Introduction to Topology

Tej Bahadur Singh

Introduction to Topology

 Springer

Tej Bahadur Singh
Department of Mathematics
University of Delhi
Delhi, India

ISBN 978-981-13-6956-8   ISBN 978-981-13-6954-4   (eBook)
https://doi.org/10.1007/978-981-13-6954-4

Library of Congress Control Number: 2019933197

This Springer imprint is published by the registered company Springer Nature Singapore Pte Ltd.
The registered company address is: 152 Beach Road, #21-01/04 Gateway East, Singapore 189721, Singapore

To the memory of my parents

Preface

This book is intended as an introductory text for senior undergraduate and beginning graduate students of mathematics.

Historically, topology has roots scattered in the nineteenth-century works on analysis and geometry. The term "topology" (actually "topologie" in German) was coined by J. B. Listing in 1836. The seminal work of Henri Poincaré ("Analysis Situs" and its five Supplements published during 1895–1909) marks the beginning of the subject of (combinatorial) topology. By the late twenties, topology has evolved as a separate discipline, and it is now a broad subject with many branches, broadly categorized as algebraic topology, general topology (or point-set topology), and geometric topology (or the theory of manifolds). Today, point-set topology is the main language for a wide variety of mathematical disciplines, while algebraic topology serves as a powerful tool for studying the problems in geometry and many other areas of mathematics. The purpose of the present work is to introduce the rudiments of general topology and algebraic topology as a part of the basic vocabulary of mathematics for higher studies.

There are some excellent textbooks and treatises on the subject, but they mostly require a level of maturity and sophistication on the part of the reader which is rather beyond what is achieved in mathematics undergraduate courses at many universities. Our objective is to make the basics of general topology and algebraic topology easily comprehensible to average students and thereby encourage them to study the subject. Of course, no claim of originality can be made in writing a book at this level. My contribution, if any, is one of presentation and selection of the material. We have included a study of elementary properties of topological groups and transformation groups because these partly geometric objects form a rich territory of interesting examples in topology and geometry and play an increasingly important role in modern mathematics and physics.

The book can be organized into three main parts. The first part comprises Chaps. 1–7 and can be considered as the core of the book. The notions of topological spaces, continuous functions, connectedness, convergence, compactness, and countability axioms are treated in these chapters. We also learn a useful method of topologizing the Cartesian product of a collection of topological spaces in Chap. 2.

The discussion of quotient spaces has been postponed until Chap. 5, for many of these spaces can easily be realized after having seen certain theorems about compactness. In fact, we study in Chap. 6 various methods of forming new spaces from old ones. The material of this part is now used in several branches of mathematics and is suitable for a one-semester first course in general topology for advanced undergraduates. The second part consists of Chaps. 8–11, where some more topics of point-set topology, specifically separation axioms, paracompactness, metrizability, completeness, and function spaces, are treated. The results of Chaps. 10 and 11 are independent of the discussions held in Chaps. 8 and 9 and important to analysts. We feel that an understanding of topological groups is advantageous to the students of several branches of mathematics. Accordingly, Chap. 12 is devoted to pretty basic information on topological groups and Chap. 13 is concerned with the action of topological groups on topological spaces and associated orbit spaces. The latter chapter also contains a brief discussion on geometric motions of Euclidean spaces. The reader is introduced to the realm of algebraic topology with the discussion of fundamental groups and covering spaces in the last two chapters (Chaps. 14–15).

We have provided enough material in the appendices to make the book self-contained. In fact, necessary background knowledge about sets, functions, countability, and ordinal numbers can be acquired from Appendix A. Some important properties of the real numbers, complex numbers, and quaternionic numbers are briefly discussed in Appendix B. However, the reader of the book who has an elementary knowledge of real analysis, group theory, and linear algebra, usually taught at the undergraduate level, would gain a deeper appreciation of the contents.

The text was originally published under the title "Elements of Topology". In this edition, we have attempted to correct errors, typographical and otherwise, and remove obscurities found in the first edition. Although the basic outline remains the same, many changes in details have been made. These consist mainly in reorganizing the contents and, where possible, simplifying the proofs; in certain cases, proofs have been completely rewritten. Certain results relegated to exercises in the first edition have been presented in the main body of the text to complement discussions on some topics. Appreciating the comments of one of the reviewers of the previous edition, we have added more entries in the index of the present edition. It is hoped that the reader will find the new treatment of the subject readily comprehensible. Solutions to all the exercises in the book are available separately to instructors.

Suggested Course Outlines

Obviously, the book has been arranged according to the author's liking. Also, several topics are independent of one another, so it is profitable to advise the reader what should be read before a particular chapter. The dependencies of chapters are roughly as follows:

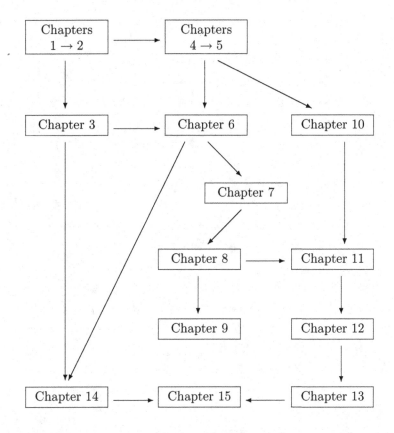

The book undoubtedly contains too much material that can be covered in a one-year course, but there is considerable flexibility for individual course design. Chapters 1–11 are suitable for a full-year course in general topology at the advanced undergraduate level. For a one-year graduate course, we suggest Chaps. 1–7 and 12–15. The subject matter of Chaps. 14 and 15 can be studied just after finishing the core part with the adaptability of turning to materials of Chaps. 12 and 13 as and when needed.

A note to the reader regarding cross-references to exercises: These have been referred to using either a single number or three numbers separated by periods. For example, the cross-reference "Exercise 4" refers to the 4th exercise of the section in which this reference has been made, while "Exercise 1.2.3" refers to Exercise 3 in Sect. 2 of Chap. 1.

Delhi, India   Tej Bahadur Singh
April 2019

Acknowledgements

Anyone who writes a book at this level merely acts as a selector of the material and owes a great deal to other people. I acknowledge my debt to previous authors of books on general topology and algebraic topology, especially those listed in the Bibliography. I would also like to acknowledge the facilities and support provided by the Harish-Chandra Research Institute, Allahabad, and the University of Delhi, Delhi, without which this book would not have been completed. I am deeply grateful to two of my teachers Prof. Shiv Kumar Gupta, West Chester University, Pennsylvania, USA, and Prof. Satya Deo Tripathi for their assistance and guidance over the years. I have had many fruitful discussions with Prof. Ramji Lal (at the University of Allahabad) and my colleagues Dr. Ratikanta Panda and Dr. Kanchan Joshi about the material presented in this book. I wish to express my great appreciation for their valuable comments and suggestions. I have been helped by several research students one way or another; among them, Mr. Sumit Nagpal deserves special thanks.

I appreciate helpful comments made by Prof. Ruchi Das and Dr. Brij Kishore Tyagi regarding improvement in the original edition of the book. Thanks are also due to the staff of Springer Nature for their cooperation in bringing out this edition.

Finally, I thank my family, especially my wife, who has endured and supported me during all these years.

Tej Bahadur Singh

Contents

About the Author

Tej Bahadur Singh is an Emeritus Professor at the Department of Mathematics, University of Delhi, India, which he joined in 1989. Earlier, he has served at the University of Allahabad, Allahabad, and Atarra P.G. College, Banda, India. He has received his Ph.D. from the University of Allahabad in 1983. With more than 40 years of teaching experience, he has written several research articles on the cohomological theory of compact transformation groups and has successfully guided eight M.Phil. and Ph.D. students. His primary research interest is algebraic topology. He has taught various courses, including point-set topology, field theory, module theory, representation of finite groups, and real analysis at graduate and Post graduate levels.

Symbols

The quantifier "there exists" is denoted by $\exists$, and the quantifier "for all" is denoted by $\forall$; this is also read as "for each." If p and q are propositions, then the logical symbol $p \Rightarrow q$ means p implies q, $p \Leftarrow q$ means p is implied by q, and $p \Leftrightarrow q$ means "$(p \Rightarrow q)$ and $(q \Rightarrow p)$," or p if and only if q.

A few particular sets frequently occur in this book; the following special symbols will be used for them.

$\varnothing$	Empty set
$\mathbb{N}$	The set of natural (or positive) integers
$\mathbb{Z}$	The set of all (positive, negative, and zero) integers
$\mathbb{Q}$	The set of all rational numbers
$\mathbb{R}$	The set (also field) of all real numbers
$\mathbb{C}$	The set (also field) of all complex numbers
$\mathbb{H}$	The set (also skew field) of all quaternions
I	The (closed) unit interval $[0, 1]$
I^n	The n-cube $I \times \cdots \times I$ (n factors)
$\mathbb{R}^n$	The set of all n-tuples $(x_1, \ldots, x_n)$ of real numbers
$\mathbb{C}^n$	The set of all n-tuples $(z_1, \ldots, z_n)$ of complex numbers
$\mathbb{H}^n$	The set of all n-tuples $(q_1, \ldots, q_n)$ of quaternions
$\langle x, y \rangle$	The standard inner product of x and y in $\mathbb{R}^n$, $\mathbb{C}^n$, or $\mathbb{H}^n$
$\|x\|$	The Euclidean norm of x
$\mathbb{D}^n$	The unit n-disc $\{x \in \mathbb{R}^n \| \|x\| \leq 1\}$
$\mathbb{S}^n$	The unit n-sphere $\left\{ (x_1, \ldots, x_{n+1}) \in \mathbb{R}^{n+1} \mid \sum_1^{n+1} x_i^2 = 1 \right\}$
ω	The well-ordered set of all nonnegative integers (also the first infinite ordinal number)
Ω	The first (or least) uncountable ordinal number
$\prod X_\alpha$	The Cartesian product of the indexed family of sets X_α, $\alpha \in A$
X^ω	The Cartesian product of countably infinite copies of X
$\sum X_\alpha$	The disjoint union of the indexed family of sets X_α, $\alpha \in A$

Chapter 1
Topological Spaces

Historically, topology has its origin in analysis and geometry and, therefore, some of the basic notions of point-set topology are abstractions of concepts which are classical in the study of analysis. We begin our study of topology by recalling the essential facts about metric spaces. In Sect. 1.2, we shall see that the properties of open sets, closed sets, or neighborhoods of points in a metric space can be used as axioms to introduce a "topology" on a given set. In Sect. 1.3, we discuss certain derived concepts such as "interior", "closure", "boundary", and "limit points" of sets in a topological space. Section 1.4 deals with the notions of "subbase" and "base" for the topology of a space X, which are families of subsets of X that generate the topology by using the operations of intersection and union or the latter one only. In Sect. 1.5, we consider the topology for subsets of a topological space induced by its topology and study the various notions for subsets of a subspace in relation to the corresponding notion in the given space.

1.1 Metric Spaces

The notions of continuity of functions and convergence of sequences are the two most fundamental concepts in analysis. Both concepts are based on the abstraction of our intuitive sense of closeness of points of a set. For example, the usual ϵ-δ definition for continuity of real- or complex-valued functions on the real line $\mathbb{R}$ (or the complex plane $\mathbb{C}$) and the definition of convergence of sequences in these spaces are based on this idea. "Closeness" of elements of a set can be measured most conveniently as distance between the elements. In any set endowed with a suitable notion of distance, one can define convergence of sequences and talk about continuity of functions between such sets. Maurice Fréchet (1906), perhaps motivated by this observation, introduced "metric spaces."

© Springer Nature Singapore Pte Ltd. 2019
T. B. Singh, *Introduction to Topology*,
https://doi.org/10.1007/978-981-13-6954-4_1

Definition 1.1.1 Let X be a (nonempty) set. A *metric* on X is a function

$$d : X \times X \to \mathbb{R}$$

such that the following conditions are satisfied for all $x, y, z \in X$:

(a) (*positivity*) $d(x, y) \geq 0$ with equality if and only if $x = y$,
(b) (*symmetry*) $d(x, y) = d(y, x)$, and
(c) (*triangle inequality*) $d(x, z) \leq d(x, y) + d(y, z)$.

The set X together with a metric d is called a *metric space*; the elements of X are called *points*. The value $d(x, y)$ on a pair of points $x, y \in X$ is called the *distance between x and y*.

Example 1.1.1 A fundamental example of a metric space is the *Euclidean n-space* $\mathbb{R}^n$. Its points are the n-tuples $x = (x_1, \ldots, x_n)$ of real numbers and the metric on this set is defined by

$$d(x, y) = \sqrt{\sum_{i=1}^{n} (x_i - y_i)^2}.$$

To see that d actually satisfies the conditions of Definition 1.1.1, we recall the definition of the "inner product" (or "scalar product") in $\mathbb{R}^n$. This is a function $(x, y) \mapsto \langle x, y \rangle$ of $\mathbb{R}^n \times \mathbb{R}^n$ into $\mathbb{R}$, where $\langle x, y \rangle = \sum_{i=1}^{n} x_i y_i$. It is linear in one coordinate when the other coordinate is held fixed (that is, it is a bilinear function). The *norm* of $x \in \mathbb{R}^n$ is defined by

$$\|x\| = \sqrt{\langle x, x \rangle} = (\sum x_i^2)^{1/2}.$$

The following properties of $\|x\|$ can be easily verified:

(a) $\|x\| > 0$ for $x \neq 0$;
(b) $\|ax\| = |a| \, \|x\|$;
(c) $|\langle x, y \rangle| \leq \|x\| \, \|y\|$ (Cauchy–Schwarz inequality); and
(d) $\|x + y\| \leq \|x\| + \|y\|$

for all $x, y \in \mathbb{R}^n$ and $a \in \mathbb{R}$.

It is now easy to check that d is a metric on $\mathbb{R}^n$, for $d(x, y) = \|x - y\|$. The Euclidean space $\mathbb{R}^1$ will be usually denoted by $\mathbb{R}$.

Example 1.1.2 Let $\mathbb{C}$ be the field of complex numbers and $\mathbb{H}$ be the skew field of quaternions. Let F denote one of these fields. Then the set F^n of ordered n-tuples $(x_1, \ldots, x_n)$, $x_i \in F$, is a vector space under the coordinatewise addition and scalar multiplication. For technical reasons (to be observed later), we will consider F^n as a right vector space over F. Let $\bar{a}$ denote the complex (resp. quaternionic) conjugate of a in $\mathbb{C}$ (resp. $\mathbb{H}$). The standard inner product on F^n is defined by $\langle x, y \rangle = \sum_1^n \bar{x}_i y_i$ for $x = (x_1, \ldots, x_n)$, $y = (y_1, \ldots, y_n)$, and it has the following properties:

(a) $\langle x, y + z \rangle = \langle x, y \rangle + \langle x, z \rangle$;

(b) $\langle x, ya \rangle = \langle x, y \rangle\, a$, $\langle xa, y \rangle = \bar{a}\,\langle x, y \rangle$;

(c) $\overline{\langle x, y \rangle} = \langle y, x \rangle$;

(d) $x \neq 0 \Rightarrow \langle x, x \rangle > 0$.

We define a function $\|\cdot\| : F^n \to \mathbb{R}$ by setting $\|x\| = \sqrt{\langle x, x \rangle}$, and refer to it as the Euclidean norm on F^n. It is easily seen that the function $\|\cdot\|$ has the properties analogous to those of the Euclidean norm on $\mathbb{R}^n$, and hence there is a metric d on F^n given by $d(x, y) = \|x - y\|$, $x, y \in F^n$. We call the metric space $\mathbb{C}^n$ the *n-dimensional complex* (or *unitary*) space, and the metric space $\mathbb{H}^n$ the *n-dimensional quaternionic* (or *symplectic*) space.

Example 1.1.3 In the standard Hilbert space ℓ_2, the points are infinite real sequences $x = (x_i)$ satisfying $\sum x_i^2 < \infty$ and its metric is defined by

$$d(x, y) = \sqrt{\sum_i (x_i - y_i)^2}.$$

We observe that $d(x, y)$ is always finite. For each positive integer n, we have

$$\sqrt{\sum_1^n (x_i - y_i)^2} \leq \sqrt{\sum_1^n x_i^2} + \sqrt{\sum_1^n y_i^2}$$

$$\leq \sqrt{\sum_i x_i^2} + \sqrt{\sum_i y_i^2} < \infty.$$

The partial sums being bounded, this monotone, nondecreasing sequence must converge and one obtains

$$\sqrt{\sum_i (x_i - y_i)^2} \leq \sqrt{\sum_i x_i^2} + \sqrt{\sum_i y_i^2}.$$

This implies that $d(x, y)$ is finite and it is indeed a metric.

Example 1.1.4 The set $\mathcal{C}(I)$ of all continuous real-valued functions on $I = [0, 1]$ with the metric

$$d(f, g) = \int_0^1 |f(t) - g(t)|\,dt$$

is an interesting metric space.

Example 1.1.5 Let X be a nonempty set and $\mathcal{B}(X)$ be the set of all bounded functions $X \to \mathbb{R}$. The metric d^* on $\mathcal{B}(X)$ given by

$$d^*(f, g) = \sup\{|f(x) - g(x)| : x \in X\}$$

is referred to as the *supremum* metric.

Definition 1.1.2 Let (X, d_X) and (Y, d_Y) be metric spaces. A function $f : X \to Y$ is *continuous* at $x \in X$ if, given $\epsilon > 0$, there exists a $\delta > 0$ such that $d_X(x, x') <$

$\delta \Rightarrow d_Y\left(f(x), f(x')\right) < \epsilon$. The function f is called *continuous* if it is continuous at each $x \in X$.

Definition 1.1.3 In a metric space (X, d), the *open r-ball with center $x \in X$ and radius $r > 0$* is the set $B(x; r) = \{y \in X \mid d(x, y) < r\}$.

In the real line $\mathbb{R}$, an open r-ball is just the open interval $(x - r, x + r)$, and an open r-ball in the plane $\mathbb{R}^2$ is a disc without its rim (see Fig. 1.2a in Sect. 1.4).

In the terminology of open ball, a function $f : X \to Y$ between two metric spaces X and Y is continuous if and only if for each open ball $B(f(x); \epsilon)$ centered at $f(x)$, there is an open ball $B(x; \delta)$ centered at x such that $f(B(x; \delta)) \subseteq B(f(x); \epsilon)$.

Example 1.1.6 Consider the function $f : \mathbb{R}^n \to \mathbb{R}^n$ given by $f(x) = x / (1 + \|x\|)$, where $\|x\| = \left(\sum x_i^2\right)^{1/2}$. We have

$$\|f(y) - f(x)\| = \frac{\|(y - x) + x(\|x\| - \|y\|) + \|x\|(y - x)\|}{(1 + \|x\|)(1 + \|y\|)}$$

$$\leq \frac{(1 + 2\|x\|)}{(1 + \|x\|)(1 + \|y\|)}\|y - x\|.$$

Therefore, f maps the open ball $B(x; \delta)$ into the open ball $B(f(x); \epsilon)$ if $\delta = \epsilon$ for $x = 0$, and $\delta = \min\{1, (1 + \|x\|)\epsilon / (1 + 2\|x\|)\}$ for $x \neq 0$. Thus f is continuous.

Definition 1.1.4 A subset U of the metric space X is called *open* if, for each point $x \in U$, there is an open r-ball with center x contained in U.

If $y \in B(x; r)$, then $B(y; r') \subseteq B(x; r)$, where $r' = r - d(x, y)$. This shows that all open balls are actually open sets.

Theorem 1.1.5 *Let X be a metric space. Then the union of any family of open sets is open, and the intersection of any finite family of open sets is also open.*

Proof The empty set $\varnothing$ and the full space X are obviously open. If $\{G_\alpha\}$ is a nonempty collection of open subsets at X, then $\bigcup_\alpha G_\alpha$ is clearly open. It remains to show that the intersection of two open subsets G_1 and G_2 is open. If $G_1 \cap G_2 = \varnothing$, we are through. So consider a point $x \in G_1 \cap G_2$. Then find positive numbers r_1 and r_2 such that $B(x; r_i) \subseteq G_i, i = 1, 2$, and put $r = \min\{r_1, r_2\}$. Then $B(x; r) \subseteq G_1 \cap G_2$, and therefore $G_1 \cap G_2$ is open. $\diamond$

We next see that the continuity of functions between metric spaces can be described completely in terms of open sets.

Theorem 1.1.6 *Let (X, d_X) and (Y, d_Y) be metric spaces. A function $f : X \to Y$ is continuous $\Leftrightarrow f^{-1}(G)$ is open in X for each open subset G of Y.*

Proof Suppose that f is continuous and $G \subseteq Y$ is open. If $f^{-1}(G) = \varnothing$, then it is open in X. Let $x \in f^{-1}(G)$ be arbitrary. Then $f(x) \in G$ and therefore there is an $\epsilon > 0$ such that $B(f(x); \epsilon) \subseteq G$. Since f is continuous at x, there exists $\delta > 0$ such that $f(B(x; \delta)) \subseteq B(f(x); \epsilon)$. This implies that $B(x; \delta) \subseteq f^{-1}(G)$; so $f^{-1}(G)$ is open.

Conversely, suppose that $x \in X$ and $\epsilon > 0$ is given. Then $f^{-1}(B(f(x); \epsilon))$ is an open subset of X containing x, by our hypothesis. Consequently, there exists a $\delta > 0$ such that $B(x; \delta) \subseteq f^{-1}(B(f(x); \epsilon)) \Rightarrow f(B(x; \delta)) \subseteq B(f(x); \epsilon)$. This implies that f is continuous at x. $\diamond$

We recall some more terminologies used in the study of metric spaces. If A and B are two nonempty subsets of a metric space (X, d), *the distance* between them is defined by

$$\operatorname{dist}(A, B) = \inf\{d(a, b) \mid a \in A, b \in B\}.$$

If $A \cap B \neq \varnothing$, then $\operatorname{dist}(A, B) = 0$. However, there exist disjoint sets with zero distance between them. When A or B is empty, we define $\operatorname{dist}(A, B) = \infty$. In particular, if $x \in X$ and $A \subseteq X$, *the distance of x from A* is $\operatorname{dist}(x, A) = \operatorname{dist}(\{x\}, A)$. The *diameter* of A, denoted by $\operatorname{diam}(A)$, is $\sup\{d(a, a') \mid a, a' \in A\}$. By convention, the diameter of the empty set is 0. A set is called *bounded* if its diameter is finite; d is a *bounded metric* if $\operatorname{diam}(X)$ is finite.

If X is a metric space and $Y \subseteq X$, then the restriction of the distance function to $Y \times Y$ is clearly a metric on Y. The set Y, with this metric, is referred to as a *subspace* of X. Thus any subset of a metric space is itself a metric space in an obvious way. This construction increases the supply of examples of metric spaces: We can now include all subsets of an Euclidean space $\mathbb{R}^n$ and the Hilbert space ℓ_2. Some subspaces of the Euclidean space $\mathbb{R}^n$, in particular, the closed unit n-disc

$$\mathbb{D}^n = \left\{ x \in \mathbb{R}^n \mid \sqrt{\textstyle\sum_1^n x_i^2} \leq 1 \right\}$$

and the unit n-cube

$$I^n = \left\{ x \in \mathbb{R}^n \mid 0 \leq x_i \leq 1, \; i = 1, \ldots, n \right\}$$

will feature in the text quite often. The subspace $I^1 = [0, 1]$ of $\mathbb{R}^1$, referred to as the unit interval, is denoted by I. Another notable subspace of $\mathbb{R}^n$ is the $(n-1)$-dimensional unit sphere

$$\mathbb{S}^{n-1} = \left\{ x \in \mathbb{R}^n \mid \sqrt{\textstyle\sum_1^n x_i^2} = 1 \right\}.$$

Notice that $\mathbb{S}^0 = \{-1, 1\}$ is a (discrete) two-point space.

Exercises

1. Given a set X, define $d(x, y) = 0$ if $x = y$, and $d(x, y) = 1$ if $x \neq y$. Check that d is a metric on X.

2. Let (X, d) be a metric space. Show that

   (a) $d'(x, y) = d(x, y)/(1 + d(x, y))$, and

   (b) $d_1(x, y) = \min\{1, d(x, y)\}$

   are bounded metrics on X.

3. • Let $F = \mathbb{R}, \mathbb{C}$ or $\mathbb{H}$, and define $\|x\| = \max\limits_{1 \leq i \leq n} |x_i|$ for every $x \in F^n$. Show that the function $\|\cdot\|$ satisfies the conditions (a), (b) and (d) described in Example 1.1.1, and hence defines a norm on F^n. This is called *the Cartesian norm* on F^n.

4. • Prove that each of the following functions defines a metric on $\mathbb{R}^n$.

   (a) $d^*((x_i), (y_i)) = \max_{1 \leq i \leq n} |x_i - y_i|$ (Cartesian metric).

   (b) $d^+((x_i), (y_i)) = \sum_1^n |(x_i - y_i)|$ (taxi-cab metric).

5. For $n = 2$ and $n = 3$, describe geometrically the open r-balls in $(\mathbb{R}^n, d)$, $(\mathbb{R}^n, d^+)$, and $(\mathbb{R}^n, d^*)$, where d is the Euclidean metric and d^+ and d^* are as in Exercise 4.

6. Verify that the functions d in Example 1.1.4, and d^* in Example 1.1.5 define metrics for $\mathcal{C}(I)$ and $\mathcal{B}(X)$, respectively.

7. • Let (Y, d) be a metric space and X be a set. Call a function $f : X \to Y$ bounded if $f(X)$ is a bounded subset of Y. Let $\mathcal{B}(X, Y)$ be the set of all bounded functions from X into Y. Show that d^* defined by

$$d^*(f, g) = \sup\{d(f(x), g(x)) \mid x \in X\}$$

   is a metric on $\mathcal{B}(X, Y)$ (This is called the *sup* metric on $\mathcal{B}(X, Y)$).

8. Show that (a) the translation function $\mathbb{R}^n \to \mathbb{R}^n$, $x \mapsto x + a$, where $a \in \mathbb{R}^n$ is fixed, and (b) the dilatation function $\mathbb{R}^n \to \mathbb{R}^n$, $x \mapsto rx$, where $r \in \mathbb{R}$ is fixed, are continuous.

9. Show that $x \mapsto \|x\|$ is a continuous function on $\mathbb{R}^n$.

10. If X is a metric space and $A \subseteq X$ is nonempty, show that the function $f : X \to \mathbb{R}$ given by $f(x) = \text{dist}(x, A)$ is continuous.

11. Show that a subset A of a metric space (X, d) is bounded if there exists a point $x \in X$ and a real number K such that $d(x, a) \leq K$ for every $a \in A$.

12. Let A, B be bounded subsets of a metric space X. Show that

    (a) $\text{diam}(A \cup B) \leq \text{diam}(A) + \text{diam}(B)$, if $A \cap B \neq \emptyset$, and

    (b) $\text{diam}(A \cup B) \leq \text{diam}(A) + \text{diam}(B) + \text{dist}(A, B)$, if they do not meet.

1.2 Topologies

In metric spaces, most of the important notions, for example, "limits," "continuity," "connectedness", "compactness," etc., may be described and many important theorems of analysis can be proved solely in terms of "open sets." This led Felix Hausdorff (1914) to abstract the basic properties of open sets and introduce a notion that is suitable for talking about these concepts and is also independent of the idea of metrics.

Definition 1.2.1 A topological structure or, simply, a topology on a set X is a collection $\mathcal{T}$ of subsets of X such that

   (a) the intersection of two members of $\mathcal{T}$ is in $\mathcal{T}$;
   (b) the union of any collection of members of $\mathcal{T}$ is in $\mathcal{T}$; and
   (c) the empty set $\varnothing$ and the entire set X are in $\mathcal{T}$.

A set X endowed with a topological structure $\mathcal{T}$ on it is called a *topological space*. The elements of X are called *points*, and the members of $\mathcal{T}$ are called the *open sets*. In general, a topological space should be denoted as a pair $(X, \mathcal{T})$. But, it is customary to use the expression "X is a topological space" or, more briefly, "X is a space" to mean $(X, \mathcal{T})$ without mentioning the topology $\mathcal{T}$ for X each time.

Example 1.2.1 Let X be any set. The family $\mathcal{D}$ of all subsets of X is a topology on X, called the *discrete* topology; the pair $(X, \mathcal{D})$ is called a *discrete space*. On the other extreme, the family $\mathcal{I} = \{\varnothing, X\}$ is also a topology on X, called the *indiscrete* or *trivial* topology; the pair $(X, \mathcal{I})$ is called an *indiscrete* or a *trivial* space.

Example 1.2.2 If $X = \{a, b\}$, then there are two topologies $\{\varnothing, \{a\}, X\}$ and $\{\varnothing, \{b\}, X\}$ on X aside from the discrete and trivial ones. The set X with one of these topologies is called the *Sierpinski space*.

Example 1.2.3 By Theorem 1.1.5, the collection of sets declared "open" in a metric space (X, d) is a topology on X; this is called the topology induced by the metric d or simply the *metric topology*. In future when a metric space is mentioned, it will be understood that the space is a topological space with the metric topology. In particular, the metric topology generated by the Euclidean metric on any subset of $\mathbb{R}^n$ will be referred to as the *usual* topology. Unless otherwise stated, a subset of $\mathbb{R}^n$ is assumed to have the usual topology. Similarly, the topologies on $\mathbb{C}^n$ and $\mathbb{H}^n$ induced by the metrics in Example 1.1.2 are referred to as the usual topologies.

Example 1.2.4 Given any set X, the family of all those subsets of X whose complements are finite together with the empty set forms a topology $\mathcal{T}_f$ on X, called the *cofinite* (or *finite complement*) topology. We call $(X, \mathcal{T}_f)$ a cofinite space. Similarly, the family of all those subsets of X whose complements are countable together with the empty set is a topology $\mathcal{T}_c$ on X, called the *cocountable* topology (or *countable complement*) topology.

We will encounter more serious examples later. It is obvious that one can assign several topological structures to a given set and these can be partially ordered by inclusion relation. If $\mathcal{T}$ and $\mathcal{T}'$ are two topologies on the same set X, we call $\mathcal{T}'$ *finer* (or *larger*) than $\mathcal{T}$ if $\mathcal{T} \subseteq \mathcal{T}'$. In this case, we also say that $\mathcal{T}$ is *coarser* (or *smaller*) than $\mathcal{T}'$. The terms "stronger" and "weaker" are also used in the literature to describe the above situation. But there is no agreement on their meaning, so we will not use these terms. It may happen that $\mathcal{T}$ is neither larger nor smaller than $\mathcal{T}'$; in this case it is said that $\mathcal{T}$ and $\mathcal{T}'$ are not comparable. Clearly, the trivial topology for a set X is the smallest possible topology on X, while the discrete topology is the largest possible topology.

Definition 1.2.2 A subset F of a topological space X is *closed* if its complement $X - F$ is open.

By De Morgan's laws, we see the following properties of closed sets in any topological space.

Proposition 1.2.3 *Let X be a space. Then*

   (a) *the union of two closed sets is a closed set;*
   (b) *the intersection of any family of closed sets is a closed set; and*
   (c) *the entire set X and the empty set $\varnothing$ are closed sets.*

Example 1.2.5 In $\mathbb{R}$, any closed interval $[a, b]$ is closed according to the above definition, for $\mathbb{R} - [a, b]$ is the union of open sets $(-\infty, a)$ and (b, ∞). The set $\mathbb{Z}$ of integers is closed, but the set $\mathbb{Q}$ of rationals is not closed.

The property (a) in Proposition 1.2.3, by iteration, implies that the union of any finite number of closed sets is closed. But it does not extend to infinite unions; for example, the union $\bigcup_{n=1}^{\infty} [1/n, 2]$ is not closed in $\mathbb{R}$.

Example 1.2.6 In a discrete space, every set is both open and closed.

Example 1.2.7 Consider the set $\mathbb{Z}$ of integers with the cofinite topology. In this topology, a finite subset of $\mathbb{Z}$ is closed but not open, $\mathbb{Z} - \{0\}$ is open but not closed, and the set $\mathbb{N}$ of positive integers is neither open nor closed.

Example 1.2.8 In the Euclidean space $\mathbb{R}^2$, $\mathbb{S}^1$ and $\mathbb{D}^2$ are closed sets, whereas the set $\{(x, y) : x \geq 0$ and $y > 0\}$ is not closed (why?).

These examples suggest that a subset can be both closed and open (called *clopen*) or it may not be either open or closed.

By the duality between the axioms for topology and the properties of closed sets, it is clear that a topology for a set X can also be described by specifying a family $\mathcal{F}$ of subsets of X satisfying the conditions in Proposition 1.2.3. In fact, the family of complements of the members of $\mathcal{F}$ is a topology for X such that $\mathcal{F}$ consists of precisely the closed subsets of X. This shows that the concept of a closed set can be taken as the primitive notion to define a topology.

Definition 1.2.4 If X is a topological space and $x \in X$, then a set $N \subseteq X$ is called a *neighborhood* (written nbd) of x in X if there is an open set U in X with $x \in U \subseteq N$.

We note that an nbd is not necessarily an open set, while an open set is an nbd of each of its points. In particular, the entire space X is an nbd of its every point. This suggests that an nbd need not be "small" as one might think. If N itself is open, we will call it an "open nbd." This is standard practice, though some mathematicians use the term "nbd" only in this sense.

Observe the following properties of the family $\mathcal{N}_x$ of all nbds of a point x in a topological space X.

(a) $\mathcal{N}_x$ is nonempty.
(b) x belongs to each N in $\mathcal{N}_x$.
(c) The intersection of two members of $\mathcal{N}_x$ is again in $\mathcal{N}_x$.
(d) If $N \in \mathcal{N}_x$ and $N \subseteq M \subseteq X$, then $M \in \mathcal{N}_x$.
(e) If $N \in \mathcal{N}_x$, then $\overset{\circ}{N} = \{y \in N \mid N \in \mathcal{N}_y\}$ is also a member of $\mathcal{N}_x$.

The last statement follows from the fact that if U is open set in X such that $x \in U \subseteq N$, then $U \subseteq \overset{\circ}{N}$.

The above properties of nbds in a space can be used to work in the opposite direction; starting with them as the axioms for "nbds", we can determine a topology.

Proposition 1.2.5 *Let X be a set and suppose that, for each $x \in X$, we are given a family $\mathcal{N}_x$ of subsets of X, satisfying the conditions (a)–(d) above. Then the collection*

$$\mathcal{T} = \{U \subseteq X \mid U \in \mathcal{N}_x \text{ for all } x \in U\}$$

is a topology on X. If (e) is also satisfied, then $\mathcal{N}_x$ is precisely the collection of all nbds of x relative to $\mathcal{T}$.

Proof The verification of the axioms of topology is routine; we prove the last statement only. If N is an nbd of the point x, then $x \in U \subseteq N$ for some open set U. Since $x \in U \in \mathcal{T}$, we have $U \in \mathcal{N}_x$. By (d), $N \in \mathcal{N}_x$. Conversely, given a set $N \in \mathcal{N}_x$, let $U = \{y \in X \mid N \in \mathcal{N}_y\}$. Clearly, $x \in U \subseteq N$. We observe that U is open. By (e), for each $y \in U$, $\overset{\circ}{N} = \{y' \in N \mid N \in \mathcal{N}_{y'}\}$ is a member of $\mathcal{N}_y$. Also, it is obvious from the definition of U that $\overset{\circ}{N} \subseteq U$. So, by (d), $U \in \mathcal{N}_y$ and we see that U is open. Therefore N is an nbd of x, as desired. $\diamond$

Clearly, a subset U of a space X is open if and only if U is an nbd of each of its points. Hence, we see that any two topologies for X, in which each point of X has the same family of nbds, must be identical. Accordingly, the topology given in the preceding proposition is unique when the condition (e) is also satisfied. It follows that even the concept of "an nbd of a point" may be used as the primitive notion and the conditions (a)–(e) in Proposition 1.2.5 as the axioms for it to define a topology.

Exercises

1. Let $X = \{a, b, c\}$ and consider the topologies $\mathcal{T}_1 = \{\varnothing, X, \{a\}, \{a, b\}\}$ and $\mathcal{T}_2 = \{\varnothing, X, \{c\}, \{b, c\}\}$ on X.

   (a) Is the union of $\mathcal{T}_1$ and $\mathcal{T}_2$ a topology for X?

   (b) Find the smallest topology containing $\mathcal{T}_1$ and $\mathcal{T}_2$, and the largest topology contained in $\mathcal{T}_1$ and $\mathcal{T}_2$.

2. Show that the intersection of any nonempty collection of topologies for a set X is a topology.

3. • Let X be an infinite set and $x_0 \in X$ be a fixed point. Show that

$$\mathcal{T} = \{G \mid \text{either } X - G \text{ is finite or } x_0 \notin G\}$$

   is a topology on X in which every point, except x_0, is both open and closed. $((X, \mathcal{T})$ is called a *Fort space*.)

4. • Decide the openness and closedness of the following subsets in $\mathbb{R}$:

   (a) $\{x \mid 1/2 < |x| \leq 1\}$, (b) $\{x \mid 1/2 \leq |x| < 1\}$,
   (c) $\{x \mid 1/2 \leq |x| \leq 1\}$, (d) $\{x \mid 0 < |x| < 1 \text{ and } (1/x) \notin \mathbb{N}\}$.

5. Find a topology on $\mathbb{R}$, different from the trivial topology and the discrete topology, so that every open set is closed and vice versa.

6. In the Euclidean space $\mathbb{R}^2$, show the following:

   (a) The first quadrant $A = \{(x, y) \in \mathbb{R}^2 \mid x, y \geq 0\}$ is closed.

   (b) $\{(x, 0) \mid -1 < x < 1\}$ is neither open nor closed.

   (c) $\{(x, 0) \mid -1 \leq x \leq 1\}$ is closed.

7. Show that $\mathbb{R}^n \times \{0\} \subset \mathbb{R}^{n+m}$ is closed in the Euclidean metric on $\mathbb{R}^{n+m}$.

8. Show that $C(I)$ is closed in the space $B(I)$ with the supremum metric (ref. Example 1.1.5).

9. Find two disjoint closed subsets of $\mathbb{R}^2$ which are zero distance apart.

10. In a metric space (X, d), for any real number $r \geq 0$, *the closed r-ball* at $x \in X$ is the set $\{y \in X \mid d(x, y) \leq r\}$.
    Show that a closed ball is always closed in the metric topology.

11. (a) What is the topology determined by the metric d on X given by $d(x, y) = 1$ if $x \neq y$ and $d(x, x) = 0$?

    (b) Let X be a set containing more than one element. Can you define a metric on X so that the associated metric topology is trivial?

12. If every countable subset of a space is closed, is the topology necessarily discrete?

13. A subset A of a space X is called G_δ (resp. F_σ) if it is the intersection (resp. union) of at most countably many open (resp. closed) sets.
    In the real line $\mathbb{R}$, prove that (a) the set $\mathbb{Q}$ of rationals is an F_σ-set, (b) the set of irrationals is a G_δ-set, and (c) a closed interval $[a, b]$ is a G_δ-set as well as an F_σ-set.

14. Show that every closed subset of metric space X is a G_δ-set. Give an example of a topological space X and a closed set $A \subset X$ which is not a G_δ-set.

15. Prove that the complement of an F_σ-set is a G_δ-set, and conversely.

16. Prove: (a) A countable intersection and a finite union of G_δ-sets are G_δ.
    (b) A countable union and a finite intersection of F_σ-sets are F_σ.

1.3 Derived Concepts

Definition 1.3.1 Let X be a space and $A \subseteq X$. The set

$$A^\circ = \bigcup \{G \mid G \text{ is open in } X \text{ and } G \subseteq A\}$$

is the largest open set contained in A; it is called the *interior* of A in X. The notation $\text{int}(A)$ is also used for A°.

Example 1.3.1 In the real line $\mathbb{R}$, $\mathbb{Q}^\circ = \varnothing = (\mathbb{R} - \mathbb{Q})^\circ$, and $[a, b]^\circ = (a, b)$.

Example 1.3.2 In the space $\mathbb{R}^2$, $\text{int}\left(\mathbb{S}^1\right) = \varnothing$, $\text{int}\left(\mathbb{D}^2\right) = B(0; 1)$.

If X is a space and $A \subseteq X$, then a point of A° is called an *interior point* of A. Obviously, a point $x \in X$ is an interior point of $A \Leftrightarrow A$ is an nbd of x. It is also clear that A is open $\Leftrightarrow A = A^\circ$.

Proposition 1.3.2 *Let X be a space. Then, for $A, B \subseteq X$, we have*

(a) $(A^\circ)^\circ = A^\circ$,
(b) $A \subseteq B \Rightarrow A^\circ \subseteq B^\circ$,
(c) $A^\circ \cap B^\circ = (A \cap B)^\circ$, *and*
(d) $A^\circ \cup B^\circ \subseteq (A \cup B)^\circ$.

We leave the simple proofs to the reader. Notice that the reverse inclusion in (d) does not hold good; this is shown by Example 1.3.1.

Definition 1.3.3 Let X be a space and $A \subseteq X$. The set

$$\overline{A} = \bigcap \{F \mid F \text{ is closed in } X \text{ and } A \subseteq F\}$$

is the smallest closed set containing A. This is called the *closure* of A, sometimes denoted by $\text{cl}(A)$. A point $x \in \overline{A}$ is referred to as an *adherent* point of A.

Example 1.3.3 In the space $\mathbb{R}$, $\overline{(a, b)} = [a, b]$ and $\overline{\mathbb{Q}} = \mathbb{R} = \overline{\mathbb{R} - \mathbb{Q}}$.

Example 1.3.4 In the space $\mathbb{R}^2$, $\overline{B(0; 1)} = \mathbb{D}^2$.

Example 1.3.5 In a cofinite space X, $\overline{A} = X$ for every infinite set $A \subseteq X$.

It is readily seen that a subset A of a space X is closed if and only if $\overline{A} = A$. We also leave the straightforward proofs of the following proposition to the reader.

Proposition 1.3.4 *Let X be a space and $A, B \subseteq X$. Then*

(a) $\overline{\overline{A}} = \overline{A}$,
(b) $A \subseteq B \Rightarrow \overline{A} \subseteq \overline{B}$,
(c) $\overline{A \cup B} = \overline{A} \cup \overline{B}$, *and*
(d) $\overline{A \cap B} \subseteq \overline{A} \cap \overline{B}$.

Note that the equality in (d) may fail, as is seen by taking $A = (-1, 0)$ and $B = (0, 1)$ in the real line $\mathbb{R}$.

Theorem 1.3.5 *Let A be subset of a space X. Then $x \in \overline{A} \Leftrightarrow U \cap A \neq \varnothing$ for every open nbd U of x.*

Proof If there exists an open set U such that $x \in U$ and $U \cap A = \varnothing$, then $F = X - U$ is a closed set which contains A but not x. Thus $x \notin \overline{A}$. Conversely, if $x \notin \overline{A}$, then $U = X - \overline{A}$ is an open nbd of x disjoint from A.   ◇

Definition 1.3.6 Let A be subset of a space X. A point $x \in X$ is a *limit point* (or *accumulation point* or *cluster point*) of A if every open nbd of x contains at least one point of $A - \{x\}$. The set A' of all limit points of A is called the *derived set* of A.

Example 1.3.6 In the space $X = \{a, b\}$ with the topology $\{\varnothing, \{a\}, X\}$, $\{a\}' = \{b\}$ and $\{b\}' = \varnothing$.

Example 1.3.7 In $\mathbb{R}$, every point of $[0, 1]$ is a limit point of $(0, 1)$, whereas the set $\mathbb{Z}$ of integers has no limit points.

Example 1.3.8 Every point of $\mathbb{R}^3$ is a limit point of the subset A of those points all of whose coordinates are rational and, at the other extreme, the subset B of points which have integer coordinates does not have any limit points.

Theorem 1.3.7 *Let A be a subset of a space X. Then $\overline{A} = A \cup A'$.*

Proof If x is neither a point nor a limit point of A, then there is an open nbd U of x such that $U \cap A = \varnothing$. Since U is a nbd of each of its points, none of these is in A'. So U is contained in the complement of $A \cup A'$, and it follows that $A \cup A'$ is closed. Hence, $\overline{A} \subseteq A \cup A'$. On the other hand, $A' \subseteq \overline{A}$, by Theorem 1.3.5. As $A \subseteq \overline{A}$ always, we find that $A \cup A' \subseteq \overline{A}$, completing the proof.   ◇

Corollary 1.3.8 *A set is closed if and only if it contains all its limit points.*

Proof A is closed $\Leftrightarrow A = \overline{A} = A \cup A' \Leftrightarrow A' \subseteq A.$ $\diamond$

Definition 1.3.9 Let A be subset of X. The *boundary* (or *frontier*) of A is defined to be the set $\partial A = \overline{A} \cap \overline{X - A}$. The notation $\mathrm{bd}(A)$ is also used for ∂A. A point $x \in \partial A$ is called a *boundary point of A*.

Obviously, ∂A is identical with $\partial(X - A)$. Also, it is clear that a point $x \in X$ is a boundary point of A if and only if each (open) nbd of x intersects both A and $X - A$.

Example 1.3.9 In $\mathbb{R}$, $\partial[0, 1] = \{0, 1\}$ and $\partial\mathbb{Q} = \mathbb{R}$.

Example 1.3.10 In $\mathbb{R}^2$, $\partial\mathbb{D}^2 = \mathbb{S}^1$ and $\partial\mathbb{S}^1 = \mathbb{S}^1$.

Example 1.3.11 Let A be the set of all points of $\mathbb{R}^3$ which have rational coordinates. Then $\partial A = \mathbb{R}^3$.

Theorem 1.3.10 *Let A be a subset of space X. Then $\overline{A} = A \cup \partial A$.*

Proof By definition, $\overline{A}$ contains both A and ∂A, and hence their union. Conversely, if $x \in \overline{A} - A$, then $x \in \overline{A} \cap (X - A) \subseteq \partial A$ and the reverse inclusion follows. $\diamond$

As an immediate consequence of this theorem, we have

Corollary 1.3.11 *A set is closed if and only if it contains its boundary.*

Definition 1.3.12 Let A be a subset of a space X. A point $a \in A$ is called *isolated* if $a \notin A'$. The set A is called *perfect* if it is closed and has no isolated points.

Example 1.3.12 Consider the closed unit interval I with the subspace topology induced from $\mathbb{R}$. Let $J_1 = \left(\frac{1}{3}, \frac{2}{3}\right)$, $J_2 = \left(\frac{1}{9}, \frac{2}{9}\right) \cup \left(\frac{7}{9}, \frac{8}{9}\right)$, $\ldots$. Notice that J_n, $n > 1$, is the union of 2^{n-1} intervals of the form $\left(\frac{1+3k}{3^n}, \frac{2+3k}{3^n}\right)$ which are contained in $I - \bigcup_{i=1}^{n-1} J_i$. The *Cantor set* is defined by $C = I - \bigcup_1^{\infty} J_n$. This is the set of all points in I whose at least one triadic expansion (base 3) contains no 1's. We observe that C is perfect. Clearly, C is closed. To see that every point of C is a limit point, let $x \in C$ be arbitrary and U be an open interval containing x. Choose a sufficiently large integer n so that U contains a closed interval $[x - 1/3^n, x + 1/3^n]$. Now, find an integer $k \geq 0$ such that x belongs to a closed interval of the form $[k/3^n, (k + 1)/3^n]$. Obviously, one end point of this interval is different from x. Thus U contains a point of C other than x, and x is a limit point of C.

We close this section with another terminology for subsets in a space, which will also appear several times in this text.

Definition 1.3.13 A subset A of a topological space X is called *dense* (or *everywhere dense*) if $\overline{A} = X$.

Example 1.3.13 In the real line $\mathbb{R}$, both the set of rational numbers and the set of irrational numbers are dense.

Example 1.3.14 If X is an infinite set with the cofinite topology, then the dense subsets of X are its infinite subsets.

Exercises

1. In a discrete space X, what are limit points of a set $A \subset X$.

2. Find the derived sets of $\{a\}$, $\{b\}$, $\{c\}$, and $\{a, c\}$ in the space $(X, \mathcal{T}_1)$ of Exercise 1.2.1

3. Describe the boundary, closure, interior, and derived set of each of the following subsets of the real line $\mathbb{R}$:

   (a) $\{(1/n) \mid n = 1, 2, \ldots\}$;   (b) $(-1, 0) \cup (0, 1)$;
   (c) $\{(1/m) + (1/n) \mid m, n \in \mathbb{N}\}$; (d) $\{(1/n) \sin n \mid n \in \mathbb{N}\}$.

   Observe that the interior operator and the closure operator do not commute.

4. Find the boundary, closure, interior, and derived set of each of the following subsets of $\mathbb{R}^2$:

   (a) $\{(x, 0) \mid x \in \mathbb{R}\}$;   (b) $\{(x, 0) \mid 0 < x < 1\}$;
   (c) $\{(x, y) \mid x \in \mathbb{Q}\}$;   (d) $\{(x, y) \mid x, y \in \mathbb{Q}\}$;
   (e) $\{(x, y) \mid 1 < x^2 + y^2 \leq 2\}$;   (f) $\{(x, y) \mid x \geq 0, y > 0\}$;
   (g) $\{(x, y) \mid x \neq 0 \text{ and } y \leq 1/x\}$; (h) $\{(x, y) \mid x \geq y^2\}$.

5. (a) Let $\{A_\alpha\}$ be an infinite family of subsets of a space X. Prove:

   (i) $\left(\bigcap A_\alpha\right)^\circ \subseteq \bigcap A_\alpha^\circ$; (ii) $\bigcup A_\alpha^\circ \subseteq \left(\bigcup A_\alpha\right)^\circ$;
   (iii) $\overline{\bigcap A_\alpha} \subseteq \bigcap \overline{A_\alpha}$;   (iv) $\bigcup \overline{A_\alpha} \subseteq \overline{\bigcup A_\alpha}$.

   (b) Give examples to show that the reverse of the inclusions in (a) may not hold good.

   (c) Prove that the equality in (a) (iv) holds if $\bigcup \overline{A_\alpha}$ is closed.

6. Let X be a space and $A \subseteq X$. Prove:

   (a) $X - \overline{A} = (X - A)^\circ$; (b) $X - A^\circ = \overline{X - A}$;
   (c) $\partial A = \overline{A} - A^\circ$;   (d) $A^\circ \cup \partial A = \overline{A}$;
   (e) $A^\circ \cap \partial A = \varnothing$;   (f) $A^\circ = A - \partial A$.

7. Let X be a space and $A \subseteq X$. Prove that A is clopen $\Leftrightarrow \partial A = \varnothing$.

8. Let X be a space, and $A, B \subseteq X$. Prove that $(A \cup B)' = A' \cup B'$. How does $\partial (A \cup B)$ relate to ∂A and ∂B?

9. Let U be an open subset of a space X. Show:

   (a) $\overline{U} = \overline{\text{int} \left(\overline{U}\right)}$. (b) $\partial U = \overline{U} - U$.
   (c) $U = \text{int} \left(\overline{U}\right)$? (d) $U \cap \overline{A} \subseteq \overline{U \cap A}$ for every $A \subseteq X$.

10. Prove that G is open in a space $X \Leftrightarrow \overline{G \cap \overline{A}} = \overline{G \cap A}$ for every subset A of X.

11. Let X be an infinite set with the cofinite topology and $A \subseteq X$. Prove that if A is infinite, then every point of X is a limit point of A and if A is finite then it has no limit points.

12. In a metric space (X, d), show the following:

   (a) x is an interior point of a subset A of X $\Leftrightarrow$ there exists an open ball $B(x; r)$ contained in A.

   (b) x is a limit point of a set $A \subseteq X$ $\Leftrightarrow$ each open ball $B(x; r)$ contains at least one point of $A - \{x\}$.

   (c) $x \in \overline{A} \Leftrightarrow dist(x, A) = 0$.

13. (a) In the Euclidean space $\mathbb{R}^n$, show that $\overline{B(x; r)}$ is the closed r-ball $B[x; r] = \{y \in X \mid d(y, x) \le r\}$.

   (b) Give an example of a metric space (X, d) in which $\overline{B(x; r)}$ is not the closed r-ball at x, and $\partial B(x; r) \ne \{y \mid d(y, x) = r\}$ for some point $x \in X$ and some real $r > 0$.

   (c) What is the relation between $\partial B(x; r)$ and $\{y \mid d(y, x) = r\}$?

14. (a) If A has no isolated points, show that $\overline{A}$ is perfect.

   (b) If a space X has no isolated points, prove that every open subset of X also has no isolated points.

15. (a) Prove that every nonempty subset of a trivial space is dense, while no proper subset of a discrete space is dense.

   (b) If no proper subset of the topological space X is dense, is the topology necessarily discrete?

   (c) What is the boundary of a subset of a discrete space? a trivial space?

16. Let D be a subset of a space X.

   (a) Prove that D is dense in X $\Leftrightarrow$ X is the only closed superset of D $\Leftrightarrow$ $X - D$ has an empty interior.

   (b) If D is dense in X, prove that $\overline{D \cap G} = \overline{G}$ for every open subset G of X.

   (c) If G and H are open subsets of a space X such that $\overline{G} = X = \overline{H}$, show that $\overline{G \cap H} = X$.

1.4 Bases

The specification of a topology by describing all of the open sets is usually a difficult task. This can often be done more simply by using the notion of a "generating family" for the topology. In this section, we will study two such concepts.

Given a set X and a family $\mathscr{S}$ of subsets of X, it is easily seen that the intersection of the collection of all topologies on X which contains $\mathscr{S}$ (certainly nonempty, for the discrete topology on X is one such topology) is a topology, denoted by $\mathscr{T}(\mathscr{S})$. Clearly, $\mathscr{T}(\mathscr{S})$ is the coarsest topology on X containing $\mathscr{S}$.

Definition 1.4.1 Let X be a space with the topology $\mathscr{T}$. A *subbasis* (or *subbase*) for $\mathscr{T}$ is a family $\mathscr{S}$ of subsets of X such that $\mathscr{T} = \mathscr{T}(\mathscr{S})$.

If $\mathscr{S}$ is a subbasis for the topology $\mathscr{T}$ on X, the members of $\mathscr{S}$ are open in X and referred to as subbasic open sets. Furthermore, the topology $\mathscr{T}$ consists of $\varnothing$, X, all finite intersections of members of $\mathscr{S}$ and all unions of these finite intersections. This can be easily ascertained by verifying that the collection of these sets is a topology for X, which contains $\mathscr{S}$ and is coarser than $\mathscr{T}$. It follows that the topology $\mathscr{T}$ is completely determined by the family $\mathscr{S}$.

As we have seen above, any family $\mathscr{S}$ of subsets of a given set X serves as a subbasis for *some* topology on X, viz., $\mathscr{T}(\mathscr{S})$. So, to define a topology on X, it suffices to specify a family $\mathscr{S}$ of subsets of X as a subbasis. The resulting topology is said to be generated by the subbasis $\mathscr{S}$. We illustrate this by introducing a topology on an ordered set $(X, \prec)$. For each pair of elements $a, b \in X$, define

$$(a, b) = \{x \in X \mid a \prec x \prec b\} \quad \text{(open interval)},$$
$$[a, b] = \{x \in X \mid a \preceq x \preceq b\} \quad \text{(closed interval)},$$

$$\left. \begin{array}{l} [a, b) = \{x \in X \mid a \preceq x \prec b\} \\ (a, b] = \{x \in X \mid a \prec x \preceq b\} \end{array} \right\} \quad \begin{array}{l} \text{(half-open or half-} \\ \text{closed intervals)}. \end{array}$$

And, for each $a \in X$, we define

$$\left. \begin{array}{l} (-\infty, a) = \{x \in X \mid x \prec a\} \\ (a, +\infty) = \{x \in X \mid a \prec x\} \end{array} \right\} \quad \begin{array}{l} \text{(open rays or one-sided} \\ \text{open intervals)}, \end{array}$$

$$\left. \begin{array}{l} (-\infty, a] = \{x \in X \mid x \preceq a\} \\ [a, +\infty) = \{x \in X \mid a \preceq x\} \end{array} \right\} \quad \begin{array}{l} \text{(closed rays or one-sided} \\ \text{closed intervals)}. \end{array}$$

It is obvious that $[a_0, a) = (-\infty, a)$ if a_0 is the first (or smallest) element, and $(a, b_0] = (a, +\infty)$ if b_0 is the last (or largest) element of X.

We ought to find a topology on X which justifies the use of adjectives closed and open here. Notice that an open interval (a, b) can be obtained as an intersection of the rays $(a, +\infty)$ and $(-\infty, b)$ so that a topology which contains these rays certainly contains (a, b).

Definition 1.4.2 Let X be an ordered set. The *order topology* (or the *interval topology*) on X is the topology generated by the subbasis consisting of all the "open rays" $(-\infty, a)$ and $(a, +\infty)$, where a ranges over X.

In the order topology on X, an open interval (a, b) is obviously open and a closed interval $[a, b]$, being the complement of $(-\infty, a) \cup (b, +\infty)$, is closed.

Example 1.4.1 Consider the set $\mathbb{R}$ of real numbers with the usual order relation on it. Since the open rays $(-\infty, a)$ and $(a, +\infty)$ $(a \in \mathbb{R})$ are open in the Euclidean topology on $\mathbb{R}$, the order topology for $\mathbb{R}$ is coarser than the Euclidean topology. On the other hand, every open ball, being an open interval, is open in the order topology. Therefore, every open subset of the real line $\mathbb{R}$ is open in the order topology, and the two topologies for $\mathbb{R}$ coincide. Thus we see that the family of all open rays in $\mathbb{R}$ is a subbase for the topology of the real line $\mathbb{R}$.

Since the operations of union and intersection both are involved in the construction of a topology from a subbasis, an important simplification occurs if the open sets are constructed only by taking unions of members of $\mathscr{S}$. This is possible, for example, if $\mathscr{S}$ is closed under the formation of finite intersections; this situation motivates the following.

Definition 1.4.3 Let $(X, \mathscr{T})$ be a space. A *basis* (or *base*) for $\mathscr{T}$ is a family $\mathscr{B} \subseteq \mathscr{T}$ such that every member of $\mathscr{T}$ is a union of members of $\mathscr{B}$.

If $\mathscr{B}$ is a basis for a topology $\mathscr{T}$ on X, then $\mathscr{T}$ is the coarsest topology on X containing $\mathscr{B}$. For, if $\mathscr{T}'$ is a topology with $\mathscr{B} \subseteq \mathscr{T}'$, then all unions of members of $\mathscr{B}$ are cetainly in $\mathscr{T}'$ and so $\mathscr{T} \subseteq \mathscr{T}'$. We say that the topology $\mathscr{T}$ is generated by the basis $\mathscr{B}$. The members of $\mathscr{B}$ are referred to as the *basic open sets* in X, and $\mathscr{B}$ is also called a basis for the space X. There is a simple characterization of bases, which some authors use as a definition.

Proposition 1.4.4 *A collection $\mathscr{B}$ of open subsets of a space X is a basis if and only if for each open subset U of X and each point $x \in U$, there exists a $B \in \mathscr{B}$ such that $x \in B \subseteq U$.*

The straightforward proof is left to the reader.

By the preceding proposition, we have a useful way to describe the open subsets of a space X when its topology is given by specifying a basis $\mathscr{B}$: A set $G \subseteq X$ is open if and only if for each $x \in G$, there is a $B \in \mathscr{B}$ such that $x \in B \subseteq G$. It follows that the topology of a space is completely determined by a basis.

Example 1.4.2 In a discrete space X, the family of all singleton sets $\{x\}$ is a basis.

Example 1.4.3 In a metric space X, the collection

$$\{B(x; r) \mid x \in X, \text{ and real number } r > 0\}$$

of open balls is a basis for the metric topology on X.

Observe that a family $\mathscr{S}$ of subsets of a space X is a subbasis if and only if the family of all finite intersections of members of $\mathscr{S}$ is a basis for X. This basis is said to be generated by $\mathscr{S}$. (We remark that some authors elude the convention that X is the intersection of the empty subfamily of $\mathscr{S}$ and require that a subbasis $\mathscr{S}$ for X must also satisfy the condition $X = \bigcup \{S \in \mathscr{S}\}$.)

It follows that the base generated by the subbasis of an ordered space X consists of all open rays, all open intervals (a, b), the empty set $\varnothing$, and the full space X. Notice that if $x \in X$ is not the first or the last element, then x belongs to an open interval in X, for there exists elements $a, b \in X$ such that $x \in (a, +\infty)$ and $x \in (-\infty, b)$. Moreover, if X has no last element, then $(a, +\infty)$ is a union of open intervals, and if X has no first element, then $(-\infty, a)$ is a union of open intervals. Accordingly, the space X has a basis consisting of the open intervals (a, b), right half-open intervals $[a_0, a) = (-\infty, a)$ (in the case a_0 is the first element of X) and left half-open intervals $(a, b_0] = (a, +\infty)$ (in the case b_0 is the last element X).

Example 1.4.4 Consider the ordering $\prec$ on $\mathbb{R}^2$ defined by $x \prec y$ if and only if either $x_1 < y_1$ or $(x_1 = y_1$ and $x_2 < y_2)$, where $x = (x_1, x_2)$ and $y = (y_1, y_2)$. This is referred to as the "dictionary ordering" for $\mathbb{R}^2$. Since there is no first or last element in the ordered set $(\mathbb{R}^2, \prec)$, the open intervals (x, y) (see Fig. 1.1 below) form a basis for the order topology on $\mathbb{R}^2$. It is easily verified that an interval (x, y), where $x_1 < y_1$, is the union of intervals of the form (x, y), where $x_1 = y_1$ (Fig. 1.1b). Therefore, the intervals of the latter type alone can generate the order topology on $\mathbb{R}^2$, and hence form a basis.

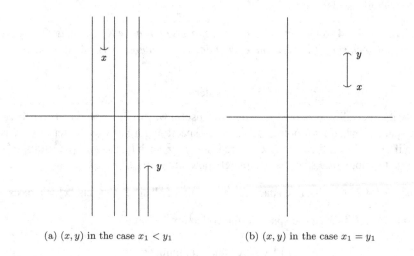

(a) (x, y) in the case $x_1 < y_1$ (b) (x, y) in the case $x_1 = y_1$

Fig. 1.1 Open intervals in the dictionary ordering on $\mathbb{R}^2$

Example 1.4.5 Recall that if x and y are two ordinal numbers, then $x < y \Leftrightarrow x \subset y$, and 0 denotes the least ordinal number (see Appendix A.7). Let Ω be the first (least) uncountable ordinal number and consider the set of all ordinal numbers $\leq \Omega$. Obviously, Ω is the last element of this set, and it is denoted by $[0, \Omega]$. By definition, the order topology for $[0, \Omega]$ is generated by the subbasis consisting of the right half-open intervals $[0, y)$, $y \leq \Omega$, and the left half-open intervals $(x, \Omega]$, $x < \Omega$. Accordingly, a basis of the ordered space $[0, \Omega]$ consists of all these intervals and the open intervals (x, y). Since each ordinal number $y < \Omega$ has an immediate successor, this topology for $[0, \Omega]$ is also generated by the basis consisting of $\{0\}$ and the left half-open intervals $(x, y]$, where $0 \leq x < y \leq \Omega$. Notice that if $x = 0$ or it has an immediate predecessor in $[0, \Omega]$, then $\{x\}$ is open.

In the above examples, we have described bases for the given topologies. But, the primary motive behind the concept of basis is to specify a topology on a *set* X by giving a basis. A natural question arises whether a given family of subsets of X would be a base for some topology on X. The answer to this question is not always positive. For example, the family $\mathscr{B} = \{\{a, b\}, \{a, c\}\}$ of subsets of $X = \{a, b, c\}$ cannot serve as a basis for a topology on X, since any topology having $\mathscr{B}$ as a basis must contain $\{a\}$ which cannot be expressed as the union of members of $\mathscr{B}$. So we ought to know when a given collection $\mathscr{B}$ of subsets of X can serve as a basis for *some* topology on X. Assume that $\mathscr{B}$ is a basis for some topology on X. Then we have $X = \bigcup\{B \mid B \in \mathscr{B}\}$, for X is open. And, for every pair of sets B_1, $B_2 \in \mathscr{B}$ and for each $x \in B_1 \cap B_2$, there exists $B_3 \in \mathscr{B}$ with $x \in B_3 \subseteq B_1 \cap B_2$, since $B_1 \cap B_2$ is open. In fact, these conditions are also sufficient, as we see below.

Theorem 1.4.5 *Let $\mathscr{B}$ be a collection of subsets of the set X such that $X = \bigcup\{B \mid B \in \mathscr{B}\}$, and for every two members B_1, B_2 of $\mathscr{B}$ and for each point $x \in B_1 \cap B_2$, there exists $B_3 \in \mathscr{B}$ with $x \in B_3 \subseteq B_1 \cap B_2$. Then there is a topology on X for which $\mathscr{B}$ is a basis.*

Proof Let $\mathscr{T}(\mathscr{B})$ be the family of all sets $U \subseteq X$ such that for each $x \in U$ there exists $B \in \mathscr{B}$ with $x \in B \subseteq U$. Then it is clear that $\mathscr{T}(\mathscr{B})$ contains $\varnothing$ (vacuously) and X (by hypothesis). If $U_\alpha, \alpha \in A$, is a family of sets in $\mathscr{T}(\mathscr{B})$ and $x \in \bigcup_\alpha U_\alpha$, then $x \in U_\beta$ for some $\beta \in A$. So there is a $B \in \mathscr{B}$ such that $x \in B \subseteq U_\beta \subseteq \bigcup_\alpha U_\alpha$, and thus $\bigcup_\alpha U_\alpha$ is a member of $\mathscr{T}(\mathscr{B})$. Next, given U_1, U_2 in $\mathscr{T}(\mathscr{B})$, if $x \in U_1 \cap U_2$, then we can find $B_i \in \mathscr{B}$, $i = 1, 2$, such that $x \in B_i \subseteq U_i$. Now, by our hypothesis, there exists a $B_3 \in \mathscr{B}$ such that $x \in B_3 \subseteq B_1 \cap B_2 \subseteq U_1 \cap U_2$, and so $U_1 \cap U_2 \in \mathscr{T}(\mathscr{B})$. Thus, we see that $\mathscr{T}(\mathscr{B})$ is a topology on X. Further, it is clear from the definition of $\mathscr{T}(\mathscr{B})$ that $\mathscr{B} \subseteq \mathscr{T}(\mathscr{B})$ and every member of $\mathscr{T}(\mathscr{B})$ is a union of some members of $\mathscr{B}$. Accordingly, $\mathscr{B}$ is a basis for $\mathscr{T}(\mathscr{B})$. $\diamond$

By way of illustration, consider the "open intervals" (a, b), $a, b \in \mathbb{R}$. Clearly, $\mathbb{R}$ is the union of the open intervals $(x - 1, x + 1)$, $x \in \mathbb{R}$, and the intersection of two open intervals is either such an interval or the empty set $\varnothing$. Hence, the family of all open intervals (a, b) is a basis for a topology on $\mathbb{R}$. The topology generated by this base is the usual topology for $\mathbb{R}$, since the (bounded) open intervals are precisely the open balls in the Euclidean metric on $\mathbb{R}$. Similarly, we see that the family of "closed intervals" $[a, b]$ is also a basis for a topology on $\mathbb{R}$. Since this family contains all the singletons $\{a\}$, $a \in \mathbb{R}$, the topology generated by this basis is the discrete topology. However, half-open intervals generate two new topologies for $\mathbb{R}$ as described below.

Example 1.4.6 Consider the family of "right half-open intervals" $[a, b)$, where $a, b \in \mathbb{R}$ and $a < b$. One can readily verify that this family satisfies the conditions for a basis. The topology generated by this basis is called the *lower limit* (or *right half-open interval*) topology for $\mathbb{R}$. The set $\mathbb{R}$ with this topology is denoted by $\mathbb{R}_\ell$, and is referred to as the "Sorgenfrey line." In this space, all the intervals $(-\infty, a)$ and $[a, +\infty)$ are both open and closed, and so is each basis element. The sets of the form (a, b) or $(a, +\infty)$ are open, for $(a, b) = \bigcup \{[x, b) \mid a < x < b\}$ and $(a, +\infty) = \bigcup \{[x, x + 1) \mid a < x\}$, but not closed.

Similarly, the family $\{(a, b] \mid a, b \in \mathbb{R} \text{ and } a < b\}$ generates a topology on $\mathbb{R}$, called the *upper limit* (or the *left half-open interval*) topology.

As another use of bases, we obtain a simple criterion for comparison of two topologies, which are specified by giving bases for them.

Proposition 1.4.6 *Let $\mathcal{T}$ and $\mathcal{T}'$ be topologies on a set X generated by the bases $\mathcal{B}$ and $\mathcal{B}'$, respectively. Then $\mathcal{T}$ is coarser than $\mathcal{T}' \Leftrightarrow$ for each $B \in \mathcal{B}$ and each $x \in B$, there exists $B' \in \mathcal{B}'$ such that $x \in B' \subseteq B$.*

Proof $\Rightarrow$: Given $B \in \mathcal{B}$, we have $B \in \mathcal{T}'$, for $\mathcal{T} \subseteq \mathcal{T}'$. If $x \in B$, then there exists $B' \in \mathcal{B}'$ with $x \in B' \subseteq B$, since $\mathcal{B}'$ is a basis for $\mathcal{T}'$.

$\Leftarrow$: Let U be $\mathcal{T}$-open. If $x \in U$, then there exists $B \in \mathcal{B}$ with $x \in B \subseteq U$, since $\mathcal{B}$ is a basis for $\mathcal{T}$. By our hypothesis, there exists $B' \in \mathcal{B}'$ such that $x \in B' \subseteq B$. So $x \in B' \subseteq U$. It follows that U is $\mathcal{T}'$-open, and so $\mathcal{T} \subseteq \mathcal{T}'$. $\diamond$

Example 1.4.7 The topology of the real line $\mathbb{R}$ is strictly smaller than that of $\mathbb{R}_\ell$ (and the upper limit topology). We already know that the open intervals (a, b) form a basis for the Euclidean topology and the half-open intervals $[a, b)$ form a basis for the lower limit topology. Also, for each $x \in (a, b)$, we have $x \in [x, b) \subset (a, b)$. So, by the preceding proposition, the Euclidean topology is coarser than the lower limit topology. But, the basis element $[a, b)$ for the lower limit topology is not open in the Euclidean topology, since there is no open interval which contains a and is contained in $[a, b)$.

Notice that distinct bases or subbases may generate the same topology; for instance, the subcollection of open intervals with rational end points is also a basis for the usual topology on $\mathbb{R}$. By way of another illustration, we give the following.

Example 1.4.8 We observe that the two topologies generated by the families of open balls in the metrics d and d^* on $\mathbb{R}^n$, given by

$$d((x_i), (y_i)) = \sqrt{\sum |x_i - y_i|^2} \quad \text{and}$$
$$d^*((x_i), (y_i)) = \max_{1 \leq i \leq n} |x_i - y_i|,$$

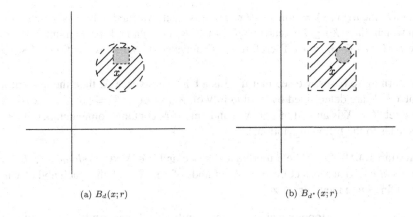

(a) $B_d(x; r)$ (b) $B_{d^*}(x; r)$

Fig. 1.2 Shapes of open balls in the metrics d and d^* on $\mathbb{R}^2$

are the same. For $n = 2$, open balls $B_d(x; r)$ (a disc without its edge) and $B_{d^*}(x; r)$ (a square without its sides) in the metrics d and d^*, respectively, are depicted in Fig. 1.2. It is intuitively clear that given a point in a disc, there is a square in the disc centered at the point. The other way round, given a point inside a square, we can find a disc in the square centered at the point. Indeed, it is easily verify that if $y \in B_d(x; r)$, then $B_{d^*}(y; r') \subset B_d(x; r)$ for $0 < r' < (r - d(y, x))/\sqrt{n}$. And, if $y \in B_{d^*}(x; r)$, then $B_d(y, r') \subset B_{d^*}(x; r)$ for $r' = r - d^*(y, x)$. So, by Proposition 1.4.6, we see that the topologies induced by d and d^* coincide.

Definition 1.4.7 Two metrics d and d' on a set X are called *equivalent* if they induce the same topology on X.

If two metrics d and d' on X are equivalent, then, for each $x \in X$ and each real number $\epsilon > 0$, the open ball $B_d(x; \epsilon)$ in the metric space (X, d) is open in the topology $\mathscr{T}_{d'}$, and the open ball $B_{d'}(x; \epsilon)$ in the metric space (X, d') is open in the topology $\mathscr{T}_d$. So there exist real numbers $\delta_1, \delta_2 > 0$ such that $B_{d'}(x; \delta_1) \subseteq B_d(x; \epsilon)$, and $B_d(x; \delta_2) \subseteq B_{d'}(x; \epsilon)$. Conversely, suppose that these conditions hold for each $x \in X$ and each real number $\epsilon > 0$. Then, given an open set U in the topology $\mathscr{T}_d$ and a point $x \in U$, there exists an open ball $B_d(x; \epsilon)$ contained in U. Now, by our assumption, there is an open ball $B_{d'}(x; \delta_2)$ contained in $B_d(x; \epsilon)$. Hence $B_{d'}(x; \delta_2) \subseteq U$, and it follows that U is also open in the topology $\mathscr{T}_{d'}$. Similarly, every open subset of the space $(X, \mathscr{T}_{d'})$ is open in the space $(X, \mathscr{T}_d)$, and the two metrics are equivalent. Thus we have proved the following.

Proposition 1.4.8 *Two metrics d and d' on a set X are equivalent if and only if for each $x \in X$ and each real number $\epsilon > 0$, there exist real numbers $\delta_1, \delta_2 > 0$ (depending upon x and ϵ) such that $B_d(x; \delta_1) \subseteq B_{d'}(x; \epsilon)$, and $B_{d'}(x; \delta_2) \subseteq B_d(x; \epsilon)$.*

Corollary 1.4.9 *Let (X, d) be a metric space. Then, for each real $\lambda > 0$, there is a metric d_λ equivalent to d such that the diameter of X in d_λ is less than λ.*

Proof Define $d_\lambda(x, y) = \min\{\lambda, d(x, y)\}$. It is easily verified that d_λ is a metric on X in which $\operatorname{diam}(X) \leq \lambda$. Clearly, $B_d(x; \epsilon) \subseteq B_{d_\lambda}(x; \epsilon)$ and, for $\delta = \min\{\lambda, \epsilon\}$, we have $B_{d_\lambda}(x; \delta) \subseteq B_d(x; \epsilon)$. Therefore d_λ is equivalent to d, by Proposition 1.4.8. $\diamond$

Returning to bases, we see that if $\mathscr{B}$ is a basis of a space X, then the nbds of a point $x \in X$ are determined by a subfamily of $\mathscr{B}$, namely, $\mathscr{B}_x = \{B \in \mathscr{B} \mid x \in B\}$. For, a set $N \subseteq X$ is an nbd of x in X if and only if N contains some member of $\mathscr{B}_x$. This leads to the following notion.

Definition 1.4.10 Let X be a topological space and $x \in X$. A *neighborhood basis* (or a *local basis*) at x is a collection $\mathscr{B}_x$ of nbds of x in X such that each nbd of x in X contains some member of $\mathscr{B}_x$.

Example 1.4.9 In a topological space, the family of all open nbds of a point is a neighborhood basis at that point.

Example 1.4.10 In a discrete space X, there is a neighborhood basis at a point x which consists of just one set, viz., $\{x\}$.

Example 1.4.11 In a metric space X, the open r-balls about a given point x for all real $r > 0$ form a neighborhood basis at x.

In a space X, it is easily seen that a collection of neighborhood bases $\mathscr{B}_x$ at different points x satisfies the following conditions:

(a) $\mathscr{B}_x$ is a nonempty set for each $x \in X$;
(b) $x \in B$ for every $B \in \mathscr{B}_x$;
(c) for every two sets $B_1, B_2 \in \mathscr{B}_x$, there exists a $B_3 \in \mathscr{B}_x$ such that $B_3 \subseteq B_1 \cap B_2$; and
(d) for each $B \in \mathscr{B}_x$, there exists a $B' \in \mathscr{B}_x$ such that B contains a member of $\mathscr{B}_y$ for every $y \in B'$.

Conversely, we have the following theorem.

Theorem 1.4.11 *Let X be a set and suppose that, for each $x \in X$, $\mathscr{B}_x$ is a family of subsets of X satisfying the conditions (a)–(d), above. Then there is a unique topology on X such that $\mathscr{B}_x$ is a neighborhood basis at x for each $x \in X$.*

Proof Consider the collection

$$\mathscr{T} = \left\{G \subseteq X \mid G \text{ contains a set } B \in \mathscr{B}_x \text{ for every } x \in G\right\}.$$

It is easily verified that $\mathscr{T}$ is a topology for X. Next, we show that each $\mathscr{B}_x$ is a neighborhood basis at x in the space $(X, \mathscr{T})$. Given a set $B \in \mathscr{B}_x$, put $U = \{y \in X \mid B \text{ contains a set } C_y \in \mathscr{B}_y\}$. It is then obvious that $x \in U \subseteq B$. Furthermore, U is open in the topology $\mathscr{T}$. For, if $y \in U$, then B contains a set $C_y \in \mathscr{B}_y$, by the definition of U. And, by (d), there exists a set $C'_y \in \mathscr{B}_y$ such that C_y contains some member of $\mathscr{B}_z$ for every $z \in C'_y$. Since $C_y \subseteq B$, we have $C'_y \subseteq U$. So U is open, and thus B is an nbd of x. Moreover, if N is an nbd of x, then there exists a $G \in \mathcal{T}$ such that $x \in G \subseteq N$. By the definition of $\mathcal{T}$, G contains a member of $\mathcal{B}_x$, and therefore so does N. Thus we see that $\mathcal{B}_x$ is a neighborhood basis at x.

Finally, to prove the uniqueness of $\mathcal{T}$, suppose that there is a topology $\mathcal{T}'$ for X such that every family $\mathcal{B}_x$ is a neighborhood basis at x. Then a set $G \subseteq X$ is open in $\mathcal{T}'$ if and only if, for each $x \in G$, there is a set $B \in \mathcal{B}_x$ such that $B \subseteq G$. Accordingly, $\mathcal{T}'$ coincides with $\mathcal{T}$. $\diamond$

The preceding theorem shows that the concept of neighborhood basis can be used as a primitive notion to define a topology on a set X.

Exercises

1. (a) What is the order topology on $\mathbb{N}$ in the usual order?

   (b) Is the ordered space $\{1, 2\} \times \mathbb{N}$ with the dictionary order discrete?

2. Show that the topology on $\mathbb{R}$ generated by the subbasis

$$\{[a, b) \mid a, b \in \mathbb{R}\} \cup \{(a, b] \mid a, b \in \mathbb{R}\}$$

   coincides with the discrete topology.

3. Describe the topology on the plane for which the family of all straight lines is a subbase.

4. Show that the sets $\{x \in \mathbb{R} \mid x > r\}$, $\{x \in \mathbb{R} \mid x < s\}$, where $r, s \in \mathbb{Q}$, form a subbasis for the Euclidean topology of $\mathbb{R}$. Is this still true if r, s are restricted to the numbers of the form $k/2^n$, where n and k are arbitrary integers?

5. Show that the sets of form $\{x \mid x \geq a\}$ and $\{x \mid x < b\}$, $a, b \in \mathbb{R}$, constitute a subbasis for the topology of the Sorgenfrey line $\mathbb{R}_\ell$.

6. • Let $A = \{1, 1/2, 1/3, \ldots\}$. Show that the collection of open intervals (a, b) and the sets $(a, b) - A$ is a basis for a topology on $\mathbb{R}$. Describe the topology generated by this base (this topology is sometimes called the *Smirnov topology* for $\mathbb{R}$), and compare it with the different topologies on $\mathbb{R}$ discussed in this section.

7. Find the boundary, closure, and interior of the set $\{1/n \mid n \in \mathbb{N}\}$ in the topology (on $\mathbb{R}$) generated by the basis $\{(a, +\infty) \mid a \in \mathbb{R}\}$. (This topology on $\mathbb{R}$ is referred to as the right order topology; a left order topology is defined similarly.)

8. (a) Determine the boundary, closure, and interior of the subsets $(0, \sqrt{2})$ and $(\sqrt{3}, 4)$ of $\mathbb{R}$ in the topology generated by the basis $\{[a, b) \mid a, b \in \mathbb{Q}\}$.

   (b) Show that the topology in (a) is strictly coarser than the lower limit topology.

9. The collection of all open intervals (a, b) together with the singletons $\{n\}$, $n \in \mathbb{Z}$, is a base for a topology on $\mathbb{R}$. Describe the interior operation in the resulting space.

10. Show that the rationals are dense in the Sorgenfrey line $\mathbb{R}_\ell$.

11. Let $\mathcal{B}$ be a basis for the topological space X and $A \subseteq X$. Show that $x \in \overline{A} \Leftrightarrow B \cap A \neq \varnothing$ for every B in $\mathcal{B}$ with $x \in B$.

12. Let $\mathcal{B}$ be a basis for the space X. Show that a subset $D \subseteq X$ is dense if and only if every nonempty member of $\mathcal{B}$ intersects D.

13. Let $\mathcal{S}$ be a subbasis for the topology of a space X. If $D \subseteq X$ and $U \cap D \neq \varnothing$ for each $U \in \mathcal{S}$, is D dense in X?

14. • Prove that Ω is a limit point of the set $[0, \Omega)$ in the ordinal space $[0, \Omega]$.

15. • Define an order relation $\preceq$ on $I \times I$ by

$$(x, y) \preceq (x', y') \Leftrightarrow y < y' \text{ or } (y = y' \text{and} x \leq x').$$

The order topology on $I \times I$ is called the *television* topology (the name is due to E.C. Zeeman). Determine the closures of the following subsets of $I \times I$:

$$A = \{(0, y) \mid 0 < y < 1\}, \qquad B = \{(0, n^{-1}) \mid n \in \mathbb{N}\},$$
$$C = \{(x, 2^{-1}) \mid 0 < x < 1\}, \qquad D = \{(2^{-1}, y) \mid 0 < y < 1\}, \text{ and}$$
$$E = \{(2^{-1}, 1 - n^{-1}) \mid n \in \mathbb{N}\}.$$

16. Let X be an ordered set with the order topology. Show that $\overline{(a, b)} \subseteq [a, b]$. Find the conditions for equality.

17. Show that the taxi-cab metric d^+ on $\mathbb{R}^n$, $d^+ ((x_i), (y_i)) = \sum |x_i - y_i|$, is equivalent to the Euclidean metric.

18. Let (X, d) be a metric space. Show that the metric d', defined by $d' = d/(1 + d)$, is equivalent to d.

19. Show that the metrics d and d^* defined on $\mathcal{C}(I)$ by

$$d(f, g) = \int_0^1 |f(t) - g(t)| dt, \quad \text{and}$$
$$d^*(f, g) = \sup \{|f(t) - g(t)| : 0 \leq t \leq 1\}$$

are not equivalent.

20. In the real line $\mathbb{R}$, show that the collection of open intervals $(x - r, x + r)$, r ranging over the set of all positive rational numbers, is a neighborhood base at x.

21. Let (a, b) be a particular point of $\mathbb{R}^2$. Show that the set of all squares with sides parallel to the axes and centered at (a, b) is a neighborhood basis at (a, b).

22. • Let X be a space and, for each $x \in X$, let $\mathcal{B}_x$ be an nbd basis at x. Show that the collection $\{\mathcal{B}_x \mid x \in X\}$ satisfies the conditions of Theorem 1.4.11.

23. Let X and $\mathscr{B}_x$ be as in Exercise 22. For any $A \subseteq X$, show the following:

    (a) x is an interior point of $A \Leftrightarrow A$ contains some member of $\mathscr{B}_x$.

    (b) x is an adherence point of $A \Leftrightarrow A$ intersects every member of $\mathscr{B}_x$.

1.5 Subspaces

A subset of a space inherits a topology from its parent space, in an obvious way. This is the simplest method of constructing a new space from a given one.

Definition 1.5.1 Let $(X, \mathscr{T})$ be a space and $Y \subseteq X$. The *relative* topology or the *subspace topology* $\mathscr{T}_Y$ on Y is the collection of all intersections of Y with open subsets of X, and Y equipped with this topology is called a *subspace* of X.

A routine verification shows that $\mathscr{T}_Y = \{Y \cap U \mid U \in \mathscr{T}\}$ is, indeed, a topology on Y. Each member H of $\mathscr{T}_Y$ is said to be *open* in Y and its relative complement $Y - H$ is *closed* in Y. We have the following

Proposition 1.5.2 *Let Y be a subspace of a space X. A subset $K \subseteq Y$ is closed in Y if and only if $K = Y \cap F$, where F is closed in X.*

Proof For any $F \subseteq X$, we have $Y - (Y \cap F) = Y \cap (X - F)$, and it follows that K is closed in Y if $K = Y \cap F$ and F is closed in X. Conversely, suppose that K is closed in Y. Then $Y - K = Y \cap G$ for an open subset G of X. This implies that $K = Y \cap (X - G)$, where $X - G$ is obviously closed in X. $\diamond$

Example 1.5.1 Let (X, d) be a metric space and $Y \subseteq X$. Then the relative topology on Y induced by the metric topology on X coincides with the metric topology determined by the restriction of d to Y. This follows from the observation that an open ball about y of radius r in Y is the intersection of Y with the open ball $B(y; r)$ in X. In particular, we see that $\mathbb{D}^n$, I^n and $\mathbb{S}^{n-1}$ have the same topological structures whether considered as subspaces of the Euclidean metric space $\mathbb{R}^n$ or those of the space $\mathbb{R}^n$ with the usual topology.

If Y is a subspace of X, then any subset of Y which is open (or closed) in X has the same property in Y. But, an open (or closed) subset of Y need not be open (or closed) in X, as shown by the following.

Example 1.5.2 In the subspace $Y = (0, 1] \cup [2, 3]$ of $\mathbb{R}$, the set $(0, 1]$ is open as well as closed. But this is not open or closed in $\mathbb{R}$.

In this regard, we have the following.

Proposition 1.5.3 *Let Y be a subspace of a space X. If Y is closed (or open) in X and A is closed (resp. open) in Y, then A is closed (resp. open) in X.*

Proof By Proposition 1.5.2, $A = Y \cap F$ for some closed subset F of X. Since the intersection of two closed subsets is closed, A is closed in X.

A similar argument applies to the "open" case.   ◇

A direct consequence of the definition of the relative topology is the following.

Proposition 1.5.4 *If Y is a subspace of X, and Z is a subspace of Y, then Z is a subspace of X.*

This property of relativization is often used without explicit mention. The proof of the next proposition is also trivial, and left to the reader.

Proposition 1.5.5 *Let X be a space and $Y \subseteq X$.*

(a) *If $\mathscr{B}$ is a basis (resp. subbasis) of X, then $\{Y \cap B \mid B \in \mathscr{B}\}$ is a basis (resp. subbasis) for the relative topology of Y.*
(b) *If $\mathscr{B}_x$ is an nbd base at $x \in X$ and $x \in Y$, then $\{B \cap Y \mid B \in \mathscr{B}_x\}$ is an nbd basis at x in Y.*

Let Y be a subspace of a space X. For any $A \subseteq Y$, we can form the boundary, closure, derived set, and interior of A using the topology of Y or X. In such situations, we need to specify the space in which the closure (boundary or interior or derived set) is taken. We shall use the notations ∂A_X, $\overline{A}_X$, A'_X, A°_X, to indicate that these operations are performed in X. The following proposition determines the various relations.

Proposition 1.5.6 *Let Y be a subspace of a space X, and $A \subseteq Y$. Then $\overline{A}_Y = \overline{A}_X \cap Y$, $A'_Y = A'_X \cap Y$, $A^\circ_Y \supseteq A^\circ_X \cap Y = A^\circ_X$ and $\partial A_Y \subseteq \partial A_X \cap Y$.*

Proof Since $\overline{A}_X$ is closed in X, $\overline{A}_X \cap Y$ is a closed subset of Y containing A. So $\overline{A}_Y \subseteq \overline{A}_X \cap Y$. On the other hand, $\overline{A}_Y$ is closed in Y, and therefore $\overline{A}_Y = Y \cap F$ for some closed subset F of X. It follows that $A \subseteq F$ whence $\overline{A}_X \cap Y \subseteq F \cap Y = \overline{A}_Y$.

The other statements follow readily from the definitions.   ◇

Example 1.5.3 Let $Y = \{(0, y) \mid y \in \mathbb{R}\}$ have the relative topology induced by the usual topology of the Euclidean space $\mathbb{R}^2 = X$. For $A = \{0\} \times [-1, +1]$, $A^\circ_Y = \{0\} \times (-1, +1)$ while $A^\circ_X = \varnothing$, and $\partial A_Y = \{0\} \times \{-1, +1\}$ while $\partial A_X = A$. This example shows that the inclusions in Proposition 1.5.6 may be strict.

Finally, in this section, we consider the behavior of relativization with order topology. Unfortunately, this is not so nice as we would like it to be. If X is an ordered set and $Y \subseteq X$, then the restriction of the simple order relation on X is a simple order relation on Y. Thus Y receives two topologies: one induced by the restricted order and the other one—the relative topology—inherited from the order topology on X. The two topologies for Y do not agree in all cases, as is seen below.

Example 1.5.4 Consider the real line with its usual order and topology. Let $Y = \{0\} \cup (1, 2)$. Then $\{0\}$ is open in the subspace Y; but, in the order topology for Y, each basic open set containing 0 contains some points greater than 1.

Example 1.5.5 Let X be an ordered space, and J be an interval in X. By Proposition 1.5.5, the sets $(-\infty, a) \cap J$ and $(a, +\infty) \cap J$, $a \in X$, form a subbasis for the relative topology on J. One observes that if $a \notin J$, then these sets are $\varnothing$ or J, and if $a \in J$, then these are obviously open rays in J. Consequently, the relative topology is coarser than the order topology of J. On the other hand, an open ray in J is $(-\infty, a) \cap J$ or $(a, +\infty) \cap J$, which is open in the relative topology for J. It follows from the definition of order topology that the relative topology is finer than the order topology on J and the two topologies for J agree.

Exercises

1. A subset Y of a space X is called *discrete* if the relative topology for Y is discrete. Prove that every subset of a discrete space is discrete.

2. (a) Show that the subset $\{1/n \mid n \in \mathbb{N}\}$ of real line $\mathbb{R}$ is discrete, while the subset $\{0\} \cup \{1/n \mid n \in \mathbb{N}\}$ is not.

   (b) Show that $\mathbb{Z}$ is a discrete subset of $\mathbb{R}$.
   (Notice that $\mathbb{Z}$ is closed in $\mathbb{R}$, while $\{1/n \mid n \in \mathbb{N}\}$ is not closed.)

3. Verify that the set $\left\{x \in \mathbb{Q} \mid -\sqrt{2} \leq x \leq \sqrt{2}\right\}$ is both open and closed in the subspace $\mathbb{Q} \subset \mathbb{R}$. (Notice that this is true for any open interval with irrational end points.)

4. Consider the subspace $X = [-1, 1]$ of the real line $\mathbb{R}$. Decide the openness and closedness in X of the sets in Exercise 1.2.4.

5. Let $\mathscr{T}$ and $\mathscr{U}$ be topologies for set X such that $\mathscr{U}$ is strictly finer than $\mathscr{T}$. For $Y \subseteq X$, what can be said about $\mathscr{T}_Y$ and $\mathscr{U}_Y$?

6. Let X be a space in which every finite subspace has the trivial topology. Show that X itself has the trivial topology. Is the corresponding assertion for the discrete topology true?

7. Let $F \subseteq X$ be closed and $U \subseteq F$ be open in F. Let V be any open subset of X with $U \subseteq V$. Prove that $U \cup (V - F)$ is open in X.

8. Let Y be a subspace of X. If A is dense in Y, show that A is dense in $\overline{Y}$.

9. Give an example of a space X which has a dense subset D and a subset Y such that $D \cap Y$ is not dense in Y.

10. Consider the sets $\mathbb{Z}$ and $\mathbb{Q}$ with the usual order relations. Show that the order topologies on $\mathbb{Z}$ and $\mathbb{Q}$ coincide with the relative topologies induced from the real line $\mathbb{R}$.

11. Consider the dictionary order on $\mathbb{R}^2$ and its restriction to I^2, $I = [0, 1]$. Is the order topology for I^2 the same as the relative topology induced from the order topology on $\mathbb{R}^2$?

12. Call a subset Y of an ordered set $(X, \preceq)$ *convex* if the interval $(a, b) \subseteq Y$ for every $a \prec b$ in Y.

   (a) Verify that an interval in X, including a ray, is convex.

   (b) Is a proper convex subset of an ordered set X an interval or a ray?

   (c) Prove that the relative topology on a convex subset Y of an ordered space X agrees with the order topology for Y.

Chapter 2
Continuity and the Product Topology

The central notion in topology is the concept of "continuity of functions" between topological spaces. A discussion of this concept and some other notions related to mappings is the object of the first section. In the second section, we consider the problem of topologizing the Cartesian products of a family of topological spaces in some natural and useful way. Here, we are mainly concerned with the "Tychonoff topology," which is the smallest topology for a product of topological spaces such that each projection is continuous.

2.1 Continuous Functions

The conditions of a topological structure have been so formulated that the definition of a continuous function can be borrowed word for word from analysis.

Definition 2.1.1 Let X and Y be spaces. A function $f : X \to Y$ is called *continuous* if $f^{-1}(U)$ is open in X for each open set $U \subseteq Y$.

Example 2.1.1 A constant function $c : X \to Y$ is obviously continuous, for $c^{-1}(U)$ is either $\varnothing$ or X for every $U \subseteq Y$.

Example 2.1.2 Every function on a discrete space is continuous.

The following theorem provides some other ways of formulating the continuity condition.

Theorem 2.1.2 *Let X and Y be topological spaces, and let $f : X \to Y$ be a function. The following conditions are equivalent:*

(a) *f is continuous.*
(b) *$f^{-1}(F)$ is closed in X for every closed set $F \subseteq Y$.*

© Springer Nature Singapore Pte Ltd. 2019
T. B. Singh, *Introduction to Topology*,
https://doi.org/10.1007/978-981-13-6954-4_2

(c) $f\left(\overline{A}\right) \subseteq \overline{f(A)}$ for every set $A \subseteq X$.

(d) $\overline{f^{-1}(B)} \subseteq f^{-1}\left(\overline{B}\right)$ for every set $B \subseteq Y$.

Proof (a) $\Leftrightarrow$ (b): This is immediate from the equality $f^{-1}(Y - B) = X - f^{-1}(B)$.

Now, we prove that (b) $\Rightarrow$ (c) $\Rightarrow$ (d) $\Rightarrow$ (b).

(b) $\Rightarrow$ (c): Since $\overline{f(A)}$ is closed, $f^{-1}\left(\overline{f(A)}\right)$ is closed, by (b). Obviously, $A \subseteq f^{-1}\left(\overline{f(A)}\right)$; so $\overline{A} \subseteq f^{-1}\left(\overline{f(A)}\right)$ and (c) holds.

(c) $\Rightarrow$ (d): Taking $A = f^{-1}(B)$ in (c), we have $f\left(\overline{f^{-1}(B)}\right) \subseteq \overline{f\left(f^{-1}(B)\right)} \subseteq \overline{B}$, which implies (d).

(d) $\Rightarrow$ (b): If F is a closed subset of Y, then (d) implies $\overline{f^{-1}(F)} \subseteq f^{-1}(F)$. But $f^{-1}(F) \subseteq \overline{f^{-1}(F)}$ always, so the equality holds and $f^{-1}(F)$ is closed. $\diamond$

If a basis (or subbasis) for a space Y is known, then the task of proving the continuity of a function into Y becomes easier, as shown by the following.

Theorem 2.1.3 *Let X and Y be topological spaces. A function $f : X \to Y$ is continuous if and only if the inverse image of every set in a basis (or subbasis) of Y is open.*

Proof In the case a basis of Y is given, the theorem follows immediately from the definitions and the equality $f^{-1}\left(\bigcup_i B_i\right) = \bigcup_i f^{-1}(B_i)$, $B_i \subseteq Y$.

To prove the second case, let $\mathscr{S}$ be a subbasis of Y. If f is continuous, then $f^{-1}(S)$ is obviously open in X for each $S \in \mathscr{S}$, for every member of $\mathscr{S}$ is open in Y. Conversely, suppose that $f^{-1}(S)$ is open in X for every $S \in \mathscr{S}$, and let $U \subseteq Y$ be open. If $f^{-1}(U) \neq X$ and $x \in f^{-1}(U)$, then there exist finitely many sets $S_1, \ldots, S_n$ in $\mathscr{S}$ (say) such that $f(x) \in \bigcap_1^n S_i \subseteq U$. So $x \in \bigcap_1^n f^{-1}(S_i) \subseteq f^{-1}(U)$, and hence $f^{-1}(U)$ is a nbd of x. Since $x \in f^{-1}(U)$ is arbitrary, we deduce that $f^{-1}(U)$ is open, and therefore f is continuous. $\diamond$

Example 2.1.3 The function $f : [0, 2\pi) \to \mathbb{S}^1$ defined by $f(t) = e^{it}$ is continuous. The collection of all open arcs of the circle is a basis for the topology on $\mathbb{S}^1$. If G is such an arc not containing $1 \in \mathbb{S}^1$, then $f^{-1}(G)$ is an open interval of the form (a, b), $0 < a < b < 2\pi$. And, if G contains 1, then $f^{-1}(G)$ has the form $[0, a) \cup (b, 2\pi)$, $a < b$. This is an open subset of $[0, 2\pi)$, and the continuity of f follows from the preceding theorem.

Theorem 2.1.4 (a) *Given a space X, the identity map $1_X : X \to X$ is continuous.*

(b) *If $f : X \to Y$ and $g : Y \to Z$ are continuous functions between spaces, then the composition $gf : X \to Z$ is also continuous.*

The straightforward proof of the theorem is left to the reader.

Let $(X, \mathscr{T})$ be a topological space and $A \subseteq X$. If $j : A \hookrightarrow X$ is the inclusion map, then $j^{-1}(U) = U \cap A$ for every $U \subseteq X$. By the definition of the relative topology, the function j is continuous in the relative topology $\mathscr{T}_A$, and it is also clear that any topology on A which makes the function j continuous must contain $\mathcal{T}_A$. So $\mathcal{T}_A$ can be characterized as the *smallest* topology on A for which j is continuous.

Let $f : X \to Y$ be a continuous function. If A is a subspace of X, then the restriction $f|A : A \to Y$ is continuous, since it is just the composite $A \overset{j}{\hookrightarrow} X \overset{f}{\to} Y$, where $j : A \hookrightarrow X$ is the inclusion map. Also, if $f(X)$ is given the relative topology, the function $g : X \to f(X)$ defined by f is continuous, for $g^{-1}(O \cap f(X)) = f^{-1}(O)$ for every $O \subseteq Y$.

As in the case of metric spaces, we also have a localized form of continuity.

Definition 2.1.5 (*Cauchy*) A function $f : X \to Y$ between spaces is said to be *continuous* at $x \in X$, if given any nbd N of $f(x)$ in Y, there exists a nbd M of x in X such that $f(M) \subseteq N$.

Since $f\left(f^{-1}(N)\right) \subseteq N$, this is the same as saying that $f^{-1}(N)$ is a nbd of x for each nbd N of $f(x)$, and for this it is sufficient that the condition holds for all members of a nbd basis of $f(x)$. When X and Y are metric spaces, this reduces to the ϵ-δ formulation.

Theorem 2.1.6 *A function $f : X \to Y$ is continuous if and only if it is continuous at each point of X.*

Proof Suppose that f is continuous, and let $x \in X$ be arbitrary. If N is a nbd of $f(x)$ in Y, then there exists an open subset U of Y such that $f(x) \in U \subseteq N$. So $x \in f^{-1}(U) \subseteq f^{-1}(N)$. By our hypothesis, $f^{-1}(U)$ is open in X, and thus $f^{-1}(N)$ is a nbd of x in X. This proves the continuity of f at x.

Conversely, suppose that f is continuous at each point of X. Let V be any open subset of Y. If $x \in f^{-1}(V)$, then V is a nbd of $f(x)$. By our assumption, there exists a nbd M of x in X such that $M \subseteq f^{-1}(V)$. This implies that $f^{-1}(V)$ is a nbd of x. Thus, being a nbd of each of its points, $f^{-1}(V)$ is open in X and f is continuous. $\Diamond$

We apply the above theorem to prove the continuity of the addition function α : $(x, y) \mapsto x + y$ and the multiplication function $\mu : (x, y) \mapsto xy$ on the Euclidean space $\mathbb{R}^2$:

As seen in Example 1.4.9, the Euclidean topology on $\mathbb{R}^2$ is also induced by the metric d^* given by $d^*(p, q) = \max\{|x - s|, |y - t|\}$ for $p = (x, y)$ and $q = (s, t)$. Let $q = (s, t)$ be an arbitrary but fixed point of $\mathbb{R}^2$, and suppose that N is a nbd of $\alpha(q) = s + t$. Then there exists a real $\epsilon > 0$ such that the open interval $(\alpha(q) - \epsilon, \alpha(q) + \epsilon) \subseteq N$. Now, for any point $p = (x, y)$ in the open ball $B_{d^*}(q; \epsilon/2)$, we have $|x - s| < \epsilon/2$ and $|y - t| < \epsilon/2$. Consequently,

$$|\alpha(p) - \alpha(q)| = |x + y - s - t| \leq |x - s| + |y - t| < \epsilon,$$

and it follows that $\alpha(p) \in (\alpha(q) - \epsilon, \alpha(q) + \epsilon) \subseteq N$. Thus α maps the nbd $B_{d^*}(q; \epsilon/2)$ of q into N, and hence α is continuous at q.

To see the continuity of the multiplication function μ at a point $q = (s, t)$ of $\mathbb{R}^2$, consider a nbd N of $\mu(q) = st$. Then find a real $\epsilon > 0$ such that the open interval $(st - \epsilon, st + \epsilon) \subseteq N$. We note that

$$|xy - st| \leq |x - s||y - t| + |x - s||t| + |y - t||s|.$$

So, for $0 < \delta < \min\{1, \epsilon/(1 + |s| + |t|)\}$, we have $|xy - st| < \epsilon$ if $|x - s| < \delta$ and $|y - t| < \delta$. It follows that $\mu(p) \in (st - \epsilon, st + \epsilon) \subseteq N$ for every $p \in B_{d^*}(q; \delta)$. Accordingly, μ maps the nbd $B_{d^*}(q; \delta)$ of q into N, and hence μ is continuous at q.

We next study the piecewise construction of continuous functions. Let X be a set and $\{A_i\}$ be a family of subsets of X such that $X = \bigcup A_i$. Suppose that, for each index i, $f_i : A_i \to Y$ is a function satisfying $f_i|(A_i \cap A_j) = f_j|(A_i \cap A_j)$ for every i and j. Then there is a function $f : X \to Y$ such that $f|A_i = f_i$ for every i. For, given $x \in X$, there is an index i such that $x \in A_i$. We set $f(x) = f_i(x)$. If $x \in A_j$ also, then $f_i(x) = f_j(x)$, by the hypothesis. So $f(x)$ is uniquely defined and $x \mapsto f(x)$ is a mapping extending each f_i. It may be noted that the function f thus defined is unique. The following result, known as the *Gluing* (or *Pasting*) *lemma*, leads to a useful process of constructing continuous functions.

Lemma 2.1.7 *Let X and Y be topological spaces, and let $A_1, \ldots, A_n$ be a finite family of closed (or open) subsets of X such that $X = \bigcup_1^n A_i$. If $f_i : A_i \to Y$ is a continuous function for every $i = 1, \ldots, n$ and $f_i|(A_i \cap A_j) = f_j|(A_i \cap A_j)$ for all i and j, then the function f defined by $f|A_i = f_i$ for every i is continuous.*

Proof Suppose first that each A_i is closed in X, and let $F \subseteq Y$ be any closed set. Then, by the continuity of f_i, $f_i^{-1}(F)$ is closed in A_i. Since A_i is closed in X for every i, we see that each $f_i^{-1}(F)$ is closed in X. Accordingly, $f^{-1}(F) = \bigcup_{i=1}^n f_i^{-1}(F)$ is closed, and f is continuous.

The proof in the case A_i's are open is quite similar. $\diamond$

Clearly, the preceding lemma extends to an arbitrary family of open sets $\{A_i\}$. However, this is not possible in the case involving infinite families of closed sets, as can be easily seen by taking one-point sets $\{x\}$ in an interval $[a, b] \subset \mathbb{R}$. To remedy the situation, we need to impose some restriction on the position of the (closed) sets A_i in X.

Definition 2.1.8 A family $\{A_i\}$ of subsets of a space X is called *locally finite* (or *nbd-finite*) if each point of X has a nbd U such that $U \cap A_i \neq \emptyset$ for at most finitely many indices i.

Proposition 2.1.9 *Let $\{A_i\}$ be a locally finite family of subsets of a space X. Then $\bigcup \overline{A_i}$ is closed in X.*

Proof Each point $x \in X$ has an open nbd U such that $U \cap A_i \neq \emptyset$ for at most finitely many indices i. The same is true of the sets $U \cap \overline{A_i}$, since $U \cap A_i = \emptyset \Rightarrow U \cap \overline{A_i} = \emptyset$. Now, if $x \notin \bigcup \overline{A_i}$ and U meets the sets $\overline{A_{i_1}}, \ldots, \overline{A_{i_n}}$ only, then the nbd $V = \bigcap_{j=1}^n \left(U \cap (X - \overline{A_{i_j}})\right)$ of x does not meet any of the $\overline{A_i}$. It follows that the $X - \bigcup \overline{A_i}$ is open, and so $\bigcup \overline{A_i}$ is closed. $\diamond$

By the preceding proposition, we see that the union of a locally finite family of closed sets is closed.

Corollary 2.1.10 *If a space X is the union of a locally finite family $\{A_i\}$ of closed sets, then a function f from X to a space Y is continuous if and only if the restriction of f to each A_i is continuous.*

Proof Suppose that $f|A_i$ is continuous for every i. If F is a closed subset of Y, then, for every i, $f^{-1}(F) \cap A_i = (f|A_i)^{-1}(F)$ is closed in A_i, and hence in X. Since $\{A_i\}$ is locally finite, so is the family $\{f^{-1}(F) \cap A_i\}$. By the preceding proposition, $f^{-1}(F) = \bigcup_i (f^{-1}(F) \cap A_i)$ is closed in X, and therefore f is continuous. The converse is obvious. ◇

We shall often set about defining a continuous function $f: X \to Y$ by cutting up X into closed (or open) subsets A_i and defining f on each A_i separately in such a way that $f|A_i$ is obviously continuous and the different definitions agree on the overlaps.

Next, we consider the notion of "equivalence" between topological spaces.

Definition 2.1.11 A *homeomorphism* between spaces X and Y is a bijective function $f: X \to Y$ such that both f and f^{-1} are continuous. Two spaces X and Y are said to be *homeomorphic*, denoted by $X \approx Y$, if there is a homeomorphism $X \to Y$.

Clearly, a continuous function $f: X \to Y$ is a homeomorphism if and only if there is a continuous function $g: Y \to X$ such that $gf = 1_X$ and $fg = 1_Y$. We must point out that a continuous bijection need not be a homeomorphism; for example, consider the identity map of a set with the discrete topology onto the same set, but equipped with a different topology.

Example 2.1.4 The Euclidean space $\mathbb{R}^n$ is homeomorphic to the open ball $B(0; 1) = \{x \in \mathbb{R}^n |\, \|x\| < 1\}$. The map $x \mapsto x/(1 + \|x\|)$ is a homeomorphism of $\mathbb{R}^n$ onto $B(0; 1)$ with $x \mapsto x/(1 - \|x\|)$ as its inverse.

Example 2.1.5 The punctured sphere $\mathbb{S}^n - \{p\}$, where $p = (0, \dots, 0, 1)$, is homeomorphic to $\mathbb{R}^n$. In fact, the mapping $f: \mathbb{S}^n - \{p\} \to \mathbb{R}^n$ defined by $f(x_0, \dots, x_n) = (x_0, \dots, x_{n-1})/(1 - x_n)$ is readily seen to be a homeomorphism with the inverse

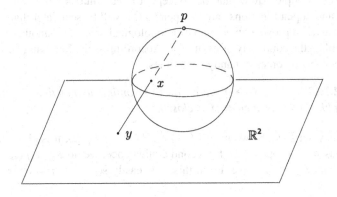

Fig. 2.1 Stereographic projection

mapping $g: \mathbb{R}^n \to \mathbb{S}^n - \{p\}$ given by $g(y) = (\lambda_y y_0, \ldots, \lambda_y y_{n-1}, 1 - \lambda_y)$, where $y = (y_0, \ldots, y_{n-1})$ and $\lambda_y = 2/(1 + \sum y_i^2)$. The homeomorphism f is called *the stereographic projection* (Fig. 2.1).

Example 2.1.6 Note that the correspondence $a + \imath b \leftrightarrow (a, b)$ between $\mathbb{C}$ and $\mathbb{R}^2$ defines a bijection $\mathbb{C}^n \leftrightarrow \mathbb{R}^{2n}$ which is clearly a distance preserving map. Therefore, the unitary space $\mathbb{C}^n$ is homeomorphic to the Euclidean space $\mathbb{R}^{2n}$. Similarly, the bijection between the symplectic space $\mathbb{H}^n$ and the Euclidean space $\mathbb{R}^{4n}$ defined by the correspondence $q \leftrightarrow (x_0, x_1, x_2, x_3)$, where q is the quaternion $x_0 + x_1 \imath + x_2 \jmath + x_3 k$, is a homeomorphism.

The general method for showing that two spaces are homeomorphic involves construction of continuous functions from one space to the other, which are inverse to each other. Unfortunately, the construction of such functions is generally not easy. We discuss below some characterizations of homeomorphisms, although these hardly make our task any easier. To this end, we first identify two types of functions between topological spaces.

Definition 2.1.12 A function $f: X \to Y$ between spaces is called *open* (resp. *closed*) if the image of each open (resp. closed) set in X is open (resp. closed) in Y.

Example 2.1.7 If Y is a discrete space, then every function $f: X \to Y$ is obviously closed and open.

Example 2.1.8 The inclusion map $(0, 2) \to \mathbb{R}$ is open but it is not closed, and the inclusion map of I into $\mathbb{R}$ is closed but not open.

Example 2.1.9 The continuous function $t \mapsto e^{\imath t}$ of $[0, 2\pi)$ onto $\mathbb{S}^1$ is neither closed nor open. For, the interval $[0, \pi)$ is open in $[0, 2\pi)$, but its image under the mapping is not open. Further, notice that the interval $[\pi, 2\pi)$ is closed in $[0, 2\pi)$, but its image under the mapping is not closed.

The above examples show that the concepts of "continuous functions," "closed functions," and "open functions" are independent. It will be seen later that, in some cases, it is easier to prove a given map of a topological space into another is closed than to establish the continuity of its inverse. Accordingly, the following description of homeomorphism comes in handy sometimes.

Theorem 2.1.13 *Let $f: X \to Y$ be a bijective continuous function. Then f is a homeomorphism $\Leftrightarrow f$ is open $\Leftrightarrow f$ is closed.*

Proof For any set $U \subseteq X$, we have $(f^{-1})^{-1}(U) = f(U)$. So, it is clear that f^{-1} is continuous $\Leftrightarrow f$ is open. For the second equivalence, we note that $f(X - U) = Y - f(U)$, since f is bijective. From this, it is easily seen that f is open $\Leftrightarrow f$ is closed. $\diamond$

By the preceding theorem, a homeomorphism $h: X \approx Y$ not only establishes a one-to-one correspondence between the points of X and Y, but also sets up a one-to-one correspondence between the open sets (resp. closed sets) in X and the open sets (resp. closed sets) in Y. Thus any assertion about X as a topological space is also valid for each homeomorph of X. For this reason, homeomorphic spaces are called *topologically equivalent*. We call a property of spaces a *topological invariant* (or a *topological property*) if it is possessed by every homeomorph of a space which has this property. Topology is often characterized as the study of topological invariants. We will see several such invariants as we go further in the book. These invariants are often used in deciding whether two given spaces are homeomorphic or not. Specifically, we show that they are not homeomorphic by obtaining different answers in the computation of some suitable invariant for the two spaces. It is interesting to notice that not all properties in a metric space are topological invariant, for example, the property of boundedness.

Closed maps and the closure operation are related by the following

Proposition 2.1.14 *Let X and Y be spaces, and let $f: X \to Y$ be a function. Then f is closed $\Leftrightarrow \overline{f(A)} \subseteq f\left(\overline{A}\right)$ for each set $A \subseteq X$.*

Proof If f is closed, then $f\left(\overline{A}\right)$ is closed. Since $f(A) \subseteq f\left(\overline{A}\right)$, we have $\overline{f(A)} \subseteq \overline{f\left(\overline{A}\right)} = f\left(\overline{A}\right)$.

Conversely, suppose that the given condition holds, and let A be any closed subset of X. Then $f(A) \subseteq \overline{f(A)} \subseteq f\left(\overline{A}\right) = f(A)$, which implies that $f(A) = \overline{f(A)}$. Thus $f(A)$ is closed. $\diamond$

Corollary 2.1.15 *A function $f: X \to Y$ between spaces is continuous and closed if and only if $f\left(\overline{A}\right) = \overline{f(A)}$ for each subset A of X.*

By the preceding corollary, we see that a homeomorphism is a bijective map which commutes with the "closure operation."

For analogous results concerning open mappings and interior operation, refer to Exercises 20 and 21. The following characterizations of open functions will be of use later.

Theorem 2.1.16 *The following properties of a function $f: X \to Y$ are equivalent:*

    (a) *f is an open map.*
    (b) *For every $x \in X$ and each nbd U of x in X, there is a nbd W of $f(x)$ in Y with $W \subseteq f(U)$.*
    (c) *For every $x \in X$ and each basic nbd B of x in X, $f(B)$ is a nbd of $f(x)$ in Y.*
    (d) *The image of each member of a basis for X is open in Y.*

The proof is easy and left to the reader.

We conclude this section with a variant of homeomorphism, called "embedding," which allows its domain to be regarded as a subspace of the codomain. Accordingly, the domain of an embedding has the topological properties which are inheritable by subspaces in common with the codomain. The importance of this fact will be realized as we go further in the book.

Definition 2.1.17 An *embedding* (or, more specifically, a *topological embedding*) of a space X into a space Y is a function $f : X \to Y$ which maps X homeomorphically onto the subspace $f(X)$ of Y.

The inclusion of a subspace in its ambient space is an embedding, and so is a closed (or open) continuous injection. However, a continuous injection may not be an embedding, as shown by the following example.

Example 2.1.10 The function $f : [0, 2\pi) \to \mathbb{R}^2$ defined by $f(t) = e^{it}$ is a continuous injection. The image of f is $\mathbb{S}^1$, the unit circle. As seen in Example 2.1.9, the function f fails to map the open neighborhoods of 0 onto the open neighborhoods of 1; accordingly, f is not an embedding.

Exercises

1. Let $\mathscr{T}_1$ and $\mathscr{T}_2$ be two topologies on a set X. Show that

    (a) the identity function $i : (X, \mathscr{T}_1) \to (X, \mathscr{T}_2)$ is continuous $\Leftrightarrow$ $\mathscr{T}_1$ is finer than $\mathscr{T}_2$, and

    (b) i is a homeomorphism $\Leftrightarrow$ $\mathscr{T}_1 = \mathscr{T}_2$.

2. Consider the set $\mathbb{R}$ of real numbers with the various topologies described in the previous chapter except the trivial topology and the discrete topology. Discuss the continuity of the identity function between each pair of spaces.

3. (a) Let $a \neq 0$ and b be fixed real numbers. Show that the affine transformation $x \mapsto ax + b$ is a homeomorphism of $\mathbb{R}$ onto itself.

    (b) Prove that any two open intervals (including the unbounded ones), any two bounded closed intervals with more than one point, and any two half-open intervals (including one-sided closed intervals) in $\mathbb{R}$ are homeomorphic.

4. Prove that the inversion function $x \mapsto x^{-1}$ on $\mathbb{R} - \{0\}$ is continuous and open.

5. Let X be an uncountable set with the cofinite (or cocountable) topology. Show that every continuous function $X \to \mathbb{R}$ is constant.

6. Give an example of a function $f : X \to Y$ between spaces and a subspace $A \subseteq X$ such that $f|A$ is continuous, although f is not continuous at any point of A.

7. Let $f : X \to Y$ be a continuous function and $A \subseteq X$. If x is a limit point of A, is $f(x)$ a limit point of $f(A)$?

8. • Let $\{U_\alpha\}$ be a family of open subsets of a space X with $X = \bigcup U_\alpha$. Show that a function f from X into a space Y is continuous if and only if $f|U_\alpha$ is continuous for each index α. (This shows that continuity of a function is a "local" property.)

9. Let $f : X \to Y$ be a function between spaces, and assume that $X = A \cup B$, where $A - B \subseteq A^\circ$, and $B - A \subseteq B^\circ$. If $f|A$ and $f|B$ are continuous, show that f is continuous.

10. Let X be a partially ordered set. Declare $U \subseteq X$ to be open if it satisfies the condition: if $y \preceq x$ and $x \in U$, then $y \in U$. Show that (a) $\{U \mid U \text{ is open}\}$ is a topology, and (b) a function $f: X \to X$ is continuous $\Leftrightarrow$ it is order preserving (i.e., $x \preceq x' \Rightarrow f(x) \preceq f(x')$).

11. Let $f, g: X \to Y$ be continuous functions, where Y has an order topology. Show that (a) the set $\{x \in X \mid f(x) \preceq g(x)\}$ is closed, and (b) the function $x \mapsto \min\{f(x), g(x)\}$ is continuous.

12. Prove that homeomorphism is an equivalence relation between spaces (equivalence classes are called *homeomorphism types*).

13. Let $f: X \to Y$ be a homeomorphism and $A \subseteq X$. Show that $f|A$ is a homeomorphism between A and $f(A)$, and $f|(X - A)$ is a homeomorphism between $X - A$ and $Y - f(A)$.

14. Let X and Y be two ordered sets with the order topology. If $f: X \to Y$ is bijective and order preserving, show that it is a homeomorphism.

15. Let Y be the two-point discrete space $\{0, 1\}$. Show that the function $f: I \to Y$ defined by $f(t) = 0$ for $t \leq 1/2$ and $f(t) = 1$ for $t > 1/2$ is closed, open, and surjective, but not continuous.

16. Let $X = [0, 1] \cup [2, 3]$ and $Y = [0, 2]$ have the relative topologies induced from the real line $\mathbb{R}$. Define a function $f: X \to Y$ by

$$f(x) = \begin{cases} x & \text{for } x \in [0, 1], \text{ and} \\ x - 1 & \text{for } x \in [2, 3]. \end{cases}$$

Show that f is a closed continuous surjection which is not open.

17. Prove that both of the addition function and the multiplication function on $\mathbb{R}^2$ are open but neither of them is closed.

18. Determine the continuity, closedness, and openness of the following functions on $\mathbb{R}$: (i) $x \mapsto x^2$ and (ii) $x \mapsto 1/(1 + x^2)$.

19. • Show that each of the following functions is open:

    (a) The function $\mathbb{R} \to [-1, 1]$, $t \mapsto \sin t$.
    (b) The function $t \mapsto \tan t$ from $(-\pi/2, \pi/2)$ onto $\mathbb{R}$ (this is actually a homeomorphism).
    (c) The exponential map $\mathbb{R}^1 \to \mathbb{S}^1$, $t \mapsto e^{it}$.

20. • Let f be a function of a space X into a space Y. Prove that f is continuous $\Leftrightarrow f^{-1}(B^\circ) \subseteq (f^{-1}(B))^\circ$ for all $B \subseteq Y \Leftrightarrow \partial(f^{-1}(B)) \subseteq f^{-1}(\partial B)$ for all $B \subseteq Y$.

22. • Let f be a function from a space X to another space Y. Show that f is open $\Leftrightarrow f(A^\circ) \subseteq f(A)^\circ$ for each set $A \subseteq X \Leftrightarrow f^{-1}(\partial B) \subseteq \partial(f^{-1}(B))$ for every $B \subseteq Y$.

23. Prove that a function f of a space X into another space Y is closed if and only if for each $y \in Y$ and each open nbd U of $f^{-1}(y)$, there exists an open nbd V of y such that $f^{-1}(V) \subseteq U$.

24. Give an example to show that a continuous open map need not send the interior of a set onto the interior of the image.

25. • Let $f: X \to Y$ be open (closed). Prove:

   (a) If $A \subseteq X$ is open (closed), $g: A \to f(A)$ defined by f is also open (closed).

   (b) If $A = f^{-1}(B)$ for $B \subseteq Y$, $g: A \to B$ defined by f is open (closed).

25. Let f and g be two functions between topological spaces such that the composition gf is defined. Prove:

   (a) If f, g are open (closed), so is gf.

   (b) If gf is open (closed), and f is continuous, surjective, then g is open (closed).

   (c) If gf is open (closed), and g is continuous, injective, then f is open (closed).

26. Let (X, d_X) and (Y, d_Y) be metric spaces, and let $f: X \to Y$ be a distance preserving map, that is, f satisfies $d_X(x, x') = d_Y(f(x), f(x')) \; \forall \; x, x' \in X$ (Such a map is called an isometry). Show that f is an embedding.

2.2 Product Spaces

There are three main methods of producing new sets out of old ones, namely, forming subsets, Cartesian products, and quotients by an equivalence relation. When these sets are constructed from topological spaces, it is natural to look at the question of topologizing them in some useful way. In the previous chapter, we have seen a simple method of topologizing subsets of topological spaces. Here, we consider the problem of topologizing products.

For simplicity, we first treat the case of finite products. Let X and Y be topological spaces. An obvious way of defining a reasonable topology on the set $X \times Y$ is to declare the sets $U \times V$ open for all open sets $U \subseteq X$ and $V \subseteq Y$. The family of all such subsets of $X \times Y$ obviously contains $\varnothing$ and $X \times Y$, and is closed under the formation of finite intersections, for $(U_1 \times V_1) \cap (U_2 \times V_2) = (U_1 \cap U_2) \times (V_1 \cap V_2)$. This family is not a topology on $X \times Y$ yet, since it is not closed under the formation of unions. However, it can be used as a basis to define a topology for $X \times Y$, by Theorem 1.4.5.

Definition 2.2.1 Let X and Y be spaces, and let $\mathscr{B}$ be the collection of sets $U \times V$, where $U \subseteq X$ and $V \subseteq Y$ are open. The topology generated by $\mathscr{B}$ as a basis is called the *product topology* for $X \times Y$, and the set $X \times Y$, when equipped with the product topology, is called the *product space*. The spaces X and Y are referred to as the *coordinate* (or *factor*) *space*.

There are natural mappings

$$p_X : X \times Y \to X, \quad (x, y) \mapsto x, \text{ and}$$
$$p_Y : X \times Y \to Y, \quad (x, y) \mapsto y,$$

referred to as the projections of $X \times Y$ onto the coordinate spaces X and Y, respectively. For $U \subseteq X$ and $V \subseteq Y$, we have $p_X^{-1}(U) = U \times Y$ and $p_Y^{-1}(V) = X \times V$. If $X \times Y$ is endowed with the product topology, then $p_X^{-1}(U)$ and $p_Y^{-1}(V)$ are open whenever $U \subseteq X$ and $V \subseteq Y$ are open. This shows that both projections p_X and p_Y are continuous in the product topology on $X \times Y$. On the other hand, suppose that there is a topology $\mathscr{T}$ for $X \times Y$ such that both p_X and p_Y are continuous. Then, for all open sets $U \subseteq X$ and $V \subseteq Y$, $U \times V = p_X^{-1}(U) \cap p_Y^{-1}(V)$ is an open set in the topology $\mathscr{T}$. Thus we see that $\mathscr{T}$ contains the basis of the product topology $\mathscr{P}$ (say) on $X \times Y$, and hence $\mathscr{T}$ is finer than $\mathscr{P}$. We summarize the above conclusions as

Proposition 2.2.2 *If $X \times Y$ has the product topology, then the projections $p_X :$ $(x, y) \mapsto x$ and $p_Y : (x, y) \mapsto y$ are continuous, and the product topology is the smallest topology for which this is true. Also, both maps p_X and p_Y are open.*

The last statement holds good because each of the projections p_X and p_Y sends the basic open sets of the product space $X \times Y$ to open sets. However, the projections of a product space are not necessarily closed functions.

Example 2.2.1 Consider $\mathbb{R}^2$ with the product topology, and let $p_1 : \mathbb{R}^2 \to \mathbb{R}^1$ be the projection onto the first factor. Soon , we shall see that this topology for $\mathbb{R}^2$ coincides with the Euclidean topology. The set $F = \{(x, y) \in \mathbb{R}^2 \mid xy = 1\}$ is closed in $\mathbb{R}^2$, since the multiplication of real numbers is continuous and points in $\mathbb{R}^1$ are closed. But $p_1(F) = \mathbb{R} - \{0\}$ is not closed in $\mathbb{R}^1$.

Notice that the collection of sets $p_X^{-1}(U)$ and $p_Y^{-1}(V)$, where $U \subseteq X$ and $V \subseteq Y$ are open, is a subbase for the product topology on $X \times Y$, since $U \times V = p_X^{-1}(U) \cap p_Y^{-1}(V)$. If bases of the coordinate spaces are known, then a basis for the product topology consisting of fewer sets can be described.

Proposition 2.2.3 *Let $\mathscr{C}$ and $\mathscr{D}$ be bases for the spaces X and Y, respectively. Then the collection $\mathscr{B} = \{C \times D \mid C \in \mathscr{C} \text{ and } D \in \mathscr{D}\}$ is a basis for the product topology on $X \times Y$.*

Proof Obviously, each set in $\mathscr{B}$ is open in the product space $X \times Y$. If G is open in $X \times Y$ and $(x, y) \in G$, then there are open nbds U of x and V of y such that $U \times V \subseteq G$. Now, we have sets $C \in \mathscr{C}$ and $D \in \mathscr{D}$ such that $x \in C \subseteq U$ and $y \in D \subseteq V$. So $(x, y) \in C \times D \subseteq U \times V \subseteq G$, and $\mathscr{B}$ is a basis of the product space $X \times Y$, by Proposition 1.4.4. $\diamond$

Example 2.2.2 The product topology on $\mathbb{R}^2$ agrees with its usual topology. We already know that the Euclidean metric d on $\mathbb{R}^2$ is equivalent to the Cartesian metric d^* given by

$$d^*(x, y) = \max\{|x_1 - y_1|, |x_2 - y_2|\}$$

(ref. to Example 1.4.8). In the metric d^*, an open ball $B_{d^*}(x; r)$ is the open square $(x_1 - r, x_1 + r) \times (x_2 - r, x_2 + r)$, where $x = (x_1, x_2)$. Since the open intervals (a, b) form a basis for the real line $\mathbb{R}^1$, the collection $\mathscr{B}$ of open rectangles $(a_1, b_1) \times (a_2, b_2)$ is a basis for the product topology on $\mathbb{R}^2 = \mathbb{R}^1 \times \mathbb{R}^1$, by Proposition 2.2.3. Obviously, every open ball $B_{d^*}(x; r)$ is a member of the base $\mathscr{B}$. On the other hand, if $x \in (a_1, b_1) \times (a_2, b_2)$, then we find a real $r > 0$ such that $B_{d^*}(x; r) \subseteq (a_1, b_1) \times (a_2, b_2)$. By Proposition 1.4.6, the topology generated by the basis $\mathscr{B}$ agrees with the topology induced by d^*.

Observe that the operation of forming the product of spaces is associative and commutative, in the sense that there are obvious canonical homeomorphisms $(X \times Y) \times Z \approx X \times (Y \times Z)$, $((x, y), z) \longleftrightarrow (x, (y, z))$, and $X \times Y \approx Y \times X$, $(x, y) \to (y, x)$. Therefore, we may define the Cartesian product of a finite number of spaces $X_1, X_2, \ldots, X_n$ by induction as the product space $X_1 \times \cdots \times X_n = (X_1 \times \cdots \times X_{n-1}) \times X_n$. Accordingly, the results for finite products follow by induction from the corresponding results for the product of two spaces. For instance, we see by Proposition 2.2.3 that a basis for the product topology on $X_1 \times \cdots \times X_n$ is the family of all products $U_1 \times \cdots \times U_n$, where each U_i is open in X_i.

Moreover, the product topology behaves well with the relativization. For, let A and B be subspaces of the spaces X and Y, respectively. Then the sets $(A \times B) \cap (U \times V)$, where $U \subseteq X$ and $V \subseteq Y$ are open, form a basis for the relative topology on $A \times B$ induced from $X \times Y$, and the collection of products $(A \cap U) \times (B \cap V)$ is a basis for the product space $A \times B$. The equality $(A \times B) \cap (U \times V) = (A \cap U) \times (B \cap V)$ implies that the two bases for $A \times B$ are the same. Hence, the relative topology on $A \times B$ coincides with the product topology. This fact increases the supply of examples of product spaces.

Example 2.2.3 As in Example 2.2.2, we see that the metric defined by the Cartesian norm and, therefore, the Euclidean norm on $\mathbb{R}^n$ induces the product topology (The same is true of $\mathbb{C}^n$ and $\mathbb{H}^n$). Therefore, the subspace I^n (the unit n-cube) of the Euclidean space $\mathbb{R}^n$ is the product of n copies of the unit interval I.

Example 2.2.4 The cylinder

$$C = \{(x, y, z) \in \mathbb{R}^3 \mid x^2 + y^2 = 1 \text{ and } 0 \leq z \leq 1\}$$

with the usual topology is the product space $\mathbb{S}^1 \times I$.

Example 2.2.5 The product space $\mathbb{S}^1 \times \mathbb{S}^1$ is called the *torus*. Intuitively, it is the surface of a solid ring or a doughnut. So, it can be considered as the surface of revolution obtained by rotating the unit circle in the xz-plane centered at $(2, 0, 0)$ about the z-axis, the plane containing the circle being always perpendicular to the xy-plane. If $P = (x, y, z)$ is a point on the circle, then its center C, the origin O, and the point $Q = (x, y, 0)$ are collinear (see Fig. 2.2). Consequently, the point P belongs to the set

$$T = \left\{ (x, y, z) \in \mathbb{R}^3 \mid \left(\sqrt{x^2 + y^2} - 2 \right)^2 + z^2 = 1 \right\}.$$

On the other hand, each point of T lies on the circle in the plane containing the point and the z-axis, with radius 1 and the center $(a,b,0)$, where $a^2 + b^2 = 4$ (verify). Thus, we see that the above surface of revolution is the subspace $T \subset \mathbb{R}^3$. We observe that $T \approx \mathbb{S}^1 \times \mathbb{S}^1$. Consider the circles $C_1 = \left\{ (x, y, -1) \in \mathbb{R}^3 \mid x^2 + y^2 = 4 \right\}$ and $C_2 = \left\{ (x, 0, z) \in \mathbb{R}^3 \mid (x - 2)^2 + z^2 = 1 \right\}$, and the function $h: T \to C_1 \times C_2$ defined by

$$h(x, y, z) = \left(\left(2x/\sqrt{x^2 + y^2},\ 2y/\sqrt{x^2 + y^2},\ -1 \right),\ \left(\sqrt{x^2 + y^2},\ 0,\ z \right) \right).$$

Then h is a homeomorphism. Since both C_1 and C_2 are homeomorphic to $\mathbb{S}^1$, it follows that T is homeomorphic to $\mathbb{S}^1 \times \mathbb{S}^1$.

We now turn to the continuity of functions into product spaces. For any function $f: Z \to X \times Y$, we have the compositions $p_X \circ f: Z \to X$ and $p_Y \circ f: Z \to Y$, where $p_X : X \times Y \to X$ and $p_Y : X \times Y \to Y$ are projections. These are referred to as the *coordinate functions* of f. If X, Y, and Z are topological spaces, there is a very useful characterization of the continuity of f in terms of its coordinate functions.

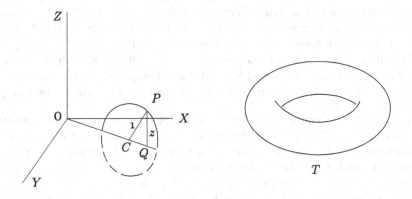

Fig. 2.2 Torus as a surface of revolution

Theorem 2.2.4 *A function* $f: Z \to X \times Y$ *is continuous if and only if the compositions* $p_X \circ f: Z \to X$ *and* $p_Y \circ f: Z \to Y$ *are continuous.*

Proof If f is continuous, then both the compositions $p_X \circ f$ and $p_Y \circ f$ are continuous, since p_X and p_Y are continuous. Conversely, suppose that the compositions $p_X \circ f$ and $p_Y \circ f$ are continuous. We show that f is continuous. By Theorem 2.1.3, we need to prove that $f^{-1}(U \times V)$ is open in Z for every open set $U \subseteq X$ and $V \subseteq Y$. It is easily checked that

$$f^{-1}(U \times V) = (p_X \circ f)^{-1}(U) \cap (p_Y \circ f)^{-1}(V).$$

By our assumption, $(p_X \circ f)^{-1}(U)$ and $(p_Y \circ f)^{-1}(V)$ are open in Z, and therefore $f^{-1}(U \times V)$ is open in Z.   ◇

Corollary 2.2.5 (a) *Given continuous functions* $f: X' \to X$, $g: Y' \to Y$, *the product function* $f \times g: X' \times Y' \longrightarrow X \times Y$ *defined by* $(f \times g)(x', y') = (f(x'), g(y'))$ *is continuous. Moreover,* $f \times g$ *is open whenever* f *and* g *are open.*

(b) *Given continuous functions* $f: Z \to X$ *and* $g: Z \to Y$, *there is a unique continuous function* $\phi: Z \to X \times Y$ *defined by* $\phi(z) = (f(z), g(z))$. *(The function* ϕ *is often denoted by* (f, g)).

The simple proofs are left to the reader.

By the preceding corollary, we see that a function $f: X \to Y$ is continuous if and only if the function $\phi: X \to X \times Y, x \mapsto (x, f(x))$, is continuous. The restriction of p_X to $\text{im}(\phi)$ is obviously the inverse of ϕ; consequently, ϕ is an embedding whenever f is continuous. The $\text{im}(\phi)$ is usually referred to as the *graph* of f. In particular, the graph of the identity map on a space X is called the *diagonal* in $X \times X$, and the (diagonal) map $x \mapsto (x, x)$ is an embedding of X into $X \times X$.

As another application of Theorem 2.2.4, we deduce that the scalar multiplication $\sigma: \mathbb{R}^1 \times \mathbb{R}^n \to \mathbb{R}^n, \sigma(a, x) = ax$ is continuous. If $\mu: \mathbb{R}^2 \to \mathbb{R}^1$ is the multiplication function and $p_j: \mathbb{R}^n \to \mathbb{R}^1$ is the jth projection, then $\mu \circ (1 \times p_j) = p_j \circ \sigma$, where 1 denotes the identity map on $\mathbb{R}^1$. So $p_j \circ \sigma$ is continuous for every j, and hence σ is continuous. Furthermore, the scalar division $(\mathbb{R}^1 - \{0\}) \times \mathbb{R}^n \to \mathbb{R}^n, (a, x) \mapsto a^{-1}x$, is the composition of the restriction of σ to $(\mathbb{R}^1 - \{0\}) \times \mathbb{R}^n$ with the product of the inverse function $a \mapsto a^{-1}$ on $\mathbb{R}^1 - \{0\}$ and the identity map on $\mathbb{R}^n$. Therefore, the scalar division is also continuous. We use this knowledge to prove the following.

Example 2.2.6 The unit n-cube I^n is homeomorphic to the unit disc $\mathbb{D}^n$. Denote the interval $[-1, +1]$ by J. Then the bijection $I \leftrightarrow J, t \leftrightarrow 2t - 1$, defines a homeomorphism $I^n \approx J^n$. Thus it suffices to show that $J^n \approx \mathbb{D}^n$. A homeomorphism between $\mathbb{D}^n$ and J^n is realized by the central projection. Specifically, for each $x = (x_1, \ldots, x_n)$ in $\mathbb{R}^n$, put $\mu(x) = \max_{1 \leq i \leq n} |x_i|$. Then $\mu(x) \leq ||x|| \leq \sqrt{n}\mu(x)$. We have functions $\phi, \psi: \mathbb{R}^n \to \mathbb{R}$ given by $\phi(x) = \frac{||x||}{\mu(x)}$ and $\psi(x) = \frac{\mu(x)}{||x||}$ for $x \neq 0$, and $\phi(0) = 0 = \psi(0)$. Clearly, ϕ and ψ are continuous. Since $|x_i|\phi(x) \leq 1 \, \forall \, x \in \mathbb{D}^n$

and $||x|| \psi(x) \leq 1 \forall x \in J^n$, ϕ and ψ determine functions $f : \mathbb{D}^n \to J^n, x \mapsto \phi(x)x$, and $g : J^n \to \mathbb{D}^n, x \mapsto \psi(x)x$. It is easily seen that f and g are continuous and inverse to each other.

Infinite Products

Now, we extend the concept of product topology to infinite Cartesian products. Let $X_\alpha, \alpha \in A$, be a family of sets. Their Cartesian product $\prod X_\alpha$ is the set of all functions $x : A \to \bigcup_\alpha X_\alpha$ such that $x(\alpha) \in X_\alpha$ for every $\alpha \in A$. We generally write $x(\alpha) = x_\alpha$ and denote x by (x_α). If any $X_\alpha = \emptyset$, then $\prod X_\alpha$ is empty. So we shall assume that $X_\alpha \neq \emptyset$ for every $\alpha \in A$. Note that, in case each of the sets X_α is the same set X, the product $\prod X_\alpha$ is just the set of all functions $A \to X$, denoted by X^A. This set may be referred to as the Cartesian product of A copies of X.

If $X_\alpha, \alpha \in A$, is a family of topological spaces, then we should like to topologize $\prod X_\alpha$ so that important theorems for finite products remain true when the indexing set A is infinite. A natural way to introduce a topology on $\prod X_\alpha$ is to adopt the method used to topologize the product of finitely many spaces; in fact, this procedure gives a topology for $\prod X_\alpha$. The collection $\left\{ \prod U_\alpha \mid U_\alpha \text{ is open in } X_\alpha \text{ for every } \alpha \in A \right\}$ is a basis for a topology on $\prod X_\alpha$. The topology generated by this basis is called the *box topology*. The map $p_\beta : \prod X_\alpha \to X_\beta, (x_\alpha) \mapsto x_\beta$, is termed projection onto βth factor. Notice that each projection $p_\beta : \prod X_\alpha \to X_\beta$ is continuous in the box topology. And, if the family $\{X_\alpha\}$ is finite, and the indexing set $A = \{1, \ldots, n\}$, then the canonical map $\eta : x \mapsto (x(1), \ldots, x(n))$ is a homeomorphism between $\prod X_\alpha$ (with the box topology) and the product space $X_1 \times \cdots \times X_n$ defined previously. But certain important theorems about finite products, such as Theorems 2.2.4 and 5.1.15, do not hold good for infinite products with the box topology.

Example 2.2.7 For each integer $n > 0$, let X_n denote the real line $\mathbb{R}$ and consider $\prod X_n$ with the box topology. Let $f : \mathbb{R} \to \prod X_n$ be the diagonal function given by $f(t) = (t, t, \ldots), t \in \mathbb{R}$. If $p_m : \prod X_n \to X_m$ is the projection map, then $p_m \circ f$ is the identity map on $\mathbb{R}$. Thus the composition of f with every projection is continuous. However, f is not continuous. For, the set $B = \prod U_n$, where $U_n = (-1/n, 1/n)$, is obviously open in the box topology, while $f^{-1}(B) = \bigcap U_n = \{0\}$ is not.

The above example shows that the box topology for the product $\prod X_\alpha$ of an arbitrary family of topological spaces contains generally too many open sets for Theorem 2.2.4 to remain valid in all the cases. Hence, we need to consider another topology for $\prod X_\alpha$, which contains fewer open sets. By Proposition 2.2.2, we see that the product topology in the finite case could equivalently have been defined as the smallest topology which makes all projections continuous. This forms the basis for the new definition of the "product topology" on a product $\prod X_\alpha$. Note that the continuity of the projection $p_\beta : \prod X_\alpha \to X_\beta$ requires $p_\beta^{-1}(U_\beta)$ to be open for every open set $U_\beta \subseteq X_\beta$.

Definition 2.2.6 The topology on $\prod X_\alpha$ generated by the subbasis $\left\{ p_\alpha^{-1}(U_\alpha) \mid U_\alpha \text{ is open in } X_\alpha, \text{ and } \alpha \in A \right\}$ is called the *product topology* (or *Tychonoff topology*). With this topology $\prod X_\alpha$ is referred to as the product space.

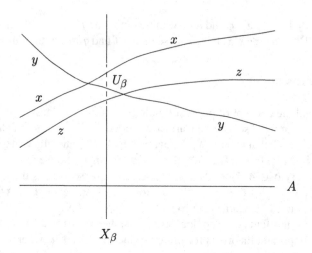

Fig. 2.3 Three elements in $p_\beta^{-1}(U_\beta)$

The definition of the product topology can be rephrased as follows.

Proposition 2.2.7 *Given a family X_α, $\alpha \in A$, of spaces, the product topology on $\prod X_\alpha$ is the smallest topology for which all projections $p_\beta : \prod X_\alpha \to X_\beta$ are continuous.*

Observe that, for $U_\beta \subseteq X_\beta$, $p_\beta^{-1}(U_\beta) = \{x \in \prod X_\alpha \mid x_\beta \in U_\beta\}$, that is, $p_\beta^{-1}(U_\beta)$ is the set of all elements of $\prod X_\alpha$ whose βth coordinates are restricted to lie in U_β (see Fig. 2.3 above). Accordingly, a finite intersection $p_{\alpha_1}^{-1}(U_{\alpha_1}) \cap \cdots \cap p_{\alpha_n}^{-1}(U_{\alpha_n})$ is the product $\prod U_\alpha$, where $U_\alpha = X_\alpha$ for $\alpha \neq \alpha_1, \ldots, \alpha_n$. Thus we see that the base generated by the subbase for the product topology on $\prod X_\alpha$ consists of the sets $\prod U_\alpha$, where $U_\alpha \subseteq X_\alpha$ are open and $U_\alpha = X_\alpha$ for all but a finite number of α's. By Theorem 2.1.16, we deduce that each projection $p_\beta : \prod X_\alpha \to X_\beta$ is open in the product topology on $\prod X_\alpha$.

We also note that each basic open set in the product topology on $\prod X_\alpha$ is also open in the box topology; therefore, the box topology is finer than the product topology. If the indexing set A is finite, then the base of the product topology on $\prod X_\alpha$ is the same as that of the box topology; hence, the two topologies for $\prod X_\alpha$ coincide. Therefore, with $A = \{1, \ldots, n\}$, we see that the product space $\prod X_\alpha$ is canonically homeomorphic to the space $X_1 \times \cdots \times X_n$ defined previously, by the map $\eta : x \mapsto (x(1), \ldots, x(n))$. Therefore, these two spaces are identified.

Clearly, if G is a nonempty open subset of $\prod X_\alpha$, then it contains a nonempty basic open set, so $p_\alpha(G) = X_\alpha$ for all but finitely many α's. Consequently, the product $\prod U_\alpha$, where each U_α is a *proper* nonempty open set in X_α, is never open in the product topology, unless the family $\{X_\alpha\}$ is finite. In contrast, any product $\prod F_\alpha$ of closed sets $F_\alpha \subseteq X_\alpha$ is closed in the product space $\prod X_\alpha$, for $\prod F_\alpha = \bigcap p_\alpha^{-1}(F_\alpha)$.

If, for every $\alpha \in A$, the topology of X_α is given by a subbasis $\mathscr{S}_\alpha$, then the family $\mathscr{S} = \{p_\alpha^{-1}(S_\alpha) \mid S_\alpha \in \mathscr{S}_\alpha, \ \alpha \in A\}$ is a subbasis for the product topology on $\prod X_\alpha$. To see this, first note that each member of $\mathscr{S}$ is open in the product space $\prod X_\alpha$. Secondly, suppose that $U_\alpha \subseteq X_\alpha$ is open and $p_\alpha(x) \in U_\alpha$. Then there exist finitely many sets, say, $S_\alpha^1, \ldots, S_\alpha^n$ in $\mathscr{S}_\alpha$ such that $x \in \bigcap_1^n p_\alpha^{-1}\left(S_\alpha^i\right) \subseteq p_\alpha^{-1}(U_\alpha)$. Since $x \in p_\alpha^{-1}(U_\alpha)$ is arbitrary, we deduce that $p_\alpha^{-1}(U_\alpha)$ is open in the topology generated by the family $\mathscr{S}$. Since the product topology for $\prod X_\alpha$ is generated by the sets $p_\alpha^{-1}(U_\alpha)$, where U_α ranges over all the open subsets of X_α and α varies over A, it follows that the two topologies for $\prod X_\alpha$ are identical.

Likewise, if the topology of each X_α is given by specifying a base, then a base for the product topology on $\prod X_\alpha$ can be described in terms of the basic open subsets of the coordinate spaces.

Proposition 2.2.8 *Let X_α, $\alpha \in A$, be a family of spaces. If $\mathscr{B}_\alpha$ is a basis for the topology of X_α for each $\alpha \in A$, then the collection $\mathscr{C}$ of sets of the form $\prod B_\alpha$, where $B_\alpha = X_\alpha$ for all but finitely many indices α and $B_\alpha \in \mathscr{B}_\alpha$ for the remaining indices, is a basis for the product topology for $\prod X_\alpha$.*

Proof Clearly, every set in $\mathscr{C}$ is open in the product topology on $\prod X_\alpha$. If G is a nonempty open subset of $\prod X_\alpha$ and $x \in G$, then there exist finitely many open sets $U_{\alpha_i} \subseteq X_{\alpha_i}$, $i = 1, \ldots, n$, say, such that $x \in \bigcap_1^n p_{\alpha_i}^{-1}\left(U_{\alpha_i}\right) \subseteq G$, where $p_{\alpha_i} : \prod X_\alpha \to X_{\alpha_i}$ is projection. Obviously, for each $i = 1, \ldots, n$, $x_{\alpha_i} \in U_{\alpha_i}$ and, since $\mathscr{B}_{\alpha_i}$ is a basis of X_{α_i}, we find a set $B_{\alpha_i} \in \mathscr{B}_{\alpha_i}$ such that $x_{\alpha_i} \in B_{\alpha_i} \subseteq U_{\alpha_i}$. Now, taking $B_\alpha = X_\alpha$ for $\alpha \neq \alpha_1, \ldots, \alpha_n$, we have $x \in \prod B_\alpha = \bigcap_1^n p_{\alpha_i}^{-1}\left(B_{\alpha_i}\right) \subseteq G$, and the proposition follows. $\diamond$

Proposition 2.2.9 *Let X_α, $\alpha \in A$, be a family of spaces and $Y_\alpha \subseteq X_\alpha$. Then the product topology for $\prod Y_\alpha$, where each Y_α has the subspace topology, coincides with the relative topology induced by the topology of the product space $\prod X_\alpha$.*

Proof It is easily checked that the subbase for the product topology on $\prod Y_\alpha$ consists of precisely the intersections of $\prod Y_\alpha$ with members of the subbase of the product topology on $\prod X_\alpha$. Since these sets form a subbase for the relative topology on $\prod Y_\alpha$, the proposition follows. $\diamond$

As we would hope, the anomaly observed in the box topology with regard to the continuity of mappings into infinite products does not occur in the product topology. In fact, we prove the following

Theorem 2.2.10 *Let X_α, $\alpha \in A$, be a family of spaces. A function f from a space Y into the product space $\prod X_\alpha$ is continuous if and only if the composition $p_\alpha \circ f$ is continuous for every $\alpha \in A$. Moreover, this property characterizes the product topology for $\prod X_\alpha$.*

Proof If f is continuous, then each $p_\alpha \circ f$, being the composition of two continuous maps, is continuous. Conversely, suppose that $p_\alpha \circ f$ is continuous for every $\alpha \in A$. Then, for any open set $U_\alpha \subseteq X_\alpha$, $f^{-1}\left(p_\alpha^{-1}(U_\alpha)\right) = (p_\alpha \circ f)^{-1}(U_\alpha)$ is open in Y.

Because the sets $p_\alpha^{-1}(U_\alpha)$, $U_\alpha \subseteq X_\alpha$ open, constitute a subbasis for the space $\prod X_\alpha$, f is continuous.

To see the last statement, denote the product space $\prod X_\alpha$ by X, and let X' denote the set $\prod X_\alpha$ with a topology such that a function f of a space Y into X' is continuous if and only if the composition $p_\alpha \circ f : Y \to X_\alpha$ is continuous for every α. Then each projection p_α of X' onto X_α is continuous, since the identity map on X' is continuous. Therefore the compositions

$$X \xrightarrow{i} X' \xrightarrow{p_\alpha} X_\alpha \quad \text{and} \quad X' \xrightarrow{i^{-1}} X \xrightarrow{p_\alpha} X_\alpha,$$

where $i : X \to X'$ is the identity function, are continuous for all α. Hence, the functions i and i^{-1} are both continuous, and thus i is a homeomorphism. It follows that the topologies of X and X' coincide. $\diamond$

For a function $f : Y \to \prod X_\alpha$, the map $p_\alpha \circ f : Y \to X_\alpha$ is referred to as the αth *coordinate function* of f. The preceding theorem is useful in the construction of continuous functions into Cartesian products. In fact, if a continuous map $f_\alpha : Y \to X_\alpha$ for each $\alpha \in A$ is given, then the function $y \mapsto (f_\alpha(y))$ of Y into $\prod X_\alpha$ is continuous. As another application, we see that if $f_\alpha : X_\alpha \to Y_\alpha$, $\alpha \in A$, is a family of continuous functions, then the product function $\prod f_\alpha : \prod X_\alpha \longrightarrow \prod Y_\alpha$, which takes (x_α) into $(f_\alpha(x_\alpha))$, is continuous.

Coming to products of metric spaces, we give an answer to the natural question if, for a given family of metric spaces X_α, $\alpha \in A$, there is a metric which induces the product topology on $\prod X_\alpha$. For finite products, the answer is in the affirmative. If (X_i, d_i), $1 \leq i \leq n$, is a finite family of metric spaces, then the function ρ defined by

$$\rho\big((x_i), (y_i)\big) = \max \{d_i (x_i, y_i) \mid 1 \leq i \leq n\}$$

is a metric on $\prod X_i$ (verify). Observe that an open ball $B(x; r)$ about $x = (x_i)$ in the metric ρ is the product of the open balls $B(x_i; r)$ in X_i, and a product of open balls $B(x_i, r_i)$, $1 \leq i \leq n$, contains the open ball $B(x; r)$, where $r = \min \{r_1, \ldots, r_n\}$. Therefore, the topology on $\prod_1^n X_i$ induced by the metric ρ coincides with the product topology, by Proposition 1.4.6.

The above definition of metric is also accessible to an infinite product of metric spaces, where the diameters of coordinate spaces are bounded by a fixed number. If (X_α, d_α), $\alpha \in A$, is a family of metric spaces each of diameter at most k (a constant), then

$$\rho\big((x_\alpha), (y_\alpha)\big) = \sup \{d_\alpha(x_\alpha, y_\alpha) \mid \alpha \in A\}$$

defines a metric on the set $\prod X_\alpha$. As in the case of finite products, we see that the topology $\mathscr{T}_\rho$ induced by the metric ρ is finer than the product topology. However, the converse is not necessarily true. In the case of countable products, we find that the induced topology $\mathscr{T}_\rho$, with a slight modification in the definition of ρ, coincides with the product topology.

Theorem 2.2.11 *Let X_n, $n = 1, 2, \ldots$, be a countable family of metric spaces. Then the product topology for $\prod X_n$ is induced by a metric.*

Proof For each n, choose a positive real number λ_n such that $\lambda_n \to 0$. Clearly, by Corollary 1.4.9, we can assume that each X_n has a metric d_n such that diam $(X_n) \leq \lambda_n$ (in the metric d_n). For $x = (x_n)$ and $y = (y_n)$ in $\prod X_n$, define $\rho(x, y) = \sup_n d_n(x_n, y_n)$. Then $\rho(x, y)$ is a nonnegative real number and it is trivial to verify that ρ is a metric on $\prod X_n$. We show that it induces the product topology. A basic open set U in the product topology can be written as $U = \prod G_n$, where G_n is open in X_n for $n = n_1, \ldots, n_k$ (say) and $G_n = X_n$ for all other indices n. Given $x \in U$, choose open balls $B(x_{n_i}; r_i) \subseteq G_{n_i}$ for $i = 1, 2, \ldots, k$, and put $r = \min \{r_1, \ldots, r_k\}$. Then $r > 0$ and if $\rho(x, y) < r$, then $y_{n_i} \in B(x_{n_i}; r_i)$ for each $i = 1, \ldots, k$. It follows that $B(x; r) \subseteq U$ and, by Proposition 1.4.6, the metric topology induced by ρ is finer than the product topology. Conversely, let $B(x; r)$ be any open ball in the metric space $\prod X_n$. Since diam$(X_n) \to 0$, there is an integer n_0 such that diam$(X_n) < r/2$, for all $n > n_0$. Let G_n be the open ball $B(x_n; r/2)$ for $n = 1, 2, \ldots, n_0$, and $G_n = X_n$, for all $n > n_0$. Then $U = \prod G_n$ is a basic open nbd of x in the product topology and we have $U \subseteq B(x; r)$. Hence, the metric topology is coarser than the product topology, and the two topologies coincide. $\diamondsuit$

Definition 2.2.12 A topological space X is said to be *metrizable* if there is a metric on X which induces its topology.

With this terminology, Theorem 2.2.11 can be rephrased as follows:
A countable product of metrizable spaces is metrizable.

It is a useful result for proving an important theorem in Chap. 8, which provides certain conditions for "metrizability" of a space. If $\{X_\alpha \mid \alpha \in A\}$ is a countably infinite family of sets, where each $X_\alpha = X$, then we usually write X^ω for X^A, since only the number of elements in A is important in determining X^A. It follows that the product space X^ω is metrizable for every metric space X; in particular, $\mathbb{R}^\omega$ and I^ω are metrizable spaces. It should, however, be noted that the product of uncountably many metric spaces, each containing more than one point, is not metrizable, since such a product fails to admit a countable local basis at any point (see Theorem 7.1.7) while the open balls of radii $1/n$, $n = 1, 2, \ldots$, about a given point of a metric space constitute a countable local basis.

Exercises

1. Let X and Y be spaces, and let $X \times Y$ have the product topology. If $A \subseteq X$ and $B \subseteq Y$, show that $(A \times B)^\circ = A^\circ \times B^\circ$, $\overline{A \times B} = \overline{A} \times \overline{B}$, $(A \times B)' = (A' \times \overline{B}) \cup (\overline{A} \times B')$, and $\partial(A \times B) = (\partial A \times \overline{B}) \cup (\overline{A} \times \partial B)$.

2. Prove that (a) $[0, 1) \times [0, 1) \approx [0, 1] \times [0, 1)$, and (b) $\mathbb{D}^m \times \mathbb{D}^n \approx \mathbb{D}^{m+n}$.

3. Let $\mathbb{R}$ be the real line and $\mathbb{R}_d$ denote the set of all real numbers with the discrete topology.

(a) Compare the product topologies on $\mathbb{R} \times \mathbb{R}$ and $\mathbb{R}_d \times \mathbb{R}$, and the dictionary order topology on $\mathbb{R} \times \mathbb{R}$.

(b) Show that $\mathbb{R} \times \mathbb{R}$ in the dictionary order topology is metrizable.

4. Compare the product topologies on $I \times I$ and $I \times I_d$, and the television topology on $I \times I$ (see Exercise 1.4.15), where I_d denotes the unit interval with the discrete topology.

5. Let X be a straight line in the plane $\mathbb{R}^2$ (e.g., $(x, y) \in X \Leftrightarrow x + y = 1$). Describe the topologies induced on X from $\mathbb{R}_\ell \times \mathbb{R}$ and $\mathbb{R}_\ell \times \mathbb{R}_\ell$.

6. Let $x_0 \in X$ and $y_0 \in Y$. Prove that the functions $x \mapsto (x, y_0)$ and $y \mapsto (x_0, y)$ are embeddings of X and Y into $X \times Y$, respectively.

7. Let (X, d) be a metric space. Show that the metric function $d: X \times X \to \mathbb{R}$ is continuous in the product topology. Also, prove that the metric topology is the smallest topology on X such that each function $f_y: X \to \mathbb{R}$, $f_y(x) = d(x, y)$, is continuous.

8 (a) Let Y be the subspace $\{(x, y) \in \mathbb{R}^2 \mid xy = 0\}$ of $\mathbb{R}^2$, and let $p: \mathbb{R}^2 \to \mathbb{R}^1$ be the map defined by $p(x, y) = x$. Show that $p|Y$ is closed but not open.

(b) Let X be the subspace of $\mathbb{R}^2$ consisting of the points of the closed right half plane $\{(x, y) \in \mathbb{R}^2 \mid x \geq 0\}$ and the x-axis $\{(x, 0) \mid x \in \mathbb{R}\}$. Show that the restriction of p to X is neither open nor closed.

9. Let X be a space and $f, g: X \to \mathbb{R}$ be continuous functions. Define the sum $f + g$, the difference $f - g$, the product $f.g$, the multiple af, $a \in \mathbb{R}$, and the quotient f/g, provided $g(x) \neq 0$ for all $x \in X$, by

$$(f \pm g)(x) = f(x) \pm g(x), \quad (af)(x) = af(x),$$
$$(f \cdot g)(x) = f(x)g(x), \quad (f/g)(x) = f(x)/g(x).$$

Show that each of these functions is continuous.

10. Let $p(x)$ be a polynomial over $\mathbb{R}$. Discuss the continuity, closedness, and openness of the function $x \mapsto p(x)$ on $\mathbb{R}$.

11. Let X be a space and $f, g: X \to \mathbb{R}$ be continuous functions. Prove that (a) the graph G_f of f is closed in $X \times \mathbb{R}$, and (b) the set of points on which f and g agree is closed in X.

12. A function f from the product of two spaces X and Y to another space Z is continuous in x if, for each $y \in Y$, the function $x \mapsto f(x, y)$ is continuous. Similarly, f is continuous in y if for each $x \in X$, the function $y \mapsto f(x, y)$ is continuous.

(a) If $f: X \times Y \to Z$ is continuous, prove that f is continuous in each variable.

(b) Consider the function $f : \mathbb{R} \times \mathbb{R} \to \mathbb{R}$ given by

$$f(x, y) = \begin{cases} 2xy/(x^2 + y^2) & \text{for } x \neq 0 \neq y, \text{ and} \\ f(0, 0) = 0. \end{cases}$$

Show that f is continuous in each variable, but it is not continuous on $\mathbb{R} \times \mathbb{R}$.

13. Prove that the subspace $\{(x_n) \in \ell_2 \mid 0 \leq x_n \leq 1/n\}$ of the Hilbert space ℓ_2 is homeomorphic to the product I^ω. (For this reason, it is called the *Hilbert cube*.)

14. Let $X_\alpha, \alpha \in A$, be a family of discrete spaces. Show that $\prod X_\alpha$ is discrete if and only if A is finite.

15. Let $X_\alpha, \alpha \in A$, be a family of topological spaces and $E_\alpha \subseteq X_\alpha$. Show:

(a) $\overline{\prod E_\alpha} = \prod \overline{E_\alpha}$.

(b) $\prod E_\alpha$ is dense in $\prod X_\alpha \Leftrightarrow$ each E_α is dense in X_α, and

(c) $(\prod E_\alpha)^\circ \neq \varnothing$ only if all but finitely many factors $E_\alpha = X_\alpha$, and then $(\prod E_\alpha)^\circ = \prod E_\alpha^\circ$.

16. • Let $X_\alpha, \alpha \in A$, be a family of spaces, and let $y = (y_\alpha)$ be a fixed point in the product $\prod X_\alpha$. Show:

(a) For each $\beta \in A$, (i) $X_\beta \approx \{x \in \prod X_\alpha \mid x_\alpha = y_\alpha \text{ for every } \alpha \neq \beta\}$, and (ii) $\prod_{\alpha \neq \beta} X_\alpha \approx \{x \in \prod X_\alpha \mid x_\beta = y_\beta\}$.

(b) The set $D = \{x \in \prod X_\alpha \mid x_\alpha = y_\alpha \text{ for all but finitely many } \alpha \in A\}$ is dense.

17. Let $f_\alpha : X_\alpha \to Y_\alpha, \alpha \in A$, be a family of open maps such that all but finitely many f_α are surjective. Show that $\prod f_\alpha : \prod X_\alpha \longrightarrow \prod Y_\alpha, (x_\alpha) \mapsto (f_\alpha(x_\alpha))$, is open.

18. Let $(X_i, d_i), 1 \leq i \leq n$, be metric spaces. Show that each of the following functions defines a metric on $\prod X_i$ which induces the product topology:

(a) $\rho((x_i), (y_i)) = \left[\sum d_i(x_i, y_i)^2\right]^{1/2}$, and

(b) $\rho^+((x_i), (y_i)) = \sum d_i(x_i, y_i)$.

19. • Show that $\rho : \mathbb{R}^\omega \times \mathbb{R}^\omega \to \mathbb{R}$ defined by

$$\rho(x, y) = \sum_n 2^{-n} |x_n - y_n| / (1 + |x_n - y_n|),$$

where x_n and y_n are the nth coordinates of x and y, respectively, is a metric, which induces the product topology on $\mathbb{R}^\omega$. The metric space $(\mathbb{R}^\omega, \rho)$ is referred to as *Fréchet space*.

20. • Let d_1^* be the metric on the set $\mathbb{R}^\omega$ defined by

$$d_1^*\big((x_n), (y_n)\big) = \sup\{d_1(x_n, y_n) \mid n = 1, 2, \ldots\},$$

where d_1 is the standard bounded metric for the set $\mathbb{R}$ of real numbers: $d_1(a, b) = \min\{1, |a - b|\}$. (This is referred to as the *sup metric* or the *uniform metric* for $\mathbb{R}^\omega$.)

(a) Show that the topology for $\mathbb{R}^\omega$ induced by the metric d_1^* is strictly finer than the product topology and coarser than the box topology.

(b) For $x, y \in \mathbb{R}^\omega$, define

$$\rho * \big((x_n), (y_n)\big) = \sup\{d_1(x_n, y_n)/n \mid n = 1, 2, \ldots\}.$$

Verify that $\rho*$ is a metric and it induces the product topology on $\mathbb{R}^\omega$.

21. Let $\mathbb{R}^\infty = \{x \in \mathbb{R}^\omega \mid x_n = 0 \text{ for almost all } n\}$. Find the closure of $\mathbb{R}^\infty$ in the box, product, and uniform topologies for $\mathbb{R}^\omega$.

Chapter 3
Connectedness

The notion of connectedness corresponds roughly to the everyday idea of an object being in one piece. One condition when we intuitively think a space to be all in one piece is if it does not have disjoint "parts". Another condition when we would like to call a space one piece is that one can move in the space from any one point to any other point. These simple ideas have had important consequences in topology and its applications to analysis and geometry. Of course, we need to formulate the above ideas mathematically. The first idea is used in Sect. 3.1 to give a precise definition of connectedness and, using the second idea, we make a slightly different formulation, path-connectedness, discussed in Sect. 3.3. If a space X is not connected, then the knowledge of its maximal connected subspaces becomes useful for description of the structure of the topology of X. These "pieces" of the space X are considered in Sect. 3.2. As we will see later, it is important for some purposes that the space satisfy one of the two connectedness conditions locally. Localized versions of both conditions are formulated and studied in Sect. 3.4.

3.1 Connected Spaces

Definition 3.1.1 A space X is called *disconnected* if it can be expressed as the union of two disjoint, nonempty, open subsets. A space which is not disconnected is called *connected*.

If $X = A \cup B$, where A and B are disjoint, nonempty, and open, then the pair $\{A, B\}$ is called a *separation* or *disconnection* of X. Notice that, in this case, A and B are also closed. Intuitively, X has a disconnection if it falls into two parts.

Example 3.1.1 The Sierpinski space and an infinite cofinite space are connected.

Example 3.1.2 An indiscrete space is connected while a discrete space having more than one point is disconnected.

© Springer Nature Singapore Pte Ltd. 2019
T. B. Singh, *Introduction to Topology*,
https://doi.org/10.1007/978-981-13-6954-4_3

Example 3.1.3 The Sorgenfrey line $\mathbb{R}_\ell$ is not connected, since the open sets $\{x \mid x < a\}$ and $\{x \mid x \geq a\}$ form a disconnection.

From the definition of connectedness, it is obvious that a space X is connected if and only if the empty set $\varnothing$ and X are the only clopen subsets of X. The following characterization of connectedness is often found convenient for establishing theorems.

Theorem 3.1.2 *A space X is connected if and only if every continuous function of X into a discrete space is constant.*

Proof $\Rightarrow$: Let $f : X \to D$ be continuous, where D is discrete. Then $f^{-1}(p)$ is clopen in X for every $p \in D$. Also, if f is not constant, then $f^{-1}(p)$ is a nonempty proper clopen subset of X for any $p \in \mathrm{im}(f)$.

$\Leftarrow$: Suppose that X contains a proper nonempty set Y which is clopen in X. Let $D = \{0, 1\}$ be the two-point discrete space. Then $f : X \to D$ defined by $f(Y) = 0$ and $f(X - Y) = 1$ is a nonconstant continuous map of X onto the discrete space D. $\diamond$

By a *connected subset* of a space X, we mean a connected subspace of X. Thus, a subset $Y \subseteq X$ is connected if there do not exist two open sets U and V in X such that $Y \subseteq U \cup V, U \cap Y \neq \varnothing \neq V \cap Y$ and $U \cap V \cap Y = \varnothing$. Note that if Y is a subspace of X, then a subset $Z \subseteq Y$ is connected if and only if Z is connected as a subset of X. Thus, connectedness is an absolute property of sets. The empty set and singletons are obviously connected sets in any space. The connected subsets of the real line $\mathbb{R}$ are just the intervals (of all kinds), as shown by the following.

Theorem 3.1.3 *A nonempty subset X of the real line $\mathbb{R}$ is connected if and only if it is an interval. In particular, $\mathbb{R}$ itself is connected.*

Proof If X is not an interval, then there exist real numbers $x, y \in X$, and a real number $z \notin X$ such that $x < z < y$. Consequently, $X = [(-\infty, z) \cap X] \cup [(z, +\infty) \cap X]$ is a disconnection of X. Therefore, X must be an interval if it is connected.

Conversely, suppose that X is an interval. We show that X is connected. Assume otherwise, and let $X = A \cup B$ be a disconnection of X. Choose $a \in A$ and $b \in B$. By relabeling, we can assume that $a < b$. Then $[a, b] \subseteq X$, for X is an interval. Consider $S = \{x \mid [a, x) \subseteq A\}$. Since A is open, there exists a $\delta > 0$ such that $(a - \delta, a + \delta) \cap X \subseteq A$. Moreover, for sufficiently small δ, we have $[a, a + \delta) \subseteq A$, since X contains $[a, b]$. Thus $a + \delta \in S$ and $S \neq \varnothing$. Clearly, $x \leq b$ for all $x \in S$, for $A \cap B = \varnothing$. Therefore $\sup(S)$ exists; put $c = \sup(S)$. Then $a < c \leq b$ and $c \in X$. We further see that $c \in A$, and hence $c < b$. For, by the choice of c, given a $\delta > 0$, there is an $x \in S$ such that $c - \delta < x$. Since $[a, x) \subseteq A$, we have $(c - \delta, c + \delta) \cap A \supseteq (c - \delta, x) \cap A \neq \varnothing$. It follows that $c \in \overline{A} \cap X = \overline{A}_X$. Since A is closed in X, we have $c \in A$. As A is also open in X, there is a $\delta > 0$ such that $(c - \delta, c + \delta) \cap X \subseteq A$. Then, for δ such that $c + \delta < b$, we have $[a, c + \delta) \subseteq A$ and therefore $c + \delta \in S$. This contradicts the choice of c, and we conclude that X is connected. $\diamond$

We call two subsets A and B of a space X *separated* if $\overline{A} \cap B = \varnothing = A \cap \overline{B}$. Notice that if $X = A \cup B$ is a separation of X, then A and B are separated. With this terminology, we have an alternative definition of connected sets, as formulated in the following.

Proposition 3.1.4 *A subset Y of a space X is connected if and only if Y is not the union of two nonempty separated sets in X.*

Proof Recall that, for any $A \subset Y, \overline{A}_Y = \overline{A}_X \cap Y$. If Y is disconnected, then $Y = A \cup B$, where $A \cap B = \varnothing$, and A and B are both nonempty closed subsets of Y. We have $\overline{A}_X \cap B = Y \cap \overline{A}_X \cap B = \overline{A}_Y \cap B = A \cap B = \varnothing$; similarly, $A \cap \overline{B}_X = \varnothing$. Thus, A and B are separated sets in X.

Conversely, suppose that $Y = A \cup B$, where A and B are nonempty and separated in X. Then

$$\overline{A}_Y = Y \cap \overline{A}_X = (A \cup B) \cap \overline{A}_X = \left(A \cap \overline{A}_X \right) \cup \left(B \cap \overline{A}_X \right) = A,$$

and therefore A is closed in Y. Similarly, B is closed in Y, and Y is disconnected. ◇

From Theorem 3.1.3, it is apparent that not every subset of a connected space is connected; for example, a disjoint of two open intervals in $\mathbb{R}$. This can also be seen by considering a finite subset (having more than one point) of the cofinite space $\mathbb{R}$.

Connectedness is a topological invariant; in fact, this is preserved by even continuous mappings.

Theorem 3.1.5 *If $f : X \to Y$ is continuous and X is connected, then $f(X)$ is connected.*

Proof Let D be a discrete space and $g : f(X) \to D$ be a continuous map. If $f' : X \to f(X)$ is the map defined by f, then the composition $gf' : X \to D$ is continuous. Since X is connected, gf' is constant. It follows that g is constant, and hence $f(X)$ is connected. ◇

As a consequence of the above theorem, we obtain a generalization of the Weierstrass intermediate value theorem.

Corollary 3.1.6 *A continuous real-valued function on a connected space X assumes all values between any two given values.*

Proof Let $f : X \to \mathbb{R}$ be a continuous function. Then, by Theorems 3.1.3 and 3.1.5, $f(X)$ is an interval. Accordingly, if $f(x) < c < f(y)$ for $x, y \in X$, then there exists a $z \in X$ with $c = f(z)$. ◇

We give two simple applications of the preceding corollary.

Corollary 3.1.7 *Let $f : I \to I$ be a continuous function. Then there exists a point $t \in I$ with $f(t) = t$.*

Proof If $f(0) = 0$ or $f(1) = 1$, then we are through. So assume that $f(0) > 0$ and $f(1) < 1$. Consequently, the continuous function $g: I \to \mathbb{R}$ defined by $g(t) = f(t) - t$ satisfies $g(1) < 0 < g(0)$. By Corollary 3.1.6, there exists a $t \in I$ with $g(t) = 0$ which implies that $f(t) = t$. $\diamond$

Corollary 3.1.8 *Let* $p: \mathbb{R} \to \mathbb{R}$ *be a polynomial function of odd degree. Then* $p(x) = 0$ *for some* $x \in \mathbb{R}$.

Proof Clearly, we can assume that $p(x)$ is monic. Then find a real $a > 0$ such that $p(x) < 0$ for $x \leq -a$, and $p(x) > 0$ for $x \geq a$. By Corollary 3.1.6, there is a point $x \in (-a, a)$ with $p(x) = 0$. $\diamond$

Theorem 3.1.9 *Let* $\{Y_\alpha\}$ *be a family of connected subsets of a space* X *such that* $Y_\alpha \cap Y_\beta \neq \varnothing$ *for every pair of indices* α, β. *Then* $\bigcup Y_\alpha$ *is also connected.*

Proof Let $f: \bigcup Y_\alpha \to D$ be a continuous function, where D is discrete. Since Y_α is connected, $f|Y_\alpha$ is constant for each α. Because any two members of the family $\{Y_\alpha\}$ intersect, it is clear that f is constant, and hence $\bigcup Y_\alpha$ is connected. $\diamond$

It is clear that the union of a family of connected subsets of a space having at least one point in common is also connected. So each open or closed ball in $\mathbb{R}^n$ is connected, for it is the union of its diameters which are homeomorphic to an interval. It is also immediate that a space in which any two points are contained in some connected subset is connected.

Corollary 3.1.10 *Let* C *be a connected subset of a space* X, *and let* $\{Y_\alpha\}$ *be a family of connected subsets of* X *such that* $C \cap Y_\alpha \neq \varnothing$ *for each* α. *Then* $\left(\bigcup_\alpha Y_\alpha\right) \cup C$ *is connected.*

Proof Set $Z_\alpha = C \cup Y_\alpha$. Then Z_α is connected, by Theorem 3.1.9. Again, the above theorem implies that $\bigcup Z_\alpha$ is connected. $\diamond$

Corollary 3.1.11 *Let* $X_1, \ldots, X_n$ *be connected spaces. Then the product space* $\prod X_i$ *is connected.*

Proof By induction, it suffices to prove the corollary for the case $n = 2$. Choose a fixed point $x_1 \in X_1$. Then $C = \{x_1\} \times X_2$ is homeomorphic to X_2, so C is connected. Similarly, $X_1 \times \{x_2\}$ is connected for each x_2 in X_2. It is obvious that $C \cap (X_1 \times \{x_2\}) \neq \varnothing$ for every $x_2 \in X_2$. So $X_1 \times X_2 = \bigcup_{x_2 \in X_2} (X_1 \times \{x_2\}) \cup C$ is connected. $\diamond$

Example 3.1.4 The unit n-cube I^n and the Euclidean space $\mathbb{R}^n$ are connected. This is immediate from Theorem 3.1.3 and Corollary 3.1.11.

Example 3.1.5 The n-sphere $\mathbb{S}^n$ is connected for $n \geq 1$. Choose two distinct points x and y of $\mathbb{S}^n$, and put $U = \mathbb{S}^n - \{x\}$ and $V = \mathbb{S}^n - \{y\}$. Then both U and V are homeomorphic to $\mathbb{R}^n$ and, therefore, connected. Obviously, we have $U \cap V \neq \varnothing$, for $n \geq 1$. Hence $\mathbb{S}^n = U \cup V$ is connected.

Example 3.1.6 The subspace $\mathbb{R}^n - \{0\}$ of the euclidean space $\mathbb{R}^n$ is connected for $n \geq 2$. The function $x \mapsto (x/\|x\|, \|x\|)$ is a homeomorphism of $\mathbb{R}^n - \{0\}$ onto $\mathbb{S}^{n-1} \times (0, \infty)$, and the latter space is connected, by Corollary 3.1.11.

The last example enables us to distinguish between the spaces $\mathbb{R}^1$ and $\mathbb{R}^n$ for $n > 1$. For, if there were a homeomorphism $h : \mathbb{R}^n \to \mathbb{R}^1$, then $h|(\mathbb{R}^n - \{0\})$ would be a homeomorphism between $\mathbb{R}^n - \{0\}$ and $\mathbb{R}^1 - \{h(0)\}$. But these are not homeomorphic, since $\mathbb{R}^n - \{0\}$ is connected while $\mathbb{R}^1 - \{h(0)\}$ is disconnected. More generally, $\mathbb{R}^m$ is not homeomorphic to $\mathbb{R}^n$ for $m \neq n$; however, the proof of this fact requires tools from algebraic topology. A similar argument shows that the unit interval I and the unit n-cube I^n, $n \geq 2$, are not homeomorphic.

Applying the same technique, we see that there is no embedding of $\mathbb{S}^1$ in $\mathbb{R}^1$. For, if $f : \mathbb{S}^1 \to \mathbb{R}^1$ is a continuous injection, then $f(\mathbb{S}^1)$, being a connected subset of $\mathbb{R}^1$, is an interval. If we remove a point from $\mathbb{S}^1$, then the resulting space is connected. But the removal of an interior point from an interval results in a disconnected space. Consequently, f cannot be a homeomorphism between $\mathbb{S}^1$ and $f(\mathbb{S}^1)$.

To see the invariance of connectedness under arbitrary products, we first prove the following simple result, which will be used for determining connectedness in some other cases, too.

Lemma 3.1.12 *Let A be a connected subset of a space X. Then any set B satisfying $A \subseteq B \subseteq \overline{A}$ is also connected; in particular, $\overline{A}$ is connected.*

Proof Let D be a discrete space and $f : B \to D$ be a continuous function. Since A is connected, $f|A$ is constant. We have $\overline{A}_B = \overline{A} \cap B = B$, so $f(B) = f(\overline{A}_B) \subseteq \overline{f(A)} = f(A)$, which implies that f is constant. Hence B is connected.   ◇

Theorem 3.1.13 *Let X_α, $\alpha \in A$, be a family of spaces. Then $\prod X_\alpha$ is connected if and only if each X_α is connected.*

Proof If $\prod X_\alpha$ is connected, then Theorem 3.1.5 implies that each X_β is connected, since the projection $p_\beta : \prod X_\alpha \to X_\beta$ is a continuous surjection.

Conversely, suppose that X_α is connected for every α and choose a fixed point z in $\prod X_\alpha$. Let Y be the union of all connected subsets of $\prod X_\alpha$ which contain z. By Theorem 3.1.9, Y is connected, and therefore $\overline{Y}$ is connected, by the preceding theorem. We claim that $\overline{Y} = \prod X_\alpha$. Consider a basic open set $B = p_{\alpha_1}^{-1}(U_{\alpha_1}) \cap \cdots \cap p_{\alpha_n}^{-1}(U_{\alpha_n})$ in $\prod X_\alpha$. Then take a point $u \in B$ such that $u_\alpha = z_\alpha$ for $\alpha \neq \alpha_1, \ldots, \alpha_n$, and define the sets

$$E_1 = \left\{ x \in \prod X_\alpha \mid x_\alpha = z_\alpha \text{ for } \alpha \neq \alpha_1 \right\},$$
$$E_2 = \left\{ x \in \prod X_\alpha \mid x_{\alpha_1} = u_{\alpha_1}, x_\alpha = z_\alpha \text{ for } \alpha \neq \alpha_1, \alpha_2 \right\},$$
$$\vdots \quad \vdots \qquad\qquad\qquad\qquad \vdots$$
$$E_n = \left\{ x \in \prod X_\alpha \mid x_{\alpha_i} = u_{\alpha_i} \text{ for } i = 1, \ldots, n-1 \text{ and} \right.$$
$$\left. x_\alpha = z_\alpha \text{ for } \alpha \neq \alpha_1, \ldots, \alpha_n \right\}.$$

Clearly, for every $i = 1, \ldots, n$, $E_i \approx X_{\alpha_i}$ (by Exercise 2.2.16) and, therefore, each E_i connected. Observe that $E_i \cap E_{i+1} \neq \varnothing$ for $1 \leq i \leq n - 1$, and hence $\bigcup_1^n E_i$ is connected. By the definition of Y, we have $\bigcup_1^n E_i \subseteq Y$, for $z \in E_1$. Also, it is obvious that $u \in E_n \subseteq Y$, so $Y \cap B \neq \varnothing$. This proves our claim.   ◇

We conclude this section with an interesting example of a connected space.

Example 3.1.7 Let $X = (0, 1]$ and $f : X \to \mathbb{R}$ be the continuous function given by $f(x) = \sin(1/x)$. The graph of f is the subspace $S = \{(x, f(x)) \mid x \in X\}$ of the euclidean space $\mathbb{R}^2$ (see Fig. 3.1 below). The set S, being a continuous image of an interval, is connected, and therefore its closure $\overline{S}$ in $\mathbb{R}^2$ is connected, by Lemma 3.1.12. We determine this subspace. Let $p_1 : X \times \mathbb{R} \to X$ and $p_2 : X \times \mathbb{R} \to \mathbb{R}$ be the projection maps. Then S is the inverse of $\{0\}$ under the function $q = p_2 - f \circ p_1$. Since q is continuous, S is closed in $X \times \mathbb{R}$. Clearly, the strip $[0, 1] \times \mathbb{R}$ is closed in $\mathbb{R}^2$ and contains S. So we have

$$\overline{S} = \left(\overline{S} \cap (X \times \mathbb{R})\right) \cup \left(\overline{S} \cap (\{0\} \times \mathbb{R})\right)$$
$$= S \cup \left(\overline{S} \cap (\{0\} \times \mathbb{R})\right).$$

It follows that all the points of $\overline{S} - S$ lie in $\{0\} \times \mathbb{R}$, the y-axis. To find them, consider first a point $(0, y)$, where $-1 \leq y \leq 1$, and let B be an open ball of radius r about $(0, y)$. Choose an integer n such that $1 < nr\pi$. Then we have $(x, y) \in B$ for all $|x| \leq 1/n\pi$. Obviously, f assumes all values between -1 and $+1$ on $[1/(n + 2)\pi, 1/n\pi]$, so we find a real $x > n\pi$ such that $y = f(x)$. Then $(x, y) \in B \cap S$ and $B \cap S \neq \varnothing$. Hence $(0, y) \in \overline{S}$. We next consider the points $(0, y)$ with $|y| > 1$. It is clear that every one of these has a nbd that does not meet S, for $|\sin 1/x| \leq 1$. Therefore $\overline{S} = S \cup \{(0, y) : |y| \leq 1\}$.

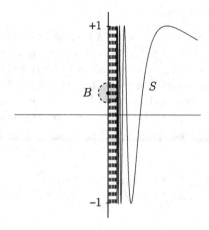

Fig. 3.1 Graph of the function $\sin(1/x)$

The subspace $\overline{S}$ of the euclidean space $\mathbb{R}^2$ is usually referred to as the "*closed topologist's sine curve*" and the subspace $S \cup \{(0, 0)\}$ is called the "*topologist's sine curve.*"

Exercises

1. Let $(X, \mathscr{T})$ be a connected space. If $\mathscr{T}'$ is a topology on X coarser than $\mathscr{T}$, show that $(X, \mathscr{T}')$ is connected.

2. Give an example of two connected subsets of a space, whose intersection is disconnected.

3. (a) Prove that every nonempty proper subset of a connected space has a nonempty boundary. Is the converse also true?

   (b) Let X be a connected metric space with an unbounded metric d. Prove that the set $\{y \mid d(x, y) = r\}$ is nonempty for each $x \in X$. Is this also true when X is disconnected?

4. Let A be a connected subset of a space X. Are A° and ∂A connected? Does the converse hold?

5. Let A_n, $n = 1, 2, \ldots$, be a countable family of connected subsets of a space X such that $A_n \cap A_{n+1} \neq \varnothing$ for each n. Prove that $\bigcup A_n$ is connected.

6. • Prove that the union of two sets $A = \{(x, y) \mid x \notin \mathbb{Q} \text{ and } 0 \leq y \leq 1\}$ and $B = \{(x, y) \mid x \in \mathbb{Q} \text{ and } -1 \leq y \leq 0\}$ is a connected subset of the euclidean space $\mathbb{R}^2$.

7. • For $n = 1, 2, \ldots$, let $X_n = \{(1/n, y) \mid -n \leq y \leq n\}$. Show that the subspace $\mathbb{R}^2 - \bigcup_n X_n$ is connected.

8. Show that $\mathbb{R}^{n+1} - \mathbb{S}^n$ is the union of two disjoint open connected sets.

9. Discuss the connectedness of the following subsets of $\mathbb{R}^n$, $n \geq 2$:

   (a) $\mathbb{R}^n - A$, where A is countable.

   (b) $\{(x_i) \in \mathbb{R}^n \mid \text{at least one } x_i \text{ is irrational}\}$.

   (c) $\{(x_i) \in \mathbb{R}^n \mid x_i \text{ is irrational for } 1 \leq i \leq n\}$.

10. Prove that $\mathbb{R}^\omega$ is disconnected in both the uniform metric topology (ref. Exercise 2.2.20) and the box topology.

11. • An ordered set X is *order complete* if each nonempty subset of X having an upper bound has a supremum. A *linear continuum* is a complete ordered set X which has no gaps, that is, if $x < y$ in X, then there exists a $z \in X$ such that $x < z < y$ (e.g., $\mathbb{R}$ with the usual order relation is a linear continuum, but the complete ordered set $[0, \Omega]$ is not.)

   (a) Prove that an ordered space X is connected $\Leftrightarrow$ it is a linear continuum.

(b) Show that $I \times I$ with the order relation, as given in Exercise 1.4.15, is a linear continuum, and therefore $I \times I$ is connected in the television topology.

12. (a) Let X be a metric space and A, B be two separated subsets of X. Prove that there exist open sets U, $V \subset X$ such that $A \subseteq U$, $B \subseteq V$ and $U \cap V = \emptyset$.

(b) Prove that a subset C of X is connected $\Leftrightarrow$ there do not exist open sets U, $V \subset X$ such that $C \subseteq U \cup V$, $C \cap U$ and $C \cap V$ are nonempty, and $U \cap V = \emptyset$.

(c) Give an example to show that (b) is not true for arbitrary topological spaces.

13. Let A and B be closed subsets of a connected space X such that $X = A \cup B$. Show that (a) A and B are connected if $A \cap B$ is connected, and (b) A or B is connected if $A \cap B$ contains at most two points.

14. ● Let A and B be separated subsets of a space, and let C be a connected subset of $A \cup B$. Show that $C \subseteq A$ or $C \subseteq B$. Give an alternative proof of Lemma 3.1.12.

15. Let C be a connected subset of a space X. If, for $A \subset X$, $C \cap A \neq \emptyset \neq C \cap (X - A)$, show that $C \cap \partial A \neq \emptyset$.

16. Let X be a connected space. Suppose that $C \subset X$ is connected and D is clopen in the subspace $X - C \subseteq X$. Show that $C \cup D$ is connected.

17. Prove that an open interval is not homeomorphic to a closed or half-open interval, and a closed interval is not homeomorphic to a half-open interval.

18. Prove that (a) every continuous open mapping $\mathbb{R} \to \mathbb{R}$ is monotonic, and (b) a monotonic bijection $\mathbb{R} \to \mathbb{R}$ is a homeomorphism.

19. Let X be the subspace $(0, \infty) \subset \mathbb{R}$. Given integer $n > 0$, prove that the function $f : X \to X$, defined by $f(x) = x^n$, is a homeomorphism. (This implies that the function $x \mapsto \sqrt[n]{x}$ on X is continuous.)

20. Let $f : \mathbb{S}^1 \to \mathbb{R}^1$ be a continuous map. Show that there exists $x \in \mathbb{S}^1$ such that $f(x) = f(-x)$.

21. Let $f : X \to Y$ be a continuous open surjection such that $f^{-1}(y)$ is connected for each $y \in Y$. Show that $C \subset Y$ is connected $\Leftrightarrow f^{-1}(C)$ is connected.

22. Prove that a connected subset of $\mathbb{R}^n$ having more than one point is uncountable.

23. Let X and Y be connected spaces, and let $A \subset X$, $B \subset Y$ be proper subsets. Prove that $(X \times Y) - (A \times B)$ is connected.

3.2 Connected Components

Definition 3.2.1 Let X be a space and $x \in X$. The *connected component*, or simply called *component*, of x in X is the union of all connected subsets of X which contain x; this will be usually denoted by $C(x)$.

By Theorem 3.1.9, the component $C(x)$ is connected. And, it is evident from its very definition that $C(x)$ is not properly contained in any connected subset of X. Thus $C(x)$ is a maximal connected subset of X; we shall see in a moment that the components of different points of X are either identical or disjoint. So we refer to them as the *components* of X.

A connected space X has only one component, viz., X itself, and at the other extreme, the components of a discrete space are one-point sets.

Proposition 3.2.2 *Let X be a space. Then:*

    (a) *The set of all components of X forms a partition of X.*
    (b) *Each component of X is closed.*
    (c) *Each connected subset of X is contained in a component of X.*

Proof (a): For every $x \in X$, the component of x in X by its definition contains x. Next, if $C(x)$ and $C(y)$ are components of x and y in X, respectively, and $C(x) \cap C(y) \neq \varnothing$, then $C(x) \cup C(y)$ is connected. So $C(x) = C(x) \cup C(y)$, by the maximality of $C(x)$, and we have $C(y) \subseteq C(x)$. Similarly, $C(x) \subseteq C(y)$ and the equality $C(x) = C(y)$ holds. Thus the components of any two points in X are either disjoint or identical.

(b): If $C(x)$ is a component of $x \in X$, then $C(x)$ is connected, and hence $\overline{C(x)}$ is also connected. By the maximality of $C(x)$, we have $C(x) = \overline{C(x)}$, for $C(x) \subseteq \overline{C(x)}$. It follows that $C(x)$ is closed in X.

(c): If A is a nonempty connected subset of X, then $A \subseteq C(a)$ for any $a \in A$; this follows from the definition of a component. $\qquad\qquad\qquad\qquad\qquad\qquad \diamondsuit$

By the above proposition, there is a decomposition of the space X into connected pieces which are closed too. However, a component of this decomposition may not be open. If the number of components is finite, then, of course, each component is open as well.

Example 3.2.1 Let $\mathbb{Q}$ be the space of rational numbers with the usual topology. If a set $X \subseteq \mathbb{Q}$ contains two points $x \neq y$, then, for any irrational number c between x and y, $X = [X \cap (-\infty, c)] \cup [X \cap (c, +\infty)]$ is a separation of X and it is disconnected. Thus every component of $\mathbb{Q}$ is an one-point set. Note that the points in $\mathbb{Q}$ are not open, for any nonempty open subset of $\mathbb{Q}$ is infinite.

If $f \colon X \to Y$ is continuous, and $C(x)$ is the component of x in X, then $f(C(x)) \subseteq C(f(x))$, the component of $f(x)$ in Y. If f is a homeomorphism, then we also have $f^{-1}(C(f(x))) \subseteq C(x)$, and therefore $f(C(x)) = C(f(x))$. This shows that the number of components of a space and structure of each component are topological invariants. Consequently, the knowledge about components can be used for distinguishing between spaces.

The next result describes the components of a product space in terms of those of its factor spaces.

Proposition 3.2.3 *Let X_α, $\alpha \in A$, be a family of spaces. Then the component of a point $x = (x_\alpha)$ in $\prod X_\alpha$ is $\prod C(x_\alpha)$, where $C(x_\alpha)$ is the component of x_α in X_α.*

Proof Let E be the component of x in $\prod X_\alpha$. By Theorem 3.1.13, $\prod C(x_\alpha)$ is connected, so we have $\prod C(x_\alpha) \subseteq E$. To prove the reverse inclusion, assume that $y \in E$. If $p_\beta : \prod X_\alpha \to X_\beta$ is the projection map, then $p_\beta(E)$ is a connected subset of X_β containing both x_β and y_β. Consequently, $y_\beta \in C(x_\beta)$, and we have $y \in \prod C(x_\alpha)$. $\diamond$

It is of some interest to note that the components of the product of discrete spaces X_α, $\alpha \in A$, are one-point sets, although $\prod X_\alpha$ is not discrete when A is infinite. As an application of the above theorem, we find the components of the cantor set.

Example 3.2.2 The components of the Cantor set C (with the subspace topology induced from $\mathbb{R}$) are singleton sets. To establish this, we show that $C \approx \prod_{n=1}^\infty X_n$, where each X_n is the discrete space $\{0, 2\}$. Note that there is at least one triadic expansion of each member in C that does not require the use of "1" (e.g. $1/3 = 0.0222\cdots$). Observe that such a representation of each element of C is unique and consider this type of representation of every element of C. Then there is a function $f : \prod X_n \to C$ defined by $f((x_n)) = \sum_1^\infty x_n/3^n$. Clearly, f is a bijection. We show that f is a homeomorphism. To see the continuity of f, consider an arbitrary point $x \in \prod X_n$. Given a real number $\epsilon > 0$, find an integer $m > 0$ so large that $3^{-m} < \epsilon$. Then the set

$$U = \{(y_n) \in \prod X_n \mid y_i = x_i \text{ for } 1 \leq i \leq m\}$$

is a basic open neighborhood of x, and $|f(y) - f(x)| < \epsilon$ for all $y \in U$. This implies that f is continuous at x. It remains to prove that f is also open. Let B be a basic open nbd of x. Then there exist finitely many integers $n_1, \ldots, n_k$, say, such that $B = \{y \in \prod X_n \mid x_{n_i} = y_{n_i} \text{ for} 1 \leq i \leq k\}$. Put $m = \max\{n_1, \ldots, n_k\}$ and let $y \in \prod X_n$ be an arbitrary point with $|f(y) - f(x)| < 1/3^{m+1}$. Then we have $y_n = x_n$ for all $1 \leq n \leq m$, for $y_n \neq x_n$ forces $\left|\sum (y_i - x_i)/3^i\right| \geq 1/3^n$, contradicting to the choice of y. So $y \in B$ and we see that $f(B)$ contains the open ball of radius 3^{-n} about $f(x)$ in C. Therefore f is open, by Theorem 2.1.16.

There is an important class of spaces typified by the Cantor set. We introduce them in the following.

Definition 3.2.4 A space X is called *totally disconnected* if its components are singleton sets.

Obviously, a discrete space is totally disconnected. More interesting examples are the Cantor set, the set $\mathbb{Q}$ of rationals, and the set $\mathbb{R} - \mathbb{Q}$ of irrationals with the relative topologies induced from $\mathbb{R}$. Note that a one-point space is both connected and totally disconnected. If X is totally disconnected and Y is a subspace of X, then Y is also totally disconnected, for a component of Y must be contained in some component of X. Also, we see from Proposition 3.2.3 that the product of a family of totally disconnected spaces is totally disconnected.

An interesting property of connected sets is that if two points x and y belong to a connected subset of a space X, then there does not exist a separation $\{A, B\}$ of X such that one point lies in A and the other lies in B (see Exercise 3.1.14). Put more succinctly, two distinct points of a connected subset of X cannot be separated by a separation of X. However, there are spaces X with a pair of distinct points, which do not lie together in a connected set, and yet they cannot be separated by any separation of X.

Example 3.2.3 Let X be the union of the line segments $J_1 = \{0\} \times [0, 1/2)$, $J_2 = \{0\} \times (1/2, 1]$ and $L_n = \{1/n\} \times I$, $n = 1, 2, \ldots$ with the relative topology induced from $\mathbb{R}^2$ (see Fig. 3.2 below). It is clear that the components of X containing the points $(0, 0)$ and $(0, 1)$ are the segments J_1 and J_2, respectively, but there is no separation of X which separates these points.

The phenomenon exhibited by the above example leads to

Definition 3.2.5 A subset K of a space X is a *quasi-component* of X if for any separation $\{A, B\}$ of X, K is contained in either A or B, and K is maximal with respect to this property (i.e., K is not a proper subset of another set $J \subseteq X$ with the same property).

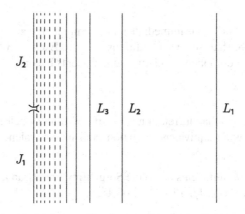

Fig. 3.2 The space X in Example 3.2.3

In Example 3.2.3, $K = J_1 \cup J_2$ is a quasi-component that is not a component. If a space has just one quasi-component, then it is clearly connected. Some more properties of quasi-components are described by the following.

Proposition 3.2.6 *Let X be a space. Then:*

(a) *Each point of X belongs to a unique quasi-component.*
(b) *The quasi-component of X containing a point x is the intersection of all clopen subsets of X which contain x, and hence is closed.*
(c) *Each component of X is contained in a quasi-component.*

Proof (a): Given a point $x \in X$, let $K(x)$ be the set of all points y in X such that there is no separation $X = A \cup B$ with $x \in A$, and $y \in B$. Then $K(x)$ is clearly a quasi-component of X with $x \in K(x)$. Now, we show that two quasi-components of X are either equal or disjoint. Suppose that K and K' are two quasi-components of X with $x \in K \cap K'$. If $X = A \cup B$ is a separation of X, then K and K', both, are contained in either A or B, according to as $x \in A$ or $x \in B$. Thus $K \cup K'$ is also contained in either A or B. By the maximality of a quasi-component, we have $K = K \cup K' = K'$. This proves the uniqueness of $K(x)$.

(b): Let K be a quasi-component of X and $x \in K$. Obviously, if A is a clopen subset of X, then $X = A \cup (X - A)$ is a separation of X. And, if $x \in A$ also, then $K \subseteq A$, and it follows that $K \subseteq \bigcap\{A \mid A$ is clopen in X and $x \in A\} = L$, say. Moreover, we see that if $X = A \cup B$ is a separation, then $L \subseteq A$ or $L \subseteq B$ according as $x \in A$ or $x \in B$. Therefore, by the maximality of K, we have $K = L$.

(c): Let C be a component of X. Choose a point $x \in C$, and denote the quasi-component of X containing x by K. If $X = A \cup B$ is a separation with $x \in A$, then $C(x) \subseteq A$, since $C(x)$ is connected. Also, $K \subseteq A$, by definition. So $C \cup K$ is contained in A. By the maximality of $K(x)$, we have $C \cup K = K$, which implies that $C \subseteq K$. $\diamond$

By Proposition 3.2.6(b), it is immediate that a component that is clopen is a quasi-component. Later, we shall see two conditions (refer to Exercise 3.4.6 and Theorem 5.1.10) under which components and quasi-components become identical.

Exercises

1. • In a space X, show that the relation $x \sim y$ if both x and y belong to a connected subset C of X is an equivalence relation, and its equivalence classes are the components.

2. • Determine the components of (a) the Sorgenfrey line $\mathbb{R}_\ell$ and, (b) the subspace $\{(x, y) \mid y = 0\} \cup \{(x, 1/x) \mid x > 0\}$ of $\mathbb{R}^2$.

3. Prove: Every nonempty connected clopen subset of a space is a component.

4. Let U be a connected open subset of a space X. Prove that U is a component of $X - \partial U$.

5. Let A be a connected subset of a connected space X. If C is a component of $X - A$, show that $X - C$ is connected.

6. Let $f : X \to Y$ be a continuous map, and let C be a component of Y. Show that $f^{-1}(C)$ is the union of components of X.

7. Find spaces X and Y such that $X \not\approx Y$, but X can be embedded in Y and Y can be embedded in X. (This shows that the analogue of the Schroeder–Bernstein theorem for topological spaces does not hold.)

8. Let $X = \bigcup_{n=0}^{\infty}[(3n, 3n + 1) \cup \{3n + 2\}]$ and $Y = (X - \{2\}) \cup \{1\}$. Give X and Y the relative topology induced from $\mathbb{R}$, and define $f : X \to Y$ by $f(x) = x$ for $x \neq 2$, and $f(2) = 1$, and $g : Y \to X$ by

$$g(x) = \begin{cases} x/2 & \text{for } 0 < x \leq 1, \\ (x - 2)/2 & \text{for } 3 < x < 4, \text{ and} \\ x - 3 & \text{for } x \geq 5. \end{cases}$$

Show that both f and g are continuous bijections, but $X \not\approx Y$.

9. Show that the subspace $A = \{0\} \cup \{1/n \mid n = 1, 2, \ldots\}$ of the real line is totally disconnected.

10. Prove that an open subset of $\mathbb{R}^n$ can have at most countably many components. Is this true for closed subsets also?

11. If $X = A \cup B$ is a separation of the space X in Example 3.2.3, show that K is contained in either A or B.

12. For each $n = 2, 3, \ldots$, let C_n denote the circle with center at the origin and radius $1 - 1/n$, and let T be the subspace of $\mathbb{R}^2$ consisting of all the C_n and the two lines $y = \pm 1$. Find the components and quasi-components of T.

13. Let X be a space, and consider the relation $\sim$ in X defined by $x \sim y$ if $f(x) = f(y)$ for every continuous map f of X into a discrete space. Show that $\sim$ is an equivalence relation and the equivalence classes are the quasi-components of X.

14. Let X and Y be spaces, and suppose that x_1, x_2 belong to a quasi-component of X, and y_1, y_2 belong to a quasi-component of Y. Show that (x_1, y_1) and (x_2, y_2) belong to a quasi-component of $X \times Y$.

3.3 Path-connected Spaces

Let X be a space. A *path* in X is a continuous function $f : I \to X$, where I is the unit interval with the usual subspace topology. The point $f(0)$ is referred to as the *origin* (or *initial point*) and the point $f(1)$ as the *end* (or *terminal point*) of f. We usually say that f is a path in X from $f(0)$ to $f(1)$ (or joining them). It is important to note that a path is a function, not the image of the function.

Definition 3.3.1 A space X is called *path-connected* if for each pair of points $x_0, x_1 \in X$, there is a path in X joining x_0 to x_1.

Example 3.3.1 An indiscrete space and the Sierpinski space are obviously path-connected.

Example 3.3.2 Given $x, y \in \mathbb{R}^n$, the set of points $(1 - t)x + ty \in \mathbb{R}^n, 0 \le t \le 1$, is called a (closed) *line segment* between x and y. A set $X \subseteq \mathbb{R}^n$ is called *convex* if it contains the line segment joining each pair of its points. Thus, if X is a convex subspace of $\mathbb{R}^n$, then, given $x, y \in X$, we have $(1 - t)x + ty \in X$ for every $t \in I$. So there is a path in X joining x to y, namely, $t \mapsto (1 - t)x + ty$. In particular, the euclidean space $\mathbb{R}^n$, the n-disc $\mathbb{D}^n$, and the open balls in $\mathbb{R}^n$ are all path-connected.

Lemma 3.3.2 *A space X is path-connected if and only if there exists an $x_0 \in X$ such that every $x \in X$ can be joined to x_0 by a path in X.*

Proof The necessity is obvious. To prove the sufficiency, suppose that $x, y \in X$ and let f and g be paths in X joining x and y to x_0, respectively. Consider the function $h: I \to X$ defined by

$$h(t) = \begin{cases} f(2t) & \text{for } 0 \le t \le 1/2, \text{ and} \\ g(2 - 2t) & \text{for } 1/2 \le t \le 1. \end{cases}$$

For $t = 1/2$, we have $f(2t) = x_0 = g(2 - 2t)$. By the Gluing lemma, h is continuous, and thus defines a path in X joining x to y. So X is path-connected. ◇

Corollary 3.3.3 *The union of a family $\{X_\alpha\}$ of path-connected subspaces of X having a common point is path-connected.*

Proof Put $Y = \bigcup X_\alpha$, and choose an $x_0 \in \bigcap X_\alpha$. Obviously, each point $y \in Y$ belongs to some X_α and there is a path $f: I \to X_\alpha$ from y to x_0, since X_α is path-connected. Then the composition $i \circ f$, where $i: X_\alpha \hookrightarrow Y$ is the inclusion map, is a path in Y joining y to x_0. Hence Y is path-connected, by Lemma 3.3.2. ◇

Example 3.3.3 Consider the subspace $A = \bigcup_{n \in \mathbb{N}} (\{1/n\} \times I) \cup (I \times \{0\})$ of $\mathbb{R}^2$. It is path-connected, for each point $(x, y) \in A$ can be joined to the point $(0, 0)$ by a path in A. Indeed, the function $f: I \to A$ defined by

$$f(t) = \begin{cases} (x, (1 - 2t)y) & \text{for } 0 \le t \le 1/2, \text{ and} \\ ((2 - 2t)x, 0) & \text{for } 1/2 \le t \le 1 \end{cases}$$

is a desired path. Consequently, $\overline{A} = A \cup (\{0\} \times I)$ is also path-connected, since A and $\{0\} \times I$ have a point in common. The closed subspace $\overline{A} \subset \mathbb{R}^2$ is called the *Comb space* (Fig. 3.3).

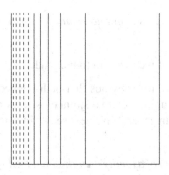

Fig. 3.3 Comb space

Example 3.3.4 The n-sphere $\mathbb{S}^n$, $n > 0$, is path-connected. For, if $H_1 = \{x \in \mathbb{S}^n \mid x_{n+1} \geq 0\}$ is the upper hemisphere and $y \in H_1$ is the north pole, then, for each $x \in H_1$, the mapping $t \to ((1 - t)x + ty) / \|(1 - t)x + ty\|$ is a path in $\mathbb{S}^n$ joining x to y. So H_1 is path-connected. Similarly, the lower hemisphere $H_2 = \{x \in \mathbb{S}^n \mid x_{n+1} \leq 0\}$ is also path-connected. Obviously, $H_1 \cap H_2 \neq \varnothing$, and hence $\mathbb{S}^n = H_1 \cup H_2$ is path-connected.

Theorem 3.3.4 *Every path-connected space is connected.*

Proof Suppose that X is path-connected, and choose a point $x_0 \in X$. For each $x \in X$, there is a path f_x in X from x_0 to x. Since I is connected, so is $\mathrm{im}(f_x)$. Clearly, X is the union of the connected subsets $\mathrm{im}(f_x)$, $x \in X$. So, by Theorem 3.1.9, X is connected. ◇

Although, the converse of the above theorem in general does not hold true, all connected subspaces of $\mathbb{R}$ are path-connected.

Example 3.3.5 As we have already seen, the closed topologist's sine curve $\overline{S}$ is connected (refer to Example 3.1.7), but it is not path-connected. Consider the points $z_0 = (0, 0)$ and $z_1 = (1/\pi, 0)$ in $\overline{S}$ and suppose that $f : I \to \overline{S}$ is a path with origin z_0 and end z_1. Let $p_i : \mathbb{R}^2 \to \mathbb{R}^1$, $i = 1, 2$, be the projection maps. Then the compositions $p_1 f$ and $p_2 f$ must be continuous. Let t_0 be the supremum of the set T of all points $t \in I$ such that $p_1 f(t) = 0$. We contend that $p_2 f$ is discontinuous at t_0, a contradiction. Since $t_0 \in \overline{T}$, we have $p_1 f(t_0) = 0$, by the continuity of $p_1 f$. So $t_0 < 1$ and we find that, for each $t_0 < t_1 \leq 1$, the image of each interval $[t_0, t_1)$ under $p_1 f$ contains the real number $2/n\pi$ for all large integers n. Consequently, $p_2 f$ takes on both values $+1$ and -1 in $[t_0, t_1)$, and it follows that $p_2 f$ fails to map any nbd of t_0 into the nbd $(y_0 - 1/3, y_0 + 1/3)$ of $y_0 = pf(t_0)$. This proves our contention.

The above example also shows that there is no analogue of Lemma 3.1.12 for path-connectedness (the set $S \subset \overline{S}$ is path-connected). But, its invariance properties are very similar to those of connectedness. For instance, we have an analogue of Theorem 3.1.5.

Theorem 3.3.5 *If* $f : X \to Y$ *is continuous and* X *is path-connected, then* $f(X)$ *is path-connected.*

We leave the simple proof of the theorem to the reader.

From the above theorem, it is obvious that path-connectedness is a topological invariant. As shown by the union of two disjoint closed (or open) intervals in $\mathbb{R}$, it is not inherited by subspaces in general. We next see its invariance under the formation of arbitrary products.

Theorem 3.3.6 *The product of a family of path-connected spaces is path-connected.*

Proof Let X_α, $\alpha \in A$, be a family of path-connected spaces, and suppose that $x, y \in \prod X_\alpha$. Then, for each $\alpha \in A$, there is a path $f_\alpha : I \to X_\alpha$ joining x_α to y_α. Define $\phi : I \to \prod X_\alpha$ by $\phi(t) = (f_\alpha(t))$, $t \in I$. If $p_\beta : \prod X_\alpha \to X_\beta$ is the projection map, then $p_\beta \phi = f_\beta$ is continuous. So ϕ is continuous, by Theorem 2.2.10. Also, we have $\phi(0) = x$ and $\phi(1) = y$, obviously. Thus x and y are joined by a path in $\prod X_\alpha$, and the theorem follows.   ◇

We now find a decomposition of a space X into path-connected pieces. By Corollary 3.3.3, for each point $x \in X$, there exists a maximal path-connected subset of X containing x, viz., the union of all path-connected subsets of X which contain x.

Definition 3.3.7 Let X be a space and $x \in X$. The *path component* of x in X is the maximal path-connected subset of X containing x.

Clearly, the path component of a point x in a space X is the set of all points $y \in X$ which can be joined to x by a path in X. If $P(x)$ and $P(y)$ are path components of two points x and y in a space X, and $z \in P(x) \cap P(y)$, then $P(x) \cup P(y)$ is path-connected. By the maximality of $P(x)$, we have $P(x) = P(x) \cup P(y)$. So $P(y) \subseteq P(x)$. Similarly, $P(x) \subseteq P(y)$ and the equality holds. It follows that the path components of the space X partition the set X, and X is path-connected if and only if it has no more than one path component. Since a path component is connected, it is contained in a component. Accordingly, each component of X is a disjoint union of its path components.

Example 3.3.6 The closed topologist's sine curve $\overline{S}$ has two path components, viz., $S = \{(x, \sin(1/x)) \mid 0 < x \le 1\}$ and $J = \{(0, y) \mid |y| \le 1\}$. Clearly, J is path-connected and so is S, being a continuous image of an interval. Moreover, by Example 3.3.5, $\overline{S} = J \cup S$ is not path-connected, hence each one of J and S is a path component of $\overline{S}$.

A Space-Filling Curve

Contrary to one's naive geometric intuition, a somewhat startling fact is that the square I^2, or more generally the unit n-cube I^n for any n, can be filled up by a curve. A *curve* in a topological space X, by definition, is the range of a path $f : I \to X$.

Theorem 3.3.8 *There exists a continuous surjection $I \to I^2$ (that is, a curve which fills up the square (with its interior)).*

Proof Let C be the Cantor set, and let X_n denote the two-point discrete space $\{0, 2\}$ for every $n = 1, 2, \ldots$. Then there is a homeomorphism $f : \prod X_n \to C$ (ref. Example 3.2.2). Similarly, there is a continuous function $g : \prod X_n \to I$ given by $g((x_n)) = \sum x_n/2^{n+1}$. Using dyadic expansion of the numbers in I, we see that g is surjective. Also, it is not difficult to see that the function $h : \prod X_n \longrightarrow \prod X_n \times \prod X_n$ defined by $h((x_n)) = ((x_{2n-1}), (x_{2n}))$ is a homeomorphism. Accordingly, we have a continuous surjection $\phi : C \to I^2$, where ϕ is the composition $(g \times g) \circ h \circ f^{-1}$. Now, let (a, b) denote an open interval removed from I in the construction of C. Since I^2 is convex, there is a line segment L in I^2 with ends $\phi(a)$ and $\phi(b)$. Let $\psi : [a, b] \to L$ be the continuous map defined by $\psi(t) = [(b - t)\phi(a) + (t - a)\phi(b)]/(b - a)$, $a \leq t \leq b$. Obviously, ψ maps a into $\phi(a)$, and b into $\phi(b)$. By the Gluing lemma, ϕ and ψ can be combined to give a continuous map $C \cup (a, b) \to I^2$. We can clearly repeat this process with other open intervals in $I - C$. Thus we obtain a continuous map $\xi : I \to I^2$ which is an extension of ϕ, and therefore surjective. $\diamond$

We conclude this section by indicating an interesting construction of a curve, due to Hilbert, which fills up the square I^2. For each integer $n > 0$, a continuous function $f_n : I \to I^2$ is defined as suggested by the following figures (Fig. 3.4). It is easily checked that the sequence $\langle f_n \rangle$ converges uniformly to a continuous function on I. The range of the limit function fills I^2. A curve in I^2, which fills up the entire space, is referred to as a "Peano space-filling curve."

Note that there cannot be a continuous bijection $I \leftrightarrow I^n, n \geq 2$. For, a continuous bijection $I \to I^n$ must be a homeomorphism but, as seen in Sect. 3.1, I and I^n are not homeomorphic.

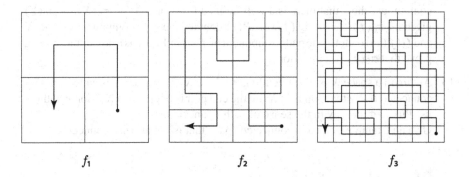

$f_1 \qquad\qquad f_2 \qquad\qquad f_3$

Fig. 3.4 Three stages of filling up the square I^2

Exercises

1. Show that a space having at most three open sets is path-connected.

2. Show that a subset of $\mathbb{R}$ is path-connected $\Leftrightarrow$ it is an interval. (Thus every connected subset of $\mathbb{R}$ is path-connected.)

3. Show that the space in Exercise 1.4.6 is connected, but not path-connected.

4. Is the space in Exercise 3.1.7 path-connected?

5. Is the union of a closed disc and an open disc in $\mathbb{R}^2$, which are externally tangent to each other, connected, path-connected?

6. If $A \subset \mathbb{R}^n$, $n > 1$, is countable, show that $\mathbb{R}^n - A$ is path-connected.

7. For each integer $n > 0$, let L_n be the line segment in $\mathbb{R}^2$ joining the origin 0 to the point $(1, 1/n)$. Is the subspace $A \cup B \subset \mathbb{R}^2$, where $A = \bigcup L_n$ and $B = [1, 2] \times \{0\}$, connected, path-connected?

8. Let A be the set as given in Example 3.3.3 and $p = (0, 1)$. Determine the components and path components of the subspace $A \cup \{p\} \subset \mathbb{R}^2$.

9. Determine the components and the path components of $I \times I$ in the television topology (refer to Exercise 3.1.11(b)).

10. Let X_α, $\alpha \in A$, be a family of spaces. Prove that the path components of $\prod X_\alpha$ are the direct product of path components of the factors.

11. Let X be a space and consider the relation $\sim$ on X defined by $x \sim y$ if there exists a path in X from x to y. Prove that (a) $\sim$ is an equivalence relation, and (b) its equivalence classes are same as the path components of X.

12. Let $\pi_0(X)$ be the set of all path components of a space X. If $f : X \to Y$ is continuous, show that there is a well-defined function $\pi_0(f) : \pi_0(X) \to \pi_0(Y)$ given by $\pi_0(f)P(x) = Q(f(x))$, $P(x)$ denotes the path component of x in X and $Q(f(x))$ denotes the path component of $f(x)$ in Y. If f is a homeomorphism, prove that $\pi_0(f)$ is a bijection.

13. An *arc* in a space X is an embedding $f : I \to X$. The space X is called *arcwise connected* if any two points in X are the end points of an arc in X. (Some authors use this terminology in the sense of path-connectedness.)
    Give an example of a path-connected space that is not arcwise connected.

3.4 Local Connectivity

As we know, the components of a space X are closed in X but need not be open (see Example 3.2.1), whereas a path component of X may not be even closed (see Example 3.3.6). We will study here two conditions, one of them implies components are open, and the other forces path components to be open, and hence closed.

Local Connectedness

Definition 3.4.1 A space X is called *locally connected at a point* $x \in X$ if, for each neighborhood U of x, there is a connected neighborhood V of x such that $V \subseteq U$. The space X is *locally* connected if it is locally connected at each of its points.

Example 3.4.1 A discrete space is locally connected, and so is an indiscrete space.

Example 3.4.2 The euclidean space $\mathbb{R}^n$ is locally connected, since its basis of open balls consists of connected sets.

Example 3.4.3 The n-sphere $\mathbb{S}^n$ is locally connected. For, by Example 2.1.5, $\mathbb{S}^n - \{p\}$, where $p = (0, \ldots, 0, 1)$, is homeomorphic to $\mathbb{R}^n$. Moreover, the *reflection mapping* $(x_0, \ldots, x_{n-1}, x_n) \mapsto (x_0, \ldots, x_{n-1}, -x_n)$ is a homeomorphism $\mathbb{S}^n - \{-p\} \approx \mathbb{S}^n - \{p\}$, and thus each point of $\mathbb{S}^n$ has an open neighborhood homeomorphic to $\mathbb{R}^n$. It is now easily seen that $\mathbb{S}^n$ is locally connected.

Example 3.4.4 The subspace $X = \{0\} \cup \{1/n \mid n = 1, 2, \ldots\}$ of the real line is not locally connected. Because the one-point sets $\{1/n\}$ are clopen in X, and any neighborhood of 0 contains them for large values of n. Thus there is no connected neighborhood of the point 0 in X.

A locally connected space need not be connected, as shown by the subspace $[0, 1/2) \cup (1/2, 1]$ of I. On the other hand, there are connected spaces which are not locally connected. In this regard, it is noteworthy that every connected subspace of the real line $\mathbb{R}$ is locally connected, while this does not hold true in $\mathbb{R}^2$.

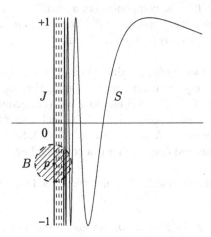

Fig. 3.5 A point of the closed topologist's sine curve, which does not have arbitrarily small connected nbds

Example 3.4.5 The closed topologist's sine curve $\overline{S} = S \cup J$, where $J = \{(0, y) \mid |y| \leq 1\}$, (refer to Example 3.1.7) is not locally connected at any point p belonging to J. Consider the open nbd $U = \overline{S} \cap B$ of p, where B is an open ball of radius less than $1/2$ and centered at p. Clearly, the boundary of B divides the wiggly part of $\overline{S}$ into infinitely many arcs, and $E = \overline{S} \cap \overline{B}$ consists of a closed segment C of J, and many arcs of S. See Fig. 3.5 above. Every arc of S lying in E is closed and connected, being a continuous image of a closed subinterval of $(0, 1]$. So, each one of these arcs is separated from C, and it follows that C is the maximal connected set containing p and contained in E. On the other hand, it is obvious that every neighborhood of p in $\overline{S}$ intersects S. Hence there exists no connected open set V in $\overline{S}$ such that $p \in V \subseteq U$, and $\overline{S}$ is not locally connected at p.

Similarly, we see that the topologist's sine curve $S \cup \{0\}$ is not locally connected at 0.

Theorem 3.4.2 *A space X is locally connected if and only if the components of each open subset of X are open.*

Proof Suppose that X is locally connected, and let U be an open subset of X. Then, given a component C of U, it is obvious that U a nbd of each point $x \in C$. By our hypothesis, there is a connected nbd V of x in X such that $V \subseteq U$. Since C is a component of U and $C \cap V \neq \varnothing$, we have $V \subseteq C$, and thus C is a nbd of x. Since $x \in C$ is arbitrary, we deduce that C is open.

Conversely, suppose that the components of each open subset of X are open. Then, given a point $x \in X$ and any open nbd U of x in X, the component V of x in U is open. Thus U contains a connected nbd of x, and it follows that X is locally connected at x.   ◇

In particular, we find that the components of a locally connected space are open as well as closed. We also see that *a space X is locally connected if and only if the family of all connected open sets in X forms a basis.*

We next observe that local connectedness is a topological invariant; in fact, the property is preserved by any continuous open mapping. For, suppose that X a locally connected space and $f: X \to Y$ is a continuous open surjection. Let U be an open subset Y and $y \in U$. Choose a point x in $f^{-1}(U)$ such that $f(x) = y$. Then there exists a connected nbd V of x in X such that $V \subseteq f^{-1}(U)$. So, we have $y \in f(V) \subseteq U$. Since f is continuous and open, $f(V)$ is a connected nbd of y in Y. Thus Y is locally connected at y.

The property of local connectedness is also preserved by continuous closed mappings.

Theorem 3.4.3 *Let $f: X \to Y$ be a continuous closed surjection. If X is locally connected, then so is Y.*

Proof Suppose that X is locally connected. By Theorem 3.4.2, we need to show that the components of an open set $U \subseteq Y$ are open. Let C be a component of U. We assert that $f^{-1}(C)$ is open in X. If $x \in f^{-1}(C)$, then there exists a connected nbd V of x in X such that $V \subseteq f^{-1}(U)$, since $f^{-1}(U)$ is a nbd of x in X, which is locally connected. So, we have $f(x) \in f(V) \subseteq C$, for $f(V)$ is connected. This implies that $x \in V \subseteq f^{-1}(C)$. Thus $f^{-1}(C)$ is a nbd of x, and the assertion follows. Now, since f is closed, $Y - C = f\left(X - f^{-1}(C)\right)$ is closed. So C is open and the theorem follows.   ◇

However, local connectedness is not preserved by continuous maps.

Example 3.4.6 Let X be the space in Example 3.4.4, and give $Y = \mathbb{N} \cup \{0\}$ the discrete topology. Obviously, the function $f : Y \to X$, defined by $f(n) = 1/n$, $n \in \mathbb{N}$, and $f(0) = 0$ is a continuous bijection. We have already seen that X is not locally connected, while Y is.

Turning to the usual questions involving products and subspaces of locally connected spaces, we note that a subspace of a locally connected space need not be locally connected (see Examples 3.4.4 and 3.4.5). However, it is obvious from the very definition that every open subset of a locally connected space is locally connected.

For products of locally connected spaces, we note that an infinite product of discrete spaces is not locally connected, for its components are one-point sets, which are not open. In this regard, we have the following theorem which, in particular, shows that local connectedness is preserved under the formation of finite products.

Theorem 3.4.4 *Let X_α, $\alpha \in A$, be a collection of spaces. Then the product $\prod X_\alpha$ is locally connected if and only if each X_α is locally connected and all but finitely many spaces X_α are also connected.*

Proof Suppose that $\prod X_\alpha$ is locally connected. Then each component C of $\prod X_\alpha$ is open. Let $p_\beta : \prod X_\alpha \to X_\beta$ denote the projection map for every index $\beta \in A$. Find a basic open set $B = \bigcap_{i=1}^{n} p_{\alpha_i}^{-1}(U_{\alpha_i})$, say, contained in C. Then, for $\alpha \neq \alpha_1, \ldots, \alpha_n$, $X_\alpha = p_\alpha(B) = p_\alpha(C)$ is connected. To see the local connectedness of X_β, let $x \in X_\beta$ be arbitrary, and U_β be an open nbd of x in X_β. Then $p_\beta^{-1}(U_\beta)$ is open in $\prod X_\alpha$ and contains a point ξ with $x = \xi_\beta$. By our assumption, there exists a connected nbd V of ξ in $\prod X_\alpha$ such that $V \subseteq p_\beta^{-1}(U_\beta)$. So, we have $x \in p_\beta(V) \subseteq U_\beta$. Clearly, $p_\beta(V)$ is a connected nbd of x in X_β, for p_β is continuous and open. This proves that X_β is locally connected at x.

To prove the converse, let $\xi = (x_\alpha) \in \prod X_\alpha$, and let $B = \bigcap_{1}^{n} p_{\alpha_i}^{-1}(U_{\alpha_i})$ be a basic open subset of $\prod X_\alpha$ containing ξ. Then, for each $i = 1, \ldots, n$, U_{α_i} is a nbd of x_{α_i}. So there exists a connected nbd V_{α_i} of x_{α_i} such that $V_{\alpha_i} \subseteq U_{\alpha_i}$, by local connectedness of X_{α_i}. Since X_α is not connected for at most finitely many indices α, we can assume that $\alpha_{n+1}, \ldots, \alpha_{n+m}$ are the indices other than $\alpha_1, \ldots, \alpha_n$ such that $X_{\alpha_{n+j}}$ is not connected. Since $X_{\alpha_{n+j}}$ is locally connected, we find a connected nbd $V_{\alpha_{n+j}}$ of $x_{\alpha_{n+j}}$

for $j = 1, \ldots, m$. Then $C = \bigcap_1^{n+m} p_{\alpha_i}^{-1} (V_{\alpha_i})$ is a nbd of ξ and is connected, by Theorem 3.1.13. Also, it is obvious that $C \subseteq B$. This shows that $\prod X_\alpha$ is locally connected at ξ, as desired.   ◇

Local Path-Connectedness

Definition 3.4.5 A space X is called *locally path-connected at a point* $x \in X$ if each neighborhood U of x contains a path-connected neighborhood V of x. The space X is called *locally path-connected* if it is locally path-connected at each of its points.

The condition on V in the above definition is equivalent to require joining each of its point to x by a path in U (see Exercise 12). It is clear that a space X is locally path-connected at $x \in X$ if and only if there exists a neighborhood basis at x consisting of path-connected sets.

Obviously, a discrete space is locally path-connected. A euclidean space $\mathbb{R}^n$ is locally path-connected, since the open balls are path-connected. The n-sphere $\mathbb{S}^n$ is locally path-connected, since each point of $\mathbb{S}^n$ has an open nbd, which is homeomorphic to an open subset of $\mathbb{R}^n$ (cf. Example 3.4.3).

A frequently used criterion for local path-connectedness is given as

Proposition 3.4.6 *A space X is locally path-connected if and only if the path components of open subsets of X are open.*

Proof $\Rightarrow$: Suppose that X is locally path-connected, and let $U \subseteq X$ be open. If P a path component of U and $x \in P$, then there exists a path-connected neighborhood V of x with $V \subseteq U$. As $x \in P \cap V$, $P \cup V$ is path-connected and contained in U. Since P is a maximal path-connected subset of U, we have $P \cup V = P$, and so $V \subseteq P$. Thus P is a neighborhood of x, and we deduce that P is open.

$\Leftarrow$: Obvious.   ◇

By the preceding proposition, it is immediate that a space X is locally path-connected if and only if it has a basis of path-connected open sets.

Observe that a locally path-connected space is locally connected, but the converse is not true (refer to Exercise 13). Furthermore, a path-connected space need not be locally path-connected.

Example 3.4.7 For each integer $n > 0$, let L_n be the line segment in $\mathbb{R}^2$ joining the origin 0 to the point $(1, 1/n)$, and L_0 be the segment $\{(s, 0) \mid 0 \leq s \leq 1\}$. Then each L_n, $n \geq 0$, is path-connected and contains the point 0. Hence $X = \bigcup L_n$ is a path-connected subset of $\mathbb{R}^2$. We observe that X is not locally path-connected at any point x of L_0 other than 0. Consider a small ball B about x such that $0 \notin B$ (see Fig. 3.6 below) and put $U = X \cap B$. Then U is an open nbd of x. We also notice that if $1/n$ is less than the radius of B, then $L_n \cap B$ is a component of U, being a clopen and connected subset of U. On the other hand, if V is a nbd of x in X, then obviously $V \cap L_n \neq \varnothing$ for all large n, and hence every nbd of x contained in U is disconnected. Thus X is not even locally connected at x.

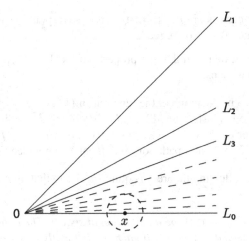

Fig. 3.6 A point of the space X in Example 3.4.7, which does not have arbitrarily small path-connected nbds

In the other direction, it is easily seen by considering the union of two disjoint intervals in the real line that a locally path-connected space is not necessarily path-connected. However, *a connected locally path-connected space is path-connected.* This is a particular case of the following more general result.

Proposition 3.4.7 *Let X be a locally path-connected space. Then each path component of X is clopen, and therefore coincides with a component of X.*

Proof By Proposition 3.4.6, it is immediate that every path component of X is open. Therefore, for each path component P of X, its complement $X - P$ is also open, being the union of the path components of X that are different from P. Hence P is closed too. We further notice that P is connected, by Theorem 3.3.4, and hence it is a component of X.   ◇

From the preceding proposition, it is immediate that a connected locally path-connected space X has only one path component, hence X must be path-connected. A discrete space with at least two points shows that the condition of connectedness in this corollary is essential. Connected locally path-connected spaces will play an important role in our discussion about "Covering Maps".

By definition, it is clear that an open subset of a locally path-connected space is locally path-connected, hence a connected open subset of $\mathbb{R}^n$ or $\mathbb{S}^n$ is path-connected.

The invariance properties of local path-connectedness are very similar to those of local connectedness. As shown by Example 3.4.7, even a closed subspace of a locally path-connected space may not be locally path-connected. From the following theorem, it is immediate that this property is a topological invariant.

Theorem 3.4.8 *Let $f : X \to Y$ be a continuous closed or open surjection. If X is locally path-connected, then so is Y.*

The proof in each case is similar to that of the corresponding result for locally connected spaces, and is left to the reader.

It should, however, be noted that the property of local connectedness is not preserved by continuous maps.

Example 3.4.8 Let S be the graph of the function $\sin(1/x), 0 < x \leq 1$, and consider the subspaces $X = S \cup \{(-1, 0)\}$ and $Y = S \cup \{(0, 0)\}$ of $\mathbb{R}^2$. Define $f : X \to Y$ by setting $f(x) = x$ for every $x \in S$, and $f(-1, 0) = (0, 0)$. Then f is a continuous surjection. Clearly, X is locally path-connected, but Y is not (see Example 3.4.5).

The proof of the following theorem is quite similar to that of Theorem 3.4.4 and is left to the reader.

Theorem 3.4.9 *Let* $X_\alpha, \alpha \in A$, *be a collection of spaces. Then the product* $\prod_\alpha X_\alpha$ *is locally path-connected if and only if each* X_α *is locally path-connected, and all but finitely many* X_α *are also path-connected.*

Exercises

1. Prove that the space in Exercise 3.1.6 is not locally connected.

2. Which of the spaces in Exercise 3.2.2 are locally connected?

3. • Discuss local connectedness and local path-connectedness of the following:

   (a) The subspace A of the comb space (see Example 3.3.3).
   (b) The deleted comb space $A \cup \{(0, 1)\}$.
   (c) The comb space.

4. Let X be the union of all line segments in $\mathbb{R}^2$ joining the point $(1, 1)$ to each point $(r, 0)$, where $r \in I$ is a rational. Show that X is locally connected only at the point $(1, 1)$. Find a path-connected subspace of $\mathbb{R}^2$ which is not locally connected at any of its points.

5. Let X be locally connected and $Y \subseteq X$. If U is a connected, open subset of Y, show that $U = Y \cap G$ for some connected open subset G of X.

6. • Let X be a locally connected space, and let $x \neq y$ be points lying in different components of X. Show that there exists a separation $X = A \cup B$ with $x \in A$ and $y \in B$ (that is, components of X are identical with its quasi-components).

7. Let X be a locally connected space, and let A and B be closed subsets of X such that $X = A \cup B$. If $A \cap B$ is locally connected, show that A and B are locally connected.

8. Let U be an open subset of a locally connected space X. Show:

   (a) If C is a component of U, then $U \cap \partial C = \varnothing$.

   (b) If $X - \overline{C} \neq \varnothing$ and X is also connected, then $\overline{C} - C \neq \varnothing$.

   (c) Even if U is connected, ∂U may fail to be connected or locally connected.

9. Prove: A space X is locally connected $\Leftrightarrow$ for each component C of a subspace $Y \subseteq X$, $\partial C \subset \partial Y$.

10. Let X be a locally connected space and $A \subseteq X$. If A is closed, and C is a component of A, prove that $\partial C = C \cap \partial A$.

11. Let X be a locally connected space. If $A \subseteq X$ with ∂A locally connected, prove that $\overline{A}$ is locally connected.

12. • Prove that a space X is locally path-connected $\Leftrightarrow$ for each point $x \in X$, and each open neighborhood U of x, there is a neighborhood V of x such that $V \subseteq U$ and any point of V can be joined to x by a path in U.

13. • Show that the space $I \times I$ with the television topology (see Exercise 1.4.15) is locally connected but not locally path-connected.

14. Let X be a space and the set $Y = X$ have the topology generated by the path components of open sets in X. Prove:

   (a) Y is locally path-connected.

   (b) The identity function $i : Y \to X$ is continuous.

   (c) A function f of a locally path-connected space Z into Y is continuous $\Leftrightarrow$ the composition $if : Z \to X$ is continuous.

Chapter 4
Convergence

We are familiar with the notion of convergence of sequences of numbers (real or complex) and its useful role in analysis. This can also be generalized to topological spaces. However, sequences in an arbitrary topological space are inadequate for certain purposes, as we shall see in Sect. 4.1. This problem is dealt with in Sect. 4.2 by means of 'nets' which are generalizations of sequences. A considerably more versatile notion, "filters", is treated in Sect. 4.3. It will be seen that uniqueness of "limits" (i.e., no sequence or net converges to more than one point) needs separation of points by disjoint open sets in the space. This condition is named after Felix Hausdorff and will be studied in Sect. 4.4 along with two other separation axioms.

4.1 Sequences

Definition 4.1.1 A *sequence* in a set X is a function $\phi : \mathbb{N} \to X$, where $\mathbb{N}$ denotes the ordered set of natural numbers.

For $n \in \mathbb{N}$, the image $\phi(n)$ of n under ϕ is referred to as the nth term of the sequence, and is usually denoted by x_n. The sequence ϕ is written in the form $\langle x_n \rangle$ or $\{x_1, x_2, \ldots\}$. If $Y \subseteq X$ and $x_n \in Y$ for every n, then $\langle x_n \rangle$ is said to be in Y. It is easy enough to generalize the notion of convergence of sequences of real numbers to sequences in any space, as we do in the following.

Definition 4.1.2 A sequence $\langle x_n \rangle$ in a space X *converges* to a point $x \in X$ if, for each open nbd U of x, there exists a positive integer n_0 such that $x_n \in U$ for every $n \geq n_0$.

If $\langle x_n \rangle$ converges to x, we write $x_n \to x$ and call x a limit of $\langle x_n \rangle$. By weakening the condition of convergence, we obtain the following.

Definition 4.1.3 Let $\langle x_n \rangle$ be a sequence in a space X. A point $x \in X$ is called a *cluster* (or *an accumulation*) point of $\langle x_n \rangle$ if each open nbd U of x contains infinitely many terms of the sequence.

© Springer Nature Singapore Pte Ltd. 2019
T. B. Singh, *Introduction to Topology*,
https://doi.org/10.1007/978-981-13-6954-4_4

In elementary analysis, we have seen a very close relation between the limit points of sets and the limits of convergent sequences. Specifically, we have the following.

Proposition 4.1.4 *Let X be a metric space and $A \subseteq X$. Then a point $x \in \overline{A}$ if and only if there exists a sequence in A converging to x.*

Proof If a sequence $\langle x_n \rangle$ in A converges to x, then each open nbd U of x contains infinitely many terms of the sequence. Hence $x \in \overline{A}$, by Theorem 1.3.5. Conversely, suppose that $x \in \overline{A}$. Then, for each integer $n > 0$, the open ball $B(x; 1/n)$ intersects A. So we can choose a point, say, x_n in $A \cap B(x; 1/n)$. We show that $x_n \to x$. Let U be an open nbd of x in X. Then there exists an open ball, say, $B(x; r)$ contained in U. Now, find an integer $n_0 > 1/r$. Then, for all $n > n_0$, we have $x_n \in B(x; 1/n) \subseteq B(x; r) \subseteq U$, and therefore $x_n \to x$. $\diamond$

This enables us to express continuity of functions between metric spaces in terms of the convergence of sequences.

Proposition 4.1.5 *Let X and Y be metric spaces. Then a function $f : X \to Y$ is continuous if and only if $x_n \to x$ in X implies that $f(x_n) \to f(x)$ in Y.*

Proof Suppose first that f is continuous and a sequence $\langle x_n \rangle$ converges to x in X. Then, given an open nbd V of $f(x)$, $f^{-1}(V)$ is an open nbd of x. So there exists an integer $n_0 > 0$ such that $x_n \in f^{-1}(V)$ for all $n \geq n_0$. This implies that $f(x_n) \in V$, and so $f(x_n) \to f(x)$ in Y.

Conversely, suppose that $f(x_n) \to f(x)$ in Y whenever $x_n \to x$ in X. We show that f is continuous. Let $A \subseteq X$ be arbitrary. If $x \in \overline{A}$, then there exists a sequence $\langle x_n \rangle$ in A converging to x, by the preceding proposition. Therefore, $f(x_n) \to f(x)$ in Y, and it follows that $f(x) \in \overline{f(A)}$. So f is continuous, by Theorem 2.1.2 . $\diamond$

If $\langle x_n \rangle$ is a sequence in X and $n_1 < n_2 < \cdots$ is a strictly increasing sequence of positive integers, then $\langle x_{n_k} \rangle$ is a *subsequence* of $\langle x_n \rangle$.

Proposition 4.1.6 *Let $\langle x_n \rangle$ be a sequence in a metric space X. Then a point $x \in X$ is a cluster point of $\langle x_n \rangle$ if and only if x is the limit of a subsequence of $\langle x_n \rangle$.*

Proof If there is a subsequence $\langle x_{n_k} \rangle$ of $\langle x_n \rangle$ converging to x, then it is clear that x is a cluster point of $\langle x_n \rangle$. Conversely, suppose that x is a cluster point of $\langle x_n \rangle$. Choose a term x_{n_1} of the sequence lying in the open ball $B(x; 1)$, and consider the open ball $B(x; 1/2)$. Since every open nbd of x contains infinitely many terms of $\langle x_n \rangle$, we can find an integer $n_2 > n_1$ such that $x_{n_2} \in B(x; 1/2)$. Continuing in this way, we find integers $n_1 < n_2 < \cdots$ such that $x_{n_k} \in B(x; 1/k)$ for every positive integer k. Clearly, $\langle x_{n_k} \rangle$ is a subsequence of $\langle x_n \rangle$ and it is easily checked that $x_{n_k} \to x$.

Unfortunately, none of the above statements is valid when considered for topological spaces in general, as shown by the following examples. $\diamond$

Example 4.1.1 Let $\mathbb{R}_c$ denote the space of real numbers with the cocountable topology. Then the complement of the range of a sequence is open in this space. Consequently, no rational number is a limit of a sequence in $\mathbb{R} - \mathbb{Q}$. On the other hand, every rational number is an adherent point of $\mathbb{R} - \mathbb{Q}$, since it intersects every nonempty open set. Next, suppose that a sequence $\langle x_n \rangle$ in $\mathbb{R}_c$ converges to a point x. Then the complement of $\{x_n \mid x_n \neq x, \text{ and } n = 1, 2, \ldots\}$ is a nbd of x. Accordingly, there exists an integer n_0 such that $x_n = x$ for all $n \geq n_0$. Thus a convergent sequence in $\mathbb{R}_c$ must be constant from some place on. It follows that the identity function $\mathbb{R}_c \to \mathbb{R}$ (the reals with the usual topology) preserves convergent sequences, although it is not continuous.

Example 4.1.2 (R. Arens) Let Y be the set of all ordered pairs of positive integers and $X = Y \cup \{(0, 0)\}$. Consider the topology on X in which every point of Y is open, and a subset U of X containing $(0, 0)$ is open if for all except a finite number of integers m, the sets $\{n \mid (m, n) \notin U\}$ are each finite. The bijective mapping $Y \to \mathbb{N}$, $(m, n) \mapsto n + \frac{1}{2}(m + n - 1)(m + n - 2)$, defines a sequence with $(0, 0)$ as a cluster point, since given an integer $k > 0$, for each m, there is an n such that $n + \frac{1}{2}(m + n - 1)(m + n - 2) > k$. We observe that no sequence in Y can converge to $(0, 0)$. If a sequence $\langle y_n \rangle$ contains points from finitely many columns only, then it is clear that it cannot converge to $(0, 0)$. Otherwise, we obtain an infinite subsequence $\langle z_n \rangle$ of $\langle y_n \rangle$, which contains at most one point from each column. Since the complement of $\langle z_n \rangle$ is a nbd of $(0, 0)$, it cannot converge to $(0, 0)$, and therefore the sequence $\langle y_n \rangle$ does not converge to $(0, 0)$.

Exercises

1. (a) What are limits of $\langle 1/n \rangle$ when $\mathbb{R}$ is assigned the cofinite topology?

   (b) Find convergent sequences and their limit(s) in the space $\mathbb{R}_\ell$.

2. Let X be space with the cofinite topology. Prove that a sequence $\langle x_n \rangle$ in X converges to a point x iff, for each $y \neq x$, the set $\{n \mid x_n = y\}$ is finite.

3. Let $\langle x_n \rangle$ be a sequence in $\mathbb{R}$ with the range $\mathbb{Q}$, the set of all rational numbers. Show that every real number is a cluster point of $\langle x_n \rangle$.

4. Let X and Y be topological spaces and let $f : X \to Y$ be a continuous function. If a sequence $\langle x_n \rangle$ converges to x in X, show that $f(x_n) \to f(x)$ in Y.

5. Show that the real-valued function f defined on $[0, \Omega]$ by $f(\alpha) = 0$ if $\alpha < \Omega$ and $f(\Omega) = 1$ is not continuous at Ω, even though it does preserve convergent sequences.

6. If a sequence $\langle x^{(n)} \rangle$ in the Hilbert space ℓ_2 converges to x, show that $x_i^{(n)} \to x_i$ for every i. Find a sequence $\langle x^{(n)} \rangle$ in ℓ_2 such that $x_i^{(n)} \to 0$ for every i but $\langle x^{(n)} \rangle$ fails to converge in ℓ_2. (This shows that the topology of ℓ_2 is different from the one it would inherit as a subspace of $\mathbb{R}^\omega$.)

7. • Let $\mathbb{R}^I$ denote the product space $\prod_{t \in I} \mathbb{R}_t$, where each $\mathbb{R}_t$ is a copy of the real line $\mathbb{R}$ and I denotes the unit interval $[0,1]$. Prove:

   (a) The point c_0 which takes value 0 everywhere is an adherent point of the set A of all functions $f : I \to \mathbb{R}$ which assume values 0 at finitely many points and 1 elsewhere, but there is no sequence in A which converges to c_0.

   (b) The sequence $\langle f_n \rangle$ in $\mathbb{R}^I$ given by $f_n(t) = t^n$ converges to the function which is zero for all points of I except 1, where it takes value 1 (accordingly, the subset $\mathcal{C}(I)$ is not closed in $\mathbb{R}^I$).

8. Let (Y, d) be a metric space and X be a set. A sequence of functions $f_n : X \to Y$ is said to *converge uniformly* to a function $f : X \to Y$ if for each $\epsilon > 0$, there exists an integer m such that $d(f_n(x), f(x)) < \epsilon$ for all $n \geq m$ and all $x \in X$.

   Prove that a sequence $\langle f_n \rangle$ in the set $\mathcal{B}(X, Y)$ of all bounded functions $X \to Y$ converges to g with respect to the sup metric d^* (ref. Exercise 1.1.7) if and only if $\langle f_n \rangle$ converges uniformly to g.

9. • Let X be a topological space and Y be a metric space. If a sequence of continuous functions $\langle f_n : X \to Y \rangle$ converges uniformly to a function $f : X \to Y$, show that f is continuous. Give an example to show that this need not be true if the convergence is not uniform.

4.2 Nets

As seen in the previous section, sequences in topological spaces are inadequate to detect limit points and continuity of functions. This difficulty is overcome by introducing a natural generalization of sequences—"nets"—in which the set of natural numbers is replaced by ordered sets. We shall use this concept to prove some important results in subsequent chapters.

Definition 4.2.1 A directed set A is a nonempty set together with a reflexive and transitive relation $\preceq$ such that, for any two elements $\alpha, \beta \in A$, there is a $\gamma \in A$ satisfying $\alpha \preceq \gamma$ and $\beta \preceq \gamma$. We say that the relation $\preceq$ directs A, and sometimes write $\beta \succeq \alpha$ for $\alpha \preceq \beta$.

Some authors require the relation $\preceq$ to be antisymmetric, too.

Example 4.2.1 The set $\mathbb{N}$ of natural numbers with its usual ordering is a directed set.

Example 4.2.2 The family of all finite subsets of a given set X is directed by inclusion $\subseteq$ (i.e., $Y \preceq Z$ if $Y \subseteq Z$). Similarly, the reverse inclusion $\supseteq$ directs any family of subsets of X which is closed under finite intersections.

Definition 4.2.2 A *net* in a space X is a function ϕ from a directed set A into X.

A net with domain the set $\mathbb{N}$ of natural numbers is a sequence. We remark that nets are often called *Moore–Smith sequences* or *generalized sequences*.

Definition 4.2.3 Let $\phi: A \to X$ be a net in the space X and $Y \subseteq X$. We say

(a) ϕ is *in* Y if $\phi(\alpha) \in Y$ for every $\alpha \in A$;
(b) ϕ is *eventually in* Y if there is an $\alpha \in A$ such that $\phi(\beta) \in Y$ for all $\beta \succeq \alpha$; and
(c) ϕ *converges* to a point $x \in X$ (written $\phi \to x$) if it is eventually in every open nbd of x.

Notice that Definition 4.2.3 (c) for sequences reduces to the one given in 4.1.2. It is common to write x_α for $\phi(\alpha)$, $\alpha \in A$, and denote the net ϕ by $\langle x_\alpha \rangle$ or $\{x_\alpha, \alpha \in A\}$. If $x_\alpha \to x$, we call x a limit of the net $\langle x_\alpha \rangle$, and occasionally write $\lim x_\alpha = x$. Limits, when they exist, are not necessarily unique. For example, every net in the space X with the trivial topology converges to every point of X. A nontrivial example is a sequence with distinct terms in a cofinite (infinite) space. It is clear that a net cannot be eventually in each of two disjoint sets. Accordingly, a convergent net has a unique limit in a space in which every pair of distinct points can be separated by open sets in the sense that if $x \neq y$, then there exist disjoint open sets U and V containing x and y, respectively. A space with this property is called a *Hausdorff space*. Interestingly, this condition is equivalent to the uniqueness of limits of nets in a topological space.

Theorem 4.2.4 *A space X is Hausdorff if and only if each net in X converges to at most one point in X.*

Proof As seen above, a net in a Hausdorff space cannot have more than one limit. To see the converse, suppose that X is not Hausdorff. Then there are two points $x \neq y$ in X which do not have disjoint open nbds. Consider the collection

$$\mathscr{P} = \{(U, V) \mid U \text{ and } V \text{ are open nbds of } x \text{ and } y, \text{ resectively}\}.$$

We direct $\mathscr{P}$ by the relation $\preceq$, where $(U, V) \preceq (U', V')$ if $U \supseteq U'$ and $V \supseteq V'$. For each (U, V) in $\mathscr{P}$, we choose a point $\phi(U, V)$ in $U \cap V$ and consider the net $\phi: \mathscr{P} \to X$. We assert that ϕ converges to both x and y, contrary to our hypothesis. To see this, let U be an open nbd of x. Then we have $(U, X) \in \mathscr{P}$. Now, if $(U', V') \succeq (U, X)$, then $U' \subseteq U$ and $\phi(U', V') \in U' \cap V' \subseteq U$. This shows that ϕ is eventually in U, and therefore it converges to x. Similarly, ϕ converges to y, and hence our assertion. $\diamond$

This theorem enables us to speak of "*the*" limit of a convergent net in a Hausdorff space. It is remarkable that the "if" part of Theorem 4.2.4 is not valid for sequences. The space $\mathbb{R}_c$ of reals with the cocountable topology is not Hausdorff, although each sequence in this space has at most one limit.

The next two theorems suggest that nets are adequate to describe all the basic topological notions, without much difficulty.

Theorem 4.2.5 *Let X be a space and $A \subseteq X$. Then $x \in \overline{A} \Leftrightarrow$ there is a net in A which converges to x.*

Proof If $x \in \bar{A}$, then each open nbd U of x intersects A. So we can pick a point $x_U \in U \cap A$. The family $\mathscr{U}_x$ of all open nbds of x is directed by the reverse inclusion (that is, $U_1 \preceq U_2$ if $U_1 \supseteq U_2$). The net $\{x_U, U \in \mathscr{U}_x\}$ obviously converges to x.

Conversely, if $\{a_\lambda, \lambda \in \Lambda\}$ is any net in A which converges to a point $x \in X$, then each open nbd U of x intersects A, and hence $x \in \bar{A}$.   ◇

By Theorem 4.2.5, a subset F of a space X is closed if and only if every convergent net in F converges to a point of F. It follows that the closed sets and hence the open sets can be described in terms of convergence of nets; in other words, the topology of the space can be completely determined by this notion.

Theorem 4.2.6 *A function* $f : X \to Y$ *between spaces is continuous if and only if for every net* ϕ *in* X *converging to a point* $x \in X$, *the net* $f \circ \phi$ *in* Y *converges to* $f(x)$.

Proof The necessary part is easy. To prove sufficiency of the condition, suppose that f is not continuous at some point $x \in X$. Then there is an open nbd V of $f(x)$ such that $f(U) \nsubseteq V$ for every open nbd U of x. Choose a point x_U in U such that $f(x_U) \notin V$. Direct the set of all open nbds U of x by the reverse inclusion and define a net ϕ by putting $\phi(U) = x_U$. If U_0 is any open nbd of x and $U \succeq U_0$, then $x_U \in U_0$ so that ϕ is eventually in U_0. Thus ϕ converges to x. But $(f\phi)(U) \notin V$ for any U; so $f \circ \phi$ does not converge to $f(x)$.   ◇

This theorem can be rephrased as

$f : X \to Y$ is continuous $\Leftrightarrow f(\lim x_\alpha) = \lim f(x_\alpha)$ for every convergent net $\langle x_\alpha \rangle$ in X.

There is a simple criterion for the convergence of nets in product spaces.

Theorem 4.2.7 *Let* X_α, $\alpha \in A$, *be a family of topological spaces. Then a net* ϕ *in* $\prod X_\alpha$ *converges to* $x = (x_\alpha)$ *if and only if* $p_\beta \circ \phi \to x_\beta$ *for each* $\beta \in A$, *where* p_β *is the projection of* $\prod X_\alpha$ *onto* X_β.

Proof If $\phi \to x$, then $p_\beta \circ \phi \to p_\alpha(x) = x_\beta$, since each p_β is continuous.

Conversely, suppose that $p_\beta \circ \phi \to x_\beta$ in X_β for every $\beta \in A$, and denote the directed set on which ϕ is defined by $(\Lambda, \preceq)$. Let $B = \bigcap_{i=1}^n p_{\alpha_i}^{-1}(U_{\alpha_i})$ be a basic nbd of $x = (x_\alpha)$ in $\prod X_\alpha$. Then, for each $i = 1, \dots, n$, U_{α_i} is a nbd of x_{α_i} in X_{α_i}. By our hypothesis, there exists a $\lambda_i \in \Lambda$ such that $(p_{\alpha_i} \phi)(\lambda) \in U_{\alpha_i}$ for all $\lambda \succeq \lambda_i$. Since Λ is directed, we find a $\lambda_0 \in \Lambda$ such that $\lambda_0 \succeq \lambda_i$ for every i. It is now obvious that $\phi(\lambda) \in B$ for all $\lambda \succeq \lambda_0$, and hence $\phi \to x$.   ◇

The sequential version of Theorem 4.2.7 is the familiar statement: A sequence $\langle x^n \rangle$ in $\prod_\alpha X_\alpha$ converges to $x^0 \Leftrightarrow x_\alpha^n \to x_\alpha^0$ for every α.

We now generalize the notions of cluster points and subsequences to nets.

Definition 4.2.8 Let X be a space and $Y \subseteq X$. A net $\phi : A \to X$ is *frequently* in Y if for each $\alpha \in A$, there is a $\beta \in A$ such that $\alpha \preceq \beta$ and $\phi(\beta) \in Y$. A point $x \in X$ is called a *cluster* (or an *accumulation*) point of ϕ if it is frequently in every nbd of x.

In the case of sequences, this notion of cluster points coincides with the one introduced previously in Definition 4.1.3.

We refer to generalizations of subsequences as *subnets*. Call a subset B of a directed set $(A, \preceq)$ *cofinal* if for each $\alpha \in A$, there is $\beta \in B$ such that $\alpha \preceq \beta$. Each cofinal subset of A is also directed by the ordering of A. Accordingly, a definition of "subnet" can be made by restricting the net to a cofinal subset of the indexing set. But this simple definition of subnet turns out to be inadequate for many purposes (ref. Example 4.1.2). Recall that subsequences are defined by precomposing the sequence with an order preserving injection, and this injection is strictly monotone increasing. Analogously, a "subnet" of a net $\phi \colon A \to X$ can be defined to be the composition $\phi\theta$, where θ is a function of a directed set B into A such that $\beta \preceq \beta' \Rightarrow \theta(\beta) \preceq \theta(\beta')$ and $\theta(B)$ is cofinal in A. This definition includes the previous case and is good for almost all purposes, yet the condition on θ is relaxed to give a more general notion of "subnet".

Definition 4.2.9 A *subnet* of a net $\phi \colon A \to X$ is a net $\psi \colon B \to X$ together with a function $\theta \colon B \to A$ such that

  (a) $\psi = \phi\theta$, and
  (b) for each $\alpha \in A$, there exists $\beta \in B$ satisfying $\alpha \preceq \theta(\beta')$ for all $\beta' \succeq \beta$.

If $\phi \colon A \to X$ is a net and $\theta \colon B \to A$ is an order preserving map such that $\theta(B)$ is cofinal in A, then $\phi\theta$ is clearly a subnet of ϕ. In particular, if B is cofinal subset of A, then the restriction of ϕ to B is a subnet of ϕ. Thus, every subsequence of a sequence $\langle x_n \rangle$ is a subnet of $\langle x_n \rangle$. It should be noted that a sequence may have subnets which are not subsequences; for example, the bijection $\mathbb{N} \leftrightarrow Y$ in Example 4.1.2 (this sequence has a subnet which converges to 0, by Theorem 4.2.10 below); see also Exercise 8.3.17.

As we would hope, the pathological behavior of subsequences observed in Example 4.1.2 is not displayed by subnets. Indeed, we have the following.

Theorem 4.2.10 *Let* $\phi \colon A \to X$ *be a net in a space* X. *A point* $x \in X$ *is a cluster point of* ϕ *if and only if there is a subnet of* ϕ *which converges to* x.

Proof Suppose first that x is a cluster point of ϕ. Let $\mathscr{U}_x$ be the family of all open nbds of x and define

$$\mathscr{D} = \big\{ (\alpha, U) \mid \alpha \in A, \ U \in \mathscr{U}_x, \text{ and } \phi(\alpha) \in U \big\}.$$

Obviously, $(\alpha, X) \in \mathscr{D}$ for any $\alpha \in A$, so $\mathscr{D}$ is nonempty. Consider the relation $\leq$ in $\mathscr{D}$ given by $(\alpha, U) \leq (\beta, V) \Leftrightarrow \alpha \preceq \beta$ and $V \subseteq U$. Then $\mathscr{D}$ is directed by $\leq$ because the intersection of two members of $\mathscr{U}_x$ is again a member of $\mathscr{U}_x$, and ϕ is frequently in each member of $\mathscr{U}_x$. Define $\theta \colon \mathscr{D} \to A$ by setting $\theta(\alpha, U) = \alpha$. Since $(\alpha, X) \in \mathscr{D}$ for every $\alpha \in A$, and $(\beta, U) \geq (\alpha, X)$ implies $\beta \succeq \alpha$, θ satisfies the condition (b) of Definition 4.2.9. So the composition of $\phi\theta$ is a subnet of ϕ. We observe that $\phi\theta$ converges to x. Given any open nbd U of x, find an $\alpha \in A$ such that $\phi(\alpha) \in U$. Then we have $(\alpha, U) \in \mathscr{D}$ and $\phi\theta((\beta, V)) = \phi(\beta) \in V \subseteq U$ for every $(\beta, V) \geq (\alpha, U)$. Accordingly, $\phi\theta$ converges to x.

Conversely, suppose that ϕ has a subnet $\psi: B \to X$ which converges to x. Let U be an open nbd of x, and $\alpha \in A$ be arbitrary. Then there exists a $\beta_1 \in B$ such that $\psi(\beta) \in U$ for all $\beta \succeq \beta_1$. Since ψ is a subnet of ϕ, there exists a mapping $\theta: B \to A$ such that $\psi = \phi\theta$ and θ satisfies the condition 4.2.9(b). So we can find a $\beta_2 \in B$ such that $\theta(\beta) \succeq \alpha$ for all $\beta \succeq \beta_2$. Since B is directed, we find a $\beta_0 \in B$ which follows both β_1 and β_2. Then $\theta(\beta_0) \succeq \alpha$ and $\phi\theta(\beta_0) = \psi(\beta_0) \in U$. It follows that ϕ is frequently in U, and thus x is its cluster point.   ◇

It is clear that if a net ϕ converges to x, then x is a cluster point of ϕ, and every subnet of ϕ converges to x. But, there are nets which have a single cluster point and yet fail to converge, for example, the sequence $\phi(n) = n + (-1)^n n$ in $\mathbb{R}$ (0 is the only cluster point of ϕ). We describe here the nets for which the converse holds.

Definition 4.2.11 A net ϕ in a set X is called a *universal net* (or an *ultranet*) if, for every $S \subseteq X$, ϕ is eventually in either S or its complement $X - S$.

As an example, a sequence which is constant from some place on (or a constant net) is a universal net. Note that a net ϕ is frequently in $U \Leftrightarrow \phi$ is not eventually in $X - U$. From this, it is immediate that a universal net in a topological space converges to each of its cluster points. The next theorem guarantees the existence of nontrivial universal nets (of course, assuming an equivalent of the Axiom of Choice, (refer to A.6.3 in Appendices)).

Theorem 4.2.12 *Every net has a universal subnet.*

Proof Let $\phi: A \to X$ be a net. Consider the collection $\mathscr{C}$ of all families $\mathscr{F}$ of subsets of X such that ϕ is frequently in each member of $\mathscr{F}$ and the intersection of any two members of $\mathscr{F}$ is also in $\mathscr{F}$. The family $\{X\}$ belongs to $\mathscr{C}$, so it is nonempty. We partially order the set $\mathscr{C}$ by the inclusion relation. If $\mathscr{D}$ is a chain in $\mathscr{C}$, then the union of the families in $\mathscr{D}$ is an upper bound for $\mathscr{D}$ in $\mathscr{C}$. Therefore Zorn's lemma applies and we obtain a maximal family $\mathscr{F}_0$ in $\mathscr{C}$. Now, consider the collection

$$\mathscr{E} = \left\{ (\alpha, F) \mid \alpha \in A, F \in \mathscr{F}_0 \text{ and } \phi(\alpha) \in F \right\}$$

and the relation $\leq$ on $\mathscr{E}$: $(\alpha, F) \leq (\beta, E) \Leftrightarrow E \subseteq F$ and $\alpha \preceq \beta$. Clearly, the relation $\leq$ on $\mathscr{E}$ is reflexive and transitive. Moreover, if (α, F) and (β, E) are in $\mathscr{E}$, then there exists a $\gamma \in A$ such that $\gamma \succeq \alpha, \beta$ and $\phi(\gamma) \in E \cap F \in \mathscr{F}_0$, by the definition of $\mathscr{C}$ and the facts that A is directed and $\mathscr{F}_0$ is a member of $\mathscr{C}$. So we have $(\gamma, E \cap F) \in \mathscr{E}$. Then $(\gamma, E \cap F) \geq (\alpha, F), (\beta, E)$ and thus the family $\mathscr{E}$ is directed by the ordering $\leq$. It is easily checked that the mapping $\theta: \mathscr{E} \to A$, $(\alpha, F) \mapsto \alpha$, satisfies the condition (b) of Definition 4.2.9, and therefore $\phi\theta$ is a subnet of ϕ. We show that $\phi\theta$ is universal. Let $S \subset X$ be any set. As noted above, if $\phi\theta$ is not eventually in S or $X - S$, then it must be frequently in S and $X - S$, both. In that case, we claim that both of these sets lie in $\mathscr{F}_0$, and consequently $\varnothing \in \mathscr{F}_0$, a contradiction. Given an $\alpha \in A$ and an $F \in \mathscr{F}_0$, we find a $\beta \succeq \alpha$ such that $\phi(\beta) \in F$. Then $(\beta, F) \in \mathscr{E}$. By our assumption, $\phi\theta$ is frequently in S; accordingly, we find a $(\gamma, E) \in \mathscr{E}$ such that $(\gamma, E) \geq (\beta, F)$ and $\phi(\gamma) = \phi\theta\,((\gamma, E)) \in S$. Thus we have $\phi(\gamma) \in E \cap S \subseteq F \cap S$, where $\gamma \succeq \alpha$. It follows that ϕ is frequently in $F \cap S$ for every $F \in \mathscr{F}_0$. By the definition of subnet $\phi\theta$, we also see that ϕ is frequently in S, and hence the family $\mathscr{F}_1$ consisting of S, all the sets in $\mathscr{F}_0$, and their intersections with S lies in $\mathscr{C}$. Then, by the maximality of $\mathscr{F}_0$, we have $\mathscr{F}_1 = \mathscr{F}_0$ so that $S \in \mathscr{F}_0$. Similarly, we deduce that $X - S \in \mathscr{F}_0$, and this proves our claim. $\diamond$

We remark that there is no analogue of this theorem for sequences, and it is equivalent to the Axiom of Choice.

Exercises

1. Prove that a net in a discrete space is convergent if and only if it is eventually constant.

2. • Let $\mathscr{S}$ be a subbasis for a space X. If a net $\langle x_\alpha \rangle$ in X is eventually in each member of $\mathscr{S}$ containing x, prove that $x_\alpha \to x$.

3. Let (X, d) be a metric space and $x_0 \in X$ be a limit point of $X - \{x_0\}$. Direct the set $X - \{x_0\}$ by the relation $x \preceq x'$ if $d(x', x_0) \leq d(x, x_0)$. Show that a net $\phi : X - \{x_0\} \to Y$, where Y is a metric space, converges to $y_0 \in Y$ if and only if $\lim_{x \to x_0} \phi(x) = y_0$ in the sense of elementary analysis.

4. Let $\mathscr{T}_1$ and $\mathscr{T}_2$ be topologies on a set X. If each net in X which converges relative to $\mathscr{T}_1$ also converges to the same point with respect to $\mathscr{T}_2$, prove that $\mathscr{T}_2 \subseteq \mathscr{T}_1$.

5. Let $(X, \leq)$ be a linearly ordered set. A net $\phi : A \to X$ is called monotone increasing (resp. decreasing) if $\alpha \preceq \beta \Rightarrow \phi(\alpha) \leq \phi(\beta)$ (resp. $\phi(\beta) \leq \phi(\alpha)$).

   Let $(X, \leq)$ be an order complete linearly ordered set. Show that each monotone increasing net in X whose range has an upper bound converges in the order topology to the supremum of its range.

6. Let X be a metric space and $\phi : [0, \Omega) \to X$ be a net. Show that $\phi \to x$ in X if and only if $\phi(\alpha)$ equals x eventually.

7. Show that there is a net in $[0, \Omega)$ which converges to Ω in the ordinal space $[0, \Omega]$, but there is no sequence in $[0, \Omega)$ converging to Ω.

8. Let X be any subset of $\mathbb{R}$, and let $\mathscr{F}$ be the family of all finite subsets of X ordered by inclusion. For $F \in \mathscr{F}$, put $\phi(F) = \sum_{x \in F} x$. Prove that the net ϕ has a limit if and only if X is a countable set and, for any enumeration $\{x_1, x_2, \ldots\}$ of X, $\sum |x_n| < \infty$. In this case, $\phi \to \sum_{n=1}^{\infty} x_n$.

9. (a) If x is a cluster point of a net which is eventually in a closed set F, show that $x \in F$.

   (b) Prove that a subset F of a space X is closed if and only if no net in F converges to a point of $X - F$.

10. Show that a cofinal subset B of a directed set $(A, \preceq)$ is also directed by $\preceq$, and the inclusion $B \hookrightarrow A$ satisfies the condition 4.2.9(b).

11. Let $\phi: A \to X$ be a net and $\theta: B \to A$ be a function of a directed set B to A, which is increasing $(\beta_1 \preceq \beta_2 \Rightarrow \theta(\beta_1) \preceq \theta(\beta_2))$ and cofinal (i.e., $\theta(B)$ is cofinal in A). Show that $\phi\theta$ is a subnet of ϕ.

12. If every subnet of a net $\phi: A \to X$ has a subnet converging to x, show that ϕ converges to x. (It remains true if we replace "nets" by sequences in this statement.)

13. Let $\phi: A \to X$ be a net in the space X and, for each $\alpha \in A$, let $Y_\alpha = \{\phi(\beta) \mid \beta \succeq \alpha$ in $A\}$. Show that x is a cluster point of ϕ if and only if $x \in \bigcap_\alpha \overline{Y}_\alpha$.

14. Let X_α, $\alpha \in A$, be a family of topological spaces, and let $p_\beta : \prod X_\alpha \to X_\beta$ be the projection onto the βth factor. Let $\phi : \Lambda \to \prod X_\alpha$ be a net with x as a cluster point. Show that for every α, $p_\alpha \circ \phi$ has $p_\alpha(x)$ for a cluster point. Give an example to show that the converse fails.

15. Show that a sequence is a universal net if and only if it is eventually constant.

16. Prove that a subnet of a universal net is universal.

17. Let $f: X \to Y$ be a function, where X and Y are sets. If ϕ is a universal net in X, prove that the composition $f\phi$ is a universal net in Y.

4.3   Filters

In the previous section, we have discussed a natural generalization of the theory of sequential convergence. However, the relation between convergence and topologies is best understood by means of the "filters". As this concept is not essential for our future discussions, we set forth the subject on a rather fast pace and relegate most properties of the filters to exercises.

Definition 4.3.1 A *filter* $\mathscr{F}$ on a set X is a nonempty family of nonempty subsets of X with the properties:

   (a) if $F \in \mathscr{F}$ and $F \subseteq F'$, then $F' \in \mathscr{F}$, and
   (b) if $F_1, F_2 \in \mathscr{F}$, then $F_1 \cap F_2 \in \mathscr{F}$.

The family of all subsets of X which contains a given nonempty set $E \subseteq X$ is a filter. If X is a topological space and $x \in X$, then the family $\mathscr{N}_x$ of all nbds of x in X is a filter on X.

For another example, consider a net $\{x_\alpha, \alpha \in A\}$ in X. Then the family $\mathscr{F}$ consisting of subsets F of X for which there exists a $\alpha \in A$ (depending on F) such that $x_\beta \in F$ for all $\beta \succeq \alpha$ is a filter on X. We call $\mathscr{F}$ the *filter generated by the net* $\langle x_\alpha \rangle$. On the other hand, given a filter $\mathscr{F}$ on X, let $\mathscr{D} = \{(x, F) \mid x \in F \in \mathscr{F}\}$. Direct

$\mathcal{D}$ by the ordering $(x, F) \preceq (x', F') \iff F' \subseteq F$. Then the function $\phi : \mathcal{D} \to X$ defined by $\phi(x, F) = x$ is a net, called the *net based on* $\mathcal{F}$.

If $\mathcal{F}$ and $\mathcal{F}'$ are two filters on X and $\mathcal{F} \subseteq \mathcal{F}'$, then we say that $\mathcal{F}'$ is *finer* than $\mathcal{F}$. A filter $\mathcal{F}$ on a topological space X *converges* to $x \in X$ (written as $\mathcal{F} \to x$) if $\mathcal{F}$ is finer than the nbd filter $\mathcal{N}_x$. In this case, we call x a limit of $\mathcal{F}$. The filter $\mathcal{F}$ is said to *accumulate at* $x \in X$ if $x \in \overline{F}$ for every $F \in \mathcal{F}$. If $\mathcal{F}$ accumulates at x, then we say that x is an accumulation point or cluster point of $\mathcal{F}$.

A *filter base* in X is a family $\mathcal{B}$ of nonempty subsets of X such that for every pair of sets B_1, B_2 in $\mathcal{B}$, there exists a set $B_3 \in \mathcal{B}$ such that $B_3 \subseteq B_1 \cap B_2$. Obviously, the family $\mathcal{U}_x$ of all open nbds of a point x in a space X is a filter base in X. If $\mathcal{B}$ is a filter base in X, then the family of supersets of members of $\mathcal{B}$ is a filter on X. This is referred to as the filter generated by $\mathcal{B}$. If $\mathcal{F}$ and $\mathcal{G}$ are two filters on X such that $F \cap G \neq \varnothing$ for all $F \in \mathcal{F}$ and $G \in \mathcal{G}$, then $\mathcal{F} \cap \mathcal{G} = \{F \cap G \mid F \in \mathcal{F}, G \in \mathcal{G}\}$ is a filter base, and the filter generated by $\mathcal{F} \cap \mathcal{G}$ is finer than both $\mathcal{F}$ and $\mathcal{G}$.

A filter base $\mathcal{B}$ in a space X is said to converge to a point $x \in X$ if the filter generated by $\mathcal{B}$ converges to x. When $\mathcal{B}$ converges to x, we write $\mathcal{B} \to x$. We say that $\mathcal{B}$ accumulates at x if the filter generated by $\mathcal{B}$ accumulates at x.

A filter $\mathcal{F}$ on a set X is called an *ultrafilter* if there is no filter on X finer than $\mathcal{F}$. In other words, an ultrafilter on X is a maximal member of the collection of filters on X partially ordered by the inclusion relation.

Proposition 4.3.2 *Every filter on a set X is contained in an ultrafilter.*

Proof Let $\mathcal{F}$ be a filter on X, and consider the collection $\mathcal{C}$ of all filters on X which contain $\mathcal{F}$. Partially order $\mathcal{C}$ by the inclusion relation. If $\mathcal{D}$ is a chain in $\mathcal{C}$, then it is easily checked that the union of members of $\mathcal{D}$ is a filter on X. Thus every chain in $\mathcal{C}$ has an upper bound. By Zorn's lemma, $\mathcal{C}$ has a maximal member $\mathcal{G}$, say. Clearly, $\mathcal{G}$ is an ultrafilter on X containing $\mathcal{F}$. $\diamond$

Given a point $x \in X$, the family of all subsets of X which contain x is an ultrafilter (called the *principal* filter generated by x). It follows that an ultrafilter on X is not unique.

(A detailed discussion on this topic can be found in the text *General Topology* by Bourbaki (see also James [5] and Willard [16]).)

Exercises

1. Which filters $\mathcal{F}$ converge to x in the trivial space X? Answer the same question in a discrete space.

2. Let X be an infinite set with cofinite topology.

   (a) Let $\mathcal{F}$ be a filter on X such that every member of $\mathcal{F}$ is infinite. Find its accumulation points.

   (b) If $\mathcal{F}$ consists of all cofinite sets, what are its limits?

3. If a filter $\mathcal{F}$ on a space X converges to x, show that x is an accumulation point of $\mathcal{F}$.

4. Prove that a filter $\mathscr{F}$ on a space X accumulates at $x \Leftrightarrow$ some filter finer than $\mathscr{F}$ converges to x.

5. Let X be a space and $E \subseteq X$. Show that $x \in \overline{E} \Leftrightarrow$ there exists a filter $\mathscr{F}$ on X such that $E \in \mathscr{F}$ and $\mathscr{F} \to x$.

6. Prove that a space X is Hausdorff if and only if $\mathscr{F} \to x$ in X implies that each accumulation point of $\mathscr{F}$ coincides with x. (Thus limits of convergent filters on a Hausdorff space are unique.)

7. Let $\mathscr{F}$ be the filter generated by a net ϕ in X. Verify

    (a) $\phi \to x \Leftrightarrow \mathscr{F} \to x$.

    (b) ϕ accumulates at $x \Leftrightarrow \mathscr{F}$ accumulates at x.

8. Let $\mathscr{F}$ be a filter on X and ϕ be the net based on $\mathscr{F}$. Prove:

    (a) $\mathscr{F} \to x$ in X if and only if $\phi \to x$.

    (b) $\mathscr{F}$ accumulates at x if and only if x is a cluster point of ϕ.

9. If ψ is a subnet of a net ϕ in X, show that the filter generated by ψ is finer than the filter generated by ϕ.

10. Let $\phi \colon A \to X$ be a net. Show that the family of subsets $B_\alpha = \{\phi(\beta) \mid \beta \succeq \alpha\}$, $\alpha \in A$, is a filter base.

11. Let $\mathscr{B}$ be a filter base in a space X. Prove:

    (a) $\mathscr{B} \to x \Leftrightarrow$ for each nbd U of x, there exists $B \in \mathscr{B}$ such that $B \subseteq U$.

    (b) $\mathscr{B}$ accumulates at $x \Leftrightarrow$ for every $B \in \mathscr{B}$ and every nbd U of x, $(U \cap B \neq \varnothing$ $\Leftrightarrow x \in \bigcap\{\overline{B} : B \in \mathscr{B}\})$.

12. Let $f \colon X \to Y$ be a function.

    (a) If $\mathscr{F}$ is a filter on X, show that $\{f(F) : F \in \mathscr{F}\}$ is a filter base in Y. (The filter on Y generated by this filter base will be denoted by $f_*(\mathscr{F})$.)

    (b) If $\mathscr{G}$ is a filter on Y and $f^{-1}(G) \neq \varnothing$ for all $G \in \mathscr{G}$, show that $\{f^{-1}(G) : G \in \mathscr{G}\}$ is a filter base in X. (The filter on X generated by this filter base will be denoted by $f^*(\mathscr{G})$.)

13. Let X and Y be spaces and $f \colon X \to Y$ be a function. Prove that f is continuous at $x_0 \in X$ if and only if for every filter $\mathscr{F} \to x_0$ in X, $f_*(\mathscr{F}) \to f(x_0)$ in Y.

14. Let X_λ, $\lambda \in \Lambda$, be a family of topological spaces. Show that a filter $\mathscr{F}$ on $\prod X_\lambda$ converges to $(x_\lambda) \Leftrightarrow (p_\mu)_*(\mathscr{F}) \to x_\mu$ for each $\mu \in \Lambda$, where $p_\mu : \prod X_\lambda \to X_\mu$ is the canonical projection map.

15. Let $f : X \to Y$ be a function.

    (a) If $\mathscr{B}$ is a filter base in X, prove that $f(\mathscr{B}) = \{f(B) \mid B \in \mathscr{B}\}$ is a filter base in Y.

    (b) If $\mathscr{B}$ is a filter base in Y and $f^{-1}(B) \neq \varnothing$ for every $B \in \mathscr{B}$, prove that $f^{-1}(\mathscr{B}) = \{f^{-1}(B) \mid B \in \mathscr{B}\}$ is a filter base in X.

16. Let $\mathscr{F}$ be a filter on X. Show that $\mathscr{F}$ is an ultrafilter if and only if for each set $S \subseteq X$, either $S \in \mathscr{F}$ or $X - S \in \mathscr{F}$.

17. Prove: A filter $\mathscr{F}$ on a set X is an ultrafilter if and only if for every pair of subsets $S, T \subset X$, $S \cup T \in \mathscr{F}$ implies that $S \in \mathscr{F}$ or $T \in \mathscr{F}$.

18. Prove: The net based on an ultrafilter is an ultranet, and the filter generated by an ultranet is an ultrafilter.

19. If a filter $\mathscr{F}$ on a set X is contained in a unique ultrafilter $\mathscr{F}_0$, show that $\mathscr{F} = \mathscr{F}_0$.

20. If $f : X \to Y$ is a surjection and $\mathscr{F}$ is an ultrafilter on X, prove that $f_*(\mathscr{F})$ is an ultrafilter on Y.

4.4 Hausdorff Spaces

As seen in Sect. 4.2, the separation of distinct points in a topological space by open sets is needed for uniqueness of "limits" (i.e., no sequence or net converges to more than one point). It is an important condition for most discussions in topology and considered to be a very mild one to be imposed on a space. We will treat it here in some more detail and also consider two other conditions obtained by making slight changes in it.

Definition 4.4.1 A space X is called *Hausdorff* (or T_2) if for every pair of distinct points x, y of X there exist disjoint open sets U and V with $x \in U$ and $y \in V$.

Example 4.4.1 A discrete space is obviously a T_2-space.

Example 4.4.2 Any metric space is Hausdorff. For, if $x \neq y$ and $2r$ is the distance between x and y, then the open balls $B(x; r)$ and $B(y; r)$ are disjoint nbds of x and y.

Example 4.4.3 An ordered space is Hausdorff: Suppose that $x \prec y$ are two points of the ordered space $(X, \preceq)$. If $(x, y) = \varnothing$, then the subbasic open sets $(-\infty, y)$ and $(x, +\infty)$ are disjoint nbds of x and y, respectively. And, if there is a point $z \in (x, y)$, then $(-\infty, z)$ and $(z, +\infty)$ satisfy the requirement.

By Theorem 4.2.4, the Hausdorff condition essentially means that "limits" are unique. Here are other useful characterizations of this property.

Proposition 4.4.2 *The following properties of a space X are equivalent:*

(a) *X is Hausdorff.*
(b) *For each $x \in X$, the intersection of all closed nbds of x is $\{x\}$.*
(c) *The diagonal $\Delta = \{(x, x) \mid x \in X\}$ is closed in $X \times X$.*

Proof (a) $\Leftrightarrow$ (b): If $y \neq x$, then there is an open nbd U of x and an open nbd V of y such that $U \cap V = \varnothing$. So $X - V$ is a closed nbd of x, for $U \subseteq X - V$. Obviously, $y \notin X - V$ and (b) holds. Conversely, if $y \neq x$, then there is a closed nbd K of x such that $y \notin K$. Choose an open set U with $x \in U \subseteq K$. Then U and $X - K$ are disjoint open nbds of x and y, respectively.

(a) $\Leftrightarrow$ (c): For any $U, V \subseteq X$, we have $U \cap V = \varnothing$ if and only if $(U \times V) \cap \Delta = \varnothing$, and the result follows immediately. $\diamond$

Corollary 4.4.3 *Let Y be a Hausdorff space.*

(a) *If $f : X \to Y$ is a continuous function, then the graph of f is closed in $X \times Y$.*
(b) *If $f, g : X \to Y$ are continuous functions, then $E = \{x \in X \mid f(x) = g(x)\}$ is closed in X.*

Proof (a): Clearly, the graph of f is the inverse image of the diagonal Δ under the continuous map

$$X \times Y \xrightarrow{f \times 1} Y \times Y, \quad (x, y) \mapsto (f(x), y).$$

Since Y is T_2, Δ is closed in $Y \times Y$, and the conclusion holds.

(b): Suppose that $f, g : X \to Y$ are continuous. Then the map $h : X \to Y \times Y$, defined by $h(x) = (f(x), g(x))$, is continuous. As Δ is closed in $Y \times Y$, $E = h^{-1}(\Delta)$ is closed in X, and the result follows. $\diamond$

By the preceding corollary, it is immediate that if two continuous functions $f, g : X \to Y$ agree on a dense subset of X, and Y is a T_2-space, then $f = g$.

The Hausdorff property is clearly a topological invariant, but it is not preserved by continuous maps or even by continuous open or continuous closed maps.

Example 4.4.4 The space $\mathbb{R}_f$ of real numbers with the cofinite topology is not Hausdorff, for there are no nonempty disjoint open sets. The identity map of the real line $\mathbb{R}$ onto $\mathbb{R}_f$ is, of course, continuous.

The reader is referred to Examples 4.4.7, 6.1.11, and 6.1.12 for the remaining cases.

A property of topological spaces is said to be *hereditary* if every subspace of a space with the property also has the property. Obviously, Hausdorff property is hereditary, and the next result shows that it is also invariant under the product formation.

Theorem 4.4.4 *Let X_α, $\alpha \in A$, be a family of Hausdorff spaces. Then the product space $\prod X_\alpha$ is Hausdorff.*

Proof Let $x = (x_\alpha)$ and $y = (y_\alpha)$ be distinct points of $\prod X_\alpha$. Then $x_\beta \neq y_\beta$ for some index $\beta \in A$. Since X_β is Hausdorff, there exist disjoint open nbds U_β and V_β of x_β and y_β in X_β, respectively. If $p_\beta : \prod X_\alpha \to X_\beta$ is the projection map, then $p_\beta^{-1}(U_\beta)$ and $p_\beta^{-1}(V_\beta)$ are clearly disjoint open nbds of x and y, respectively. $\diamond$

It is remarkable that the converse of this theorem is also true (provided $X_\alpha \neq \varnothing$ for all α), because each X_α is homeomorphic to a subspace of $\prod X_\alpha$.

We now discuss two weaker conditions, viz., T_0 and T_1 which are sufficient for some purposes, and will be talked about occasionally.

Definition 4.4.5 A space X is called T_1 if for each pair of distinct points $x, y \in X$, there is an open set containing x but not y, and another open set containing y but not x.

Definition 4.4.6 A space X is called T_0 if for any two distinct points of X, there is an open set which contains one point but not the other.

The axiom T_1 is also referred to as the *Fréchet* condition, and axiom T_0 as the *Kolmogorov* condition. Evidently, a T_2-space is T_1, and a T_1-space is T_0. But the converse is false in either case.

Example 4.4.5 The Sierpinski space satisfies the axiom T_0 but not T_1.

Example 4.4.6 The space $\mathbb{R}_f$ in Example 4.4.4 satisfies T_1-axiom but not T_2.

The assertion in the preceding example follows at once from the following.

Proposition 4.4.7 *A space X is $T_1 \Leftrightarrow \{x\}$ is closed for every $x \in X \Leftrightarrow$ the intersection of all nbds of $x \in X$ is $\{x\}$.*

Proof Suppose that X is a T_1-space. Then, for each $y \in X - \{x\}$, there is an open nbd U_y of y such that $x \notin U_y$. So $X - \{x\} = \bigcup \{U_y \mid y \in X \text{ and } y \neq x\}$ is open. Conversely, if every one-point set in X is closed and $x \neq y$, then the open sets $X - \{y\}$ and $X - \{x\}$ are nbds of x and y, respectively, not containing the other point.

To see the second equivalence, let x_0 be a fixed element of X. If $\{x\}$ is closed for every $x \in X$ and $\mathcal{N}_x$ denotes the family of all nbds of x in X, then $X - \{x\} \in \mathcal{N}_{x_0}$ for all $x \neq x_0$. So we have $\bigcap \{N \mid N \in \mathcal{N}_{x_0}\} = \{x_0\}$. Conversely, if $\bigcap \{N \mid N \in \mathcal{N}_x\} = \{x\}$ for every $x \in X$, then, for every $x \neq x_0$, $X - \{x_0\}$ contains an $N \in \mathcal{N}_x$. This implies that $X - \{x_0\}$ is open and $\{x_0\}$ is closed in X. $\diamond$

It is immediate from the preceding proposition that every finite set in a T_1-space is closed. As another consequence, we see that the image of a T_1-space under a closed function is T_1. But this property is not necessarily preserved by a continuous function or even a continuous open function. This can be seen by the following.

Example 4.4.7 Denote the Sierpinski space $\{0, 1\}$ with $\{0\}$ open by Y, and consider the mapping $f : I \to Y$ defined by $f(t) = 0$ for all $t < 1$ and $f(1) = 1$. Then f is clearly a continuous open mapping of the T_2-space I onto Y, where Y is not T_1.

It is also clear that subspaces and products of T_1-spaces are T_1.

Proposition 4.4.8 *Let X be a T_1-space and $A \subseteq X$. Then a point $x \in X$ is a limit point of A if and only if each open nbd of x contains infinitely many points of A.*

Proof The sufficiency is obvious. To prove the other half, assume that there is an open nbd U of x such that $A \cap U$ is finite. Then the set $F = A \cap U - \{x\}$ is closed in X, since X is T_1. Thus $U - F$ is an open nbd of x, which contains no point of A except possibly x itself. So x is not a limit point of A, and the proposition follows. ◇

It follows that if x is a limit point of a subset A of a T_1-space, then there are infinitely many points of A which are arbitrarily close to x. As an application of the above proposition, we show that an infinite Hausdorff space contains a large number of open sets.

Theorem 4.4.9 *An infinite Hausdorff space has infinitely many disjoint open sets.*

Proof Let X be an infinite T_2-space. If there are no limit points in X, then X is a discrete space and the theorem is obvious. So assume that y is a limit point of X. Choose a point $x_1 \in X$ different from y. Then there exist disjoint open sets U_1 and V_1 in X such that $x_1 \in U_1$, $y \in V_1$. As y is a limit point, $V_1 - \{y\}$ is nonempty. So we can find a point $x_2 \in V_1$ different from y. By our hypothesis, there are disjoint open sets U_2 and V_2 contained in V_1 such that $x_2 \in U_2$, $y \in V_2$. Notice that $U_1 \cap U_2 = \varnothing$, for $U_2 \subseteq V_1$. Clearly, the above argument is inductive because every nbd of y contains infinitely many points. Thus we obtain a sequence of open sets U_n and V_n such that $U_n \cap V_n = \varnothing$, and both U_n and V_n are contained in V_{n-1}. Since $V_1 \supseteq V_2 \supseteq \cdots$, U_n is disjoint from every U_i for $1 \leq i < n$. ◇

Exercises

1. Prove that space X in Example 4.1.2 is T_2.

2. Let X be any uncountable set and $\mathscr{T}_c$ be the cocountable topology for X. Show that $(X, \mathscr{T}_c)$ is a T_1-space which is not T_2.

3. On the set $\mathbb{R}$ of real numbers, consider the topology

$$\mathscr{T} = \{\varnothing\} \cup \{\mathbb{R} - F \mid F \text{ is closed and bounded in } \mathbb{R}\}.$$

   Show that $(\mathbb{R}, \mathscr{T})$ is T_1 but not T_2.

4. Let X be a T_2-space and $f : X \to Y$ be a closed bijection. Show that Y is T_2.

5. Let Y be a T_2-space.

   (a) If $f: X \to Y$ is a continuous injection, show that X is T_2.
       (This implies that a topology finer than a Hausdorff topology is Hausdorff.)

   (b) Let $f: X \to Y$ and $g: Y \to X$ be continuous with $gf = 1_X$. Show that X is T_2 and $f(X)$ is closed in Y.

6. Let $f: X \to Y$ be a continuous open surjection. Show that Y is Hausdorff if and only if the set $\{(x, x') \mid f(x) = f(x')\}$ is closed in $X \times X$.

7. Let X be an infinite Hausdorff space. Prove:

   (a) X contains a countably infinite discrete subspace.

   (b) X contains a strictly decreasing sequence of closed sets.

8. Show that any finite T_1-space is discrete.

9. For a set X, show that cofinite topology is the smallest topology for X satisfying T_1-axiom.

10. If X is a T_1-space with more than one point, show that a base which contains X as an element remains a base if X is dropped.

11. Let X be a T_1-space and $A \subseteq X$. Prove that A' is a closed set.

12. If A is a subset of the T_1-space X, show that the intersection of all the nbds of A in X is A itself.

13. Suppose that X is a T_1-space such that the intersection of every family of open sets is open. Show that X is discrete.

14. If a T_1-space X has no isolated points, show that every dense set in X also has no isolated points. Is the condition of T_1-ness on X necessary?

15. Prove: A T_1-space in which each point has a local base consisting of clopen sets is totally disconnected.

16. Show that a connected subset of a T_1-space having more than one point is infinite.

17. Give an example of a countable connected Hausdorff space.

Chapter 5
Compactness

The concept of compactness is undoubtedly the most important idea in topology. We have seen the importance of closed and bounded subsets of a Euclidean space $\mathbb{R}^n$ in real analysis; these sets, in topological parlance, are "compact". This term has been used to describe several (related) properties of topological spaces before its current definition was accepted as the most satisfactory. It was initially coined to describe the property of a metric space in which every infinite subset has a limit point. However, this sense of compactness failed to give some desirable theorems for topological spaces, especially the invariance under the formation of topological products. On the other hand, it was found that, in metric spaces, this property is equivalent to the Borel–Lebesgue property: Each "open cover" has a finite "subcover". After Tychonoff proved that the latter property is inherited by products from their factors, it has been universally adopted now as the definition of compactness. We will study this formulation of compactness in Sect. 5.1 and discuss its other variants in Sect. 5.2. In the case of metric spaces, which were then the most widely studied objects, different notions of compactness turn out to be equivalent to the present definition. This will be seen in Sect. 5.3. It is remarkable that the class of compact spaces is rather restricted; it does not include even Euclidean spaces. Interestingly, there are many spaces which satisfy a localized version of the property. Some useful results can be proved for spaces with this property, for they can be embedded into compact spaces. The next section concerns with such spaces. In the last section, we shall deal with "perfect maps"; these maps preserve several topological properties that are not invariant under continuous closed maps.

5.1 Compact Spaces

We begin this section with some terminology. A *covering* of a set X is a family $\mathcal{G}$ of subsets of X such that $X = \bigcup \{G : G \in \mathcal{G}\}$; in this case, we also say that $\mathcal{G}$ covers X. A *subcovering* of $\mathcal{G}$ is a subfamily $\mathcal{H}$ of $\mathcal{G}$ such that $\mathcal{H}$ covers X. If X is a topological space and every member of $\mathcal{G}$ is open (resp. closed), then $\mathcal{G}$ is called an

© Springer Nature Singapore Pte Ltd. 2019
T. B. Singh, *Introduction to Topology*,
https://doi.org/10.1007/978-981-13-6954-4_5

open (resp. closed) covering of X. For a subset Y of X, we sometimes consider a family $\mathscr{G}$ of subsets of X whose union contains Y. In this case, we say that $\mathscr{G}$ is a covering of Y by subsets of X.

Example 5.1.1 Let $r > 0$ be a real number, and let $\mathscr{G}$ be the family of intervals $(x - r, x + r)$ for every $x \in \mathbb{R}$. Then $\mathscr{G}$ is an open cover of $\mathbb{R}$. Similarly, in any metric space X, the family of all open balls $B(x; r)$, $x \in X$, is an open cover of X.

Example 5.1.2 Let X be a metric space and $x_0 \in X$ be a fixed point. The family of open balls $B(x_0; n)$, $n \in \mathbb{N}$, is an open cover of X.

Example 5.1.3 The family of intervals $(1/n, 1]$, n a positive integer, is an open cover of $(0, 1]$.

Definition 5.1.1 A space X is said to be *compact* if every open covering of X has a finite subcover.

In the literature, the term "bicompact" has also been used by some mathematicians to describe the above property of compactness.

Example 5.1.4 A cofinite space is compact.

Example 5.1.5 A Fort space (see Exercise 1.2.3) is compact.

Example 5.1.6 An infinite discrete space is not compact.

Example 5.1.7 The Euclidean space $\mathbb{R}^n$ is not compact, since the open covering $\{B(x; 1) \mid$ every coordinate of x is an integer$\}$ does not have a finite subcovering.

We will have more examples of compact spaces after the following.

Definition 5.1.2 A *subset* Y of a space X is said to be compact if every covering of Y by sets open in X has a finite subcovering.

Since a set $U \subseteq Y$ is open in the relative topology on Y if and only if $U = Y \cap G$ for some open set $G \subseteq X$, it follows that Y is compact if and only if it is compact as a *subspace* of X. Hence, we see that a set $K \subseteq Y$ is a compact subset of the subspace Y if and only if K is a compact subset of X. Contrast this with the property of being open or closed.

Example 5.1.8 If $\langle x_n \rangle$ is a convergent sequence in a space X with limit x, then $\{x\} \cup \{x_n \mid n = 1, 2, \ldots\}$ is a compact subset of X.

An important example of compact set is given by the following theorem.

Theorem 5.1.3 (Heine–Borel Theorem) *A closed interval $J = [a, b]$ in the real line $\mathbb{R}$ is compact.*

Proof Let $\mathcal{G}$ be a family of open subsets of $\mathbb{R}$ covering J, and consider the set

$$X = \{x \in J \mid [a, x] \text{ is covered by a finite subfamily of } \mathcal{G}\}.$$

Obviously, a belongs to some member G of $\mathcal{G}$. So there exists a real number $\epsilon > 0$ such that $(a - \epsilon, a + \epsilon) \subseteq G$. Then, for any element $x \in J$ satisfying $x < a + \epsilon$, we have $[a, x] \subset G$, and therefore X is nonempty. Moreover, X is bounded above by b, so $c = \sup X$ exists. As $a < c \leq b$, there exists a set G_0 in $\mathcal{G}$ such that $c \in G_0$. Since G_0 is open in $\mathbb{R}$, we find a real number $\delta > 0$ such that $(c - \delta, c + \delta) \subseteq G_0$. By the choice of c, there exists an $x \in X$ such that $c - \delta < x$. Then, by the definition of X, there are finitely many sets, say $G_1, \ldots, G_n$, in $\mathcal{G}$ such that $[a, x] \subseteq \bigcup_1^n G_i$. It is now obvious that the family $\mathcal{H} = \{G_0, G_1, \ldots, G_n\}$ covers $[a, c]$, and thus $c \in X$. Were $c < b$, then we would find a number $d \in J$ such that $c < d < c + \delta$. It is then clear that $[a, d]$ is covered by $\mathcal{H}$ forcing d in X. This contradicts the choice of c, and therefore $b = c$. Thus, J is covered by $\mathcal{H}$, completing the proof. $\diamond$

On the other hand, an open interval or a half-closed interval in $\mathbb{R}$ is not compact. For example, consider the half-closed interval $(0, 1]$. Clearly, its open covering $\{(1/n, 1] \mid n = 2, 3, \ldots\}$ has no finite subcovering.

An argument similar to the one given in Theorem 5.1.3 shows that every closed interval in an order complete space X is compact. For the ordinal space $[0, \Omega]$, we have an alternate proof.

Example 5.1.9 The ordinal space $[0, \Omega]$ is compact. Given an open covering $\mathcal{U}$ of $[0, \Omega]$, consider an open nbd $V(\Omega) = (x_1, \Omega]$ of Ω contained in some member of $\mathcal{U}$. Since, for each ordinal $x < \Omega$, the open sets $(y, x] \subseteq (0, \Omega]$ form a nbd basis at x, we find a sequence of open nbds $V(x_1) = (x_2, x_1], V(x_2) = (x_3, x_2], \ldots$ such that each $V(x_n)$ is contained in some $U \in \mathcal{U}$. As $x_1 > x_2 > \cdots$, we must have $x_n = 0$ for some integer $n > 0$, by the well-ordering property of $[0, \Omega)$. Now, we choose sets $U_i \in \mathcal{U}, i = 0, 1, \ldots, n$, such that $U_0 \supseteq V(\Omega)$, $U_i \supseteq V(x_i)$ for $1 \leq i \leq n - 1$, and $U_n \supseteq \{0\}$. Then the family $\{U_i \mid 0 \leq i \leq n\}$ is a subcovering of $\mathcal{U}$, and therefore $[0, \Omega]$ is compact.

Taking complements and applying De Morgan's rules, we obtain a useful formulation of compactness in terms of closed sets.

Theorem 5.1.4 *A space X is compact if and only if every family $\{F_\alpha\}_{\alpha \in A}$ of closed subsets of X with $\bigcap_{\alpha \in A} F_\alpha = \varnothing$ contains a finite subfamily $\{F_{\alpha_1}, \ldots, F_{\alpha_n}\}$ such that $\bigcap_{i=1}^n F_{\alpha_i} = \varnothing$.*

Definition 5.1.5 A collection $\mathcal{C}$ of subsets of a given set X is said to have the *finite intersection property* if the intersection of any finite subcollection of $\mathcal{C}$ is nonempty.

With this terminology, Theorem 5.1.4 can be restated as

Theorem 5.1.6 *A space X is compact if and only if each family of closed subsets of X with the finite intersection property has a nonempty intersection.*

Although compactness is not a hereditary property, it is inherited by closed subsets.

Theorem 5.1.7 *A closed subset of a compact space is compact.*

Proof Let X be a compact space and $Y \subseteq X$ be closed. Let $\{F_\alpha \mid \alpha \in A\}$ be any family of closed subsets of Y with the finite intersection property. Since Y is closed in X, the F_α are closed in X, too. By the direct part of Theorem 5.1.6, $\bigcap F_\alpha \neq \varnothing$, and its converse part implies that Y is compact. $\diamond$

We can now describe all compact subsets of the real line $\mathbb{R}$ in more familiar terms. From Theorems 5.1.3 and 5.1.7, it follows that a closed and bounded subset of the real line $\mathbb{R}$ is compact. The converse of this statement also holds good. For, if $X \subset \mathbb{R}$ is compact, then there exist finitely many points, say $x_1, \ldots, x_n$, in X such that $X \subseteq \bigcup_1^n (x_i - 1, x_i + 1)$. Accordingly, $\operatorname{diam}(X) \leq M + 2$, where $M = \max_{i,j} |x_i - x_j|$, and X is bounded. The closedness of X is an immediate consequence of an important property of Hausdorff spaces described as

Theorem 5.1.8 *A compact subset of a Hausdorff space is closed.*

Proof Let X be a Hausdorff space and suppose that $A \subset X$ is compact. If $x \in X - A$, then, for each $a \in A$, there exist open sets U_a and V_a such that $x \in U_a$, $a \in V_a$ and $U_a \cap V_a = \varnothing$. Obviously, the family $\{V_a \mid a \in A\}$ covers A. Since A is compact, there are finitely many points $a_1, \ldots, a_n$ in A such that $A \subseteq \bigcup_1^n V_{a_i} = V$. If $U = \bigcap_1^n U_{a_i}$, then clearly U does not intersect V. So we have $x \in U \subseteq X - V \subseteq X - A$, and x is an interior point of $X - A$. Since $x \in X - A$ is arbitrary, we conclude that $X - A$ is open, as desired. $\diamond$

We remark that the Hausdorff condition in the preceding theorem cannot be relaxed, for an infinite proper subset of a cofinite space is compact but not closed. Further, it follows from Theorems 5.1.7 and 5.1.8 that a subset of a compact Hausdorff space is compact if and only if it is closed. By the proof of the preceding theorem, we see that if A is a compact subset of a Hausdorff space X and $x \in X - A$, then there are disjoint open sets U and V such that $x \in U$ and $A \subseteq V$. Using this result, we prove the following.

Proposition 5.1.9 *If A and B are disjoint compact subsets a Hausdorff space X, then there exist two disjoint open one containing A and the other containing B.*

Proof As noted above, for each $b \in B$, there are disjoint open sets U_b and V_b in X such that $b \in U_b$ and $A \subseteq V_b$. Obviously, the family $\{U_b \mid b \in B\}$ covers the compact set B, and so there are finitely many points $b_1, \ldots, b_n$ in B such that $B \subseteq \bigcup_1^n U_{b_i} = U$. Then the set $V = \bigcap_1^n V_{b_i}$ is open and contains A. Since $U_{b_i} \cap V_{b_i} = \varnothing$ for every i, we have $U \cap V = \varnothing$, and the proposition follows. $\diamond$

As an application of the above proposition, we prove the following

Theorem 5.1.10 *Every quasi-component of a compact Hausdorff space is a component.*

Proof Let X be a compact Hausdorff space and $x \in X$. Denote the component and the quasi-component of x by $C(x)$ and $K(x)$, respectively. Then $C(x) \subseteq K(x)$, by Proposition 3.2.6(c). To prove the reverse inclusion, it is clearly enough to show that $K(x)$ is connected. Assume that $K(x) = E \cup F$, where E and F are nonempty closed subsets of $K(x)$ and $E \cap F = \varnothing$. Since $K(x)$ is closed in X, E and F are also closed in X. So, by Proposition 5.1.9, there exist disjoint open sets U and V in X such that $E \subseteq U$ and $F \subseteq V$. Moreover, by Proposition 3.2.6(b), we have $K(x) = \bigcap C_\alpha$, where the C_α are clopen nbds of x in X. We observe that some $C_\alpha \subseteq U \cup V = W$, say. Put $K_\alpha = C_\alpha - W$ for every α. Then each K_α is closed in X and $\bigcap K_\alpha = \varnothing$. Since X is compact, there are finitely many indices $\alpha_1, \ldots, \alpha_n$, say, such that $\bigcap_1^n K_{\alpha_i} = \varnothing$. Obviously, $\bigcap_1^n C_{\alpha_i} = C_{\alpha_0}$ for some index α_0. Then $K_{\alpha_0} = \varnothing \Rightarrow C_{\alpha_0} \subseteq W$. Consequently, we have separations

$$X = (U \cap C_{\alpha_0}) \cup [(V \cap C_{\alpha_0}) \cup (X - C_{\alpha_0})]$$
$$= (V \cap C_{\alpha_0}) \cup [(U \cap C_{\alpha_0}) \cup (X - C_{\alpha_0})]$$

of X. If $x \in E$, then we must have $K(x) \subseteq U \cap C_{\alpha_0}$ which forces $F = \varnothing$. Similarly, $x \in F$ implies $E = \varnothing$. Thus, we have a contradiction in either case, and therefore $K(x)$ is connected. This finishes the proof. $\diamond$

By the preceding theorem and Proposition 3.2.6, we conclude that the component of a compact Hausdorff space X containing a point x is the intersection of all clopen nbds of x in X.

We next observe that compactness is a topological property; in fact, it is preserved by continuous mappings.

Theorem 5.1.11 *Let $f: X \to Y$ be a continuous map. If X is compact, then so is $f(X)$.*

Proof Suppose that X is compact and let $\mathscr{U} = \{U_\alpha \mid \alpha \in A\}$ be an open covering of $f(X)$. Then, for each $\alpha \in A$, there exists an open set $G_\alpha \subseteq Y$ such that $U_\alpha = f(X) \cap G_\alpha$. Clearly, the open sets $f^{-1}(G_\alpha)$ cover X. Since X is compact, we find finitely many sets G_{α_i}, $i = 1, \ldots, n$, in the family $\{G_\alpha\}_{\alpha \in A}$ such that $X = \bigcup_i f^{-1}(G_{\alpha_i})$. This implies that $f(X) = \bigcup_i U_{\alpha_i}$, as desired. $\diamond$

Applying Theorems 5.1.7, 5.1.8 and 5.1.11, we obtain the following.

Proposition 5.1.12 *Every continuous map of a compact space into a Hausdorff space is closed.*

Corollary 5.1.13 *A one-to-one continuous map of a compact space onto (resp. into) a Hausdorff space is a homeomorphism (resp. embedding).*

For invariance of compactness under products, we have the following

Theorem 5.1.14 (Tychonoff Theorem) *An arbitrary product of compact spaces is compact.*

This theorem is the main reason for adopting the product (Tychonoff) topology for the Cartesian product of an infinite family of topological spaces. For the moment, we defer a proof of the theorem in full generality and treat the finite case. The reason for this priority is the fact that proof in this case is quite easy and the theorem only for finite products is needed for immediate applications.

Theorem 5.1.15 *If X and Y are compact spaces, then so is $X \times Y$.*

Proof Let $\mathscr{G}$ be an open cover of $X \times Y$. Let $x \in X$ be arbitrary but fixed. Then, for each $y \in Y$, there exist open sets $U_y \subseteq X$ and $V_y \subseteq Y$ such that $(x, y) \in U_y \times V_y$ and $U_y \times V_y$ is contained in some member of $\mathscr{G}$. Since Y is compact, the open covering $\{V_y \mid y \in Y\}$ has a finite subcover. Accordingly, there are finitely many points $y_1, \ldots, y_{n(x)}$ in Y such that $Y = \bigcup_1^n V_{y_i}$. Put $U_x = \bigcap_1^{n(x)} U_{y_i}$ and, for each i, choose a member G_x^i of $\mathscr{G}$ such that $U_{y_i} \times V_{y_i} \subseteq G_x^i$. Then U_x is an open nbd of x and $U_x \times V_{y_i} \subseteq G_x^i$. Consequently, $U_x \times Y \subseteq \bigcup_1^{n(x)} G_x^i$. Now, consider the open covering $\{U_x \mid x \in X\}$ of X. Since X is compact, there exist finitely many points $x_1, \ldots, x_m$ in X such that $X = \bigcup_1^m U_{x_j}$. It follows that $X \times Y = \bigcup_{j=1}^m (U_{x_j} \times Y) = \bigcup_{j=1}^m \bigcup_{i=1}^{n(x_j)} G_{x_j}^i$, and thus the family $\{G_{x_j}^i \mid 1 \leq i \leq n(x_j) \text{ and } 1 \leq j \leq m\}$ is a finite subcovering of $\mathscr{G}$. $\diamond$

By induction, we see that a finite product of compact spaces is compact. In particular, by the Heine–Borel theorem, the unit n-cube I^n is compact. The unit disc $\mathbb{D}^n$, being homeomorphic to I^n (see Example 2.2.6), is also compact. More generally, we have the following

Theorem 5.1.16 (Generalized Heine–Borel Theorem) *A closed and bounded subset A of the Euclidean space $\mathbb{R}^n$ is compact, and conversely.*

Proof For each $i = 1, \ldots, n$, let $p_i : \mathbb{R}^n \to \mathbb{R}^1$ denote the projection onto the ith factor. Then $p_i(A)$ is a bounded subset of $\mathbb{R}^1$; consequently, there is a closed interval $[a_i, b_i]$ such that $p_i(A) \subseteq [a_i, b_i]$. It follows that $A \subseteq [a_1, b_1] \times \cdots \times [a_n, b_n] = X$, say. Clearly, X is homeomorphic to I^n and, therefore, compact. So, being a closed subset of X, A is compact. Conversely, if A is compact, then it is closed, by Theorem 5.1.8, and its boundedness follows from the fact that it can be covered by finitely many open balls, each of a fixed radius 1. $\diamond$

A straightforward application of this theorem shows that the n-sphere $\mathbb{S}^n$ is compact. For, it is obviously bounded and closed, being the inverse image of the real number 1 under the continuous function $\mathbb{R}^{n+1} \to \mathbb{R}^1$, $x \mapsto \|x\|$.

Turning to a proof of the Tychonoff theorem in full generality, we remark that there is no way to generalize its proof in the finite case to infinite products. Fortunately, our task is made considerably easier by the following characterizations of compactness.

Theorem 5.1.17 *In any space X, the following statements are equivalent:*

  (a) *X is compact.*

  (b) *Every universal net in X converges.*

  (c) *Every net in X has a convergent subnet.*

Proof (a) $\Rightarrow$ (b): Let $\phi \colon A \to X$ be a universal net, and assume that ϕ fails to converge in X. Then, for each $x \in X$, there exists an open nbd U_x of x such that ϕ is eventually in $X - U_x$. Accordingly, there is an element $\alpha_x \in A$ such that $\phi(\alpha) \in X - U_x$ for all $\alpha \succeq \alpha_x$. Since X is compact, there exist finitely many points $x_1, \ldots, x_n$ in X such that $X = \bigcup_{i=1}^{n} U_{x_i}$. As A is directed, there is an $\alpha \in A$ such that $\alpha \succeq \alpha_{x_i}$ for every $i = 1, \ldots, n$. So $\phi(\alpha) \in \bigcap_{i=1}^{n} (X - U_{x_i}) = X - \bigcup_{i=1}^{n} U_{x_i} = \varnothing$, a contradiction.

(b) $\Rightarrow$ (c): This follows from Theorem 4.2.12.

(c) $\Rightarrow$ (a): Given a family $\mathscr{F}$ of closed subsets of X with the finite intersection property, let $\mathscr{E}$ be the family of all members of $\mathscr{F}$ and their finite intersections. Direct $\mathscr{E}$ by the reverse inclusion, that is, $E \preceq E' \Leftrightarrow E \supseteq E'$. Notice that every member of $\mathscr{E}$ is nonempty, so we can choose a point $x_E \in E$ for every E in $\mathscr{E}$. By our hypothesis, the net $\langle x_E \rangle$ has a convergent subnet, say, $\{x_{f(\alpha)}, \alpha \in \mathscr{D}\}$. Suppose that $x_{f(\alpha)} \to x$ in X. By definition, $\mathscr{D}$ is a directed set and $f \colon \mathscr{D} \to \mathscr{E}$ is a function with the property that, for each $E \in \mathscr{E}$, there exists a $\beta \in \mathscr{D}$ such that $f(\gamma) \subseteq E$ for every $\gamma \succeq \beta$. This shows that the net $\langle x_{f(\alpha)} \rangle$ is eventually in E. Since E is closed and $x_{f(\alpha)} \to x$, we have $x \in E$. Accordingly, $x \in \bigcap \{E \in \mathscr{E}\} \subseteq \bigcap \{F \in \mathscr{F}\}$. Now, by Theorem 5.1.6, we deduce that X is compact, and thus (a) holds. $\diamond$

Proof of the Tychonoff theorem: Let $X_\lambda, \lambda \in \Lambda$, be a family of compact spaces, and consider a universal net $\phi \colon A \to \prod X_\lambda$. If $p_\mu \colon \prod X_\lambda \to X_\mu$ is the projection map, then $\prod X_\lambda - p_\mu^{-1}(S) = p_\mu^{-1}(X_\mu - S)$ for every $S \subseteq X_\mu$. So ϕ is eventually in either $p_\mu^{-1}(S)$ or $p_\mu^{-1}(X_\mu - S)$; accordingly, $p_\mu \circ \phi$ is eventually in either S or $X_\mu - S$. Thus, for each $\mu \in \Lambda$, the composition $p_\mu \circ \phi$ is a universal net in X_μ. By Theorem 5.1.17, $p_\mu \circ \phi \to x_\mu$ in X_μ, say. Then $\phi \to (x_\lambda)$, and hence $\prod X_\lambda$ is compact. $\diamond$

The above proof of the Tychonoff theorem appears much easier than others. But this is not actually the case; here the entire difficulty has been subsumed in the result about universal nets, used in establishing the characterization of compactness (Theorem 5.1.17) crucial for proving the Tychonoff theorem. This result is known to be equivalent to the axiom of choice. Furthermore, notice that the converse of the theorem also holds, since each projection of $\prod X_\lambda$ is a continuous surjection.

The Tychonoff theorem is unquestionably the most useful result in point-set topology; we will see several important applications of the theorem as we go further in the book.

When one factor of the product of two spaces is compact, the following result often comes in useful.

Lemma 5.1.18 (The Tube Lemma) *Let Y be compact and X be any space. Let N be an open nbd of the slice $\{x\} \times Y$ in $X \times Y$. Then there is an open nbd U of x such that $U \times Y \subseteq N$.*

We leave the easy proof along the lines of Theorem 5.1.15 to the reader.

The product $U \times Y$ in the above lemma is referred to as a *tube* about $\{x\} \times Y$ (see Fig. 5.1).

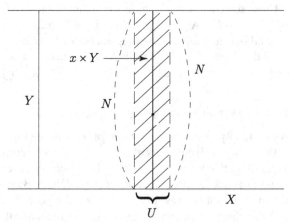

Fig. 5.1 A tube about $x \times Y$ contained in N

As an application, we show that if X is compact and Y is any space, then the projection $p \colon X \times Y \to Y$, $(x, y) \mapsto y$, is closed. Let $F \subseteq X \times Y$ be closed. Let $y \in Y - p(F)$. Then $(X \times \{y\}) \cap F = \varnothing$ so that $(X \times Y) - F$ is nbd of $X \times \{y\}$. By Lemma 5.1.18, there exists an open nbd U of y such that $X \times U \subseteq (X \times Y) - F$. Consequently, $U \cap p(F) = \varnothing$ which implies that $U \subseteq Y - p(F)$. Thus, $Y - p(F)$ is a nbd of y. Since $y \in Y - p(F)$ is arbitrary, $Y - p(F)$ is open, and therefore $p(F)$ is closed.

Interestingly, the above property characterizes compact spaces.

Theorem 5.1.19 *A space X is compact if and only if the projection $p \colon X \times Y \to Y$, $(x, y) \mapsto y$, is closed for all spaces Y.*

Proof Clearly, we need to prove the converse. Suppose that the mapping $p \colon X \times Y \to Y$, $p(x, y) = y$, is closed for every space Y. Let $\mathcal{U}$ be an open cover of X. Choose an element $\infty \notin X$, and let $Y = X \cup \{\infty\}$. Give Y the topology generated by the subbase which consists of one-point sets $\{x\}$, $x \in X$, and sets $Y - U$, $U \in \mathcal{U}$. Consider the subset $D = \{(x, x) \mid x \in X\}$ of $X \times Y$. By our assumption, the projection $p \colon X \times Y \to Y$ is closed; so $p\overline{D} = \overline{pD} = \overline{X}$. For each $x \in X$, there exists $U \in \mathcal{U}$ with $x \in U$. Then $U \times (Y - U)$ is a nbd of (x, ∞) and $[U \times (Y - U)] \cap D = \varnothing$. So (x, ∞) is not an adherent point of D, and $\overline{D} = D$. It follows that $\overline{X} = X$ and the singleton set $\{\infty\}$ is open in Y. So there exist finitely many sets $U_1, \ldots, U_n$ in $\mathcal{U}$ such that $\{\infty\} = \bigcap_1^n (Y - U_i)$, which implies that $X = \bigcup_1^n U_i$. Thus, $\mathcal{U}$ has a finite subcovering, and X is compact. $\diamondsuit$

Exercises

1. Show that the interval $[0,1]$ is not compact in either of the topologies induced by the cocountable topology, the lower limit topology and the topology described in Exercise 1.4.6 for $\mathbb{R}$.

2. Prove that a space X is compact $\Leftrightarrow$ every open covering consisting of members of a base for X has a finite subcovering. (The result also holds if the condition is satisfied by subbasic open coverings.)

3. Prove that a space X is compact if and only if each open covering of X has a nbd-finite subcovering.

4. (a) Prove that a finite union of compact subsets of a space is compact.

   (b) Give an example of two compact subsets K_1 and K_2 of a space such that $K_1 \cap K_2$ is not compact.

5. Give an example of a compact set whose closure is not compact.

6. If A is a compact subset of a Hausdorff space, prove that its derived set A' is compact. Is it true in any topological space?

7. Give an example of a T_1-space which contains a compact set that is not closed.

8. If every subset of a T_2-space X is compact, show that X is discrete.

9. Let X be an ordered space. If every closed interval of X is compact, show that X has the least upper bound property.

10. Generalize Theorem 5.1.3 to a closed interval in any order complete space.

11. (a) If X is order complete, prove that a closed subset of X with both lower and upper bounds is compact.

    (b) Prove that $I \times I$ with the television topology is compact.

12. Let $\mathscr{T}$ be a compact Hausdorff topology on a set X. Show that a topology strictly larger than $\mathscr{T}$ is not compact, and the one strictly smaller than $\mathscr{T}$ is not Hausdorff.

13. • Let X be a compact Hausdorff space, and $f : X \to Y$ be a continuous surjection. Show that Y is Hausdorff $\Leftrightarrow f$ is closed.

14. Give an example of a continuous injection of a compact space into a Hausdorff space that is not open.

15. Let X be a compact space, and let $\{f_n\}$ be a sequence of continuous real-valued functions on X such that $f_n(x) \le f_{n+1}(x)$ (or $f_n(x) \ge f_{n+1}(x)$) for all n and for all $x \in X$. If $f_n(x) \to f(x)$ for each $x \in X$ and f is continuous, show that $f_n \to f$ uniformly.

16. Prove (without using universal nets) that every net in a compact space has a cluster point (and hence a convergent subnet).

17. Prove: A space X is compact $\Leftrightarrow$ each filter on X accumulates at some point $x \in X \Leftrightarrow$ each ultrafilter on X is convergent.

18. Let $\mathscr{F}$ be a filter on a compact space X. Prove:

    (a) If E is the set of accumulation points of $\mathscr{F}$, then each nbd of E is a member of $\mathscr{F}$.

    (b) If there is only one accumulation point of $\mathscr{F}$ in X, then $\mathscr{F}$ is convergent.

19. • Let $\{K_\alpha\}$ be a family of compact subsets of a Hausdorff space X.

    (a) Prove that every open nbd U of $\bigcap K_\alpha$ contains a finite intersection of members of $\{K_\alpha\}$.

    (b) If the intersection of every finite subfamily of $\{K_\alpha\}$ is nonempty, show that $\bigcap K_\alpha$ is nonempty and compact.

20. Let X and Y be Hausdorff spaces and $f : X \to Y$ a continuous function. If $\{F_n\}$ is a decreasing sequence of compact subsets of X, show that $f\left(\bigcap F_n\right) = \bigcap f(F_n)$.

21. Let X be a compact Hausdorff space and $f : X \to X$ be a continuous function. Prove that there exists a nonempty closed set $A \subseteq X$ such that $A = f(A)$.

22. Let f be a function from a space X into a compact Hausdorff space Y. Show that f is continuous if and only if its graph G_f is closed in $X \times Y$.

23. • Let X and Y be spaces, and $A \subseteq X$ and $B \subseteq Y$ be compact sets. If W is a nbd of $A \times B$ in $X \times Y$, prove that there exist nbds U of A and V of B such that $U \times V \subseteq W$.

24. If a compact Hausdorff space X has no isolated points, prove that X is uncountable. Deduce that the Cantor set is uncountable.

25. A compact connected set in a space is called a *continuum*. Show that a Hausdorff continuum having more than one point is uncountable.

26. Let X be a compact Hausdorff space.

    (a) Suppose that $\mathscr{C}$ is a nonempty collection of closed connected subsets that is simply ordered by inclusion. Prove that $\bigcap\{C : C \in \mathscr{C}\}$ is connected.

    (b) Let $\mathscr{C}$ be a family consisting of closed connected subsets of X such that the intersection of any two members of $\mathscr{C}$ contains a member of $\mathscr{C}$. Show that the intersection of all members of $\mathscr{C}$ is a closed connected set.

27. Prove that the intersection of a (nonempty) family of continua in a Hausdorff space X directed by the reverse inclusion is a continuum.

28. Let X be a Hausdorff continuum and $A \subset X$ be closed. Show that there exists a closed connected subset $B \supseteq A$ such that no proper closed connected subset of B contains A.

29. Prove that a compact locally connected space has a finite number of components.

30. Prove that a compact Hausdorff space X is locally connected $\Leftrightarrow$ each open covering $\{U_\alpha\}$ of X has a finite subcovering $\{C_i\}$, where each C_i is connected and open.

31. Prove: A compact Hausdorff space X is totally disconnected if and only if for every pair x, y of distinct points, there exists a clopen set $A \subset X$ such that $x \in A$ and $y \notin A$.

32. Let X be a compact Hausdorff space and C be a component of X. Show that for each open set $U \subseteq X$ with $C \subseteq U$, there exists an open set V such that $\partial V = \emptyset$ and $C \subseteq V \subseteq U$.

33. Let X be a Hausdorff continuum. If C is a component of an open subset $U \subseteq X$, show that C has a limit point in $\overline{U} - U$.

34. Let X be a compact Hausdorff space. Consider the relation $\sim$ on X defined by $x \sim y$ if for every continuous function $f : X \to \mathbb{R}$ with $f(x) = 0$, $f(y) = 1$, there is a $z \in X$ such that $f(z) = 1/2$. Show that $\sim$ is an equivalence relation and its equivalence classes are the components of X.

5.2 Countably Compact Spaces

For a long time, a space X was said to be compact if it had the **Bolzano–Weierstrass property** (briefly, B-W property): Every infinite subset of X has a limit point. Some authors call this property "countable compactness", others term it the "limit point compactness". We begin this section by showing that B-W property is a generalization of compactness.

Theorem 5.2.1 *A compact space has the B-W property.*

Proof Suppose that X is a compact space and $A \subseteq X$ is infinite. If A has no limit point, then each point $x \in X$ has an open nbd G_x such that $A \cap G_x - \{x\} = \emptyset$. The family $\{G_x\}$ forms an open covering of X. Since X is compact, there exist finitely many points $x_i \in X$, $1 \leq i \leq n$, such that $X = \bigcup_1^n G_{x_i}$. This implies that $A = \bigcup_1^n \left(A \cap G_{x_i} \right)$ is finite, a contradiction. $\diamond$

The converse of the theorem is not true, as shown by the following example.

Example 5.2.1 On the set $X = \mathbb{R} - \mathbb{Z}$, consider the topology generated by the open intervals $(n - 1, n)$. If a subset A of X contains a point $x \in (n - 1, n)$, then every point of $(n - 1, n)$ except possibly x is a limit point of A. But X is not compact, since the open covering consisting of the basis elements is not reducible to a finite subcovering.

Clearly, the B-W property is a topological invariant but, unlike compactness, it is not invariant under continuous mappings (ref. Example 5.2.2 below). Of course, the image of a T_1-space X with the B-W property under a continuous map has the property. For, if $f: X \to Y$ is a continuous surjection and $B \subseteq Y$ is an infinite set, then we can construct an infinite subset of X by choosing a point $x \in f^{-1}(y)$ for every $y \in B$. Then the set A of all such points x has a limit point p, say. If G is an open nbd of $f(p)$ in Y, then $f^{-1}(G)$ is an open nbd of p. Since X is T_1, $f^{-1}(G)$ contains infinitely many points of A. Consequently, $G \cap B$ is infinite, and hence $f(p)$ is a limit point of B.

Example 5.2.2 Let $\mathbb{Z}$ have the discrete topology, and let X be the space of Example 5.2.1. The function $f: X \to \mathbb{Z}$, given by $f(x) = n$, for $n - 1 < x < n$, is continuous and surjective. But $\mathbb{Z}$ does not have the B-W property.

For almost all topological spaces, the B-W property coincides with the following condition.

Definition 5.2.2 A space X is called countably compact if every countable open covering of X has a finite subcovering.

Theorem 5.2.3 *A T_1-space X is countably compact if and only if it has the Bolzano–Weierstrass property.*

Proof Assume that some infinite subset $A \subseteq X$ has no limit point. Then we find a countably infinite set $B \subseteq A$. Since no point of X is a limit point of B, for each $x \in X$, there exists an open nbd U_x of x such that $U_x \cap B = \varnothing$ for $x \notin B$, and $U_x \cap B = \{x\}$ for $x \in B$. Then the family $\{X - B\} \cup \{U_x : x \in B\}$ is a countable open covering of X, which clearly has no finite subcovering.

Conversely, if X is not countably compact, then there exists a countable open covering $\{U_n \mid n = 1, 2, \ldots\}$ of X which contains no finite subcovering. Now, for each n, we choose a point $x_n \in X$ such that $x_n \notin \bigcup_1^n U_i$ and $x_n \neq x_i$ for every $1 \leq i \leq n - 1$. This is possible since $X - \bigcup_1^n U_i$ is infinite for every n. Since each $x \in X$ belongs to some U_n and $x_i \notin U_n$ for all $i > n$, the infinite set $\{x_n \mid n = 1, 2, \ldots\}$ has no limit point in X (for X is T_1). ◇

Countable compactness is characterized in terms of convergence of *subnets* of sequences.

Theorem 5.2.4 *A space X is countably compact if and only if each sequence in X has a cluster point (equivalently, every sequence has a convergent subnet).*

Proof $\Rightarrow$: Let $\langle x_n \rangle$ be a sequence in X and put $T_n = \{x_n, x_{n+1}, \ldots\}$. If $\bigcap \overline{T_n} = \varnothing$, then $\{X - \overline{T_n}\}$ is a countable open covering of X. Since X is countably compact, this open cover has a finite subcovering. Accordingly, there exist finitely many integers $n_1, \ldots, n_k$ such that $\bigcap_1^k \overline{T_{n_i}} = \varnothing$. This is plainly not true and, therefore, we have a point $x \in \bigcap \overline{T_n}$. Then, given a nbd U of x in X and an integer $n > 0$, we have $U \cap T_n \neq \varnothing$. Consequently, there is an $m \geq n$ such that $x_m \in U$, and thus x is a cluster point of $\langle x_n \rangle$.

$\Leftarrow$: If there is a countable open covering $\{U_n\}$ of X, which contains no finite subcovering, then, as in the proof of Theorem 5.2.3, we obtain a sequence $\langle x_n \rangle$ in X such that $x_n \notin U_i$ for every $1 \le i \le n$. As each $x \in X$ belongs to some U_n and $x_i \notin U_n$ for all $i \ge n$, x is not a cluster point of $\langle x_n \rangle$. $\diamondsuit$

By definition, every compact space is countably compact, but the converse is not true in general.

Example 5.2.3 The ordinal space $X = [0, \Omega)$ is not compact, since it is not closed in the Hausdorff space $[0, \Omega]$. We next show that it is countably compact. Let $\{U_n : n = 1, 2, \ldots\}$ be a countable open covering of $[0, \Omega)$. For each $y < \Omega$, the closed interval $[0, y]$ is compact, being a closed subset of the compact space $[0, \Omega]$ (see Example 5.1.9). Therefore, $[0, y]$ can be covered by finitely many U_n's. If the covering $\{U_n\}$ has no finite subcovering, then, we can find $x_n \in X - \bigcup_{i=1}^{n} U_i$ for each n. Put $y = \sup x_n$. Then $y < \Omega$ and the closed interval $[0, y]$ can't be covered by finitely many U_n's, a contradiction. Therefore, the covering $\{U_n\}$ has a finite subcovering, and X is countably compact.

The invariance properties of countable compactness are relegated to the exercises.

By Theorems 5.1.3 and 5.2.1, we conclude the Bolzano–Weierstrass theorem: A bounded infinite subset of real numbers has a limit point. It is equivalent to the property: A bounded sequence of real numbers has a convergent subsequence. We next generalize this property of the real line to introduce another version of compactness for topological spaces.

Definition 5.2.5 A space X is called sequentially compact if every sequence in X contains a convergent subsequence.

Clearly, every sequence in a sequentially compact space has at least one cluster point. Therefore, a sequentially compact space X is also countably compact, by Theorem 5.2.4. On the other hand, there are compact spaces which are not sequentially compact (see Exercises 14 and 8.3.17). However, the notions of countable compactness and sequential compactness agree for a wide class of spaces, as we shall soon see.

Definition 5.2.6 A topological space is said to satisfy the *first axiom of countability* (or be *first countable*) if it has a countable neighborhood basis at each point.

Recall that a nbd basis at a point x of a space X is a family $\mathscr{B}_x$ of nbds of x (in X) such that each nbd of x contains some member of $\mathscr{B}_x$.

Example 5.2.4 A metric space is obviously first countable, for the open balls of radii $1/n, n = 1, 2, \ldots$, about a given point of the space constitute a countable nbd basis.

Example 5.2.5 An uncountable space X with cofinite topology is not first countable. For, if $\mathscr{B}_x$ is a local basis at a point $x \in X$ and $y \ne x$, then there is a member of $\mathscr{B}_x$ not containing y. Hence, we see that $\bigcap\{B : B \in \mathscr{B}_x\} = \{x\}$, and so $X - \{x\} = \bigcup\{X - B : B \in \mathscr{B}_x\}$. If $\mathscr{B}_x$ were countable, then $X - \{x\}$ would be countable, being a countable union of finite sets. This is contrary to the assumption about X, and therefore there is no countable nbd basis at x.

It is worth noticing that a first countable space X admits a countable nbd basis $\{U_n\}$ at each point x such that $U_n \supseteq U_{n+1}$ for every $n = 1, 2, \ldots$. In fact, given a countable nbd basis $\{B_n\}$ at x, one needs to define $U_n = B_1 \cap \cdots \cap B_n$.

We now prove the following theorem.

Theorem 5.2.7 *A countably compact space satisfying the first countability axiom is sequentially compact.*

Proof Let $\langle x_n \rangle$ be a sequence in X. Since X is countably compact, the sequence $\langle x_n \rangle$ has a cluster point $x \in X$, by Theorem 5.2.4. We show that $\langle x_n \rangle$ has a subsequence which converges to x. By the first axiom of countability, we have a monotonically decreasing countable nbd base $\{B_i\}$ at x. Now, given an integer $m > 0$, each B_i contains a term x_n for some $n \geq m$. Consequently, we obtain a subsequence $\langle x_{n_i} \rangle$ of $\langle x_n \rangle$ such that $x_{n_i} \in B_i$. This subsequence clearly converges to x, and the theorem follows. $\diamond$

It follows that sequential compactness is equivalent to the B-W property for first countable T_1-spaces. In the next section, we shall see that the two properties are actually equivalent to compactness for metric spaces.

Exercises

1. Is the real line $\mathbb{R}$ is countably compact?

2. Find a space in which every uncountable subset has a limit point but no countable subset has a limit point.

3. Prove that the following are equivalent:

   (a) A space X is countably compact.
   (b) Each countable family of closed sets in X having the finite intersection property has a nonempty intersection.
   (c) Every decreasing sequence $C_1 \supseteq C_2 \supseteq \cdots$ of nonempty closed sets in X has a nonempty intersection.

4. Let X be a T_1-space. Prove that a space X is countably compact $\Leftrightarrow$ every infinite open cover of X has a proper subcover.

5. Is the subspace $[0, 1]$ of the Sorgenfrey line $\mathbb{R}_\ell$ countably compact?

6. Let X be a T_2-space. Show that X is countably compact $\Leftrightarrow$ each discrete closed subset of X is finite.

7. Prove that a closed subset of a countably compact space is countably compact.

8. Give an example of an open subspace of a countably compact space that is not countably compact.

9. Show that the continuous image of a countably compact space is countably compact.

10. Let X be compact and Y be a countably compact space. Show that $X \times Y$ is countably compact.

11. Prove that a closed subset of a sequentially compact space is sequentially compact.

12. Prove that the sequential compactness is preserved by continuous functions.

13. Prove that a countable product of sequentially compact spaces is sequentially compact.

14. • Show that the product of uncountably many copies of the unit interval I is not sequentially compact.

15. Give an example of a sequentially compact space that is not compact.

5.3 Compact Metric Spaces

Notice that a compact metric space X is bounded. For, the covering of X by the open balls $B(x_0; n)$, where $x_0 \in X$ is a fixed point and n ranges over $\mathbb{N}$, has a finite subcovering so that $X = B(x_0; n)$ for some n. In fact, X satisfies a stronger condition described in the following definition.

Definition 5.3.1 A metric space X is called *totally bounded* (or *precompact*) if for each real $\epsilon > 0$, there exists a finite set $F \subseteq X$ such that the open balls $B(x; \epsilon)$, $x \in F$, cover X. The subset F is called an $\epsilon - net$ for X.

If a metric space X is totally bounded, then it is also bounded. For, if F is an ϵ-net for X, then $\text{diam}(X) \leq \text{diam}(F) + 2\epsilon$. However, a bounded metric space is not necessarily totally bounded; for example, the set $\mathbb{R}$ of reals with the bounded metric $d(x, y) = \min\{1, |x - y|\}$ is not totally bounded.

It is clear that a compact metric space is totally bounded. This is true even for countably compact spaces, as shown by the following.

Proposition 5.3.2 *Every countably compact metric space is totally bounded.*

Proof Let (X, d) be a countably compact metric space. If X is not totally bounded, then there is a real number $\epsilon > 0$ for which X has no ϵ-net. Choose a point $x_1 \in X$. By our assumption, $X \neq B(x_1; \epsilon)$. So we can choose a point x_2 outside $B(x_1; \epsilon)$. Then $d(x_1, x_2) \geq \epsilon$. Since $\{x_1, x_2\}$ is not an ϵ-net for X, we find x_3 outside $\bigcup_1^2 B(x_i; \epsilon)$. Then $d(x_i, x_3) \geq \epsilon$, for i = 1, 2. Proceeding inductively, suppose that we have chosen points $x_1, \ldots, x_n$ in X such that $d(x_i, x_j) \geq \epsilon$ for $i \neq j$. Again, since $\{x_1, \ldots, x_n\}$ is not an ϵ-net for X, we can choose $x_{n+1} \in X - \bigcup_{i=1}^n B(x_i; \epsilon)$. Then $d(x_i, x_{n+1}) \geq \epsilon$ for $i = 1, \ldots, n$. By induction, we have a sequence of points $x_n \in X$ such that $d(x_i, x_j) \geq \epsilon$ for $i \neq j$. Since X is countably compact, the infinite set $E = \{x_n \mid n = 1, 2, \ldots\}$ must have a limit point $x \in X$. Then the open ball

$B(x; \epsilon/2)$ contains an infinite number of points of E, for X is T_1. On the other hand, $B(x; \epsilon/2)$ does not contain more than one point of E, since $d(x_i, x_j) \geq \epsilon$ for $i \neq j$. Therefore, X must be totally bounded. $\diamond$

Now, we can prove the following.

Theorem 5.3.3 *In a metric space X, the following conditions are equivalent:*

    (a) *X is compact;*
    (b) *X is countably compact; and*
    (c) *X is sequentially compact.*

Proof As noted above, (b) and (c) are equivalent for X and $(a) \Rightarrow (b)$ is always true. Accordingly, we need only to prove the implication (b) $\Rightarrow$ (a). To see this, suppose that X is a countably compact metric space and let $\mathcal{G}$ be an open covering of X. Then X is totally bounded, by the preceding proposition. So, for each positive integer n, we have an $1/n$-net A_n in X. Put $A = \bigcup A_n$ and let D be the collection of all pairs $(a, m) \in A \times \mathbb{N}$ such that the open ball $B(a; 1/m)$ is contained in some member of $\mathcal{G}$. Then, for each pair (a, m) in D, select one set in $\mathcal{G}$ containing $B(a; 1/m)$ and denote it by $G(a, m)$. Then $\mathcal{H} = \{G(a, m)|(a, m) \in D\}$ is a countable subcollection of $\mathcal{G}$. We observe that $\mathcal{H}$ covers X. Given a point $x \in X$, we find a set $G \in \mathcal{G}$ containing x. Since G is open, there exists a real number $r > 0$ such that $B(x; r) \subseteq G$. Then, choose an integer $n > 2/r$. Now, by the definition of A_n, there exists an $a \in A_n$ such that $x \in B(a; 1/n)$. It is easily checked that $B(a; 1/n) \subseteq G$. Therefore, $(a, n) \in D$ and $x \in G(a, n)$, as desired. Since X is countably compact, there is a finite subcollection of $\mathcal{H}$ covering X. This finite subcollection of $\mathcal{H}$ is a finite subcovering of $\mathcal{G}$, and thus X is compact. $\diamond$

Moving on to compact subsets of metrizable spaces, we see that every compact subset of a metric space is closed and bounded. For, if C is a compact subset of the metric space X, then it is closed, by Theorem 5.1.8. It is also bounded, since the open covering $\{B(c; 1) \mid c \in C\}$ reduces to a finite subcovering $\{B(c_i; 1) \mid 1 \leq i \leq n\}$ and so $\mathrm{diam}(C) \leq M + 2$, where $M = \max\{d(c_i, c_j) \mid 1 \leq i < j \leq n\}$. Although we have proved the converse in the case of Euclidean spaces (see the Generalized Heine–Borel Theorem), it does not hold true in general. For example, the unit sphere S in the Hilbert space ℓ_2, which consists of all points $x \in \ell_2$ such that $\sum x_n^2 = 1$, is clearly closed and bounded. But it is not compact, since the sequence $\langle x^{(n)} \rangle$ in S, where $x^{(n)}$ is the element of ℓ_2 with its nth coordinate 1 and all other coordinates 0, has no convergent subsequence.

Considering compact subsets of Euclidean spaces, we notice that many familiar objects such as a cube, a disc, a tetrahedron, etc., are convex, too. In Chap. 2, we have seen that two such objects, namely, the unit n-cube I^n and the unit n-disc $\mathbb{D}^n$ are homeomorphic. This is actually true of the most compact convex subsets of $\mathbb{R}^n$. To see this, we first prove the following.

Lemma 5.3.4 *Let $A \subseteq \mathbb{R}^n$ be a compact convex set with the origin 0 as an interior point. Then each ray from 0 intersects ∂A (the boundary of A) in exactly one point.*

Proof Suppose that $\mathscr{R}$ is a ray from the origin, and let $u \in \mathbb{R}^n$ be a unit vector in the direction of $\mathscr{R}$. Then $\mathscr{R} = \{tu \mid t \geq 0 \text{ real}\}$. Since A is bounded, the set $T = \{t > 0 \mid tu \notin A\}$ is nonempty and bounded below. Put $t_0 = \inf T$. We observe that $t_0 u \in \partial A$. By the choice of t_0, $tu \in A$ for every $0 \leq t < t_0$. Since A is closed, we have $t_0 u \in A$. The convexity of A forces that $tu \notin A$ for all $t > t_0$; consequently, $t_0 u \notin A^\circ$. Thus, $t_0 u \in A - A^\circ = \partial A$. To establish the uniqueness of $t_0 u$, suppose that there is $t_1 > t_0$ with $t_1 u \in \partial A$. Let $B(0; r)$ be a ball centered at 0 and contained in A. Let V be the geometric cone over $B(0; r)$ with the vertex $t_1 u$ (that is, V is the union of all line segments joining $t_1 u$ to the points of $B(0; r)$) (see Fig. 5.2 below). Then $V \subseteq A$, since A is convex. If $r_1 = r(t_1 - t_0)/t_1$, then $B(t_0 u; r_1) \subseteq V \subseteq A$ so that $t_0 u \in A^\circ$, a contradiction. Similarly, there is no $t_1 < t_0$ with $t_1 u \in \partial A$.   ◇

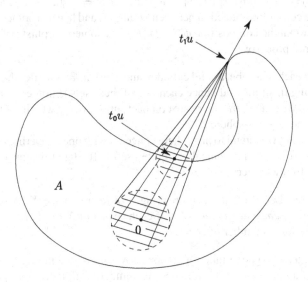

Fig. 5.2 Proof of Lemma 5.3.4

Theorem 5.3.5 *Let K be a compact, convex subset of $\mathbb{R}^n$ with $K^\circ \neq \varnothing$. Then there is a homeomorphism of K with $\mathbb{D}^n$ carrying ∂K onto $\mathbb{S}^{n-1}$.*

Proof Let $w \in K^\circ$, and $\tau : \mathbb{R}^n \to \mathbb{R}^n$ be the translation $x \mapsto x - w$. Then τ is a homeomorphism and preserves convex combinations. So $A = \tau(K)$ is also compact and convex, and the origin $0 \in A^\circ$. We observe that $A \approx \mathbb{D}^n$ by a homeomorphism which carries ∂A onto $\mathbb{S}^{n-1}$. Consider the mapping $h : \partial A \to \mathbb{S}^{n-1}$ defined by $h(x) = x/\|x\|$. By the preceding lemma, h is a bijection. The function h is the restriction of the continuous map $\mathbb{R}^n - \{0\} \to \mathbb{S}^{n-1}$, $x \mapsto x/\|x\|$, and therefore continuous. Since ∂A is compact and $\mathbb{S}^{n-1}$ is T_2, h is a homeomorphism. Let $f : \mathbb{S}^{n-1} \to \partial A$ denote the inverse of h. We extend f to a function $\phi : \mathbb{D}^n \to A$ by setting

$$\phi(x) = \begin{cases} \|x\| f\,(x/\|x\|) & \text{for } x \neq 0, \text{ and} \\ 0 & \text{for x=0.} \end{cases}$$

We show that ϕ is a homeomorphism. If $x \neq 0$, then ϕ is clearly continuous at x. The continuity of ϕ at $x = 0$ follows from the inequality $\|\phi(x)\| \leq b\|x\|$, where b is a bound for $\{\|x\| : x \in A\}$. To see the injectivity of ϕ, suppose $\phi(x) = \phi(y)$. Since $0 \in A^\circ$ and $f\,(x/\|x\|) \in \partial A, \phi(x) = 0$ implies $x = 0$. Accordingly, $\phi(x) = \phi(y) = 0 \Rightarrow x = y = 0$. If $x \neq 0 \neq y$, then $f\,(y/\|y\|) = \frac{\|x\|}{\|y\|} f\,(x/\|x\|)$ and both $f\,(x/\|x\|)$ and $f\,(y/\|y\|)$ are on the same ray from 0. Because these points belong to ∂A, we must have $f\,(x/\|x\|) = f\,(y/\|y\|)$. Then, by our assumption, $\|x\| = \|y\|$. Since f is injective, we deduce that $x = y$. Next, to see that ϕ is surjective, let $0 \neq v \in A$. If $v \in \partial A$, then there is an $x \in \mathbb{S}^{n-1} \subset \mathbb{D}^n$ such that $v = f(x) = \phi(x)$. If $v \in A^\circ$, then $w = f(v/\|v\|) \in \partial A$ satisfies $v/\|v\| = w/\|w\|$. So $x = v/\|w\| \in \mathbb{D}^n$, and $\phi(x) = v$. Thus, ϕ is a continuous bijection between $\mathbb{D}^n$ and A, and hence a homeomorphism. It is now clear that the composition $\phi^{-1} \circ (\tau|K)$ is a homeomorphism of K onto $\mathbb{D}^n$ with the desired property. $\diamond$

Notice that each of a cube, a solid cylinder, and a disc in $\mathbb{R}^n$ satisfies the hypothesis of the preceding theorem, and hence each one of these sets is homeomorphic to the unit n-disc $\mathbb{D}^n$. We remark that there are compact subsets of $\mathbb{R}^n$ which are not convex, for example, the $(n-1)$-sphere $\mathbb{S}^{n-1}$.

The other result we intend to prove here concerns with open coverings of compact metric spaces. It is rather technical but quite useful. To state this conveniently, we introduce the following terminology.

Definition 5.3.6 Let $\mathcal{G}$ be an open covering of a metric space X. A real number $r > 0$ is called a *Lebesgue number* of $\mathcal{G}$ if each subset of X of diameter less than r is contained in some member of $\mathcal{G}$.

Obviously, a Lebesgue number of an open covering $\mathcal{G}$ of a metric space X is not unique. For, if a number $r > 0$ is a Lebesgue number of $\mathcal{G}$, then every positive real number less than r is also a Lebesgue number of $\mathcal{G}$. It is also clear that $\mathcal{G}$ has a Lebesgue number if and only if there exists a real $r > 0$ such that, for each point $x \in X$, the open ball $B(x; r)$ is contained in a member of $\mathcal{G}$.

Example 5.3.1 Any number $0 < r \leq 1/2$ is a Lebesgue number of the open covering $\{(n, n+2) \mid n \in \mathbb{Z}\}$ of the real line $\mathbb{R}$.

Example 5.3.2 The open covering $\{(1/n, 1) \mid n = 1, 2, \ldots\}$ of the subspace $(0,1)$ of $\mathbb{R}$ has no Lebesgue number.

Lemma 5.3.7 (Lebesgue Covering Lemma) *Every open covering of a compact metric space has a Lebesgue number.*

Proof Let $\mathcal{G}$ be an open covering of the compact metric space (X, d). Given $x \in X$, there is a set $G \in \mathcal{G}$ with $x \in G$. Choose $r_x > 0$ such that the open ball $B(x; r_x) \subseteq G$. Consider the open covering $\{B\,(x; r_x/2) \mid x \in X\}$ of X. Since X is compact, this open covering has a finite subcovering. Accordingly, there exist finitely many points

$x_1, \ldots, x_n$ such that the family $\left\{ B\left(x_i; r_{x_i}/2\right) \mid i = 1, \ldots, n \right\}$ covers X. We observe that $\lambda = (1/2) \min \left\{ r_{x_1}, \ldots, r_{x_n} \right\} > 0$ is a Lebesgue number. Let A be any subset of X with $\mathrm{diam}(A) < \lambda$. Choose a point $a \in A$. Then $a \in B\left(x_i; r_{x_i}/2\right)$ for some i; consequently, $d(a', x_i) \leq d(a', a) + d(a, x_i) < \lambda + r_{x_i}/2 \leq r_{x_i}$ for every $a' \in A$. This implies that $A \subseteq B\left(x_i; r_{x_i}\right)$. Since the latter set is contained in some member of $\mathscr{G}$, the result follows. $\diamond$

We will find many applications of the above lemma in this work. As the first case, we generalize a well-known result in analysis: A continuous real-valued function on a closed interval is uniformly continuous.

Definition 5.3.8 Let (X, d_x) and (Y, d_y) be metric spaces. A function $f : X \to Y$ is *uniformly continuous* if for each $\epsilon > 0$, there exists a $\delta > 0$ (depending only on ϵ) such that $d_x(x, x') < \delta \Rightarrow d_y(f(x), f(x')) < \epsilon$ for all $x, x' \in X$.

Theorem 5.3.9 *Let f be a continuous function of a compact metric space (X, d_X) into a metric space (Y, d_Y). Then f is uniformly continuous.*

Proof Given $\epsilon > 0$, consider the open ball $B(y; \epsilon/2)$ for each point $y \in Y$. Since f is continuous, the family $\left\{ f^{-1}\left(B(y; \epsilon/2)\right) \mid y \in Y \right\}$ is an open covering of X. Since X is compact, this open covering has a Lebesgue number δ. If $x, x' \in X$ and $d_X(x, x') < \delta$, then there exists a $y \in Y$ such that both $f(x)$ and $f(x')$ belong to $B(y; \epsilon/2)$. So $d_Y(f(x), f(x')) \leq d_Y(f(x), y) + d_Y(y, f(x')) < \epsilon$, and f is uniformly continuous. $\diamond$

Note that the condition of compactness on X in the preceding theorem is essential, as shown by the function $f : \mathbb{R} \to \mathbb{R}$ given by $f(x) = x^2$.

Exercises

1. Let X be a metric space. If the projection $q : X \times \mathbb{R} \to \mathbb{R}$ is a closed map, show that X is bounded.

2. Show that a continuous function of a compact space into a metric space is bounded.

3. • (a) Show that a real-valued continuous function f on a compact space X attains its bounds, that is, there exist two points x_0 and y_0 in X such that $f(x_0) = \sup f(X)$, and $f(y_0) = \inf f(X)$.

   (b) If $f(x) > 0$ for every $x \in X$, show that there exists an $r > 0$ such that $f(x) > r$ for all $x \in X$.

4. Do as in Exercise 3 for real-valued continuous functions on a countably compact space.

5. Let A be a (nonempty) compact subset of a metric space (X, d). Show that there exist points a, b in A such that $\mathrm{diam}(A) = d(a, b)$.

6. Let X and Y be spaces, and let $f : X \times Y \to \mathbb{R}$ be a continuous function. If X is compact, prove that the function $g, h : Y \to \mathbb{R}$ defined by $g(y) = \sup \{ f(x, y) \mid x \in X \}$ and $h(y) = \inf \{ f(x, y) \mid x \in X \}$ are continuous.

7. Let (X, d) be a metric space, and let $A, B \subset X$ be nonempty, disjoint closed sets. Show:

   (a) If B is compact, then there exists $b \in B$ such that $\text{dist}(A, B) = \text{dist}(A, b) > 0$.

   (b) If both A and B are compact, then there exist $a \in A$, $b \in B$ such that $\text{dist}(A, B) = d(a, b)$.

   (c) The conclusion fails if A and B are not compact.

8. Let X be a metric space and $A \subseteq X$. Prove:

   (a) $\text{diam}(A) = \text{diam}(\overline{A})$. (Hence, A is bounded if and only if $\overline{A}$ is bounded.)

   (b) A is totally bounded if and only if $\overline{A}$ is totally bounded.

   (c) A is totally bounded if X is so.

9. Prove that a bounded subspace of $\mathbb{R}^n$ is totally bounded.

10. Give an example of a bounded metric space which is not totally bounded.

11. Give an example to show that total boundedness is a property of metrics (i.e., this is not a topological property).

12. (a) Prove that the Hilbert cube $\{(x_n) \in \ell_2 \mid 0 \le x_n \le 1/n$ for every $n\}$ is sequentially compact (and hence compact).

    (b) Is every closed bounded subspace of ℓ_2 sequentially compact?

13. Let (X, d) be a compact metric space and $f : X \to X$ be a mapping such that $d(x, y) = d(f(x), f(y))$. Prove that f is onto.

14. A space X is *pseudocompact* if and only if every continuous real-valued function on X is bounded.

    (a) Prove that every countably compact space is pseudocompact.

    (b) Give an example of a pseudocompact space which is not countably compact.

5.4 Locally Compact Spaces

There are many spaces, the most important being the Euclidean spaces, which are not compact, but have instead a localized version of the property. Following an established practice, we adopt the following definition.

Definition 5.4.1 A space X is locally compact at $x \in X$ if it has a compact neighborhood of x. X is called locally compact if it is locally compact at each of its points.

Since a space is a nbd of each of its points, a compact space is locally compact. The converse of this is not true, as shown by the following.

Example 5.4.1 An infinite discrete space is locally compact that is not compact.

Example 5.4.2 A Euclidean space $\mathbb{R}^n$ is locally compact, for the closed unit ball centered at $x \in \mathbb{R}^n$ is a compact neighborhood of x.

Example 5.4.3 The Hilbert space ℓ_2 is not locally compact. To see this, consider the point $x_0 \in \ell_2$ all of whose coordinates are zero. If N is a nbd of x_0 in ℓ_2, then there exists a closed ball $\{x \in \ell_2 \mid \|x\| \leq r\} \subseteq N$. For each positive integer n, let $x^{(n)} \in \ell_2$ be the point given by $x_i^{(n)} = 0$ for $i \neq n$ and $x_n^{(n)} = r$. Then $x^{(n)} \in N$ for every n, and the sequence $\langle x^{(n)} \rangle$ does not have a convergent subsequence, for $d\left(x^{(n)}, x^{(m)}\right) = \sqrt{2}r$ when $n \neq m$. Hence, N is not compact.

Although the above formulation of local compactness is contrary to the general spirit of local properties, it is equivalent to a truly local condition for a very wide class of spaces.

Theorem 5.4.2 *Let X be a Hausdorff space. The following conditions are equivalent:*

(a) *X is locally compact.*
(b) *For each $x \in X$, there is an open nbd W of x such that $\overline{W}$ is compact.*
(c) *For each $x \in X$, and each nbd U of x, there is an open set V such that $x \in V \subseteq \overline{V} \subseteq U$ and $\overline{V}$ is compact (that is, the closed compact nbds of x form a nbd basis at x).*
(d) *X has a basis consisting of open sets with compact closures.*

Proof (a) $\Rightarrow$(b): Let N be a compact nbd of x in X. Since X is Hausdorff, N is closed in X. Let W be the interior of N. Then $x \in W$ and $\overline{W} \subseteq N$ is compact.

(b) $\Rightarrow$ (c): Let $x \in X$ be arbitrary and U be any nbd of x. Obviously, it suffices to prove (c) when U is open. By our hypothesis, there exists an open nbd W of x with $\overline{W}$ compact. Then $F = \overline{W} - U$ is a closed set with $x \notin F$. Since $\overline{W}$ is compact Hausdorff, there exist disjoint sets G and H open in $\overline{W}$ such that $x \in H$ and $F \subseteq G$ (see the remarks following Theorem 5.1.8). So $H \subseteq \overline{W} - G \subseteq \overline{W} - F \subseteq U$ and we have $\overline{H}_{\overline{W}} \subseteq U$. By definition, there exists an open subset of $O \subseteq X$ such that $H = \overline{W} \cap O$. Put $V = O \cap W$. Then V is an open nbd of x in X, and $V \subseteq H \cap W$, obviously. Now, we have $\overline{V} \subseteq \overline{H} \cap \overline{W} = \overline{H}_{\overline{W}} \subseteq U$. As $\overline{W}$ is compact, so is $\overline{V}$. Thus, V is the desired nbd of x.

The implications (c) $\Rightarrow$ (d) $\Rightarrow$ (a) are trivial. $\diamond$

The property of local compactness is a topological invariant; in fact, it is preserved by continuous open mappings. For, if $f: X \to Y$ is a continuous open map and U is a compact nbd of x in X, then $f(U)$ is a compact nbd of $f(x)$ in Y. But, unlike compactness, this property is not invariant under continuous mappings, as shown by the identity map $\ell_2 \to \ell_2$, where the domain is assigned the discrete topology. Even continuous closed maps fail to preserve the property (see Exercise 6.1.12).

Local compactness is not inherited by subspaces in general, as shown by the following.

Example 5.4.4 The subspace $\mathbb{Q} \subset \mathbb{R}$ is not locally compact. For, if K is a compact subset of $\mathbb{Q}$, then it is a compact subset of $\mathbb{R}$ and therefore must be closed in $\mathbb{R}$, by the Heine–Borel theorem. If K also contains an open subset of $\mathbb{Q}$, then some irrational numbers must be its limit points, a contradiction.

However, every closed subset of a locally compact space X is locally compact. To see this, suppose that F is a closed subset of X. If $x \in F$ and $K \subseteq X$ is a compact nbd of x, then $F \cap K$ is obviously a nbd of x in F and, being a closed subset of K, it is compact. Thus, F is locally compact at x. If X is also Hausdorff, then every open subset of X is locally compact, by Theorem 5.4.2. In this case, we further notice that if $F \subseteq X$ is closed and $G \subseteq X$ is open, then $F \cap G$, being an open subset of the locally compact subspace F, is locally compact. Conversely, we have

Theorem 5.4.3 *A locally compact subspace of a Hausdorff space is the intersection of a closed set and an open set.*

Proof Suppose that A is a locally compact subspace of a Hausdorff space X. For $a \in A$, let U be an open nbd of a in A such that $\overline{U}_A$ is compact. Then there exists an open nbd V of a in X with $U = A \cap V$. So we have $\overline{U}_A = \overline{U} \cap A = \overline{A \cap V} \cap A$; accordingly, $\overline{A \cap V} \cap A$ is closed in X, being a compact subset of a Hausdorff space. This implies that $\overline{A \cap V} \subseteq A$. Also, one easily checks that $\overline{A} \cap V \subseteq \overline{A \cap V}$. Thus, we see that A contains an open nbd of a in $\overline{A}$, and hence A is open in $\overline{A}$. Then, by the definition of relative topology, there exists an open set $G \subseteq X$ such that $A = G \cap \overline{A}$, as desired. ◇

To describe the intersection of a closed set and an open set more succinctly, we introduce the following terminology.

Definition 5.4.4 A subset A of a topological space X is called *locally closed* if each point of A belongs to an open subset G of X such that $G \cap A$ is closed in G.

Proposition 5.4.5 *A subset A of a space X is locally closed if and only if $A = F \cap G$ for some closed set F and open set G in X.*

Proof Suppose that A is locally closed. Then for each $a \in A$, there exists an open nbd O_a of a such that $O_a \cap A$ is closed in O_a. So $O_a \cap A = \overline{O_a \cap A} \cap O_a = O_a \cap \overline{A}$. Taking $G = \bigcup_{a \in A} O_a$, we have $G \cap \overline{A} = \bigcup_{a \in A} (O_a \cap \overline{A}) = \bigcup_{a \in A} (O_a \cap A) = G \cap A = A$. Obviously, G is open in X, and the necessity follows. The converse is obvious. ◇

It follows that locally compact subspaces of a locally compact Hausdorff space are precisely its locally closed subsets.

The property of local compactness is invariant under the formation of finite products. More generally, we prove

Theorem 5.4.6 *The product $\prod X_\alpha$ of a family of spaces X_α, $\alpha \in A$, is locally compact if and only if each X_α is locally compact and all but finitely many X_α are compact.*

Proof If $\prod X_\alpha$ is locally compact, then each X_α is locally compact because the projections are continuous and open. To see the second statement, choose a compact nbd K of some point $x \in \prod X_\alpha$. Let V be a basic nbd of x contained in K, and suppose that $V = \bigcap_1^n p_{\alpha_i}^{-1}\left(U_{\alpha_i}\right)$, where U_{α_i} is open in X_{α_i}. Then, for $\alpha \neq \alpha_1, \ldots, \alpha_n$, we have $p_\alpha(K) = X_\alpha$, so X_α is compact.

Conversely, suppose that the given conditions hold, and let $\alpha_1, \ldots, \alpha_n$ be the indices for which X_{α_i} is not compact. Given $x \in \prod X_\alpha$, choose a compact nbd U_{α_i} of x_{α_i} in X_{α_i} for $i = 1, \ldots, n$. Then $\bigcap_{i=1}^n p_{\alpha_i}^{-1}\left(U_{\alpha_i}\right)$ is a compact nbd of x, by the Tychonoff theorem. $\diamond$

One-Point Compactification

The study of a non-compact space X is often made easier by constructing a compact space which contains X as a subspace. There are many ways of producing such spaces. We will learn here one method of doing this, which is quite similar to the construction of a Riemann sphere (identified with the extended complex plane), and discuss another one in Chap. 8. In the construction of the extended complex plane, an ideal point ∞ is adjoined to the complex plane $\mathbb{R}^2$ and the complements of the closed and bounded subsets of $\mathbb{R}^2$ (which are, in fact, compact subsets of $\mathbb{R}^2$) are taken to be the neighborhoods of ∞. We shall generalize this construction to a topological space.

Definition 5.4.7 Let X be a space. The (Alexandroff) *one-point compactification* of X is the space $X^* = X \cup \{\infty\}$, where ∞ is an element that is not in X, and its topology $\mathcal{T}^*$ consists of all open subsets of X and all subsets U of X^* such that $X^* - U$ is a closed and compact subset of X.

Of course, we must check that the above specification does give a topology for X^*. We do this verification now. Since the empty set $\varnothing$ is trivially compact, X^* and $\varnothing$ are members of $\mathcal{T}^*$. Suppose $U, V \in \mathcal{T}^*$. If both U and V are contained in X, then $U \cap V$ is open in X. If $U \subseteq X$ and $\infty \in V$, then $U \cap V$ is open in X, for $V \cap X = X - (X^* - V)$ is open in X. And, if $\infty \in U \cap V$, then $X^* - (U \cap V) = (X^* - U) \cup (X^* - V)$ is closed and compact in X. So $U \cap V \in \mathcal{T}^*$. Next, let $\{U_\alpha\}$ be a family of members of $\mathcal{T}^*$. If each $U_\alpha \subseteq X$, then $\bigcup U_\alpha$ is open in X. And, if ∞ belongs to some member U_β of $\{U_\alpha\}$, then $X^* - \bigcup U_\alpha = \bigcap(X - U_\alpha)$ is closed in X. This is also compact, being a closed subset of the compact set $X^* - U_\beta$. So $\bigcup U_\alpha \in \mathcal{T}^*$, and $\mathcal{T}^*$ is a topology for X^*.

The point ∞ is often referred to as the *point at infinity* in X^*. If X is already compact, then ∞ is an isolated point of X^*; so, in this case, the one-point compactification is uninteresting. If X is not compact, then ∞ is a limit point of X so that X is dense in X^*. Note that this point may have just the trivial nbd in X^*. Of course, any space contains compact subsets (e.g., singletons), but they need not be closed.

Theorem 5.4.8 *Let X be a space and X^* be its one-point compactification. Then the inclusion $X \hookrightarrow X^*$ is an embedding, and X^* is compact. Furthermore, X^* is Hausdorff if and only if X is Hausdorff and locally compact, and in this case, any homeomorphism between X and the complement of a single point of a compact Hausdorff space Y extends to a homeomorphism between X^* and Y.*

Proof We first show that the relative topology on X induced from X^* is the same as the given topology on X. If $U \subseteq X$ is open, then U is open in X^*. Conversely, for any $U \subseteq X^*$, $U \cap X = X - (X^* - U)$ or U, according as U contains or does not contain ∞. Thus, $U \cap X$ is open in X for every open $U \subseteq X^*$.

To see that X^* is compact, let $\mathcal{U}$ be any open cover of X^*. Then there is a member U_∞ of $\mathcal{U}$ which contains ∞. By definition, $X^* - U_\infty$ is a compact subset of X. Hence, there is a finite subfamily $\{U_1, \ldots, U_n\}$ of $\mathcal{U}$ such that $X^* - U_\infty \subseteq \bigcup_1^n U_i$. It follows that $\{U_1, \ldots, U_n, U_\infty\}$ is a finite subcovering of $\mathcal{U}$, and X^* is compact.

Next, suppose that X is locally compact and T_2. Let x, y be distinct points of X^*. If x and y both belong to X, then there are disjoint open sets U and V in X such that $x \in U$ and $y \in V$. By the definition of topology for X^*, U and V are also open in X^*. If $y = \infty$, then $x \in X$ and there is an open set $U \subseteq X$ such that $x \in U$ and $\overline{U}$ (in X) is compact. So $V = X^* - \overline{U}$ is an open nbd of ∞, which is obviously disjoint from U. Conversely, suppose that X^* is T_2. Then X is T_2, for X is a subspace of X^*. To see that X is locally compact, let $x \in X$ be arbitrary. By our hypothesis, there are disjoint open sets U and V in X^* such that $x \in U$ and $\infty \in V$. Then $X^* - V$ is a compact nbd of x, for $x \in U \subseteq X^* - V$. Thus, X is locally compact at x.

To establish the last statement, suppose that X^* is T_2, and let h be a homeomorphism of X into a compact Hausdorff space Y such that $Y - h(X)$ is the singleton $\{y_0\}$. Define $h^* : X^* \to Y$ by $h^*(x) = h(x)$ for all $x \in X$ and $h^*(\infty) = y_0$. Then h^* is obviously bijective. We observe that h^* is open. Let U be any open subset of X^*. If $\infty \notin U$, then U is open in X so that $h^*(U) = h(U)$ is open in Y, for $\{y_0\}$ is closed. And, if $\infty \in U$, then $X^* - U$ is a compact subset of X. So $h(X^* - U)$ is compact, and hence closed in Y. As $h(X^* - U) = Y - h^*(U)$, $h^*(U)$ is open in Y. Thus, h^* is open. By symmetry, $h^{*-1} : Y \to X^*$ is also open, and therefore h^* is a homeomorphism. $\diamond$

The last statement of the theorem implies that the topology of the one-point compactification X^* of a locally compact Hausdorff space X is the only topology which makes X^* a compact Hausdorff space such that $X \subset X^*$ as a subspace. Also, it enables us to determine the one-point compactification of spaces.

Example 5.4.5 The subspace $(0, 1]$ of the real line $\mathbb{R}$ is locally compact and its one-point compactification is the closed interval $[0, 1]$.

Example 5.4.6 The one-point compactification of $\mathbb{N}$ (with the discrete topology) is the subspace $\{1/n \mid n \in \mathbb{N}\} \cup \{0\}$ of $\mathbb{R}$.

Example 5.4.7 The one-point compactification of the Euclidean space $\mathbb{R}^n$ is the n-sphere $\mathbb{S}^n$, since $\mathbb{S}^n$ is compact Hausdorff and the stereographic projection $\mathbb{S}^n - \{\text{north pole}\} \to \mathbb{R}^n$ is a homeomorphism.

We shall come across several applications of this construction as we proceed further in the book.

k-Spaces

In Analysis, we have seen that a function on the real line is continuous if and only if its restriction to each closed interval is continuous. Locally compact Hausdorff and many other spaces enjoy this property. Such spaces are important to our discussion about function spaces because continuous functions on them are precisely those which behave well on compact subsets. The topologies of these spaces are determined by the families of all their compact subsets in the following sense.

Definition 5.4.9 A Hausdorff space X is called a k-space if a set $F \subseteq X$ such that $F \cap C$ is closed (in C) for every compact subset C of X is closed.

A k-space is also said to be *compactly generated*. It is clear that a subset U of a k-space X is open if and only if the intersection of U with each compact set $C \subseteq X$ is open in C.

There is a very wide class of k-spaces, which includes all metric spaces and Hausdorff locally compact spaces.

Proposition 5.4.10 *Every metric space is a k-space.*

Proof Let X be a metric space. If $A \subseteq X$ is not closed, then there is a point $x \in \overline{A} - A$. Consequently, we can find a sequence $\langle a_n \rangle$ in A such that $a_n \to x$. The set $C = \{x\} \cup \{a_n \mid n = 1, 2, \ldots\}$ is compact, but $A \cap C$ is not closed. It follows that X satisfies the required condition for a k-space.   ◇

It follows, in particular, that every Euclidean space is a k-space. The above proposition can be easily generalized to Hausdorff first countable spaces (by applying Theorem 7.1.2). Next, we prove

Proposition 5.4.11 *Every locally compact Hausdorff space is a k-space.*

Proof Let X be a locally compact T_2-space and $A \subseteq X$. Assume that $A \cap C$ is closed in C for each compact set $C \subseteq X$. We show that $X - A$ is open. Let $x \in X - A$ be arbitrary. Let N be a compact nbd of x. By our hypothesis, $(X - A) \cap N$ is open in N; accordingly, $(X - A) \cap N^\circ = G$ is open in N°. Since N° is open in X, G is an open nbd of x contained in $X - A$. Hence, $X - A$ is open, and X is a k-space.   ◇

There is a partial but useful converse of the preceding proposition (see Theorem 6.4.2).

Moving on to inheritance property, we have

Proposition 5.4.12 *A closed subspace of a k-space is a k-space.*

Proof Suppose that X is a k-space and $F \subseteq X$ is closed. Let $A \subseteq F$ be such that $A \cap K$ is closed in K for every compact subset K of F. If $C \subseteq X$ is compact, then $K = F \cap C$ is also compact. By our assumption, $A \cap C = A \cap K$ is closed in K which is obviously closed in C. Since X is a k-space, A is closed in X and hence in F.   ◇

In Sect. 6.4, we will prove a similar result for open subspaces, too.

For product of k-spaces, we content ourselves here with the following.

Proposition 5.4.13 *If X is a k-space and Y is a compact Hausdorff space, then $X \times Y$ is a k-space.*

Proof Suppose that F is a subset of $X \times Y$ such that $F \cap C$ is closed in C for every compact subset C of $X \times Y$. We show that F is closed in $X \times Y$. Consider a point (x, y) in the complement of F in $X \times Y$. Since $F \cap (x \times Y)$ is closed, there exists an open nbd V of y in Y such that $F \cap (x \times V) = \varnothing$. If $p : X \times Y \to X$ is the projection map and $x \in \overline{p(F)} - p(F)$, then there exists a compact set $K \subseteq X$ such that $p(F \cap (K \times Y)) = p(F) \cap K$ is not closed. Since p is a closed map, we deduce that $F \cap (K \times Y)$ is not closed, contrary to our hypothesis. Hence, $p(F)$ is closed in X, and thus $U = X - p(F)$ is an open nbd of x. Obviously, $F \cap (U \times V) = \varnothing$, and therefore $(X \times Y) - F$ is a nbd of (x, y). Since (x, y) is an arbitrary point of $(X \times Y) - F$, we see that F is closed, as desired.   ◇

Later, in Sect. 6.4, it will be seen that the above proposition remains valid for the product of a k-space and a locally compact Hausdorff space. However, it should be noted that the product of two k-spaces in general need not to be a k-space.

We conclude this section with a test for the continuity of mappings on k-spaces.

Theorem 5.4.14 *A mapping of a k-space X into a space Y is continuous if and only if its restriction to every compact subset of X is continuous.*

Proof The necessity of the condition is obvious. To prove sufficiency, suppose that $f : X \to Y$ is a function such that $f|C$ is continuous for every compact subset C of X. Then, for any closed set $F \subseteq Y$, $f^{-1}(F) \cap C$ is closed in C. Since X is a k-space, we conclude that $f^{-1}(F)$ is closed in X. Thus, f is continuous.   ◇

Exercises

1. Discuss local compactness of the following:

   (a) The Sorgenfrey line $\mathbb{R}_\ell$, (b) the metric spaces $\mathcal{B}(I)$ and $\mathcal{C}(I)$ with the sup metric d^* (refer to Example 1.1.5).

2. Prove that every open subset of a compact metric space is locally compact.

3. Prove that a locally compact dense subset of a Hausdorff space is open.

4. Let X be a locally compact Hausdorff space. Show:

   (a) If $K \subseteq X$ is a compact set and $U \subseteq X$ is an open set with $K \subseteq U$, then there exists an open set V such that $K \subseteq V \subseteq \overline{V} \subseteq U$ and $\overline{V}$ is compact.

   (b) If K_1, K_2 are disjoint compact subsets of X, then they have disjoint nbds with compact closures.

5. Let X be a connected, locally connected and locally compact Hausdorff space, and let $x, y \in X$. Show that there is a compact connected set C in X which contains both x and y.

6. Prove that a connected, locally compact, Hausdorff space X is locally connected $\Leftrightarrow$ for each compact set $K \subset X$ and each open set $U \supseteq K$, all but a finite number of components of $X - K$ lie in U.

7. Prove that in a locally compact Hausdorff space, every compact quasi-component is a component, and every compact component is a quasi-component.

8. Prove that one-point compactification of a T_1-space is a T_1-space.

9. • Show that the ordinal space $[0, \Omega)$ is locally compact Hausdorff and find its one-point compactification.

10. (a) If two locally compact Hausdorff spaces are homeomorphic, show that their one-point compactifications are also homeomorphic.

    (b) Find two spaces X and Y such that $X \not\approx Y$ but $X^* \approx Y^*$.

11. Let f be a continuous open mapping of a locally compact T_2-space X onto a T_2-space Y. Given a compact subset $B \subseteq Y$, show that there exists a compact subset $A \subseteq X$ such that $f(A) = B$.

12. If X is a compact space, show that $\{\infty\}$ is a component of X^*.

13. • (a) If X is connected, is X^* necessarily connected?

    (b) If the one-point compactification X^* of a space X is connected, is X connected?

14. Let X be a k-space and $\langle f_n \rangle$ be a sequence of continuous functions from X into metric space Y. Suppose that $\langle f_n \rangle$ converges to $f : X \to Y$ uniformly on each compact $K \subseteq X$. Show that f is continuous.

5.5 Proper Maps

Definition 5.5.1 A continuous map $f : X \to Y$ is called *proper* (or *perfect*) if, for all spaces Z, the product map $f \times 1 : X \times Z \longrightarrow Y \times Z$ is closed.

If $f : X \to Y$ is proper, then, for any $B \subseteq Y$, the function $g : f^{-1}(B) \to B$ defined by f is also proper. This follows from the equality

$$(B \times Z) \cap (f \times 1)(K) = (f \times 1)\left(\left(f^{-1}(B) \times Z\right) \cap K\right)$$

for all $K \subseteq X \times Z$. In particular, we see that the constant map $f^{-1}(y) \to y$ is proper. Therefore, by Theorem 5.1.19, $f^{-1}(y)$ is compact. Also, by taking Z to be a one-point space, we see that f is closed. The following theorem shows that these conditions are sufficient, too.

Theorem 5.5.2 *Let* $f: X \to Y$ *be continuous. Then* f *is proper if and only if* f *is closed and* $f^{-1}(y)$ *is compact for every* $y \in Y$.

Proof (Sufficiency) Let Z be any space and consider the function $f \times 1: X \times Z \longrightarrow Y \times Z$. Given a closed set $F \subseteq X \times Z$, suppose that $(y, z) \notin (f \times 1)(F)$. If $y \notin f(X)$, then $(Y - f(X)) \times Z$ is a nbd of (y, z) disjoint from $(f \times 1)(F)$. So assume that $y = f(x)$ for some $x \in X$. Then $f^{-1}(y) \times \{z\} \subseteq (X \times Z) - F$. Since F is closed and $f^{-1}(y)$ is compact, there exists an open nbd U of $f^{-1}(y)$ and V of z such that $U \times V \subseteq (X \times Z) - F$ (ref. Exercise 5.1.23). So $F \subseteq (X \times Z) - (U \times V) = X \times (Z - V) \cup (X - U) \times Z = E$, say. It is obvious that $(y, z) \notin (f \times 1)(E)$ and $(f \times 1)(F) \subseteq (f \times 1)(E)$. Since f is closed, $(f \times 1)(E) = [f(X) \times (Z - V)] \cup [f(X - U) \times Z]$ is closed. Therefore, $Y \times Z - (f \times 1)(F)$ is a nbd of (y, z), and it follows that $(f \times 1)(F)$ is closed. Thus, $f \times 1$ is a closed map, and the proof is complete. $\diamond$

Corollary 5.5.3 *Let* X *be a compact space.*

(a) *If* Y *is a Hausdorff space, then any continuous map* $f: X \to Y$ *is proper.*
(b) *For any space* Y, *the projection* $X \times Y \to Y$ *is proper.*

The simple proofs are omitted.

An important property of proper mappings is given by

Proposition 5.5.4 *Let* $f: X \to Y$ *be a proper map. If* Y *is compact, then so is* X.

Proof Let $\{G_\alpha \mid \alpha \in A\}$ be an open cover of X. By Theorem 5.5.2, $f^{-1}(y)$ is compact for every $y \in Y$ and f is closed. Consequently, there exists a finite set $B_y \subseteq A$ such that $f^{-1}(y) \subseteq \bigcup \{G_\alpha \mid \alpha \in B_y\} = N_y$, say. As $f(X - N_y)$ is closed in Y, we find an open nbd V_y of y such that $f^{-1}(V_y) \subseteq N_y$. If Y is compact, then there are finitely many points $y_1, \ldots, y_n$ in Y such that $Y = \bigcup_1^n V_{y_i}$. It follows that $X = \bigcup_1^n N_{y_i}$ and the family $\{G_\alpha \mid \alpha \in B_{y_i}, i = 1, \ldots, n\}$ is a finite subcovering of $\{G_\alpha \mid \alpha \in A\}$. Thus, X is compact. $\diamond$

By the proof of the preceding proposition, we see that if $f: X \to Y$ is a proper map, then $f^{-1}(C)$ is compact for every compact subset C of Y. The converse holds under the conditions given by

Theorem 5.5.5 *Let* X *be a Hausdorff space and* Y *be a locally compact Hausdorff space. Then a continuous map* $f: X \to Y$ *is proper if and only if* $f^{-1}(C)$ *is compact for every compact* $C \subseteq Y$.

Proof As observed above, the given condition is necessary. To prove sufficiency, suppose that $f^{-1}(C)$ is compact for every compact subset C of Y. Then $f^{-1}(y)$ is obviously compact for each $y \in Y$. So it is enough to show that f is closed, by Theorem 5.5.2. To this end, we first observe that X is locally compact. Given a point $x \in X$, there is a compact nbd N of $f(x)$ in Y, since Y is locally compact. By our hypothesis, $f^{-1}(N)$ is a compact nbd of x in X, and thus X is locally compact. Denote the one-point compactifications of X and Y by X^* and Y^*, respectively, and define a function $f^* : X^* \to Y^*$ by setting $f^*(x) = f(x)$ for every $x \in X$ and $f^*(\infty_x) = \infty_y$. We assert that f^* is continuous. Let U be an open set subset of Y^*. If $U \subseteq Y$, then U is open in Y. So $f^{*-1}(U) = f^{-1}(U)$ is open in X, and hence in X^*. If $\infty_y \in U$, then $C = Y^* - U$ is compact, and so $f^{-1}(C)$ is compact, by our assumption. Since X is T_2, $f^{-1}(C)$ is closed in X. Thus, $f^{*-1}(U) = X^* - f^{-1}(C)$ is open in X^*, and hence the assertion. Now, if F is a closed subset of X, then $F \cup \{\infty_x\}$ is closed in X^* and, therefore, compact. It follows that $f^*(F \cup \{\infty_x\})$ is compact, and hence closed in Y^*. So $f(F) = Y \cap f^*(F \cup \{\infty_x\})$ is closed in Y, and the proof is complete. $\diamond$

We remark that the preceding theorem can be proved more generally for mappings into k-spaces (see Exercise 7). It should also be noted that some mathematicians take this property as the defining condition for "proper maps".

As we proceed in the text, it will be seen that a number of topological properties are preserved by proper mappings. We end this section with the following illustration.

Proposition 5.5.6 Let $f : X \to Y$ be a proper surjection. If X is Hausdorff, then so is Y.

Proof Let y_1, y_2 be two distinct points of Y. Since f is proper, both $f^{-1}(y_1)$ and $f^{-1}(y_2)$ are compact subsets of X. So, by Proposition 5.1.9, there exist disjoint open nbds U_1 and U_2 of $f^{-1}(y_1)$ and $f^{-1}(y_2)$, respectively. Set $V_i = Y - f(X - U_i)$, $i = 1, 2$. Then each V_i is open in Y, since f is closed, and $y_i \in V_i$. Also, we have $f^{-1}(V_i) \subseteq U_i$. Since $U_1 \cap U_2 = \varnothing$ and f is surjective, $V_1 \cap V_2 = \varnothing$. Thus, V_1 and V_2 are disjoint nbds of y_1 and y_2, respectively, and Y is Hausdorff. $\diamond$

Exercises

1. Give an example of a continuous surjection $f : X \to Y$ such that $f^{-1}(y)$ is compact for all $y \in Y$, but f is not closed.

2. Let $f : X \to Y$ be a continuous injection. Show that f is closed $\Leftrightarrow$ f is proper $\Leftrightarrow$ f is an embedding and im(f) is closed in Y.

3. Let $f : X \to Y$, $g : Y \to Z$ be continuous surjections. Prove:

    (a) If f is proper and $A \subseteq X$ is closed, then $f|A$ is proper.

    (b) If f and g are proper, then gf is proper.

    (c) If gf is proper, then g is proper.

    (d) If gf is proper and g is injective or Y is T_2, then f is proper.

4. Let $f : X \to Y$ be a proper surjection. If Y is countably compact, prove that so is X.

5. Let f be a proper map of a Hausdorff space X onto a space Y. Prove that X is locally compact $\Leftrightarrow$ Y is locally compact (and Hausdorff).

6. Let X and Y be locally compact Hausdorff spaces, and let $f: X \to Y$ be a continuous surjection. Show that f is proper $\Leftrightarrow$ f extends to a continuous function $f^*: X^* \to Y^*$ with $f^*(\infty_x) = \infty_y$, where X^* denotes the one-point compactification of X, etc.

7. Let $f: X \to Y$ be a continuous surjection. If Y is a k-space and $f^{-1}(C)$ is compact for every compact $C \subseteq Y$, show that f is proper.

Chapter 6
Topological Constructions

The main object of this chapter is the construction of "quotient spaces". The motivation for this comes from the construction of models of some geometrical objects such as circle, cylinder, Möbius band, etc., by gluing things together. In Sect. 6.1, we translate the process of gluing into precise mathematical language by using the notion of equivalence relation and study the "quotient topology" for the set of equivalence classes of a topological space with an equivalence relation. Of course, quotient spaces could have been discussed just after studying subspaces and products; however, comprehending the discussion becomes much easier after having seen some results about compactness. Moreover, it is often more convenient to describe quotient spaces by means of "identification maps". This approach will be adopted in Sect. 6.2. In the next three sections, we develop several interesting techniques for producing new spaces from old ones by combining different constructions. These methods of generating new spaces are of great importance in the study of algebraic topology. In the last section, we will study topology induced by functions from a given set into a collection of topological spaces and its dual notion. In particular, we will see the concept of topology generated by subspaces of a given space.

6.1 Quotient Spaces

Let X be a space with an equivalence relation $\sim$, and let $X/\sim$ be the set of equivalence classes. Let $\pi: X \to X/\sim$ be the natural projection which takes $x \in X$ to its equivalence class $[x]$. Then the family $\mathscr{T}(\sim)$ of subsets $U \subseteq X/\sim$ such that $\pi^{-1}(U)$ is open in X is a topology on $X/\sim$ because the inverse of an intersection (or a union) of members of $\mathscr{T}(\sim)$ is the intersection (resp. union) of the inverses.

Definition 6.1.1 The set $X/\sim$ together with the topology

$$\left\{ U \mid \pi^{-1}(U) \text{ is open in } X \right\}$$

© Springer Nature Singapore Pte Ltd. 2019
T. B. Singh, *Introduction to Topology*,
https://doi.org/10.1007/978-981-13-6954-4_6

is called the quotient space of X by the relation $\sim$, and this topology for $X/\sim$ is referred to as the quotient topology.

We think of the quotient space $X/\sim$ as the one obtained from X by identifying each of the equivalence classes $[x]$, $x \in X$, to a single point. Observe that a set $F \subseteq X/\sim$ is closed if and only if $\pi^{-1}(F)$ is closed in X. Also, notice that the quotient topology for $X/\sim$ is the largest topology such that the mapping π is continuous. With this topology on $X/\sim$, we refer to the projection $\pi\colon X \to X/\sim$ as a quotient map. It is clear that mapping π is closed (resp. open) if and only if $\pi^{-1}(\pi(F))$ is closed (resp. open) in X for every closed (resp. open) subset $F \subseteq X$. We further remark that the quotient mapping π need not be open or closed (refer to Exercise 1).

For a subset $A \subseteq X$, the set

$$\pi^{-1}(\pi(A)) = \{x \in X \mid x \sim a \text{ for some } a \in A\}$$

is called the *saturation* of A. We say that the set A is *saturated* if $A = \pi^{-1}(\pi(A))$. With this terminology, we have the following.

Proposition 6.1.2 *A necessary and sufficient condition for the quotient map $\pi\colon X \to X/\sim$ to be a closed (resp. open) mapping is that the saturation of every closed (resp. open) set is closed (resp. open).*

To realize a quotient space $X/\sim$ of a space X, we usually need to find a homeomorphism between the space $X/\sim$ and a known space. Obviously, this entails the construction of a continuous function on $X/\sim$. With this end in view, we first suppose that a continuous function $g\colon (X/\sim) \to Y$ is given. Then the composition $g\pi\colon X \to Y$, where $\pi\colon X \to X/\sim$ is the quotient map, is continuous. Conversely, if $g\pi$ is continuous and $U \subseteq Y$ is open, then $\pi^{-1}\left(g^{-1}(U)\right) = (g\pi)^{-1}(U)$ is open in X. So $g^{-1}(U)$ is open in $X/\sim$, by the definition of its topology, and g is continuous. Now, if a function $f\colon X \to Y$ is constant on each equivalence class $[x]$, $x \in X$, then we can define a function $\bar{f}\colon (X/\sim) \to Y$ by $\bar{f}([x]) = f(x)$, $x \in X$. Notice that the composition

$$X \xrightarrow{\pi} (X/\sim) \xrightarrow{\bar{f}} Y$$

is the function f, that is, $\bar{f} \circ \pi = f$. Therefore, if f is also continuous, then $\bar{f}$ is continuous. We summarize this observation in the following.

Proposition 6.1.3 *Let X be a space with an equivalence relation $\sim$. If $f\colon X \to Y$ is a continuous function which is constant on each equivalence class, then the induced function $\bar{f}\colon [x] \mapsto f(x)$ on the quotient space $X/\sim$ is continuous. Furthermore, $\bar{f}$ is open if and only if $f(U)$ is open for each saturated open set $U \subseteq X$. A similar statement obtained by replacing "open" with "closed" everywhere holds good.*

Proof To prove the second statement, suppose that $f(U)$ is open for every saturated open set $U \subseteq X$. Clearly, for each open set $V \subseteq X/\sim$, $\pi^{-1}(V)$ is an open saturated set. So $\bar{f}(V) = f\left(\pi^{-1}(V)\right)$ is open, and thus $\bar{f}$ is open. Conversely, if $\bar{f}$ is open, and $U \subseteq X$ is an open saturated set, then $f(U) = \bar{f}\pi(U)$ is open, since $\pi(U)$ is open.

A similar proof applies in the "closed" case. ◇

Example 6.1.1 Intuitively, a circle is obtained from an interval by gluing its end points together. Let I be the unit interval with its usual topology and consider the equivalence relation $\sim$ on it, where $0 \sim 1$ and $x \sim x$ for $x \neq 0, 1$. The quotient space $I/\sim$ is homeomorphic to the unit circle $\mathbb{S}^1$ by the mapping $[x] \mapsto e^{2\pi i x}$.

Example 6.1.2 Intuitively, a cylinder is obtained from a rectangle (which is homeomorphic to the square $I \times I$) by gluing a pair of its opposite sides together (Fig. 6.1). So it can be considered as the quotient space $(I \times I)/\sim$, where $\sim$ is the equivalence relation $(s, 0) \sim (s, 1)$, $0 \leq s \leq 1$. Indeed, this quotient space is homeomorphic to the product space $\mathbb{S}^1 \times I$; a homeomorphism is given by $[(s, t)] \mapsto (e^{2\pi i s}, t)$. In Example 2.2.4, we have also seen that the unit cylinder is the product space $\mathbb{S}^1 \times I$.

Example 6.1.3 A *Möbius band* (or *strip*) is a surface in $\mathbb{R}^3$ generated by moving a line segment of finite length in such a way that its middle point glides along a circle, and the segment remains normal to the circle and turns uniformly through a total angle of 180°. A model of a Möbius band can be constructed by gluing the ends of a rectangular strip of paper after giving it a half twist (of 180°). (See Fig. 6.2). The quotient space $(I \times I)/\sim$, where the equivalence relation $\sim$ is given by $(0, t) \sim (1, 1 - t)$, $0 \leq t \leq 1$ is homeomorphic to a Möbius band.

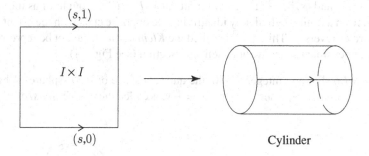

Fig. 6.1 Construction of a cylinder

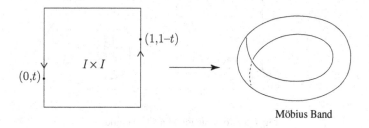

Fig. 6.2 Construction of a Möbius band

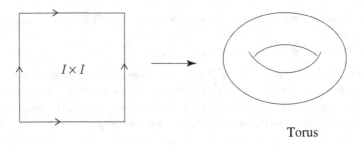

Torus

Fig. 6.3 Construction of a torus

Example 6.1.4 In Example 2.2.5, we have defined the *torus* as the product space $\mathbb{S}^1 \times \mathbb{S}^1$. A model of torus can be constructed from a rectangle by first gluing a pair of opposite edges to get a cylinder and then the circular ends are glued together. This identification is described by the equivalence relation on $I \times I$ given by $(s, 0) \sim (s, 1)$ and $(0, t) \sim (1, t)$ for all $s, t \in I$. The quotient space $(I \times I)/\sim$ is homeomorphic to torus; in fact, the function $I \times I \longrightarrow \mathbb{S}^1 \times \mathbb{S}^1, (s, t)] \mapsto \left(e^{2\pi i s}, e^{2\pi i t}\right)$, induces a desired homeomorphism (Fig. 6.3).

Example 6.1.5 The quotient space of $I \times I$ by another equivalence relation $\sim$, where $(0, t) \sim (1, t)$ and $(s, 0) \sim (1 - s, 1)$ for all $s, t \in I$, can be thought of as the space obtained from a (finite) cylinder by identifying the opposite ends with the orientation of two circles reversed. This space is called the *Klein bottle*. It cannot be represented in three-dimensional space without self-intersection (see Fig. 6.4).

Example 6.1.6 For any integer $n \geq 0$, the quotient space of the n-sphere $\mathbb{S}^n$ by the equivalence relation $\sim$, where $x \sim y$ if $x = -y$, is called the $n - dimensional\ real$

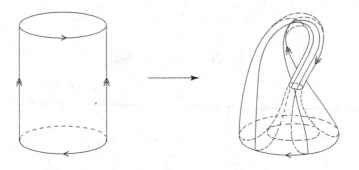

Fig. 6.4 Construction of a Klein bottle

projective space, denoted by $\mathbb{RP}^n$. There are other equivalent descriptions of the space $\mathbb{RP}^n$, given in the exercises.

Next, we formalize the construction of a sphere from a disc by collapsing its boundary to a point.

Definition 6.1.4 Let X be a space and $A \subseteq X$. The quotient space of X by the equivalence relation $\sim$ given by $a \sim a'$ for every $a, a' \in A$ is usually denoted by X/A, and called the quotient of X by A.

We think of X/A as the space X with the subspace A identified (collapsed) to a point. If A is open (resp. closed), then the projection map $\pi : X \to X/A$ is open (resp. closed). In this case, it is also easily seen that $X - A$ is homeomorphic to $X/A - [A]$, where $[A]$ is the equivalence class of any point of A.

Example 6.1.7 The quotient space $\mathbb{D}^2/\mathbb{S}^1$, where $\mathbb{S}^1 = \partial\mathbb{D}^2$, is homeomorphic to $\mathbb{S}^2$. To see this, we first construct a homeomorphism between $\mathrm{int}(\mathbb{D}^2)$ and $\mathbb{S}^2 - \{p\}$. Clearly, the mappings $h : \mathrm{int}(\mathbb{D}^2) \to \mathbb{R}^2$, $x \mapsto x/(1 - \|x\|)$, and $g : \mathbb{S}^2 - \{p\} \longrightarrow \mathbb{R}^2$, $(x_0, x_1, x_2) \mapsto \left(\frac{x_0}{1-x_2}, \frac{x_1}{1-x_2}\right)$, where $p = (0, 0, 1) \in \mathbb{S}^2$ are homeomorphisms. So $g^{-1}h : \mathrm{int}(\mathbb{D}^2) \to \mathbb{S}^2 - \{p\}$ is a desired homeomorphism. Now, we define a function $f : \mathbb{D}^2 \to \mathbb{S}^2$ by

$$f(x) = \begin{cases} g^{-1}h(x) & \text{for } x \in \mathbb{D}^2 - \mathbb{S}^1 \text{ and} \\ p & \text{for } x \in \mathbb{S}^1. \end{cases}$$

(The function f maps a concentric circle of radius r in $\mathbb{D}^2$ homeomorphically onto a parallel at the latitude $(2r - 1)/(1 - 2r + 2r^2)$ in $\mathbb{S}^2$ and takes the radii in $\mathbb{D}^2$ onto the meridians running from the south pole to the north pole. See Fig. 6.5.) Clearly, f induces a bijection $\bar{f} : \mathbb{D}^2/\mathbb{S}^1 \to \mathbb{S}^2$ such that $f = \bar{f}\pi$, where $\pi : \mathbb{D}^2 \to \mathbb{D}^2/\mathbb{S}^1$ is the quotient map. We show that $\bar{f}$ is a homeomorphism. By Proposition 6.1.3, we need to prove that f is continuous and closed. To see the continuity of f, consider a point $x \in \mathbb{D}^2$, and let $U \subset \mathbb{S}^2$ be an open nbd of $f(x)$. If $x \in \mathbb{D}^2 - \mathbb{S}^1$, then $V = f^{-1}(U - \{p\})$ is a nbd of x such that $f(V) \subseteq U$. And, if $x \in \mathbb{S}^1$, then we choose

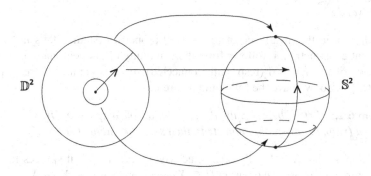

Fig. 6.5 Construction of a sphere from a disc

a small open ball $B(p; \epsilon) \subseteq U$. It is easily checked that $V = f^{-1}(B(p; \epsilon))$ is the annulus $\{x \in \mathbb{D}^2 \mid r < \|x\| \leq 1\}$, where $r^2 = (4 - \epsilon^2)/(4 + 2\epsilon\sqrt{4 - \epsilon^2})$. Thus, V is an open nbd of x in $\mathbb{D}^2$ with $f(V) \subseteq U$, and the continuity of f follows. Now, it is immediate that f is closed, since $\mathbb{D}^2$ is compact and $\mathbb{S}^2$ is Hausdorff.

Similarly, we see that the quotient space $\mathbb{D}^n/\mathbb{S}^{n-1}$ is homeomorphic to $\mathbb{S}^n$.

We remark that a relation $\sim$ on X is often regarded as its graph

$$G(\sim) = \{(x, x') \mid x, x' \in X \text{ and } x \sim x'\}.$$

We call the relation $\sim$ open (resp. closed) if its graph $G(\sim)$ is open (resp. closed) in the product space $X \times X$. Accordingly, if an equivalence relation $\sim$ is open (resp. closed), then every equivalence class $[x]$ is open (resp. closed) in X, because $[x]$ is the image of $(\{x\} \times X) \cap G(\sim)$ under the homeomorphism $\{x\} \times X \approx X$. So, for an open equivalence relation $\sim$, the quotient topology for $X/\sim$ is discrete and the quotient map $\pi\colon X \to X/\sim$ is open. However, a similar statement for a closed relation does not hold good, as shown by Example 6.1.8. On the other hand, if the quotient map $\pi\colon X \to X/\sim$ is closed (resp. open), the relation $\sim$ is not necessarily closed (resp. open) in $X \times X$. Incidentally, we would like to mention that the terms open relation and closed relation are used in a different sense too in the literature.

Example 6.1.8 Consider the equivalence relation $\sim$ on $\mathbb{R}$ defined by $x \sim y$ if $x = y$ or $x \neq 0$ and $y = 1/x$. This relation is obviously closed in $\mathbb{R}^2$, but the quotient map $\pi\colon \mathbb{R} \to \mathbb{R}/\sim$ is not closed (the saturation of the closed set $[1, \infty)$ is $(0, \infty)$).

Example 6.1.9 Let X be the Sierpinski space with the identity relation ($x \sim y \Leftrightarrow x = y$). Then the quotient map $X \to X/\sim$ is obviously closed, while $G(\sim)$ is not closed in $X \times X$ (cf. Exercise 8.1.14).

Example 6.1.10 In the real line $\mathbb{R}$, consider the equivalence relation $x \sim y$ if $x - y$ is rational. The quotient space $\mathbb{R}/\sim$ has the trivial topology, and the quotient map $\pi\colon \mathbb{R} \to \mathbb{R}/\sim$ sends each open interval (a, b) onto $\mathbb{R}/\sim$, and hence π is open. But the relation $G(\sim)$ is not open in $\mathbb{R}^2$, since it does not contain any rectangle $(a, b) \times (c, d)$.

Moving on to the invariance of topological properties (studied thus far) under the present construction, it follows from the continuity of quotient map that a quotient space of a connected (resp. path-connected) space is connected (resp. path-connected). Also, we have the following theorem.

Theorem 6.1.5 *Let $\sim$ be an equivalence relation on a space X. If X is locally connected (resp. locally path-connected), then so is the quotient space $X/\sim$.*

Proof Suppose that X is locally connected. By Theorem 3.4.2, it suffices to show that the components of each open set $U \subseteq X/\sim$ are open. Let $\pi\colon X \to X/\sim$ be the quotient map and C be a component of U. We observe that $\pi^{-1}(C)$ is open. Let $x \in \pi^{-1}(C)$ be an arbitrary element and V be the component of $\pi^{-1}(U)$ containing x. Then V is open, since X is locally connected and $\pi^{-1}(U)$ is open iln X. It is obvious that $\pi(x) \in \pi(V) \subseteq C$, for $\pi(V)$ is connected. So $x \in V \subseteq \pi^{-1}(C)$, and $\pi^{-1}(C)$ is a neighborhood of x. It follows that $\pi^{-1}(C)$ is open, and hence C is open, by the definition of the topology on $X/\sim$.

Using Proposition 3.4.6, a similar argument can be given in the case of local path-connectedness. ◇

In the next section, it will be shown that a closed (or open) map differs from a quotient map by a homeomorphism. Hence, the preceding theorem can be considered as the unification of the cases of closed maps and open maps in 3.4.3 and in 3.4.8.

Like connectedness, compactness is also transmitted to quotient spaces. It should, however, be noted that an analogous statement about local compactness is not true (see Exercise 12).

Coming to T_1-axiom, we see that a quotient space $X/\sim$ is T_1 if and only if each equivalence class of the space X is closed. Said differently, Axiom T_1 certainly passes down to quotient space if the quotient map is closed. In particular, if X satisfies the T_1-axiom and $A \subseteq X$ is closed, then X/A is a T_1-space. However, a quotient space of a T_1-space need not be T_1, even if the quotient map is open.

Example 6.1.11 Let $X = [0, 2]$ be the subspace of the real line $\mathbb{R}$ and $A = (1, 2]$. It is easy to see that the projection map $\pi: X \to X/A$ is open. But X/A fails to satisfy T_1-axiom. For, every saturated open nbd of point 1 contains the set A; so there is no nbd of the point $\pi(1)$, which excludes the point $\pi(2)$.

Observe that a quotient space of a space X is Hausdorff if and only if any two distinct equivalence classes of X have disjoint saturated open nbds. From Example 6.1.11, it is clear that Hausdorff property is not preserved even by open maps. The same is true of closed maps, as shown by the following.

Example 6.1.12 Let $A = \{1/n \mid n = 1, 2, \ldots\}$ and X be the space of real numbers with the topology

$$\mathcal{T} = \{U - B \mid U \text{ is open in the real line } \mathbb{R} \text{ and } B \subseteq A\}$$

(ref. Exercise 1.4.6). Then A is closed in X, and X is Hausdorff, since the topology of X is finer than the usual topology of $\mathbb{R}$. It follows that the projection map $X \to X/A$ is closed, but the points $[0]$ and $[A]$ in X/A cannot be separated by open sets, since every open set containing A intersects every nbd of 0 in X. The quotient space X/A satisfies the T_1-axiom, however.

In this regard, the following result is interesting.

Proposition 6.1.6 *If X is a compact Hausdorff space and $A \subseteq X$ is closed, then X/A is (compact and) Hausdorff.*

Proof Suppose that X is a compact Hausdorff space and $A \subseteq X$ is closed. Let π: $X \to X/A$ be the quotient map. If $x \neq y$ are in $X - A$, then there exist disjoint open sets U and V in X with $x \in U$ and $y \in V$. Since A is closed in X, we can assume that $U \cap A = \varnothing = V \cap A$. Then $\pi(U)$ and $\pi(V)$ are disjoint nbds of $\pi(x)$ and $\pi(y)$ in X/A. If $x \notin A$ and $y \in A$, then we find disjoint open sets U and V in X with $x \in U$ and $A \subseteq V$, by Theorem 5.1.8. Now, $\pi(U)$ and $\pi(V)$ are disjoint nbds of $\pi(x)$ and $\pi(y)$ in X/A. Thus, X/A is a Hausdorff space.   ◇

The preceding result is valid for a wider class of spaces X (refer to Exercise 8.1.8). For checking the Hausdorff property of quotient spaces, the next proposition sometimes comes in useful.

Proposition 6.1.7 *Let X be a space with an equivalence relation $\sim$ and $\pi: X \to X/\sim$ be the natural projection. If the quotient space $X/\sim$ is Hausdorff, then $G(\sim)$ is closed in $X \times X$. Conversely, if $G(\sim)$ is closed in $X \times X$ and π is open, then $X/\sim$ is Hausdorff.*

Proof Assume that $X/\sim$ is T_2 and $x \not\sim y$. Then there exist disjoint open nbds U of $\pi(x)$ and V of $\pi(y)$. We have $(x, y) \in \pi^{-1}(U) \times \pi^{-1}(V) \subseteq (X \times X) - G(\sim)$, where $G(\sim) = \{(x, x') \in X \times X \mid x \sim x'\}$. This shows that $(X \times X) - G(\sim)$ is open; so $G(\sim)$ is closed.

Conversely, suppose that π is open and $G(\sim)$ is closed in $X \times X$. If $\pi(x) \neq \pi(y)$ for $x, y \in X$, then $(x, y) \in (X \times X) - G(\sim)$. Since $G(\sim)$ is closed in $X \times X$, there exist nbds N of x and M of y such that $N \times M \subseteq (X \times X) - G(\sim)$. This implies that $\pi(N) \cap \pi(M) = \varnothing$. Since π is open, $\pi(N)$ and $\pi(M)$ are disjoint nbds of $\pi(x)$ and $\pi(y)$, respectively.   ◇

We remark that the graph of an equivalence relation $\sim$ is not necessarily closed, even if the projection $\pi: X \to X/\sim$ is open (cf. Exercise 10).

We now turn to the behavior of quotient topology under the formation of subspaces and product spaces.

Consider an equivalence relation $\sim$ on a space X and let $A \subset X$. Then, by restricting $\sim$ to A, we obtain an equivalence relation $\sim_A$ on A. The canonical mapping $\lambda: A/\sim_A \to X/\sim$, which maps the equivalence class of $a \in A$ under $\sim_A$ into its equivalence class under $\sim$, is obviously injective with the range $\pi(A)$, where π is the projection $X \to X/\sim$. So we can identify $A/\sim_A$ with $\pi(A)$, and regard $A/\sim_A$ as a subset of $X/\sim$. Notice that if A is saturated, then $A/\sim_A$ is actually a subset of $X/\sim$. Thus, $A/\sim_A$ receives two topologies, one as the quotient space of A and the other as the subspace of the quotient space $X/\sim$. The latter topology for $A/\sim_A$ is coarser than the former, in general (see Example 6.1.13). We shall see later that if A is open (or closed), then the two topologies for $A/\sim_A$ agree (cf. Proposition 6.2.7).

Example 6.1.13 Consider the equivalence relation $\sim$ on I as given in Example 6.1.1, and let $A = [0, 1/2) \cup (2/3, 1)$. The relation $\sim_A$ on A obtained by restricting $\sim$ to A is the identity relation. So the quotient space $A/\sim_A$ can be identified with A itself. Thus, $[0, 1/4)$ is open in the quotient topology for $A/\sim_A$, while a nbd of 0 in the relative topology induced from $I/\sim$ must contain a set $[0, \epsilon) \cup (1 - \epsilon, 1)$ for some $\epsilon > 0$.

To see a relationship between the quotient topology and the product topology, consider two spaces X_1 and X_2 with the equivalence relations $\sim_1$ and $\sim_2$, respectively. Then an equivalence relation R on $X_1 \times X_2$ can be defined by $(x_1, x_2) R (y_1, y_2) \iff x_1 \sim_1 y_1$ and $x_2 \sim_2 y_2$. The relation R on $X_1 \times X_2$ is usually denoted by $\sim_1 \times \sim_2$. There is a canonical map

$$\phi: (X_1 \times X_2)/R \longrightarrow (X_1/\sim_1) \times (X_2/\sim_2)$$

which makes the diagram

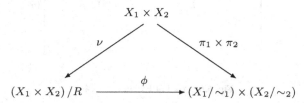

commutative, where ν, π_1 and π_2 are quotient maps. The mapping ϕ is clearly a continuous bijection, and if both π_1 and π_2 are open, then it is a homeomorphism, since the mapping $\pi_1 \times \pi_2$ is open, by Corollary 2.2.5. Thus, we have the following.

Proposition 6.1.8 *If the quotient maps $\pi_i: X_1 \to X_i/\sim_1$, $i = 1, 2$, are open, then the canonical bijection*

$$\phi: (X_1 \times X_2)/R \longrightarrow (X_1/\sim_1) \times (X_2/\sim_2)$$

is a homeomorphism, where the equivalence relation R and the map ϕ are defined as above.

However, ϕ is not open or, equivalently, closed in general.

Example 6.1.14 Let $\sim$ be the equivalence relation on $\mathbb{Q}$ which identifies $\mathbb{Z}$ to a point, and let $=$ denote the identity relation on $\mathbb{Q}$. Consider the relation S on $\mathbb{Q} \times \mathbb{Q}$ given by $(x, y)S(x', y') \iff x, x' \in \mathbb{Z}$ and $y = y'$. Denote the quotient map $\mathbb{Q} \to \mathbb{Q}/\sim$ by π. We show that the canonical mapping

$$\phi: (\mathbb{Q} \times \mathbb{Q})/S \longrightarrow (\mathbb{Q}/\sim) \times \mathbb{Q}, \quad [(x, y)] \mapsto (\pi(x), y)$$

is not a homeomorphism. For each positive integer n, choose an irrational number α_n such that $\alpha_n \to 0$. Let $G_n \subseteq [n, n+1] \times \mathbb{R}$ be an open set such that $\overline{G}_n$ meets $\{n, n+1\} \times \mathbb{R}$ only in the points (n, α_n) and $(n+1, \alpha_{n+1})$ (see Fig. 6.6 below). Set $F_n = \overline{G}_n \cap (\mathbb{Q} \times \mathbb{Q})$ for every n. Then $\{F_n\}$ is a locally finite family of closed subsets of $\mathbb{Q} \times \mathbb{Q}$, and hence $A = \bigcup F_n$ is closed. Since A does not meet $\mathbb{Z} \times \mathbb{Q}$,

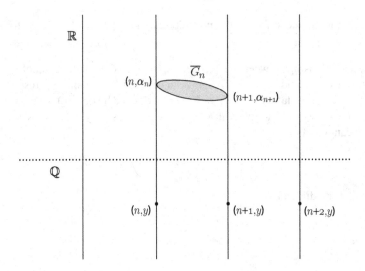

Fig. 6.6 Proof of Example 6.1.14

it can be regarded as a closed subset of $(\mathbb{Q} \times \mathbb{Q})/S$. But $B = \phi(A)$ is not closed in $(\mathbb{Q}/\sim) \times \mathbb{Q}$; in fact, we observe that the point $p = ([0], 0)$ is a limit point of B which lies outside it. Since $A = \phi^{-1}(B)$ does not contain any $(n, 0)$, it is obvious that p does not belong to B. To see that p is a limit point of B, consider a basic nbd $O = N \times M$ of p in $(\mathbb{Q}/\sim) \times \mathbb{Q}$. Then there exists a $\delta > 0$ such that $(-\delta, \delta) \cap \mathbb{Q} \subseteq M$. Since $\alpha_n \to 0$, we find an integer $n > 0$ such that $|\alpha_n| < \delta$. Then there is an $\epsilon > 0$ such that $(n - \epsilon, n + \epsilon) \cap \mathbb{Q} \subseteq \pi^{-1}(N)$, for N is a nbd of $[0]$ in $\mathbb{Q}/\sim$. Clearly,

$$[(n - \epsilon, n + \epsilon) \times (-\delta, \delta)] \cap (\mathbb{Q} \times \mathbb{Q}) \cap G_n \neq \varnothing,$$

and the image of this set under $\pi \times 1$ is contained in $O \cap B$. Hence, $p \in \overline{B}$.

Under the condition of Proposition 6.1.8, $\pi_1 \times \pi_2$ is clearly a quotient map. This holds true without any condition on the mappings π_1 and π_2 if both spaces X_1 and X_2 are locally compact Hausdorff (see Corollary 6.2.9).

We end this section by describing a method of constructing continuous functions between quotient spaces. Let X and Y be spaces with the equivalence relations $\sim$ and $\simeq$, respectively. Let $f: X \to Y$ be a relation-preserving function, that is, it satisfies the condition: $f(x) \simeq f(x')$ whenever $x \sim x'$. Then the equivalence class of $f(x)$ is independent of the choice of the representative x of $[x]$, for $[x] = [x']$ in $X/\sim$ implies that $x \sim x'$ which, in turn, implies that $f(x) \sim f(x')$. So there is a function $\bar{f}: (X/\sim) \longrightarrow (Y/\simeq)$ defined by $\bar{f}([x]) = [f(x)]$. If $\pi: X \to X/\sim$ and $\varpi: Y \to Y/\simeq$ are the natural projections, then we have $(\bar{f}\pi)(x) = (\varpi f)(x)$ for all $x \in X$, that is, the diagram

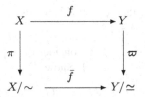

commutes. Since π is surjective, $\bar{f}$ is uniquely determined by the equation $\bar{f}\pi = \varpi f$. Moreover, if f is continuous, then so also is $\bar{f}$. We summarize this discussion as

Proposition 6.1.9 *If $f : X \to Y$ is a relation-preserving continuous map, then the mapping $\bar{f} : (X/\sim) \longrightarrow (Y/\simeq)$, $[x] \mapsto [f(x)]$, is continuous, where $X/\sim$ and $Y/\simeq$ have the quotient topologies.*

Exercises

1. ● In the unit interval I, consider the equivalence relation such that $x \sim y$ if both are less than $1/2$ or both are $\geq 1/2$. Show that the quotient space $I/\sim$ is homeomorphic to the Sierpinski space. Also, prove that the quotient map $I \to I/\sim$ is neither open nor closed.

2. (a) Does there exist an equivalence relation on I so that the quotient space is homeomorphic to a two-point discrete space?

    (b) Consider the equivalence relation $\sim$ on I defined by $x \sim y$ if both x and y are either rational or irrational. Determine the quotient space $I/\sim$.

3. Show that topology of the quotient space $\mathbb{R}/\sim$ in Example 6.1.10 is trivial. What is the cardinality of $\mathbb{R}/\sim$?

4. Show that (a) $\mathbb{RP}^0$ is the one-point space, and (b) $\mathbb{RP}^1 \approx \mathbb{S}^1$.

5. Show that $(I \times I)/\partial(I \times I)$ is homeomorphic to $\mathbb{S}^2$.

6. Let $A = \{1/n : n \in \mathbb{N}\}$ and $\pi : I \to I/A$ be the projection map. Let $G = [0, 1/2] \cup \{1\}$. Show that G is a saturated nbd of 0, but $\pi(G)$ is not a nbd of $\pi(0)$.

7. Let $X = I \times \mathbb{N} \subset \mathbb{R}^2$ and $A = \{0\} \times \mathbb{N}$. Find a continuous bijection from the quotient space X/A on to the subspace $Y \subset \mathbb{R}^2$ which consists of the closed line segments joining the origin to the points $(1, 1/n)$, $n \in \mathbb{N}$. Are they homeomorphic?

8. For each integer $n = 1, 2, \ldots$, let U be the subspace of $\mathbb{R}^2$ consisting of circles $(x - n)^2 + y^2 = n^2$, X be the subspace of $\mathbb{R}^2$ which is the union of circles $(x - 1/n)^2 + y^2 = (1/n)^2$, and Y denote the quotient space $\mathbb{R}/\mathbb{Z}$. Show that no two of U, X and Y are homeomorphic.

9. Find an example showing that if $A \subset X$ is not open or closed, then $X - A$ need not be homeomorphic $X/A - [A]$.

10. • On the interval $J = [-1, 1] \subset \mathbb{R}$, consider the equivalence relation $\sim$ which identifies x with $-x$ for $-1 < x < 1$. Show:

   (a) The quotient map $\pi \colon J \to J/\sim$ is open, but the graph $G(\sim)$ is not closed in $J \times J$.

   (b) The quotient space $J/\sim$ is T_1 but not T_2.

11. Show that the quotient space X/A in Example 6.1.12 is connected but not path-connected.

12. • Show that the quotient map $\mathbb{R} \to \mathbb{R}/\mathbb{Z}$ is closed, but the quotient space $\mathbb{R}/\mathbb{Z}$ is not locally compact. Is it Hausdorff?

13. Let $\sim$ be an equivalence relation on a locally compact Hausdorff space X such that the quotient map $\pi \colon X \to X/\sim$ is closed and each equivalence class is compact. Prove that $X/\sim$ is also locally compact.

14. Let $A \subseteq B$ be subsets of a space X and A be closed in X. Prove that B/A is a subspace of X/A.

15. Let $\pi \colon X \to X/\sim$ be open or closed. If $A \subseteq X$ meets every equivalence class of X, show that the induced map $A/\sim \to X/\sim$ is a homeomorphism.

16. Show that $\mathbb{R}P^n$ is homeomorphic to each of the following quotient spaces:

   (a) $\left(\mathbb{R}^{n+1} - \{0\} \right) / \sim$, where $x \sim y$ if $y = \lambda x$ for some $0 \neq \lambda \in \mathbb{R}$.

   (b) $\mathbb{D}^n/\sim$, where $x \sim y$ if $x = -y$ for $x, y \in \mathbb{S}^{n-1} = \partial\mathbb{D}^n$.

17. Show that $\mathbb{R}P^2$ is homeomorphic to $(I \times I)/\sim$, where $(s, 0) \sim (1 - s, 1)$ and $(0, t) \sim (1, 1 - t)$, for all $s, t \in I$.

18. Describe the space obtained by identifying the boundary (edge) of the Möbius band to a point.

19. Let X be a compact Hausdorff space. If A is closed in X, prove that X/A is homeomorphic to the one-point compactification of $X - A$.

20. (a) Determine the one-point compactification of the subspace $\mathbb{R}^2 - \{0\}$ of the Euclidean space $\mathbb{R}^2$.

   (b) Prove that the one-point compactification of the open Möbius band (i.e., without its boundary) is homeomorphic to $\mathbb{R}P^2$.

21. Let X be a space with an equivalence relation $\sim$. Prove that the quotient map $X \to X/\sim$ is closed if and only if for each element $[x]$ of $X/\sim$ and each open set $U \subseteq X$ containing the equivalence class $[x]$, there is a saturated open set $V \subseteq X$ such that $[x] \subseteq V \subseteq U$. (Such an equivalence relation is called *upper semicontinuous*.)

6.2   Identification Maps

In this section, we continue with the study of quotient topology in a more general setting.

Definition 6.2.1 Let X and Y be spaces. A function $f : X \to Y$ is called an *identification map* if it is surjective and Y has the largest topology such that f is continuous.

Clearly, a surjection $f : X \to Y$ is an identification map if and only if (a set U is open in $Y \Leftrightarrow f^{-1}(U)$ is open in X). If we substitute closed sets for open sets, we obtain another equivalent definition. When f is an identification map, the space Y is referred to as an *identification* space of f.

Given a function $f : X \to Y$, we say that a subset $A \subseteq X$ is f-saturated if $A = f^{-1}f(A)$. With this terminology, a continuous surjection $f : X \to Y$ is an identification if the open (resp. closed) subsets of Y are precisely the images of the f-saturated open (resp. closed) subsets of X. If $\sim$ is an equivalence relation on a space X, then the natural projection of X onto the quotient space $X / \sim$ is an identification map. The following result provides another two useful classes of identification maps.

Proposition 6.2.2 *A closed (or open) continuous surjection is an identification map.*

We leave the straightforward proof to the reader.

By Corollary 5.1.12 and the preceding proposition, we deduce the following.

Corollary 6.2.3 *Let $f : X \to Y$ be a continuous surjection. If X is compact and Y is Hausdorff, then f is an identification.*

We remark that there are identification maps which are neither open nor closed (refer to Exercise 5 or 6.1.1). It should be also noted that not every continuous surjection is an identification map, for example, the identity map of $\mathbb{R}$, where the domain is assigned the discrete topology and the range has the usual topology. But the range of every surjection f of a given space X can be assigned a particular topology so that it becomes an identification map. In fact, the family $\mathscr{T}(f)$ of all subsets U of $Y = \text{im}(f)$ such that $f^{-1}(U)$ is open in X is a topology for Y, since the function f^{-1} behaves well with the operations of taking intersection and union. Moreover, f is an identification map with the topology $\mathscr{T}(f)$ on Y. This topology for Y is called the identification topology or the quotient topology determined by f.

The next result is a generalization of Proposition 6.1.3 to identification maps.

Theorem 6.2.4 *Let $f : X \to Y$ be an identification map, and, for a space Z, let $h : X \to Z$ be a continuous map which is constant on each set $f^{-1}(y)$, $y \in Y$. Then there is a continuous map $g : Y \to Z$ such that $gf = h$. Also, g is an open (closed) map if and only if $h(U)$ is open (closed) for every f-saturated open (closed) set $U \subseteq X$.*

Proof Given $y \in Y$, there is $x \in X$ such that $f(x) = y$. We set $g(y) = h(x) \Leftrightarrow$ $f(x) = y$. If $f(x) = f(x')$, our hypothesis implies that $h(x) = h(x')$. Thus, g is a single-valued function satisfying $gf(x) = h(x)$, for all $x \in X$. By the proof of Proposition 6.1.3, it is immediate that g is continuous. For the last statement, we note that $g(V)$, for each open (resp. closed) set $V \subseteq Y$, is the image under h of an f-saturated open (resp. closed) subset of X. $\diamond$

We use the preceding theorem to show that an identification map f differs from a quotient map by a homeomorphism. A function $f: X \to Y$ between topological spaces defines an equivalence relation on X: $x \sim x'$ if $f(x) = f(x')$. The equivalence classes of X under this relation are nonempty sets $f^{-1}(y)$, $y \in Y$, and the associated quotient space, denoted by $\widetilde{X}$, will be referred to as the *decomposition space* of f. The function f induces an injective function $\tilde{f}: \widetilde{X} \to Y$, which maps the equivalence class $[x]$ into $f(x)$, and there is a factorization $f = \tilde{f} \circ \pi$ that is, the following triangle

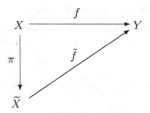

commutes, where π is the natural projection. Moreover, if f is continuous, then $\tilde{f}$ is a continuous injection, by Theorem 6.2.4. For $\tilde{f}$ to be a homeomorphism, a necessary and sufficient condition is provided by

Theorem 6.2.5 *Let X and Y be spaces and $f: X \to Y$ be a function. Then the mapping $\tilde{f}: \widetilde{X} \to Y$ is a homeomorphism if and only if f is an identification.*

Proof Assume that f is an identification. Then, as seen above, $\tilde{f}$ is a continuous bijection. Also, if $O \subseteq \widetilde{X}$ is open, then $f^{-1}\tilde{f}(O) = \pi^{-1}(O)$ is open in X. By our assumption, $\tilde{f}(O)$ is open in Y, and $\tilde{f}$ is a homeomorphism. Conversely, assume that $\tilde{f}$ is a homeomorphism. Then $f = \tilde{f} \circ \pi$ is obviously a continuous surjection. Now suppose that for $U \subseteq Y$, $f^{-1}(U)$ is open in X. Then $\tilde{f}^{-1}(U)$ is open in $\widetilde{X}$, so $U = \tilde{f}(\tilde{f}^{-1}(U))$ is open in Y. Thus, f is an identification. $\diamond$

It follows that the identification space Y of a map f can be always regarded as a quotient space of its domain via the equivalence relation whose equivalence classes are the sets $f^{-1}(y)$, $y \in Y$. Accordingly, an identification map is usually referred to as a *quotient map*. The preceding theorem is often used to compare different descriptions of the same space.

Now, we come to the main property of identification maps, which is often convenient for deciding whether a given map is an identification. If $f: X \to Y$ is an identification map and $g: Y \to Z$ is a function such that $gf: X \to Z$ is continuous, then g is continuous, by the proof of Theorem 6.2.4. Conversely, suppose that $f: X \to Y$

is a continuous surjection such that for any space Z the continuity of a function $g: Y \to Z$ follows from that of the composition $gf: X \to Z$. Let Z be the decomposition space $\widetilde{X}$ of f, and let $\pi: X \to \widetilde{X}$ be the projection map and $\tilde{f}: \widetilde{X} \to Y$ be the continuous bijection, $[x] \mapsto f(x)$. Then $\tilde{f}^{-1} \circ f = \pi$ is continuous. By our hypothesis, $\tilde{f}^{-1}$ is continuous. Therefore, $\tilde{f}: \widetilde{X} \to Y$ is a homeomorphism, and hence an identification map. Being a composition of two identification maps, f is an identification map. Thus, we have established the following.

Theorem 6.2.6 *Let $f: X \to Y$ be a continuous surjection. Then f is an identification if and only if, for any space Z, the continuity of a function $g: Y \to Z$ follows from that of $gf: X \to Z$.*

Turning to the behavior of identification maps under various operations, we first note that the composition of two identification maps is an identification. In particular, we see that a quotient space of a quotient space of X is a quotient space of X; in other words, taking quotient is a transitive operation:

Let $\sim$ be an equivalence relation on a space X and $\simeq$ be an equivalence relation on $Y = X/\sim$. If $\pi: X \to X/\sim$ and $\varpi: Y \to Y/\simeq$ are the projection maps, then the composition $X \xrightarrow{\pi} Y \xrightarrow{\varpi} Y/\simeq$ is an identification map. Therefore, $Y/\simeq$ is homeomorphic to the decomposition space of the composition $\varpi \circ \pi$.

Next, consider an identification map $f: X \to Y$ and let A be a subspace of X. As shown by Example 6.1.13, the map $g: A \to f(A)$ defined by f need not be an identification. However, in certain situations, it turns out that g is an identification map.

Proposition 6.2.7 *Let $f: X \to Y$ be an identification map, and let $A \subseteq X$ be an f-saturated set. Then each of the following conditions implies that $g: A \to f(A)$, $a \mapsto f(a)$, is an identification map:*

(a) *f is open (or closed).*
(b) *A is open (or closed).*

Proof If $V \subseteq Y$ and $U = f(A) \cap V$, then $g^{-1}(U) = A \cap f^{-1}(V)$. This shows that $g^{-1}(U)$ is open in A whenever U is open in $f(A)$.

Conversely, let U be a subset of $f(A)$ such that $g^{-1}(U)$ is open in A. We need to prove that U is open in $f(A)$. Observe that $g^{-1}(U) = f^{-1}(U)$, since A is f-saturated.

Case (a): Suppose that f is open. We have $f^{-1}(U) = A \cap G$ for some open set $G \subseteq X$. We observe that $U = f(A) \cap f(G)$, which implies that U is open in $f(A)$, since f is open. If $f(x) \in f(A)$, then $x \in f^{-1}f(A) = A$; consequently, $f(A) \cap f(G) \subseteq f(A \cap G)$. The reverse inclusion is obvious, and the equality holds. The case f being closed is proved similarly.

Case (b): If A is open in X, then $f^{-1}(U)$ is open in X. Since f is an identification, U is open in Y and, therefore, in $f(A)$. The proof for closed A is similar. $\diamond$

More generally, it can be seen that g is an identification map if each g-saturated set which is open (resp. closed) in A is the intersection of A with an f-saturated set open (resp. closed) in X. The converse also holds.

As seen in the previous section, if $f: X \to Y$ and $g: Z \to T$ are identification maps, then the mapping $f \times g: X \times Z \longrightarrow Y \times T$ may not be an identification (Example 6.1.14). We conclude this section with the following useful theorem, due to J.H.C. Whitehead, in this regard.

Theorem 6.2.8 *Let $f: X \to Y$ be an identification map, and let Z be locally compact Hausdorff. Then $f \times 1: X \times Z \longrightarrow Y \times Z$ is an identification map.*

Proof By Theorem 6.2.6, it suffices to prove the continuity of a given function $g: Y \times Z \to T$ for which the composition

$$X \times Z \xrightarrow{f \times 1} Y \times Z \xrightarrow{g} T$$

is continuous. Write $h = g \circ (f \times 1)$, and let (y_0, z_0) be a point in $g^{-1}(O)$, where $O \subseteq T$ be open. Since f is surjective, we find an $x_0 \in X$ such that $f(x_0) = y_0$. Then, by the continuity of h, there exist open sets $U \subseteq X$ and $V \subseteq Z$ such that $(x_0, z_0) \in U \times V \subseteq h^{-1}(O)$. Since Z is locally compact Hausdorff, V contains a compact nbd N of z_0. Let $W = \{y \in Y \mid g(y, z) \in O \text{ for all } z \in N\}$. Then we have $y_0 \in W$ and $W \times N \subseteq g^{-1}(O)$. We further observe that W is open in Y. For,

$$f^{-1}(W) = \{x \in X \mid h(x, z) = g(f(x), z) \in O \text{ for all } z \in N\}$$
$$= \{x \in X \mid h(\{x\} \times N) \subseteq O\},$$

and therefore

$$X - f^{-1}(W) = \{x \in X \mid h(\{x\} \times N) \cap (T - O) \neq \varnothing\}$$
$$= p\left(h^{-1}(T - O) \cap (X \times N)\right),$$

where $p: X \times N \to X$ is the projection map. Since N is compact, the mapping p is closed, and hence we deduce from the above equation that $f^{-1}(W)$ is open in X. Therefore, W is open in Y, since f is an identification map. Accordingly, $W \times N$ is a nbd of (y_0, z_0), and thus g is continuous at the point (y_0, z_0).   ◇

Corollary 6.2.9 *Let $f: X \to Y$ and $g: Z \to T$ be identification maps. If both X and T are locally compact Hausdorff, then $f \times g$ is an identification.*

Proof Obviously, the mapping $f \times g: X \times Z \longrightarrow Y \times T$ is the composition $X \times Z \xrightarrow{1 \times g} X \times T \xrightarrow{f \times 1} Y \times T$. If X and T are locally compact Hausdorff, then both mappings $1 \times g$ and $f \times 1$ are identifications, by the preceding theorem. Hence, $f \times g$ is an identification.   ◇

Exercises

1. Prove that an injective identification map is a homeomorphism.

2. ● (a) Let $f: X \to Y$ be a continuous map. If f admits a continuous right inverse, prove that it is an identification map.

(b) Let $f: X \to Y$ and $g: Y \to Z$ be continuous surjections such that gf is an identification. Show that g is an identification.

3. Prove that the function $\bar{f}$ in Proposition 6.1.9 is an identification map if f is so.

4. Prove that an identification map $f: X \to Y$ is closed (resp. open) if and only if the saturation of every closed (resp. open) subset of X is closed (resp. open).

5. • Let $\chi: [0, 2] \to \{0, 1\}$ be the characteristic function of the set $[1, 2]$. Determine the identification topology on $\{0, 1\}$ induced by χ.

6. Determine the quotient space $\mathbb{R}^2/\sim$, where the equivalence relation $\sim$ is given by

   (a) $(x, y) \sim (x', y')$ if $y = y'$;
   (b) $(x, y) \sim (x', y')$ if $x + y^2 = x' + y'^2$; and
   (c) $(x, y) \sim (x', y')$ if $x^2 + y^2 = x'^2 + y'^2$.

7. Describe each of the following spaces:

   (a) The cylinder with each of its boundary circles identified to a point.

   (b) The torus with the subset consisting of a meridional and a longitudinal circle identified to a point.

8. Let $X = [0, 1] \cup [2, 3]$ and $Y = [0, 2]$ have the subspace topology from the real line $\mathbb{R}$. Show that the mapping $f: X \to Y$ defined by

$$f(x) = \begin{cases} x & \text{for } 0 \le x \le 1, \text{ and} \\ x - 1 & \text{for } 2 \le x \le 3 \end{cases}$$

is an identification map. Is it open? Is the restriction of f to the subspace $A = [0, 1] \cup (2, 3]$ an identification map?

9. Let X be the closed topologist's sine curve (ref. Example 3.1.7) and $f: X \to I$ be the projection map $(x, y) \mapsto x$. Show that f is an identification map which is not open.

10. Let $f: \mathbb{R} \to [-1, 1]$ be the function $f(x) = \sin 1/x$ for $x \ne 0$ and $f(0) = 0$. Give $[-1, 1]$ the identification topology induced by f. Prove that the subspace $[-1, 0) \cup (0, 1]$ has its usual topology but the only nbd of 0 is $[-1, 1]$.

11. Show that the quotient space of a space X by the equivalence relation in Exercise 3.2.1 is totally disconnected.

12. Let $f: X \to Y$ be an identification such that $f^{-1}(y)$ is connected for each $y \in Y$. Prove that X is connected $\Leftrightarrow Y$ is connected.

13. Let $f: X \to Y$ be an identification map. If Y is T_2, show that the decomposition space $\tilde{X}$ of f is also T_2.

14. Let X be a compact Hausdorff space and $f : X \rightarrow Y$ be an identification. Show
    that Y is Hausdorff $\Leftrightarrow$ the set $\{(x_1, x_2) \mid f(x_1) = f(x_2)\}$ is closed in $X \times X$.

6.3 Cones, Suspensions, and Joins

Definition 6.3.1 Given a space X, the quotient space $(X \times I) / \sim$, where the equivalence relation $\sim$ is given by $(x, 0) \sim (x', 0)$ for all $x, x' \in X$, is called the *cone* on
X and is usually denoted by CX.

Notice that the cone on X can be also described as the quotient space $(X \times I) /$
$(X \times \{0\})$. The identified point $[X \times \{0\}]$ is called the *apex* or *vertex* of the cone. The
equivalence class of $(x, t) \in X \times I$ will be denoted by $[x, t]$. Intuitively, the cone
on a space X is obtained from the cylinder $X \times I$ (over X) by pinching the bottom
$X \times \{0\}$ to a single point. The mapping $x \mapsto [x, 1]$ is an embedding of X into CX;
so X can be regarded as a subspace of CX. We refer to the subspace $X \subset CX$ as the
base of the cone (see Fig. 6.7).

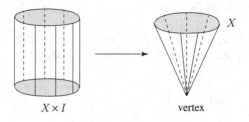

$X \times I$ vertex

Fig. 6.7 Cone on X

If $X \subseteq \mathbb{R}^n \subset \mathbb{R}^{n+1}$ and $v \in \mathbb{R}^{n+1} - \mathbb{R}^n$, then the subspace $TX = \{tx + (1 - t)v \mid$
$x \in X$, and $0 \leq t \leq 1\}$ is referred to as the *geometric cone* on X. Clearly, TX is
obtained by joining each point of X to v by a line segment. In general, CX has more
open sets than the geometric cone TX. Yet, the cone CX over a space X is intuitively
considered as the union of all line segments joining points of X to a point outside it.

If $f : X \rightarrow Y$ is a continuous function, then the function $(x, t) \mapsto (f(x), t)$ of
$X \times I$ into $Y \times I$ is continuous and relation preserving. Hence, there is an induced
continuous map $Cf : CX \rightarrow CY$ taking a point $[x, t]$ into $[f(x), t]$.

Example 6.3.1 The cone $C\mathbb{S}^n$ over the n-sphere $\mathbb{S}^n$ is homeomorphic to the disc $\mathbb{D}^{n+1}$.
To see this, consider the function $f : \mathbb{S}^n \times I \rightarrow \mathbb{D}^{n+1}$ defined by $f(x, t) = tx$. This
is clearly a continuous surjection and sends $\mathbb{S}^n \times 0$ to the point $0 \in \mathbb{D}^{n+1}$. Therefore,
f induces a continuous bijection $\tilde{f} : C\mathbb{S}^n \rightarrow \mathbb{D}^{n+1}$. Since $\mathbb{S}^n \times I$ is compact and
$\mathbb{D}^{n+1}$ is Hausdorff, the induced mapping $\tilde{f}$ is a homeomorphism.

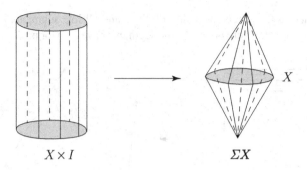

$$X \times I \qquad\qquad\qquad \Sigma X$$

Fig. 6.8 Suspension of X

Definition 6.3.2 Given a space X, the quotient space $(X \times I)/\sim$, where the equivalence relation $\sim$ is defined by $(x, 0) \sim (x', 0)$, $(x, 1) \sim (x', 1)$ for all $x, x' \in X$, is called the *suspension* of X.

The suspension of a space X is denoted by ΣX. The space ΣX can be alternatively described as the quotient space CX/X. Intuitively, the suspension of a space X is obtained from the cylinder $X \times I$ by pinching the bottom $X \times \{0\}$ to a point and the top $X \times \{1\}$ to another point. The identified points $[X \times \{0\}]$ and $[X \times \{1\}]$ are the vertices of ΣX. The subspace $\left[X \times \left\{\frac{1}{2}\right\}\right]$ is homeomorphic to X and is referred to as the base of the suspension (see Fig. 6.8).

As in the case of the cone over a space, a continuous function $f : X \to Y$ induces a continuous function $\Sigma f : \Sigma X \to \Sigma Y$, since it is constant on each equivalence class.

Example 6.3.2 The suspension of the n-sphere $\mathbb{S}^n$ is homeomorphic to $\mathbb{S}^{n+1}$. To see this, consider the mapping $f : \mathbb{S}^n \times I \to \mathbb{S}^{n+1}$ defined by $f(x, t) = (x \sin \pi t, \cos \pi t)$. It is clearly a continuous surjection and maps $\mathbb{S}^n \times 0$ and $\mathbb{S}^n \times 1$ into the points p and $-p$, respectively, where $p = (0, \ldots, 0, 1)$. Hence, it induces a continuous bijection $\tilde{f} : \Sigma \mathbb{S}^n \to \mathbb{S}^{n+1}$. Since $\mathbb{S}^n \times I$ is compact and $\mathbb{S}^{n+1}$ is Hausdorff, the induced mapping $\tilde{f}$ is a homeomorphism.

Definition 6.3.3 Given two spaces X and Y, their *join* $X * Y$ is the quotient space of $X \times Y \times I$ by the equivalence relation $(x, y, 0) \sim (x, y', 0)$ for all $y, y' \in Y$ and $(x, y, 1) \sim (x', y, 1)$ for all $x, x' \in X$.

Under the equivalence relation $\sim$, each set $\{x\} \times Y \times \{0\}$, $x \in X$, is identified to a point, and a similar remark is true about each of the sets $X \times \{y\} \times \{1\}$, $y \in Y$. See Fig. 6.9 below. The mappings

$$x \longmapsto [\{x\} \times Y \times \{0\}] \quad \text{and} \quad y \longmapsto [X \times \{y\} \times \{1\}]$$

are embeddings of X and Y into $X * Y$, respectively. Their images are usually identified with X and Y, and called the *bases* of the join. Then, intuitively speaking, $X * Y$

can be regarded as the space consisting of X, Y and all line segments joining each $x \in X$ to every $y \in Y$, where no two of the segments have interior points in common.

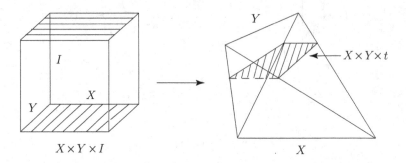

Fig. 6.9 Join of X and Y

Example 6.3.3 For any space X, $X * \{p\} \approx CX$, where $\{p\}$ is a one-point space. Obviously, the mapping $(x, p, t) \longmapsto (x, 1 - t)$ is a homeomorphism of $X \times \{p\} \times I$ with $X \times I$. By passing to the quotients, we obtain $X * \{p\} \approx CX$.

Example 6.3.4 For any space X, $X * \mathbb{S}^0 \approx \Sigma X$. Indeed, the mapping $f \colon X \times \mathbb{S}^0 \times I \longrightarrow X \times I$ defined by $f(x, -1, t) = (x, (1 - t)/2)$ and $f(x, +1, t) = (x, (1 + t)/2)$, $x \in X$ and $t \in I$, is a continuous closed surjection. Therefore, its composition with the natural projection $X \times I \to \Sigma X$ is an identification $\phi \colon X \times \mathbb{S}^0 \times I \longrightarrow \Sigma X$, say. Clearly, the decomposition space of ϕ is $X * \mathbb{S}^0$, and hence $X * \mathbb{S}^0 \approx \Sigma X$.

The join operation $*$ is commutative, in the sense that $X * Y \approx Y * X$. A desired homeomorphism is induced by the bijective mapping $X \times Y \times I \longleftrightarrow Y \times X \times I$, $(x, y, t) \leftrightarrow (y, x, 1 - t)$. However, the join operation is not associative with this topology. Also, this topology for the join is not convenient for maps *into* the join. Therefore, we introduce another topology for the join, which enjoys these properties and agrees with the aforesaid topology for a wide class of spaces.

Let Δ^1 be the subspace $\{(1 - t, t) \mid 0 \leq t \leq 1\}$ of $\mathbb{R}^2$. Then $I \approx \Delta^1$, a desired homeomorphism is given by the mapping $t \mapsto (1 - t, t)$. Consider the equivalence relation $\sim$ on $X \times Y \times \Delta^1$ generated by $(x, y, 1, 0) \sim (x, y', 1, 0)$, $(x, y, 0, 1) \sim (x', y, 0, 1)$. The canonical homeomorphism $(x, y, t) \longleftrightarrow (x, y, 1 - t, t)$ between $X \times Y \times I$ and $X \times Y \times \Delta^1$ clearly preserves the equivalence relations. Hence, there is a homeomorphism of $X * Y$ onto $(X \times Y \times \Delta^1)/\sim$, given by the mapping $[x, y, t] \longmapsto [x, y, 1 - t, t]$. Accordingly, the two spaces are topologically same, and the same notation $X * Y$ may be used to denote the space $(X \times Y \times \Delta^1)/\sim$ as well. For notational convenience, we will write the element $[x, y, 1 - t, t]$ as $(1 - t)x + ty$. Then $(1 - t')x' + t'y' = (1 - t)x + ty$ if $t = t'$, $x = x'$ for $t \neq 1$ and $y = y'$ for $t \neq 0$. In particular, note that $1x + 0y = 1x + 0y'$ even if $y \neq y'$, and

$0x + 1y = 0x' + 1y$ even if $x \neq x'$. The spaces X and Y are embedded into $X * Y$ via the mappings $x \mapsto 1x + 0y$ and $y \mapsto 0x + 1y$, respectively.

With the above definition of $X * Y$, we have four functions

$$\rho: X * Y \to I, \quad (1-t)x + ty \mapsto 1 - t;$$
$$\sigma: X * Y \to I, \quad (1-t)x + ty \mapsto t;$$
$$\xi: \rho^{-1}(0, 1] \to X, \quad (1-t)x + ty \mapsto x;$$
$$\psi: \sigma^{-1}(0, 1] \to Y, \quad (1-t)x + ty \mapsto y.$$

These functions are referred to as the *coordinate functions*.

Definition 6.3.4 The *coarse topology* for $X * Y$ is the smallest topology which makes the four coordinate functions ρ, σ, ξ and ψ continuous, where the domains of ξ and ψ have the relative topologies.

In other words, the coarse topology for $X * Y$ is generated by the subbase consisting of the inverse images of the open subsets of X under ξ, of Y under ψ and of I under ρ and σ. We observe that if for a function $f: Z \to X * Y$, the compositions of f with the coordinate functions ρ, σ, ξ and ψ (defined on the appropriate domains) are continuous in the coarse topology for $X * Y$, then f is continuous. Note that the domains of the compositions $f\xi$ and $f\psi$ are the subspaces $Z - f^{-1}(Y)$ and $Z - f^{-1}(X)$ of Z, respectively, and these sets are open in Z, by the continuity of $f\sigma$ and $f\rho$. Now, we see that the inverse image of every subbasic open set in $X * Y$ under f is open, and the assertion follows. In particular, the natural projection $\varpi: X \times Y \times \Delta^1 \longrightarrow X * Y$ is continuous in the new topology for $X * Y$, since its compositions with the coordinate functions are obviously continuous. Therefore, the coarse topology for $X * Y$ is indeed coarser than the quotient topology. This justifies the choice of the term for the new topology on $X * Y$. By contrast, the quotient topology on $X * Y$ is referred to as the "fine topology". Note also that the mappings $X \to X * Y$, $x \mapsto 1x + 0y$, and $Y \to X * Y$, $y \mapsto 0x + 1y$, remain closed embeddings even with the coarse topology on $X * Y$.

The fine topology and the coarse topology for $X * Y$ are identical for many nice spaces X and Y. We content ourselves by establishing the case wherein both the bases of $X * Y$ are compact Hausdorff. To this end, we first prove

Proposition 6.3.5 *If X and Y are Hausdorff spaces, then so is $X * Y$ (with either of the two topologies).*

Proof Clearly, it suffices to prove the proposition in the case $X * Y$ has the coarse topology. Suppose that X and Y are Hausdorff spaces, and let $\zeta = (1-t)x + ty$ and $\zeta' = (1 - t')x' + t'y'$ be distinct elements of $X * Y$. If $t \neq t'$, then we find disjoint open sets V and V' in I with $t \in V$ and $t' \in V'$. By the definition of the coarse topology for $X * Y$, $\sigma^{-1}(V)$ and $\sigma^{-1}(V')$ are open subsets of $X * Y$. It is obvious that $\sigma^{-1}(V)$ and $\sigma^{-1}(V')$ are disjoint and contain ζ and ζ', respectively. If $t = t' \neq 0$ and $y \neq y'$, then there exist disjoint open nbds U and U' of y and y', respectively, in Y. Consequently, $\psi^{-1}(U)$ and $\psi^{-1}(U')$ are disjoint open nbds of ζ and ζ', respectively. If $t = t' = 0$ or $y = y'$, then we must have $x \neq x'$, and a similar argument applies. $\diamond$

If the spaces X and Y are compact Hausdorff, then so is the product space $X \times Y \times \Delta^1$, for $\Delta^1 \approx I$. By Corollary 6.2.3 and the preceding proposition, the natural projection $\varpi: X \times Y \times \Delta^1 \longrightarrow X * Y$,

$$(x, y, 1 - t, t) \longmapsto (1 - t)x + ty,$$

is an identification with the coarse topology on $X * Y$. By Theorem 6.2.5, the range of ϖ is homeomorphic to its decomposition space, and this is obviously the space $X * Y$ with the fine topology. Therefore, the two topologies for $X * Y$ are same.

The commutativity of the join operation remains valid in the coarse topology, too. Moreover, the associativity also holds good.

Proposition 6.3.6 *The join operation $*$ is associative in the coarse topology, that is, there is a natural homeomorphism $(X * Y) * Z \approx X * (Y * Z)$ for all spaces X, Y, and Z.*

Proof Consider $X * Y$ with the coarse topology and define the mapping η_{XY}: $X * Y \to CX \times CY$ by $\eta_{XY}((1 - t)x + ty) = ([x, 1 - t], [y, t])$. Clearly, η_{XY} is a continuous injection, so $\eta_{XY}^{-1}: \mathrm{im}(\eta_{XY}) \longrightarrow X * Y$ is continuous. Also, it is easily seen that the compositions of η_{XY}^{-1} with the coordinate functions ρ, σ, ξ and ψ (defined on the appropriate domains) are continuous. Therefore, η_{XY}^{-1} is continuous, and hence η_{XY} is an embedding. For $[x, r] \in CX$, $[y, s] \in CY$ and $t \in I$, define $t([x, r], [y, s]) = ([x, tr], [y, ts])$. Then we have a continuous injection $\eta_{(XY)Z}: (X * Y) * Z \longrightarrow CX \times CY \times CZ$ given by

$$\eta_{(XY)Z}((1 - t)u + tz) = ((1 - t)\eta_{XY}(u), [z, t]),$$

where $u \in X * Y$. Obviously,

$$\mathrm{im}(\eta_{(XY)Z}) = \{([x, r], [y, s], [z, t]) \mid r + s + t = 1\}$$

and the function $\eta_{(XY)Z}^{-1}: \mathrm{im}(\eta_{(XY)Z}) \longrightarrow (X * Y) * Z$ sends the element $([x, r], [y, s], [z, t])$ to $(1 - t)u + tz$, where $u = \frac{r}{1-t}x + \frac{s}{1-t}y$ for $t \neq 1$, and $u \in X * Y$ is arbitrary for $t = 1$. The compositions of $\eta_{(XY)Z}^{-1}$ with the coordinate functions on $(X * Y) * Z$ are continuous, and therefore $\eta_{(XY)Z}^{-1}$ is continuous. It follows that $\eta_{(XY)Z}$ is an embedding. Similarly, the function $\eta_{(YZ)X}: (Y * Z) * X \longrightarrow CY \times CZ \times CX$ is an embedding. The canonical homeomorphism $CX \times CY \times CZ \approx CY \times CZ \times CX$ maps $\mathrm{im}(\eta_{(XY)Z})$ onto $\mathrm{im}(\eta_{(YZ)X})$. Hence,

$$(X * Y) * Z \approx (Y * Z) * X \approx X * (Y * Z).$$ $\diamond$

It follows that, if the spaces X, Y, and Z are compact Hausdorff, $(X * Y) * Z \approx X * (Y * Z)$ in the fine topology for joins, too. Moreover, the preceding proposition enables us to discuss with the coarse topology the multiple joins of a finite number of spaces, without using parentheses. As an immediate consequence of this, we have

$$X * \mathbb{S}^n \approx X * \Sigma \mathbb{S}^{n-1} \approx X * \mathbb{S}^{n-1} * \mathbb{S}^0 \approx X * \mathbb{S}^0 * \mathbb{S}^{n-1} \approx \Sigma X * \mathbb{S}^{n-1},$$

since $X * \mathbb{S}^0 \approx \Sigma X$ (ref. Example 6.3.4). By induction, we deduce that $X * \mathbb{S}^n$ is homeomorphic to the $(n + 1)$-fold suspension of X. In particular, we have $\mathbb{S}^m * \mathbb{S}^n \approx \mathbb{S}^{m+n+1}$.

We conclude this section by showing a relationship between the constructions of "cones" and "joines".

Proposition 6.3.7 *With the fine topology for $X * Y$, $C (X * Y) \approx CX \times CY$ for any two spaces X and Y.*

Proof There is a homeomorphism $h \colon \Delta^1 \times I \longrightarrow I \times I$ given by

$$h ((1 - t, t), s) = \begin{cases} (s, st/(1 - t)) & \text{for } 0 \leq t \leq 1/2, \\ (s(1 - t)/t, s) & \text{for } 1/2 \leq t \leq 1. \end{cases}$$

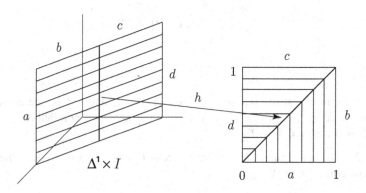

Fig. 6.10 The homeomorphism h in Proposition 6.3.7

Consequently, the composition

$$X \times Y \times \Delta^1 \times I \xrightarrow{1_X \times 1_Y \times h} X \times Y \times I \times I \xrightarrow{1_X \times g \times 1_Y} X \times I \times Y \times I$$

is a homeomorphism, where g is the homeomorphism $Y \times I \approx I \times Y$, $(y, u) \mapsto (u, y)$. Denote this composite by η. Then

$$\eta(x, y, 1-t, t, s) = \begin{cases} (x, s, y, st/(1-t)) & \text{for } 0 \leq t \leq 1/2, \\ (x, s(1-t)/t, y, s) & \text{for } 1/2 \leq t \leq 1. \end{cases}$$

If $p \colon X \times I \to CX$ and $q \colon Y \times I \to CY$ are the natural projections, then the composite

$$X \times Y \times \Delta^1 \times I \xrightarrow{\eta} X \times I \times Y \times I \xrightarrow{p \times q} CX \times CY$$

is an identification. It is easily checked that the decomposition space of the map $(p \times q) \circ \eta$ is $C(X * Y)$. Hence, $C(X * Y) \approx CX \times CY$.   ◇

Observe that the homeomorphism in the preceding proposition maps the base $X * Y$ of the cone $C(X * Y)$ onto $(CX \times Y) \cup (X \times CY)$, where X and Y are the bases of the cones CX and CY, respectively. Thus, with the fine topology, $X * Y \approx (CX \times Y) \cup (X \times CY)$. Of course, this result as well as the above proposition is also valid in the coarse topology when both the spaces X and Y are compact Hausdorff.

By way of an application, we determine the join of two discs.

Example 6.3.5 $\mathbb{D}^m * \mathbb{D}^n$ is homeomorphic to $\mathbb{D}^{m+n+1}$. We know that $\mathbb{D}^n \approx C\mathbb{S}^{n-1}$ (Example 6.3.1) and, by Example 6.3.3, $C\mathbb{S}^{n-1} \approx \mathbb{S}^{n-1} * \{pt\}$, etc. Accordingly, we have

$$\begin{aligned} \mathbb{D}^m * \mathbb{D}^n &\approx \left(\mathbb{S}^{m-1} * \{pt\}\right) * \left(\mathbb{S}^{n-1} * \{pt\}\right) \\ &\approx \left(\mathbb{S}^{m-1} * \{pt\} * \mathbb{S}^{n-1}\right) * \{pt\} \\ &\approx C\left(\mathbb{S}^{m-1} * \{pt\} * \mathbb{S}^{n-1}\right) \\ &\approx C\mathbb{S}^{m-1} \times C\{pt\} \times C\mathbb{S}^{n-1} \\ &\approx \mathbb{D}^m \times I \times \mathbb{D}^n \approx \mathbb{D}^{m+n+1}. \end{aligned}$$

Exercises

1. Let X be the subspace $\{(n, 0) \in \mathbb{R}^2 \mid n = 1, 2, \ldots\}$ and TX be the geometric cone on X with vertex $v = (0, 1)$. Show that there is a continuous bijection $CX \to TX$, but $CX \not\approx TX$.

2. Let $X \subseteq \mathbb{R}^n$ be compact. Prove that the cone CX is homeomorphic to the geometric cone TX on X.

3. If X is T_2, show that CX is also T_2.

4. Let X be a compact Hausdorff space. Show that the one-point compactification of $X \times (0, 1]$ is homeomorphic to the cone CX on X.

5. Show that the cone CX and the suspension ΣX of any space X are path-connected.

6. Prove that CX can be regarded as a closed subspace of ΣX.

7. If the cone CX over a space X is locally compact Hausdorff, show that X must be compact.

8. Prove that $X * Y$ is path-connected for any two spaces X and Y.

9. Show that $(X * Y) * Z \approx X * (Y * Z)$ in the fine topology if (a) X and Y, or (b) Y and Z are compact Hausdorff, or (c) X and Z are locally compact Hausdorff.

10. If $f : X \to X'$ and $g : Y \to Y'$ are continuous, show that there is a continuous map $f * g : X * Y \to X' * Y'$ in either topology for joins.

11. If $X \approx X'$ and $Y \approx Y'$, show that $X * Y \approx X' * Y'$.

12. (a) For any two subspaces $A \subseteq X$ and $B \subseteq Y$, show that $A * B$ can be identified with a subspace of $X * Y$ in the coarse topology. (b) If both A and B are closed, show that $A * B$ can be identified with a closed subspace of $X * Y$ in the fine topology.

13. Let Y be a compact T_2-space, and let $A_i, i = 1, 2, \ldots, n$, be compact subsets of a Hausdorff space X. Prove:

    (a) $\left(\bigcup_1^n A_i \right) * Y \approx \bigcup_1^n (A_i * Y)$, and

    (b) $\left(\bigcap_1^n A_i \right) * Y \approx \bigcap_1^n (A_i * Y)$.

14. Show that the quotient space of $X * Y$ (with the fine topology) obtained by identifying its bases to points is homeomorphic to $\Sigma (X \times Y)$.

6.4 Topological Sums

There is a simple way of topologizing the disjoint union of a family of topological spaces. Clearly, if a space Y is the union of a family of open subspaces $\{X_\alpha\}$, then a set $U \subseteq Y$ is open if and only if $U \cap X_\alpha$ is open in X_α for every α. Thus, we see that the topology of Y is determined by the topologies of the X_α's. This observation gives us a cue to introduce a topology on the union Y of a family of disjoint topological spaces $\{X_\alpha\}$, where the topologies of the spaces X_α are not necessarily induced from one and the same space. Indeed, the collection of subsets $U \subseteq Y$ such that $U \cap X_\alpha$ is open in X_α for every α is a topology for Y. Note that, with this topology on Y, each X_α becomes a clopen subspace of Y. Consequently, a subset $F \subseteq Y$ is closed if and only if $F \cap X_\alpha$ is closed in X_α for each index α.

Definition 6.4.1 Let $\{X_\alpha\}$ be an indexed family of disjoint topological spaces. Their *topological sum* is the space $Y = \bigcup X_\alpha$ with the topology such that a set $U \subseteq Y$ is open (resp. closed) if and only if $U \cap X_\alpha$ is open (resp. closed) in X_α for every α.

We denote the space Y in the above definition by $\sum X_\alpha$. In case the family $\{X_\alpha\}$ is finite and indexed by the numbers $1, \ldots, n$, we also use the notation $X_1 + \cdots + X_n$ for $\sum X_\alpha$.

The above approach can be easily extended to form the sum of an indexed family of spaces $\{X_\alpha\}$, which may not be pairwise disjoint. In this case, we construct the sets

$X'_\alpha = X_\alpha \times \{\alpha\}$ which are obviously pairwise disjoint and consider a topology for each X'_α so that it is homeomorphic to X_α. Clearly, the bijection $X'_\alpha \to X_\alpha, (x, \alpha) \mapsto x$, induces a topology on X'_α and, with this topology, $X'_\alpha \approx X_\alpha$. Now, a topology for $\bigcup X'_\alpha$ is defined, as above. So, we can define the *topological sum* (or *disjoint union* or *free union*) of the family of spaces $\{X_\alpha\}$ to be the space $\sum X'_\alpha$. For each β, there is an injection $i_\beta \colon X_\beta \to \sum X'_\alpha, x \mapsto (x, \beta)$. Clearly, i_β is an embedding and its range $X_\beta \times \{\beta\}$ is clopen in $\sum X_\alpha$. In practice, we generally identify X_β with $X_\beta \times \{\beta\}$, and treat X_β as a subspace of $\sum X'_\alpha$. With this identification, $\sum X'_\alpha$ is a disjoint union of the clopen subsets X_α and will be written as $\sum X_\alpha$.

To define a continuous function from $\sum X_\alpha$ into a space Y, we simply need a family of continuous functions $\{f_\alpha \colon X_\alpha \to Y\}$. For, if $x \in \sum X_\alpha$, then there is a unique X_α containing x. Therefore, we have a function $\phi \colon \sum X_\alpha \to Y$ defined by $\phi(x) = f_\alpha(x)$. Since $\phi|X_\alpha = f_\alpha$ is continuous for every α, it is easily seen that ϕ is continuous.

The natural questions about the topological properties of sums have easy answers and are relegated to exercises.

We can now establish a closer link between locally compact Hausdorff spaces and k-spaces (cf. Proposition 5.4.11).

Theorem 6.4.2 (D.E. Cohen) *A Hausdorff space X is a k-space if and only if it is homeomorphic to a quotient space of a locally compact space.*

Proof Let X be a k-space and $\{C_\alpha\}$ be the family of all compact subsets of X. By the definition of a k-space, a set $F \subseteq X$ is closed if and only if $F \cap C_\alpha$ is closed for every α, and hence the mapping $\phi \colon \sum C_\alpha \to X$ defined by the inclusion maps $C_\alpha \hookrightarrow X$ is an identification. Moreover, it is obvious that the topological sum $\sum C_\alpha$ is locally compact.

Conversely, suppose that X is a quotient space of a locally compact space Y and let $\pi \colon Y \to X$ be the quotient map. Let F be a subset of X such that $F \cap C$ is closed in C for every compact $C \subseteq X$. We need to show that F is closed. Since π is a quotient map, F is closed in X if and only if $\pi^{-1}(F)$ is closed in Y. So it suffices to prove that $\pi^{-1}(F)$ is closed in Y. Assume otherwise, and choose a point $y \in \overline{\pi^{-1}(F)} - \pi^{-1}(F)$. Since Y is locally compact, we find a compact nbd N of y. Then it is easily verified that $\pi(y) \in \overline{\pi(N) \cap F}$. Thus, we see that $\pi(N) \cap F$ is not closed, contrary to the choice of F. Hence, $\pi^{-1}(F)$ is closed, and the theorem follows. $\diamond$

The above theorem has several interesting consequences.

Corollary 6.4.3 *If X is a k-space and $\sim$ is an equivalence relation on X such that the quotient space $X/\!\sim$ is Hausdorff, then it is also a k-space.*

Proof This is immediate from the preceding theorem, since the composition of two quotient maps is a quotient map. $\diamond$

By Proposition 5.4.11 and the above corollary, we see that a Hausdorff quotient of a locally compact space is a k-space; although it may not be locally compact.

Next, we derive an analogue of Proposition 5.4.12 for open subspaces.

Corollary 6.4.4 *An open subspace of a k-space is a k-space.*

Proof Let G be an open subspace of a k-space X. Then, by Theorem 6.4.2, there exists an identification map $f : Y \to X$, where Y is a locally compact Hausdorff space. Clearly, $f^{-1}(G)$ is locally compact, being an open subspace of Y, and so it is a k-space. Also, it is easily checked that the mapping $g : f^{-1}(G) \to G$ defined by f is an identification. Then the converse part of the above theorem implies that G is a k-space. $\diamond$

As yet another consequence, we have the following.

Corollary 6.4.5 *The product of a k-space and a locally compact Hausdorff space is a k-space.*

Proof Let X be a k-space and Y a locally compact Hausdorff space. Then, by Theorem 6.4.2, there exists an identification map f from a locally compact Hausdorff space Z to X. Theorem 6.2.8 shows that the mapping

$$f \times 1 : Z \times Y \longrightarrow X \times Y, \quad (z, x) \mapsto (f(z), y),$$

is an identification and, by Theorem 5.4.6, $Z \times Y$ is locally compact. Hence, $X \times Y$ is a k-space. $\diamond$

Wedge Sums and Smash Products

The technique of forming wedge of a collection of topological spaces involves gluing prespecified points of the spaces together.

Definition 6.4.6 A *pointed space* is a pair (X, x_0), where X is a topological space and x_0 is a fixed point of X; the distinguished point x_0 is called the *base point*.

Definition 6.4.7 Let $\left(X_\alpha, x_\alpha^0\right)$, $\alpha \in A$, be a family of pointed spaces. The quotient space of the sum $\sum X_\alpha$ obtained by identifying all the base points x_α^0 is called the *wedge* (or the *one-point union* or the *bouquet*) of the spaces X_α and is denoted by $\bigvee X_\alpha$. The identified point $[x_\alpha^0]$ is taken as the base point of $\bigvee X_\alpha$.

Clearly, for each $\beta \in A$, the composition $X_\beta \xrightarrow{i_\beta} \sum X_\alpha \xrightarrow{\pi} \bigvee X_\alpha$ is an embedding, denoted by j_β. Also, note that each point x of $\bigvee X_\alpha$ other than the base point lies in a unique X_α. So there are natural maps $q_\beta : \bigvee X_\alpha \to X_\beta$ defined by

$$q_\beta(x) = \begin{cases} x & \text{if } x \in X_\beta - \left[x_\alpha^0\right], \text{ and} \\ x_\beta^0 & \text{otherwise.} \end{cases}$$

It is obvious that $q_\beta \circ j_\beta = 1_{X_\beta}$ and $q_\beta \circ j_{\beta'}$ is the constant map if $\beta \neq \beta'$. It is easily seen that each q_β is continuous. Let X_β' denote the slice parallel to X_β through the point $\left(x_\alpha^0\right)$ of $\prod X_\alpha$. Then the subspace $X_\beta' \subset \prod X_\alpha$ is homeomorphic to X_β, and the correspondence $x \leftrightarrow (q_\alpha(x))$ is a bijection between $\bigvee X_\alpha$ and $\bigcup X_\alpha'$. The mapping $\bigvee X_\alpha \to \prod X_\alpha$, $x \mapsto (q_\alpha(x))$, is clearly continuous but it need not be a homeomorphism, even if the family $\{X_\alpha\}$ is finite. Notice that if the base point x_β^0 is closed in X_β for every $\beta \in A$, then each slice X_β' is closed in $\prod X_\alpha$. In this case, we see that the mapping $\bigvee X_\alpha \longrightarrow \bigcup X_\alpha', x \mapsto (q_\alpha(x))$, is a homeomorphism for a finite family of pointed spaces $\{X_\alpha\}$. If the indexing set $A = \{1, 2, \ldots, n\}$, then the wedge of the spaces X_α is also written as $X_1 \vee \cdots \vee X_n$. Thus, for pointed T_1-spaces $X_1, \ldots, X_n$, $X_1 \vee \cdots \vee X_n$ is homeomorphic to the subspace $X_1' \cup \cdots \cup X_n'$ of the product space $X_1 \times \cdots \times X_n$. Accordingly, the wedge $X_1 \vee \cdots \vee X_n$ is regarded as the subspace $\bigcup_1^n X_\alpha' \subset \prod_1^n X_\alpha$.

Example 6.4.1 As seen above, the wedge $\mathbb{S}^1 \vee \mathbb{S}^1$ of two circles is homeomorphic to the subspace $\{(z_1, z_2) \in \mathbb{S}^1 \times \mathbb{S}^1 \mid z_1 = 1 \text{ or } z_2 = 1\}$ of the torus. This space can also be viewed as the subspace of the plane $\mathbb{R}^2$ which is the union of two unit circles centered at $(-1, 0)$ and $(1, 0)$, or as the quotient space of the unit circle $\mathbb{S}^1$ obtained by identifying the points -1 and 1, that is, $\mathbb{S}^1 / \{-1, 1\}$. We usually call it the *figure 8 space* (see Fig. 6.11).

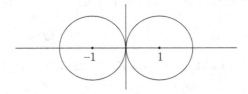

Fig. 6.11 Wedge of two circles

Example 6.4.2 For each positive integer n, consider the circle C_n in the plane $\mathbb{R}^2$ having radius $1/n$ and center $(1/n, 0)$. Notice that each C_n touches the y-axis at the origin. The subspace $X = \bigcup C_n$ of $\mathbb{R}^2$ is usually referred to as the *Hawaiian earring*. Consider the origin as the base point of each C_n, and let Y be the wedge $\bigvee C_n$ (see Fig. 6.12). Although there exists a continuous bijection from Y to X, they are not homeomorphic. For, the set $F = \{(2/n, 0) \mid n = 1, 2, \ldots\}$ is closed in Y (since $F \cap C_n$ is closed in C_n for every n), while it is not closed in X (since the origin is its limit point).

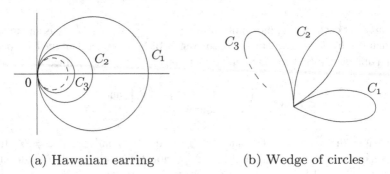

(a) Hawaiian earring   (b) Wedge of circles

Fig. 6.12 Spaces X and Y in Example 6.4.2

Turning to the behavior of topological properties (studied until now), we note that the wedge of any family of connected (resp. path-connected) spaces is connected (resp. path-connected). A similar result holds for Hausdorff property under this construction, as we will see below. However, compactness and local compactness do not behave as nicely, but it is easily verified that the wedge of a finite family of compact (resp. locally compact) spaces is compact (resp. locally compact).

Proposition 6.4.8 *If (X_α, x_α), $\alpha \in A$, are Hausdorff pointed spaces, then $\bigvee X_\alpha$ is also Hausdorff.*

Proof Let $x_0 = [x_\alpha]$ denote the base point of $\bigvee X_\alpha$. Then it is obviously closed in $\bigvee X_\alpha$. If x, y are two distinct points in $\bigvee X_\alpha - \{x_0\}$, then they have disjoint nbds in $\bigvee X_\alpha$, since each $X_\alpha - \{x_\alpha\}$ is Hausdorff and open in $\bigvee X_\alpha$. If $x = x_0 \neq y$, then there exists a unique index β such that $y \in X_\beta - \{x_\beta\}$. So we have disjoint open sets U_β and V_β in X_β such that $x_\beta \in U_\beta$ and $y \in V_\beta$. For each index $\alpha \neq \beta$, find an open nbd U_α of x_α in X_α and put $U = \sum U_\alpha$ and $V = V_\beta$. Then $\pi(U)$ and $\pi(V)$, where $\pi: \sum X_\alpha \to \bigvee X_\alpha$ is the identification map, are disjoint open nbds of x and y, respectively.   ◇

Smash Product

Recall that the wedge of two pointed spaces (X, x_0) and (Y, y_0) is essentially the subspace $(X \times \{y_0\}) \cup (\{x_0\} \times Y) \subset X \times Y$, and so the quotient space $(X \times Y) / (X \vee Y)$ is defined. This leads to the following.

Definition 6.4.9 Let (X, x_0) and (Y, y_0) be pointed spaces. The quotient space $(X \times Y) / (X \vee Y)$ is called the *smash product* of X and Y, and is denoted by $X \wedge Y$ (some authors write $X \# Y$).

We remark that $X \wedge Y$ depends on the choice of the base points x_0 and y_0. For example, consider the pointed space $(I, 0)$. It is easily seen that $I \wedge I$ is homeomorphic to $\mathbb{D}^2$. On the other hand, if we choose $1/2$ as the base point of I, then $I \wedge I$ is homeomorphic to the wedge of four copies of $\mathbb{D}^2$.

For any pointed space (X, x_0), we have $X \vee \mathbb{S}^0 = (X \times \{-1\}) \cup \{(x_0, 1)\}$, so $X \wedge \mathbb{S}^0 \approx X \times \{1\} \approx X$.

To see $X \wedge \mathbb{S}^1$, we regard $\mathbb{S}^1$ as the quotient space $I/\partial I$ with the base point $z_0 = [\partial I]$. Since the projection $p: I \to I/\partial I$ is a proper map, the function $1 \times p: X \times I \longrightarrow X \times \mathbb{S}^1$ is closed. Hence, the composition

$$X \times I \xrightarrow{1 \times p} X \times \mathbb{S}^1 \xrightarrow{\pi} X \wedge \mathbb{S}^1,$$

where π is the quotient map, is an identification. Clearly, the inverse of the subspace $X \vee \mathbb{S}^1 = (X \times \{z_0\}) \cup (\{x_0\} \times \mathbb{S}^1) \subset X \times \mathbb{S}^1$ under $1 \times p$ is the subspace $(X \times \{0\}) \cup (X \times \{1\}) \cup (\{x_0\} \times I)$ of $X \times I$. Therefore,

$$X \wedge \mathbb{S}^1 = (X \times \mathbb{S}^1) / [(X \times \{z_0\}) \cup (\{x_0\} \times \mathbb{S}^1)]$$
$$\approx (X \times I) / [(X \times \partial I) \cup (\{x_0\} \times I)].$$

The quotient space $(X \times I)/[(X \times \{0\}) \cup (X \times \{1\}) \cup (\{x_0\} \times I)]$ is called the *reduced suspension* of the pointed space (X, x_0) and is denoted by SX. Thus, $X \wedge \mathbb{S}^1 \approx SX$. Later, in this section, we shall determine $X \wedge \mathbb{S}^n$ for $n > 1$.

Example 6.4.3 The reduced suspension $S\mathbb{S}^{n-1}$ is homeomorphic to $\mathbb{S}^n$. Denote the base point of $\mathbb{S}^{n-1}$ by x_0. Then

$$X = S\mathbb{S}^{n-1} - \{[x_0, 0]\} \approx \left(\mathbb{S}^{n-1} - \{x_0\}\right) \times (0, 1) \approx \mathbb{R}^{n-1} \times \mathbb{R} \approx \mathbb{R}^n.$$

So their one-point compactifications X^* and $(\mathbb{R}^n)^*$ are homeomorphic. Since $S\mathbb{S}^{n-1}$ is compact Hausdorff, $X^* \approx S\mathbb{S}^{n-1}$. So $S\mathbb{S}^{n-1} \approx (\mathbb{R}^n)^* \approx \mathbb{S}^n$.

It is clear that the operation of taking a smash product is commutative: $X \wedge Y \approx Y \wedge X$. The associativity of smash product holds under certain conditions; we will establish this for locally compact Hausdorff spaces. To this end, we first prove the following.

Proposition 6.4.10 *If both of the spaces X and Y are Hausdorff, then so is $X \wedge Y$.*

Proof Let x_0 and y_0 be base points of X and Y, respectively, and let $*$ denote the base point of $X \wedge Y$. It is clear that $X \vee Y$ is closed in $X \times Y$, and hence $X \wedge Y - \{*\}$ is an open subspace of $X \times Y$. Consequently, every pair of distinct points in $X \wedge Y - \{*\}$ have disjoint open nbds in $X \wedge Y$. To complete the proof, we need to separate the point $*$ and any other point (x, y) in $X \wedge Y$. Since X is T_2, there exist disjoint open sets U_{x_0} and U_x in X with $x_0 \in U_{x_0}$ and $x \in U_x$. For the same reason, there are disjoint open sets V_{y_0} and V_y in Y with $y_0 \in U_{y_0}$ and $y \in U_y$. Then $\left(U_{x_0} \times Y\right) \cup \left(X \times V_{y_0}\right)$ is a nbd of $X \vee Y$, and $U \times V$ is a nbd of (x, y) in $X \times Y$. These sets are disjoint and saturated, and therefore their images in $X \wedge Y$ separate the points $*$ and (x, y). $\diamond$

Theorem 6.4.11 *If any two of the pointed spaces X, Y, and Z are compact Hausdorff or X and Z are locally compact Hausdorff, then $X \wedge (Y \wedge Z) \approx (X \wedge Y) \wedge Z$.*

Proof Let $*$ denote all base points, and let π denote the canonical projection of the product of two spaces onto their smash product. If X is locally compact Hausdorff, then $X \times Y \times Z \xrightarrow{1 \times \pi} X \times (Y \wedge Z)$ is an identification, by Theorem 6.2.8. If Y and Z are compact Hausdorff, then $\pi: Y \times Z \to Y \wedge Z$ is a proper map, by Corollary 5.5.3. Thus, $1 \times \pi$ is a closed map, and hence an identification. Let ϕ denote the composition

$$X \times Y \times Z \xrightarrow{1 \times \pi} X \times (Y \wedge Z) \xrightarrow{\pi} X \wedge (Y \wedge Z).$$

Then ϕ, being a composition of two identifications, is an identification. Similarly, if Z is locally compact Hausdorff or X and Y are compact Hausdorff, then the composition

$$X \times Y \times Z \xrightarrow{\pi \times 1} (X \wedge Y) \times Z \xrightarrow{\pi} (X \wedge Y) \wedge Z$$

is an identification, denoted by ψ, say. Observe that the decomposition space of each of ϕ and ψ is the quotient space $(X \times Y \times Z)/A$, where $A = (\{*\} \times Y \times Z) \cup (X \times \{*\} \times Z) \cup (X \times Y \times \{*\})$. So, by Theorem 6.2.5,

$$(X \wedge Y) \wedge Z \approx (X \times Y \times Z)/A \approx X \wedge (Y \wedge Z). \qquad \diamond$$

As an application of the preceding result, we compute $X \wedge \mathbb{S}^n$ for every n. For $n = 1$, we have already seen that $X \wedge \mathbb{S}^n \approx SX$. So

$$X \wedge \mathbb{S}^2 \approx X \wedge (\mathbb{S}^1 \wedge \mathbb{S}^1) \approx (X \wedge \mathbb{S}^1) \wedge \mathbb{S}^1 \approx SX \wedge \mathbb{S}^1 \approx S(SX).$$

By Example 6.4.3, $\mathbb{S}^n \approx \mathbb{S}^{n-1} \wedge \mathbb{S}^1$. Therefore, induction applies, and we deduce that $X \wedge \mathbb{S}^n$ is the n-fold reduced suspension of X. In particular, we note that $\mathbb{S}^m \wedge \mathbb{S}^n \approx \mathbb{S}^{m+n}$.

Exercises

1. Let X and Y be disjoint subspaces of a space Z such that $Z = X \cup Y$. Show that the following conditions are equivalent:

    (a) $Z = X + Y$,
    (b) X, Y are both open in Z,
    (c) X is clopen in Z, and
    (d) $\overline{X} \cap Y = \varnothing = \overline{Y} \cap X$.

2. Prove that a space X is connected if and only if X is not the sum of two nonempty disjoint subspaces.

3. Show that there are canonical homeomorphisms:

    (a) $(X + Y) \times Z \approx (X \times Z) + (Y \times Z)$ and
    (b) $Y \times \sum X_\alpha \approx \sum (Y \times X_\alpha)$ for any family of spaces X_α, $\alpha \in A$.

4. Let X_α, $\alpha \in A$, be a family of spaces each of which is homeomorphic to a space X. Prove that $\sum X_\alpha \approx X \times A$, where A has the discrete topology.

5. Let $\{X_\alpha\}$ be a family of T_i-spaces, $i = 1, 2$. Show that $\sum X_\alpha$ is a T_i-space.

6. Prove that any sum of locally compact spaces is locally compact.

7. Let X_α, $\alpha \in A$, be a family of metrizable spaces and let d_α be the standard bounded metric on X_α. Verify that δ given by

$$\delta(x, y) = \begin{cases} d_\alpha(x, y) & \text{if } x, y \in X_\alpha \text{ and} \\ 1 & \text{if } x \in X_\alpha, y \in X_\beta \text{ and } \alpha \neq \beta \end{cases}$$

is a metric on $\sum X_\alpha$ and the topology induced by δ is the topology of $\sum X_\alpha$.

8. Show that the space Y in Example 6.4.2 is homeomorphic to the quotient space of $\mathbb{R}$ obtained by collapsing $\mathbb{Z}$ to a point. Conclude that the wedge sum of infinitely many locally compact spaces need not be locally compact.

9. Describe each of the following spaces: (a) $\mathbb{S}^2$ with the equator identified to a point, and (b) $\mathbb{R}^2$ with each of the circles centered at the origin and of integer radii identified to a point.

10. Let X_α, $\alpha \in A$, be a family of pointed spaces. If each X_α is connected (resp. path-connected), prove that so is $\bigvee X_\alpha$.

11. Prove that $S(X \vee Y) \approx SX \vee SY$ for any pointed T_1-spaces X and Y.

12. If (X, x_0) and (Y, y_0) are compact Hausdorff pointed spaces, prove that $X \wedge Y$ is the one-point compactification of $(X - \{x_0\}) \times (Y - \{y_0\})$.

6.5 Adjunction Spaces

We shall study here a method of "attaching two spaces via a continuous map".

Definition 6.5.1 Let X and Y be spaces, $A \subseteq X$ and let $f : A \to Y$ be a continuous function. The quotient space $(X + Y)/\sim$, where the equivalence relation $\sim$ on $X + Y$ is generated by the relation $a \sim f(a)$, $a \in A$, is denoted by $X \cup_f Y$. The space $X \cup_f Y$ is called the *adjunction space* obtained by adjoining (or attaching) X to Y via f (see Fig. 6.13). The function f is called the *attaching map*.

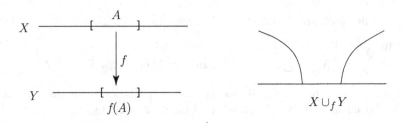

Fig. 6.13 The space formed by attaching X to Y via f

Example 6.5.1 Let $A \subseteq X$ be closed and Y be a one-point space $\{p\}$. Then, for the obvious map $f : A \to Y$, $X \cup_f Y \approx X/A$. Since $a \sim p$ in $X + Y$ for all $a \in A$, the inclusion map $\lambda : X \to X + Y$ is relation preserving. Hence, it induces a continuous bijection $\bar{\lambda} : X/A \to X \cup_f Y$. If $\pi : X + Y \to X \cup_f Y$ and $\phi : X \to X/A$ are the natural projections, then we have $\pi^{-1}\bar{\lambda}(F) = \phi^{-1}(F) + \{p\}$ if $[A] \in F$, and $\pi^{-1}\bar{\lambda}(F) = \phi^{-1}(F)$ if $[A] \notin F$. Therefore, $\bar{\lambda}$ is also a closed mapping, and thus a homeomorphism.

Example 6.5.2 The torus $\mathbb{S}^1 \times \mathbb{S}^1$ can be obtained by attaching a 2-disc to the figure 8 space via a mapping of the boundary of the disc. It has already been realized in Example 6.1.4 as the quotient space of the square $I \times I$ obtained by identifying its opposite sides. Under this identification, all four vertices of the square are identified, and the two pairs of identified sides result in the figure 8 space, essentially $\mathbb{S}^1 \vee \mathbb{S}^1$. If $\pi: I \times I \longrightarrow (I \times I)/\sim$ is the quotient map, then the image of $\partial\left(I^2\right)$ under π is clearly $\mathbb{S}^1 \vee \mathbb{S}^1$. See Fig. 6.14.

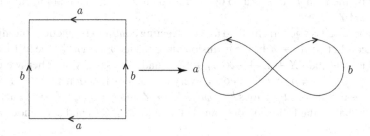

Fig. 6.14 Image of ∂I^2 under the quotient map π

Let $h: \mathbb{D}^2 \to I^2$ be a homeomorphism. Then h sends $\mathbb{S}^1 = \partial\left(\mathbb{D}^2\right)$ onto $\partial\left(I^2\right)$, and thus we have a continuous function $f: \mathbb{S}^1 \to \mathbb{S}^1 \vee \mathbb{S}^1$ defined by the composition πh. We observe that the adjunction space $\mathbb{D}^2 \cup_f \left(\mathbb{S}^1 \vee \mathbb{S}^1\right)$ is homeomorphic to the torus $(I \times I)/\sim$. Clearly, the mapping

$$g: \mathbb{D}^2 + \left(\mathbb{S}^1 \vee \mathbb{S}^1\right) \longrightarrow (I \times I)/\sim$$

given by $g(x) = \pi h(x)$ for $x \in \mathbb{D}^2$ and $g(y) = y$ for $y \in \mathbb{S}^1 \vee \mathbb{S}^1$ is an identification. As $f = g|\mathbb{S}^1$, it is easily verified that the decomposition space of g is the space $\mathbb{D}^2 \cup_f \left(\mathbb{S}^1 \vee \mathbb{S}^1\right)$, hence our assertion.

Proposition 6.5.2 *Suppose that X is attached to Y via a map $f: A \to Y$, and let $\pi: X + Y \longrightarrow X \cup_f Y$ be the quotient map. Then*

(a) *$\pi|Y$ is an embedding of Y onto a closed subspace of $X \cup_f Y$, and*
(b) *$\pi|(X - A)$ is an embedding of $X - A$ onto an open subspace of $X \cup_f Y$.*

Proof (a): If $F \subseteq Y$ is closed, then $f^{-1}(F)$ is closed in A and hence in X. So $\pi^{-1}(\pi(F)) = f^{-1}(F) + F$ is closed in $X + Y$ which, in turn, implies that $\pi(F)$ is closed in $X \cup_f Y$. Thus, $\pi|Y$ is a closed mapping; in particular, $\pi(Y)$ is closed in $X \cup_f Y$. Also, $\pi|Y$ is a continuous injection, and therefore $Y \approx \pi(Y)$.

(b): If $G \subseteq X - A$ is open, then $\pi^{-1}(\pi(G)) = G$ is open in $X + Y$. So $\pi(G)$ is open in $X \cup_f Y$, and it follows that $\pi|(X - A)$ is an open mapping. In particular, $\pi(X - A)$ is open in $X \cup_f Y$. Obviously, $\pi|(X - A)$ is a continuous injection of $X - A$ into $X \cup_f Y$, and thus it is a homeomorphism of $X - A$ onto $\pi(X - A)$. $\diamond$

By identifying Y with $\pi(Y)$, it is usually regarded as a closed subspace of $X \cup_f Y$. Similarly, $X - A$ may be considered as an open subspace of $X \cup_f Y$. It is clear that $\pi(X - A)$ and $\pi(Y)$ are disjoint, and $X \cup_f Y$ equals their union. Therefore, $X \cup_f Y$ can be considered as the disjoint union of $X - A$ and Y glued together by a topology defined by the map f. If A is also open in X, then $X \cup_f Y$ is a topological sum of $X - A$ and Y.

We now discuss a general method for constructing continuous functions on adjunction spaces. Let $j: A \to X$ be the inclusion map, and let ρ_x and ρ_y be the restrictions of the quotient map $X + Y \longrightarrow X \cup_f Y$ to X and Y, respectively. Then we obviously have $\rho_x \circ j = \rho_y \circ f$. Moreover, every continuous function $\phi: X \cup_f Y \to Z$ determines continuous functions $k = \phi \rho_x: X \to Z$ and $g = \phi \rho_y: Y \to Z$ such that $kj = gf$, that is, the following diagram of topological spaces and continuous maps commutes:

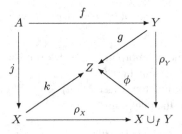

The other way around, we have

Proposition 6.5.3 *Suppose that A is a closed subset of a space X, and $f: A \to Y$, $g: Y \to Z$ and $k: X \to Z$ are continuous maps such that $kj = gf$, where $j: A \hookrightarrow X$ is the inclusion mapping. Then there is a unique continuous map $\phi: X \cup_f Y \to Z$ such that $\phi \rho_x = k$ and $\phi \rho_y = g$, where the notations ρ_x and ρ_y have the above meaning.*

Proof Consider the map $F: X + Y \to Z$ defined by

$$F(x) = \begin{cases} k(x) & \text{for } x \in X \text{ and} \\ g(y) & \text{for } y \in Y. \end{cases}$$

Obviously, F is continuous. Moreover, if $x \in X$, $y \in Y$ and $x \sim y$, then $x \in A$ and $y = f(x)$. Consequently, $g(y) = gf(x) = kj(x) = k(x)$, and therefore F induces a function $\phi: X \cup_f Y \to Z$ such that $\phi \circ \rho_x = F|X = k$, and $\phi \circ \rho_y = F|Y = g$. By the definition of the topology on $X \cup_f Y$, it is clear that ϕ is continuous. To see the uniqueness of ϕ, suppose that there is a mapping $\psi: X \cup_f Y \to Z$ with $\psi \circ \rho_x = k$ and $\psi \circ \rho_y = g$. Then $\psi \circ \pi = F$, where π is the quotient map on $X + Y$, and hence $\psi = \phi$.   ◇

Some topological properties such as connectedness, path-connectedness, and compactness behave nicely under the construction of adjunction spaces. Moreover, it is easily checked that if both X and Y are T_1-spaces and A is a closed subset of X, then $X \cup_f Y$ is a T_1-space for any continuous map $f : A \to Y$. However, a similar result for the Hausdorff property does not hold true. In this regard, the following result is quite useful.

Theorem 6.5.4 *Suppose X and Y are Hausdorff spaces, and $A \subseteq X$ is compact. Then, for any continuous function $f : A \to Y$, $X \cup_f Y$ is Hausdorff.*

Proof Let $\pi : X + Y \to X \cup_f Y$ be the quotient map. If $F \subseteq X + Y$ is closed, then $F \cap A$ is closed in A, and hence compact. Consequently, $f(F \cap A)$ is closed in Y, and it follows that

$$\pi^{-1}\pi(F) = F \cup f(F \cap A) \cup f^{-1}((F \cap Y) \cup f(F \cap A))$$

is closed. Thus, $\pi(F)$ is closed and π is a closed map. Moreover, for a point z in $X + Y$, the equivalence class of z is an one-point set if $z \in (X - A) \cup (Y - f(A))$, and $\{z\} \cup f^{-1}(z)$ if $z \in f(A)$. This shows that $\pi^{-1}\pi(z)$ is compact and π is a proper map. By Proposition 5.5.6, we see that $X \cup_f Y$ is Hausdorff.   ◇

The technique involved in the construction of adjunction spaces can be used to form two new spaces associated with a continuous map: "Mapping Cylinder" and "Mapping Cone". These constructs find many applications in algebraic topology. Let $f : X \to Y$ be a continuous map, and consider the cylinder $X \times I$, where I is the unit interval. Then the set $X \times \{0\}$ is closed in $X \times I$ and we have a continuous map $\phi : X \times \{0\} \to Y$ defined by f, i.e., $\phi((x, 0)) = f(x)$. The adjunction space of the map ϕ is called the *mapping cylinder* of f, and is denoted by M_f (Fig. 6.15 below).

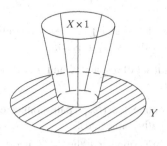

Fig. 6.15 Mapping cylinder of f

Let $\pi: (X \times I) + Y \longrightarrow M_f$ be the quotient map. By Proposition 6.5.2, $\pi|Y$ is a homeomorphism of Y onto a closed subspace of M_f. By means of this embedding, Y is considered as a closed subspace of M_f. We also note that the map $i: X \to M_f$ given by $i(x) = \pi((x, 1))$, $x \in X$, is a closed embedding. Accordingly, X is embedded as a closed subspace of M_f, too. The quotient space M_f/X is called the *mapping cone* of f, and denoted by C_f.

Observe that if Y is a one-point space, then M_f is CX (the cone over X), and C_f is ΣX (the suspension of X).

Exercises

1. Let $X = [0, 1]$, $Y = [2, 3]$ be the subspaces of the real line $\mathbb{R}$ and $A = \{0, 1\}$. Let $f: A \to Y$ be the map given by $f(0) = 3$, $f(1) = 2$. Show that $X \cup_f Y$ is homeomorphic to $\mathbb{S}^1$.

2. Suppose that X and Y are closed subsets of a space Z such that $Z = X \cup Y$. Let $A = X \cap Y$ and $f: A \to Y$ be the inclusion map. Show that $Z = X \cup_f Y$.

3. Suppose that $A \subseteq X$ is closed and $f: A \to Y$ is a continuous map. Let $\pi: X + Y \longrightarrow X \cup_f Y$ be the natural projection. Show:

   (a) For a closed set $F \subseteq X$, $\pi(F)$ is closed in $X \cup_f Y$ if and only if $f(F \cap A)$ is closed in Y.

   (b) If $U \subseteq X$ and $V \subseteq Y$ are open sets such that $f^{-1}(V) = A \cap U$, then $\pi(U + V)$ is open in $X \cup_f Y$.

4. Show that $\mathbb{RP}^2 \approx \mathbb{D}^2 \cup_f \mathbb{S}^1$ for a continuous map $f: \mathbb{S}^1 \to \mathbb{S}^1$.

5. Let X, Y be spaces and $A \subset X$ be closed. Let $f: A \to Y$ be an identification. Prove that the composition $X \to X + Y \longrightarrow X \cup_f Y$ is also an identification.

6. Give an example to show that the adjunction space of two Hausdorff spaces need not be Hausdorff.

7. Suppose that $A \subseteq X$ is closed and $f: A \to Y$ is continuous. If both spaces X and Y are compact, show that $X \cup_f Y$ is so.

8. Let A be a nonempty subset of a space X and $f: A \to Y$ be a continuous map. Prove:

   (a) If X and Y are connected (resp. path-connected), then $X \cup_f Y$ is also connected (resp. path-connected).

   (b) If A and $X \cup_f Y$ are connected, then Y is also connected.

9. Show that there is a continuous map $f: \mathbb{S}^1 \to \mathbb{S}^1 \vee \mathbb{S}^1$ such that the Klein bottle (Example 6.1.5) is homeomorphic to $\mathbb{D}^2 \cup_f (\mathbb{S}^1 \vee \mathbb{S}^1)$.

10. Prove that $\mathbb{S}^{m+n+1} \approx (\mathbb{D}^{m+1} \times \mathbb{S}^n) \cup (\mathbb{S}^m \times \mathbb{D}^{n+1})$. Use it to show that $\mathbb{S}^m \times \mathbb{S}^n$ is homeomorphic to $\mathbb{D}^{m+n} \cup_f (\mathbb{S}^m \vee \mathbb{S}^n)$ for a continuous map $f: \mathbb{S}^{m+n-1} \longrightarrow \mathbb{S}^m \vee \mathbb{S}^n$.

11. Let A be a closed subset of a space X and $f : A \to Y$ be a continuous function. Suppose that $A \subseteq S \subseteq X$ and $f(A) \subseteq T \subseteq Y$. If $g : A \to T$ is the map defined by f, show that $S \cup_g T$ is a subspace of $X \cup_f Y$.

12. Let A be a closed subspace of X, and let $f : A \to Y$ and $g : Y \to Z$ be continuous maps. Prove that $(X \cup_f Y) \cup_g Z \approx X \cup_{gf} Z$.

13. Let $A \subseteq B$ be closed subspaces of X. If $f : A \to Y$ is a continuous map, prove that $X \cup_f Y \approx X \cup_g (B \cup_f Y)$, where g is the restriction of the quotient map $B + Y \longrightarrow B \cup_f Y$ to B.

14. If X, Y are Hausdorff spaces, and $f : X \to Y$ is a continuous map, prove that M_f and C_f are Hausdorff.

15. Show that $X * Y$ with the fine topology is homeomorphic to the space obtained by attaching $X \times Y \times I$ to the topological sum $X + Y$ by the mapping $\phi : X \times Y \times \{0, 1\} \longrightarrow X + Y$ given by $\phi(x, y, 0) = x$ and $\phi(x, y, 1) = y$, for all $x \in X$, $y \in Y$.

6.6 Induced and Coinduced Topologies

In this section, we first consider a straightforward generalization of the product topology, and then dualize this construction to introduce the concept of "coinduced topology" on a set X determined by functions from a collection of topological spaces into X. Moreover, we shall study the topology of a space generated by its certain subspaces; this makes precise the intuitive idea of "pasting" spaces together along prespecified subsets.

Induced Topologies

Definition 6.6.1 Let X be a set and Y_α, $\alpha \in A$, be a family of spaces with a function $f_\alpha : X \to Y_\alpha$ for each α. The *topology induced on* X by the functions f_α is the smallest topology on X such that each f_α is continuous.

Some authors use the term "initial topology" or "weak topology" to mean what we call induced topology.

Notice that a subbasis for the induced topology consists of the sets $f_\alpha^{-1}(U_\alpha)$, where U_α runs through the open subsets of Y_α and α ranges over A. Of course, the subbase for the topology induced on X by a single function $f : X \to Y$ (Y a topological space) itself turns out to be a topology and therefore, in this case, the induced topology on X consists of precisely the inverse images of the open subsets of Y under f. In particular, if $X \subseteq Y$, then the topology on X induced by the inclusion map $X \hookrightarrow Y$ is simply the relative topology. More generally, if $f : X \to Y$ is an embedding, then X has the topology induced by f, and conversely, if $f : X \to Y$ is an injection (or a bijection), then the induced topology on X makes f an embedding (or a homeomorphism). As another instance, we see that the product topology on the cartesian product $X = \prod Y_\alpha$ of the spaces Y_α is simply the topology induced on X

by the projections $p_\alpha \colon X \to Y_\alpha$. Thus, we see that the notion of induced topology unifies the study of relative topology and product topology.

Observe that the relative topology on the product $\prod X_\alpha$ of subspaces $X_\alpha \subseteq Y_\alpha$ agrees with the topology induced by the projections $\prod X_\alpha \to X_\beta$. This is essentially Proposition 2.2.9. The proof of the following theorem goes through entirely as that of Theorem 2.2.10.

Theorem 6.6.2 *If X has the topology induced by a collection of functions $f_\alpha \colon X \to Y_\alpha$, $\alpha \in A$, then a function g of a space Z into X is continuous $\Leftrightarrow f_\alpha \circ g \colon Z \to Y_\alpha$ is continuous for every α. Moreover, this property characterizes the induced topology for X.*

There is a simple criterion to decide whether the topology of a space is induced by a given family of continuous functions on the space.

Proposition 6.6.3 *Let $f_\alpha \colon X \to Y_\alpha$, $\alpha \in A$, be a family of continuous functions satisfying the condition that for each point $x \in X$ and each closed set $F \subset X$ not containing x, there is an $\alpha \in A$ such that $f_\alpha(x) \notin \overline{f_\alpha(F)}$. Then the topology of X coincides with the topology induced by the functions f_α.*

Proof Obviously, $f_\alpha^{-1}(V)$ is open in X for every open set $V \subseteq Y_\alpha$, and this is true for all α. Consequently, the topology $\mathcal{T}'$ induced on X by the functions f_α is coarser than the topology of X. Conversely, suppose that $U \subseteq X$ is open and $x \in U$. Then $F = X - U$ is close and $x \notin F$. By our hypothesis, $f_\alpha(x) \notin \overline{f_\alpha(F)}$ for some α. Therefore, there exists an open set $V \subseteq Y_\alpha$ such that $f_\alpha(x) \in V$ and $V \cap f_\alpha(F) = \varnothing$. Hence, $x \in f_\alpha^{-1}(V) \subseteq U$, and it follows that the topology of X coarser than $\mathcal{T}'$. Thus, the two topologies for X are the same. $\diamond$

A family of functions $f_\alpha \colon X \to Y_\alpha$, $\alpha \in A$, from a fixed space X to the various spaces Y_α is said to *separate points and closed sets* if for each closed set $F \subset X$ and each point $x \in X - F$, there is an index α such that $f_\alpha(x) \notin \overline{f_\alpha(F)}$.

With this terminology, Proposition 6.6.3 can be rephrased as: A space X has the topology induced by a family of continuous functions $f_\alpha \colon X \to Y_\alpha$ if the functions f_α separate points and closed sets.

The converse of this proposition is not true.

Example 6.6.1 Let $p_1, p_2 \colon \mathbb{R}^2 \to \mathbb{R}^1$ be the projection maps. The usual topology of $\mathbb{R}^2$ is induced by the family $\Phi = \{p_1, p_2\}$. Obviously, Φ does not separate the point $0 = (0, 0)$ and the closed set $F = \{(x, y) \mid xy = 1\}$.

Coinduced Topologies

In this subsection, we generalize the concept of identification topology. Let X_α, $\alpha \in A$, be a family of spaces and Y be a set. Given functions $f_\alpha \colon X_\alpha \to Y$, one for each α, we wish to find a topology for Y such that each f_α is continuous. Obviously, the trivial topology on Y satisfies this requirement, but it is not interesting. If $\mathcal{T}$ is any topology on Y which makes each f_α continuous, then $f_\alpha^{-1}(U)$ is open in X_α for all $U \in \mathscr{T}$ and $\alpha \in A$. As we would like, the family

$$\mathscr{U} = \left\{ U \subseteq Y \mid f_\alpha^{-1}(U) \text{ is open in } X_\alpha \text{ for each } \alpha \in A \right\}$$

is a topology on Y having the desired property. Obviously, $\mathscr{U}$ is finer than any topology $\mathscr{T}$ for Y with this property. Thus, $\mathscr{U}$ is the largest topology on Y such that each f_α is continuous.

Definition 6.6.4 Let X_α, $\alpha \in A$, be a family of spaces and Y be a set. Given functions $f_\alpha \colon X_\alpha \to Y$, one for each α, the largest (or finest) topology on Y which makes each f_α continuous is called the *topology coinduced* by the collection $\{f_\alpha\}$. This is also referred to as the *final topology* with respect to $\{f_\alpha\}$.

Since $f_\alpha^{-1}(Y - U) = X_\alpha - f_\alpha^{-1}(U)$, it is clear that a set $F \subseteq Y$ is closed in the topology coinduced by $\{f_\alpha\}$ if and only if $f_\alpha^{-1}(F)$ is closed in X_α for every $\alpha \in A$.

It is obvious that the topology coinduced by a function f of a space X onto a set Y is the identification topology. Observe also that the topology of the sum of a family of spaces X_α is coinduced by the injections $i_\beta \colon X_\beta \to \sum X_\alpha$. As another instance, we see that the topology of the adjunction space $X \cup_f Y$ is coinduced by the canonical mappings $\rho_x \colon X \to X \cup_f Y$ and $\rho_y \colon Y \to X \cup_f Y$.

The following theorem characterizes the topology on Y coinduced by the functions $f_\alpha \colon X_\alpha \to Y$, and is dual of Theorem 6.6.2.

Theorem 6.6.5 *Let X_α, $\alpha \in A$, be a family of spaces for each index α. Then the topology of a space Y is coinduced by the functions $f_\alpha \colon X_\alpha \to Y$ if and only if, for any space Z, a function $g \colon Y \to Z$ is continuous $\Leftrightarrow g \circ f_\alpha \colon X_\alpha \to Z$ is continuous for every $\alpha \in A$.*

Proof Suppose that Y has the topology coinduced by $\{f_\alpha\}$. If g is continuous, then $g \circ f_\alpha$ is continuous for every α, since each f_α is continuous. Conversely, if $g \circ f_\alpha$ is continuous for all α, then $f_\alpha^{-1}\left(g^{-1}(U)\right) = (g \circ f_\alpha)^{-1}(U)$ is open in X_α for every open $U \subseteq Z$. So $g^{-1}(U)$ is open in Y and g is continuous. This proves the necessity of the condition.

To prove the sufficiency, suppose that the space Y satisfies the given condition. Then each $f_\alpha \colon X_\alpha \to Y$ is continuous, since the identity map on Y is continuous. If Y' is the set Y together with the topology coinduced by $\{f_\alpha\}$, then the functions $f_\alpha \colon X_\alpha \to Y'$ are continuous. Thus, for the identity function $i \colon Y \to Y'$, each composition $X_\alpha \xrightarrow{f_\alpha} Y \xrightarrow{i} Y'$ is continuous. So i is continuous, by our hypothesis. Since the topology of Y' is coinduced by $\{f_\alpha\}$ and, as seen above, each composition $X_\alpha \xrightarrow{f_\alpha} Y' \xrightarrow{i^{-1}} Y$ is continuous, i^{-1} is continuous. It follows that $i \colon Y \to Y'$ is a homeomorphism, and the spaces Y and Y' have the same topology. $\diamond$

Coherent Topologies

The notion of coinduced topology is especially useful when all the spaces X_α are subsets of (a set) Y. Assume that the set Y is already given a topology, and each X_α

is a subspace of Y. Consider the topology on Y coinduced by the inclusion maps $X_\alpha \hookrightarrow Y$. Then the sets open (or closed) in the original topology of Y remain open (or closed) in the new topology. So the new topology on Y is, in general, finer than the given topology of Y. We obtain many interesting results for Y when its original topology coincides with the new one. In this situation, the topology of Y is described by using the adjective "coherent" instead of "coinduced".

Definition 6.6.6 Let Y be a space and $\{X_\alpha\}$, $\alpha \in A$, be a collection of subspaces of Y. The topology of Y is said to be *coherent* with the collection $\{X_\alpha \mid \alpha \in A\}$ if it is coinduced from the subspaces X_α by the inclusion maps $X_\alpha \hookrightarrow Y$.

The coherent topology for Y relative to the collection $\{X_\alpha\}$ is also called the *weak topology*. (We remark that this term is also used in the literature to mean something quite different). Of course, there is only one topology on Y coherent with a given collection $\{X_\alpha\}$ of subspaces $X_\alpha \subseteq Y$, and this is the largest topology on Y which induces the initial topology back on each X_α. Thus, the topology of a space Y coherent with a given family of subsets of Y preserves the predetermined topologies of the members. We can also describe the open (closed) sets of a topology coherent with a collection of spaces $\{X_\alpha\}$ directly in terms of the open (closed) sets of X_α.

Proposition 6.6.7 *A necessary and sufficient condition that a space Y has a topology coherent with a collection of its subspaces X_α is that a set $U \subseteq Y$ be open (or closed) if and only if $U \cap X_\alpha$ is open (or closed) in X_α for every index α.*

The simple proof of the proposition is left to the reader.

From the preceding proposition, it is clear that if a space Y is the union of a collection $\{U_\alpha\}$ of open sets, then its topology is coherent with $\{U_\alpha\}$. As another case, suppose that a space Y is the union of a locally finite collection $\{F_\alpha\}$ of closed sets. Then the topology of Y is coherent with $\{F_\alpha\}$, by Proposition 2.1.9. We also see that a k-space is the coherent union of its compact sets.

Example 6.6.2 The canonical embedding $\mathbb{R}^n \to \mathbb{R}^{n+1}$, $(x_1, \ldots, x_n) \mapsto (x_1, \ldots, x_n, 0)$ permits us to identify $\mathbb{R}^n$ with the subspace of $\mathbb{R}^{n+1}$ consisting of all points having their last coordinate zero. Thus, we have inclusions $\mathbb{R}^1 \subset \mathbb{R}^2 \subset \cdots$. Set $\mathbb{R}^\infty = \bigcup_{n \geq 1} \mathbb{R}^n$, and topologize this set by the metric $d(x, y) = \max\{|x_n - y_n| : n \in \mathbb{N}\}$. The metric space $\mathbb{R}^\infty$ is called a *generalized Euclidean space*, and contains $\mathbb{R}^n$ as a subspace. The topology coherent with the subspaces $\mathbb{R}^n$ is called the *inductive* topology for $\mathbb{R}^\infty$. This topology is distinct from the metric topology of $\mathbb{R}^\infty$ because the function $f: \mathbb{R}^\infty \to \mathbb{R}^1$ defined by $f(x) = \sum_1^\infty n x_n$ is continuous in the inductive topology, but discontinuous in the metric topology.

If the topology of a space Y is coherent with a collection of subspaces X_α, and $Y = \bigcup X_\alpha$, then we often say that Y is a *coherent union* of the X_α. In this case, intuitively speaking, we can think of the space Y as obtained by the "pasting on" of the subspaces X_α. This is evident from the following.

Proposition 6.6.8 *Let Y be a space and X_α, $\alpha \in A$, be a family of subspaces of Y with $Y = \bigcup X_\alpha$. Then the topology of Y is coherent with the collection $\{X_\alpha\}$ if and only if the mapping $f : \sum X_\alpha \to Y$, $(x, \alpha) \mapsto x$, is an identification map. In this case, Y is homeomorphic to the decomposition space of f.*

Proof Obviously, f is surjective. The composition of f with each injection i_β: $X_\beta \to \sum X_\alpha, x \mapsto (x, \beta)$, is the inclusion map $X_\beta \subseteq Y$. Since the topology of $\sum X_\alpha$ is coinduced by the injections i_β, $\beta \in A$, f is continuous, by Theorem 6.6.5. For any set $U \subseteq Y$, we obtain $f^{-1}(U) \cap (X_\beta \times \{\beta\}) = i_\beta (U \cap X_\beta)$. Since i_β is a homeomorphism between X_β and $X_\beta \times \{\beta\}$, $f^{-1}(U)$ is open in $\sum X_\alpha$ if and only if $U \cap X_\beta$ is open in X_β for every $\beta \in A$. It follows that if $f^{-1}(U)$ is open, and Y has a topology coherent with $\{X_\alpha\}$, then U is open in Y. So f is an identification. Conversely, if f is an identification, and $U \cap X_\alpha$ is open in X_α for every α, then U is open in Y, and Y has a topology coherent with $\{X_\alpha\}$.

The last statement follows from Theorem 6.2.5. $\diamond$

An important result about the continuity of functions on the spaces with coherent topologies is given by the following proposition.

Proposition 6.6.9 *Suppose that a space Y has the topology coherent with $\{X_\alpha\}$. Then, for any space Z, a function $f : Y \to Z$ is continuous if and only if each restriction $f|X_\alpha$ is continuous.*

Proof This is clear from the definition of the topology on Y and Theorem 6.6.5. $\diamond$

It should be noted that if $\{X_\alpha\}$ is a family of spaces with each $X_\alpha \subseteq Y$, then there is in general no topology on Y such that each X_α is a subspace of Y (see Exercise 16). However, there is a simple condition the family $\{X_\alpha\}$ which guarantees the existence of a topology on Y such that it induces the preassigned topology back on each X_α.

Theorem 6.6.10 *Let $\{X_\alpha\}$ be a family of subsets of a set Y. Suppose that each X_α has a topology such that (a) the relative topologies on $X_\alpha \cap X_\beta$ induced from X_α and X_β agree, and (b) $X_\alpha \cap X_\beta$ is closed (or open) in X_α and in X_β for each pair of indices α and β. Then there is a topology on Y in which each X_α is a closed (resp. open) subspace of Y and which is coherent with $\{X_\alpha\}$.*

Proof We show that the topology on Y coinduced by the inclusion maps $X_\alpha \hookrightarrow Y$ has the required properties. Clearly, it suffices to establish the first statement, that is, each X_α, as a subspace of Y, retains its original topology and is closed in Y. By definition of the coinduced topology, $F \cap X_\alpha$ is closed in X_α for every closed set $F \subseteq Y$. Conversely, suppose that $E \subseteq X_\alpha$ is closed. Then, for any index β, $E \cap X_\beta = E \cap X_\alpha \cap X_\beta$ is closed in $X_\alpha \cap X_\beta$. By our hypothesis, $X_\alpha \cap X_\beta$ is a closed subspace of X_β. So $E \cap X_\beta$ is closed in X_β, and therefore E is closed in Y. It follows that the relative topology on X_α induced from Y coincides with its original topology. Also, taking $E = X_\alpha$, we see that X_α itself is closed in Y.

The theorem in the other case is proved just by replacing the term "closed" with "open" in the above proof. $\diamond$

Exercises

1. Let $f : X \to Y$ be a mapping of a set X onto a space Y. If X has the topology induced by f, prove that f is closed and open.

2. Determine the topology induced by the function $f : \mathbb{R} \to \mathbb{R}$, where

   (a) $f(x) = 0$ if $x \le 0$, and $f(x) = 1$ if $x > 0$,

   (b) $f(x) = -1$ if $x < 0$, $f(0) = 0$, and $f(x) = 1$ if $x > 0$.

3. Let $\mathscr{F}$ be a collection of real-valued functions on $\mathbb{R}$. Describe the topology on $\mathbb{R}$ induced by $\mathscr{F}$, if

   (a) $\mathscr{F}$ consists of all constant functions,

   (b) $\mathscr{F}$ consists of all functions which are continuous in the usual topology on $\mathbb{R}$, and

   (c) $\mathscr{F}$ consists of all bounded functions which are continuous in the usual topology for $\mathbb{R}$.

4. Determine the topology on X such that every real-valued function on X is continuous.

5. Let $\mathscr{F}$ be a family of real-valued functions on a set X. Show that a basis for the topology on X induced by $\mathscr{F}$ is given by the sets of the form $\{y \in X : \exists\, x \in X$, a real $\epsilon > 0$ and finitely many functions $f_1, \ldots, f_n$ in $\mathscr{F}$ such that $|f_i(y) - f_i(x)| < \epsilon\}$.

6. Let $\{X_\alpha\}$ be a family of topological spaces, all with the same underlying set X. Describe the topology on X induced by the identity functions $i_\alpha : X \to X_\alpha$. Show that X (with the induced topology) is homeomorphic to the subspace $\{x \in \prod X_\alpha \mid x_\alpha = x_\beta$ for all indices $\alpha, \beta\}$ of the product space $\prod X_\alpha$.

7. • Suppose that a space X has the topology induced by a collection of functions $f_\alpha : X \to Y_\alpha, \alpha \in A$. If $Z \subseteq X$, show that the relative topology on Z is induced by the restrictions of the f_α to Z.

8. Let X be a set and Y_α, $\alpha \in A$, be a family of spaces. Given functions $f_\alpha : X \to Y_\alpha$ for every α, let $f : X \to \prod Y_\alpha$ be defined by $f(x) = (f_\alpha(x))$. Prove:

   (a) The topologies induced by f and the family $\{f_\alpha\}$ coincide.

   (b) If for each pair of distinct points x and x' in X, there is an index $\alpha \in A$ such that $f_\alpha(x) \ne f_\alpha(x')$, then f is an embedding.

9. Let $\mathscr{F}$ be a family of real-valued functions on a set X, and assign X the topology induced by $\mathscr{F}$. Show that X is $T_2 \Leftrightarrow$ for each pair of distinct points x and y in X, there exists an $f \in \mathscr{F}$ with $f(x) \ne f(y)$.

10. Let Y have the topology coinduced by $f : X \to Y$. Prove that $f : X \to f(X)$ is an identification map.

11. Show that the topology on $\mathbb{R}$ coinduced by the inclusion maps $i : \mathbb{Q} \hookrightarrow \mathbb{R}$ and $j : \mathbb{R} - \mathbb{Q} \hookrightarrow \mathbb{R}$, where $\mathbb{Q}$ and $\mathbb{R} - \mathbb{Q}$ have their usual topologies, is strictly finer than the usual topology of $\mathbb{R}$.

12. Let $\{X_\alpha\}$ be a family of spaces each having the same underlying set Y. What is the topology coinduced by the identity functions $X_\alpha \to Y$?

13. Suppose that $\{X_\alpha\}$ is a family of subspaces of a space $(Y, \mathcal{U})$, and $Z \subseteq Y$. Let $\mathcal{T}$ be the topology on Y coinduced by the inclusions $X_\alpha \hookrightarrow Y$, and let $\mathcal{T}^Z$ be the topology on Z coinduced by the inclusions $Z \cap X_\alpha \hookrightarrow Z$. Show:

    (a) The identity map from $(Z, \mathcal{T}^Z)$ into $(Z, \mathcal{T}_Z)$ is continuous.

    (b) If Z and each X_α are closed in Y, then $\mathcal{T}^Z = \mathcal{T}_Z$, and Z is a closed subspace of $(Y, \mathcal{T})$.

14. Let $X_\alpha, \alpha \in A$, be a family of spaces and Y be a set. Given a function $f_\alpha : X_\alpha \to Y$ for each $\alpha \in A$, there is a function $f : \sum X_\alpha \to Y$ defined by $f((x, \alpha)) = f_\alpha(x)$, $x \in X_\alpha$. Prove that the topology on Y coinduced by the functions f_α coincides with the one coinduced by f.

15. Prove that coinduced topologies are transitive: Let $Y_\lambda, \lambda \in \Lambda$, be a family of spaces and suppose that, for each $\lambda \in \Lambda$, the topology of Y_λ is coinduced by the functions $f_{\lambda,\alpha} : X_{\lambda,\alpha} \to Y_\lambda$, $\alpha \in A_\lambda$. Then, given a set Z and a family of functions $g_\lambda : Y_\lambda \to Z$, $\lambda \in \Lambda$, the topology on Z coinduced by the functions g_λ coincides with the topology coinduced by the compositions $g_\lambda \circ f_{\lambda,\alpha}$.

16. ● Give an example to show that without condition (b) in Theorem 6.6.10, there may not be a topology on Y such that each X_α, as a subspace of Y, retains its original topology.

17. Let Y be the coherent union of the subspaces $\{X_\alpha\}$. Show that if Z is a closed (or open) subspace of Y, then Z is the coherent union of its subspaces $\{Z \cap X_\alpha\}$.

18. Suppose that Y and Y' have topologies coherent with the collections $\{X_\alpha\}$ and $\{X'_\beta\}$, respectively. Let $\mathcal{T}$ be the topology on $Y \times Y'$ coinduced by the inclusion $X_\alpha \times X'_\beta \hookrightarrow Y \times Y'$ and $\mathcal{P}$ be the product topology on $Y \times Y'$. Show:

    (a) The identity map $i : (Y \times Y', \mathcal{T}) \to (Y \times Y', \mathcal{P})$ is continuous.

    (b) If $Y = \bigcup X_\alpha$, $Y' = \bigcup X'_\beta$, where each X_α is open in Y and each X'_β is open in Y', then the mapping ι in (a) is a homeomorphism.

    (c) The conclusion in (b) holds if each $y \in Y$ lies in the interior of some X_α, and each $y' \in Y'$ lies in the interior of some X'_β.

    (d) The conclusion in (b) also holds if $\{X_\alpha\}$ and $\{X'_\beta\}$ are locally finite collections of closed subsets of Y and Y', respectively, such that $Y = \bigcup X_\alpha$, $Y' = \bigcup X'_\beta$.

19. Suppose that a space X is a coherent union of the subspaces X_α. If each X_α is a T_1-space, prove that X is T_1.

20. If a space X is a coherent union of the subspaces X_α, and Y is a locally compact Hausdorff, prove that the topology of $X \times Y$ is coherent with the subspaces $X_\alpha \times Y$.

21. Let X be a T_2-space. Show that there exists a k-space Y and a continuous bijection $f : Y \to X$ such that f is a homeomorphism on every compact subset of Y and X.

Chapter 7
Countability Axioms

In Sect. 5.2, we have seen that the notions of countable compactness and sequential compactness coincide for topological spaces which have countable local bases. A more important but restricted class of spaces consists of the ones which have countable bases. The first section concerns with such spaces. There are two other properties, namely, separability and Lindelöfness which are described by using the notion of countability, and each of them guarantees the existence of countable bases in metrizable spaces. These are treated in the second section.

7.1 First and Second Countable Spaces

We have already introduced the first axiom of countability: A topological space is first countable if it has a countable local basis at each of its points. Clearly, any space which admits a countable open basis is first countable.

Definition 7.1.1 A topological space is said to satisfy the *second axiom of countability* (or be *second countable*) if its topology has a countable basis.

Since the collection of finite intersections of members of a countable family $\mathscr{S}$ of subsets of a set S is countable (ref. Theorem A.5.12), a space is second countable if it has a countable subbase.

Example 7.1.1 The open intervals with rational end points form a countable basis for the real line $\mathbb{R}$. More generally, the Euclidean space $\mathbb{R}^n$ is second countable for every $n \geq 1$. The cubes $(a_1, b_1) \times \cdots \times (a_n, b_n)$, where a_i and b_i are rationals for all i, form a countable base.

Example 7.1.2 As seen in Example 5.2.4, every metric space is a first countable space. But an uncountable set X with the discrete metric, $[d(x, y) = 1$ if $x \neq y$, and $d(x, x) = 0]$, is not second countable, since any basis of a discrete space must contain all one-point sets.

© Springer Nature Singapore Pte Ltd. 2019
T. B. Singh, *Introduction to Topology*,
https://doi.org/10.1007/978-981-13-6954-4_7

Example 7.1.3 The Sorgenfrey line $\mathbb{R}_\ell$ satisfies the first axiom of countability, since the intervals $[x, x + 1/n), n = 1, 2, \ldots$, form a nbd basis at x, for each $x \in \mathbb{R}$. But it is not second countable. Given a basis $\mathscr{B}$ of $\mathbb{R}_\ell$, choose $B_x \in \mathscr{B}$ for each $x \in \mathbb{R}$ such that $x \in B_x \subseteq [x, x + 1)$. Then $x \mapsto B_x$ is an injective mapping $\mathbb{R} \to \mathscr{B}$; consequently, $\mathscr{B}$ is uncountable. Thus, $\mathbb{R}_\ell$ fails to satisfy the second axiom of countability.

A useful fact about first countable spaces is that the theory of sequential convergence is adequate to express most topological concepts in these spaces. This can be deduced from the following three results.

Theorem 7.1.2 *Let X be a first countable space and $A \subseteq X$. Then $x \in \overline{A} \Leftrightarrow$ there is sequence in A which converges to x.*

Proof Suppose that $x \in \overline{A}$. Since X satisfies the first axiom of countability, there is a nbd basis $\{U_n \mid n = 1, 2, \ldots\}$ at x such that $U_1 \supseteq U_2 \supseteq \ldots$. Since $x \in \overline{A}$, we can select a point $a_n \in U_n \cap A$ for every n. Then $a_n \to x$. Conversely, if there is a sequence in A converging to x, then every open nbd of x intersects A, and $x \in \overline{A}$, by Theorem 1.3.5. $\Diamond$

Theorem 7.1.3 *Let X be a first countable space and Y be any space. A function $f: X \to Y$ is continuous $\Leftrightarrow x_n \to x$ in X implies $f(x_n) \to f(x)$.*

Proof Suppose that $f: X \to Y$ is continuous, and $x_n \to x$ in X. For any open nbd U of $f(x)$, $f^{-1}(U)$ is an open nbd of x. So there is an integer n_0 such that $x_n \in f^{-1}(U)$ for all $n \geq n_0$. This implies that $f(x_n) \in U$ for all $n \geq n_0$, and $f(x_n) \to f(x)$.

Conversely, assume that $f(x_n) \to f(x)$ in Y whenever $x_n \to x$ in X. We show that f is continuous by showing that $f\left(\overline{A}\right) \subseteq \overline{f(A)}$ for every $A \subseteq X$. If $x \in \overline{A}$, then there is a sequence $\langle a_n \rangle$ in A which converges to x, by Theorem 7.1.2. Our hypothesis implies that $f(a_n) \to f(x)$, and hence $f(x) \in \overline{f(A)}$. This completes the proof. $\Diamond$

Theorem 7.1.4 *A first countable space X is Hausdorff if and only if each convergent sequence in X has a unique limit.*

Proof The necessity is clear, for a sequence cannot be eventually in each of two disjoint sets. To prove the sufficiency, assume that the Hausdorff condition fails for two points x and y in X. Let $\{U_n\}$ and $\{V_n\}$ be monotonically decreasing countable nbd bases at x and y, respectively. By our assumption, $U_n \cap V_n \neq \varnothing$ for every $n = 1, 2, \ldots$. So we can choose a point z_n in $U_n \cap V_n$ for every n. Clearly, the sequence $\langle z_n \rangle$ converges to both x and y. $\Diamond$

It should be noted that there are spaces (for example, the cocountable space $\mathbb{R}_c$) which are not Hausdorff, though every convergent sequence in those spaces has a unique limit.

Both the properties are topological invariants; in fact, they are preserved by continuous open mappings.

Proposition 7.1.5 *The continuous open image of a first countable (resp. second countable) space is first countable (resp. second countable).*

The proofs are simple and we leave these to the reader.

We remark that this proposition does not hold good for continuous closed maps, as shown by the following.

Example 7.1.4 Consider the real line $\mathbb{R}$ with the subspace $\mathbb{Z}$ of all integers pinched to a point. Let $\{U_n\}$ be a countable family of open nbds of $[\mathbb{Z}]$ in the quotient space $Y = \mathbb{R}/\mathbb{Z}$. Then each set $\pi^{-1}(U_n)$ is a nbd of $\mathbb{Z}$, where $\pi : \mathbb{R} \to Y$ is the quotient map. So, for each integer n, we can find a real r_n such that $1/2 > r_n > 0$ and $(n - r_n, n + r_n) \subseteq \pi^{-1}(U_n)$. The image of the open set $\bigcup (n - r_n/2, n + r_n/2)$ under π is an open nbd G of $[\mathbb{Z}]$, and it is clear that G contains no U_n. Therefore, Y does not satisfy the first axiom of countability.

This example also shows that quotients of a first countable or a second countable space need not be so. However, we have the following proposition.

Proposition 7.1.6 *Let $f : X \to Y$ be a proper surjection. If X is a second countable, then so is Y.*

Proof Let $\{B_n\}$ be a countable basis for X. Then the family $\mathscr{G}$ of all finite unions of B_n is countable. So $\mathscr{G}$ can be indexed by the positive integers; let $\mathscr{G} = \{G_n\}$. Since f is closed, $U_n = Y - f(X - G_n)$ is open for every n. We observe that $\{U_n\}$ is a countable basis for Y. Let $V \subseteq Y$ be open, and $y \in V$ be arbitrary. Then $f^{-1}(V)$ is open in X and contains $f^{-1}(y)$. For each $x \in f^{-1}(y)$, there is an integer n such that $x \in B_n \subseteq f^{-1}(V)$. Varying x over $f^{-1}(y)$, we obtain a covering of $f^{-1}(y)$ consisting of basis elements each of which is contained in $f^{-1}(V)$. By Theorem 5.5.2, $f^{-1}(y)$ is compact, so this open cover reduces to a finite subcover. Consequently, we find a set G_n in $\mathscr{G}$ such that $f^{-1}(y) \subseteq G_n \subseteq f^{-1}(V)$. Then $X - f^{-1}(V) \subseteq X - G_n \subseteq X - f^{-1}(y) = f^{-1}(Y - y)$, which implies that $y \in U_n \subseteq f(G_n) \subseteq V$. Thus, V is a union of sets in $\{U_n\}$ and the proposition holds.   ◇

From Proposition 1.5.5, it is clear that both properties of the first countability and the second countability are inherited by subspaces in general. However, neither of them is invariant under the formation of arbitrary products, as shown by the following.

Example 7.1.5 Consider a discrete space X having more than one point. Let A be an uncountable set (e.g., $A = [0, 1]$) and, for each $\alpha \in A$, let X_α denote the space X. Obviously, X_α is first countable for each $\alpha \in A$. Assume that $\{B_n\}$ is a countable local basis at a point $\xi \in \prod X_\alpha$. Then, for each integer n, there is a finite set $F_n \subset A$ such that $p_\beta(B_n) = X$ for every $\beta \notin F_n$, where p_β is the βth projection of $\prod X_\alpha$. As A is uncountable, we can choose a $\beta \in A$ such that $\beta \notin F_n$ for all n. Then $U = p_\beta^{-1}(\xi_\beta)$ is a nbd of ξ, which contains no B_n. This contradicts our assumption, and thus $\prod X_\alpha$ is not first countable.

The above example also shows that an uncountable product of nontrivial metric spaces is not metrizable, for every metric space is first countable. The next result gives a necessary and sufficient condition for the product space to be the first (or the second) countable.

Theorem 7.1.7 *A product of topological spaces is first countable (resp. second countable) if and only if each factor is first countable (resp. second countable) and all but a countable number of the spaces are trivial.*

Proof Let X_α, $\alpha \in A$, be a family of spaces such that $\prod X_\alpha$ is first countable. Since the projections $p_\beta : \prod X_\alpha \to X_\beta$ are continuous open surjections, each X_α is first countable, by Proposition 7.1.5. Let $B = \{\alpha \in A \mid X_\alpha$ is a nontrivial space$\}$. For each $\beta \in B$, we choose a point $c_\beta \in X_\beta$ which has a proper nbd in X_β and consider a point $x \in \prod X_\alpha$ with $x_\beta = c_\beta$. Let $\mathscr{M}$ be a countable nbd basis at x. By the definition of the product topology, for every nbd N of x, $p_\alpha(N) = X_\alpha$ for all but finitely indices α. So, for each $M \in \mathscr{M}$, there exists a finite set $A_M \subset A$ such that $p_\alpha(M) = X_\alpha$ for all $\alpha \in A - A_M$. Since $\mathscr{M}$ is countable, $\Gamma = \bigcup \{A_M \mid M \in \mathscr{M}\}$ is also countable. We assert that $B \subseteq \Gamma$ which implies that B is countable. By the choice of x, for each index $\beta \in B$, there exists a proper open subset $U_\beta \subset X_\beta$ such that $p_\beta(x) \in U_\beta$. Then there is a member M in $\mathscr{M}$ such that $p_\beta(M) \subseteq U_\beta$, and $\beta \in \Gamma$. This proves our assertion.

Conversely, let X_α, $\alpha \in A$, be a family of first countable spaces, and B be a countable subset of A such that X_α is trivial for every $\alpha \in A - B$. Given a point $x = (x_\alpha) \in \prod X_\alpha$, let $\mathscr{M}_\alpha$ be a countable local base at x_α. Note that $\mathscr{M}_\alpha = \{X_\alpha\}$ for every $\alpha \in A - B$. Let $\mathscr{V}$ denote the family

$$\{p_\beta^{-1}(M) \mid M \in \mathscr{M}_\beta,\ \beta \in B\}.$$

Then $\mathscr{V}$ is clearly a countable family of nbds of x, and so is the family $\mathscr{N}$ of all finite intersections of members of $\mathscr{V}$. We claim that $\mathscr{N}$ is a local base at x. For a proper nbd N of x, we have $p_\alpha(N) = X_\alpha$ for all but finitely many indices $\alpha_1, \ldots \alpha_n$, say. Then the indices $\alpha_1, \ldots, \alpha_n$ are in B and, for each $i = 1, \ldots, n$, there exists an open subset $U_i \subset X_{\alpha_i}$ such that $x \in \bigcap_1^n p_{\alpha_i}^{-1}(U_i) \subseteq N$. Now, we find a set $M_i \in \mathscr{M}_{\alpha_i}$ such that $M_i \subseteq U_i$ for every i. Clearly, we have $\bigcap_1^n p_{\alpha_i}^{-1}(M_i) \subseteq N$, and hence our claim.

In the case of second countability, the proof of the necessity is quite similar to the above case. In fact, if $\prod X_\alpha$ is second countable, then each X_α is second countable, by Proposition 7.1.5 again. Moreover, there are at most countably many nontrivial factors, since $\prod X_\alpha$ is first countable. Conversely, for each $\alpha \in A$, let $\mathscr{U}_\alpha$ be a countable base of X_α. Then, by our hypothesis, there is a countable set $B \subseteq A$ such that $\mathscr{U}_\alpha = \{X_\alpha\}$ for all $\alpha \notin B$. It is not difficult to show that the family $\{p_\beta^{-1}(U) \mid U \in \mathscr{U}_\beta,\ \beta \in B\}$ is a countable subbase for the topology on $\prod X_\alpha$, and hence it is second countable. $\diamond$

Corollary 7.1.8 *If the product of a family of spaces $\{X_\alpha\}$ is metrizable, then each X_α is metrizable, and all but countably many of these are trivial.*

Proof Since X_β, $\beta \in A$, is homeomorphic to a subspace of $\prod X_\alpha$, it is metrizable. The second statement follows from the preceding theorem, since every metric space is first countable. ◇

For metrizable spaces, there are several properties which imply the axiom of second countability. As a first evidence, we have the following proposition.

Proposition 7.1.9 *Every compact metric space is second countable.*

Proof Let X be a compact metric space. Then, for each positive integer n, we find a finite set A_n in X such that the open balls $B(a; 1/n)$, $a \in A_n$, cover X. We observe that the family $\mathscr{B} = \{B(a; 1/n) \mid a \in A_n, n = 1, 2, \ldots\}$ is a base for X. Let G be a nonempty open set in X and $x \in G$ be arbitrary. Then there exists a real number $r > 0$ such that $B(x; r) \subseteq G$. Now, choose an integer $n > 2/r$. Then, by the definition of A_n, there exists an $a \in A_n$ such that $x \in B(a; 1/n)$. We clearly have $B(a; 1/n) \subseteq G$, proving our claim. Thus, $\mathscr{B}$ is a countable basis for X, and it is second countable. ◇

One might guess that the one-point compactification X^* of a locally compact metric space X is metrizable. This is not true unless X is second countable. For, if X^* were a compact metric space, then it would be second countable. So X, being a subspace of X^*, must be second countable. Conversely, it is true that X^* is metrizable if X is also second countable (see Exercise 8.2.28). To this end, the following result will come in useful.

Theorem 7.1.10 *The one-point compactification of a second countable locally compact Hausdorff space X is second countable.*

Proof By Theorem 5.4.2, there is a countable basis $\{U_n\}$ of X such that $\overline{U}_n$ is compact. Put $V_n = X^* - \bigcup_1^n \overline{U}_i$ for $n = 1, 2, \ldots$. We assert that $\{V_n\}$ is a countable local base at ∞, the point at infinity in X^*. Obviously, each V_n is an open nbd of ∞. If O is an open set in X^* containing ∞, then $X^* - O$ is a compact subset of X. As $\{U_n\}$ covers $X^* - O$, there are finitely many sets, say, $U_{n_1}, \ldots, U_{n_k}$ in $\{U_n\}$ such that $X^* - O \subseteq \bigcup_1^k U_{n_i}$. If $m = \max\{n_1, \ldots, n_k\}$, then $V_m \subseteq O$, and our assertion follows. Since X is open in X^*, the family $\{U_n\} \cup \{V_n\}$ is a countable basis for X^*. ◇

The converse of the above theorem is also true, for the property of being second countable is hereditary.

We close this section with the remark that an analogous result for the first countability is not true. For example, the ordinal space $[0, \Omega)$ is first countable, but its one-point compactification $[0, \Omega]$ is not (see Exercises 7.2.16 and 7.2.17).

Exercises

1. Show that none of the following spaces is first countable:

   (a) An uncountable space with the cocountable topology.

   (b) An uncountable Fort space (Exercise 1.2.3).

2. Let X be a first countable T_1-space. Show that every one-point set is a G_δ-set. What about the converse?

3. • Let X be a first countable space. Prove:

   (a) A point $x \in X$ is a cluster point of a sequence $\langle x_n \rangle$ in X if and only if $\langle x_n \rangle$ has a subsequence which converges to x.

   (b) For $A \subseteq X$, a point $x \in A'$ if and only if there is a sequence in $A - \{x\}$ which converges to x.

   (c) A subset F of X is closed if and only if the cluster points of each sequence in F belong to F.

   (d) A subset G of X is open if and only if each sequence which converges to a point of G is eventually in G. (A space with this property is called *sequential*.)

   (e) A subset G of X is open if and only if $G \cap A$ is open in A for every countable subset A of X.

4. Give an example of a sequential space which does not satisfy the axiom of first countability.

5. • Let X be the Euclidean plane and $A \subseteq X$ be the x-axis. Show that the projection map $X \to X/A$ is closed, but X/A is not first countable.

6. Prove: (a) A Hausdorff first countable space is a k-space.

   (b) A sequential T_2-space is a k-space.

7. If A is a countably compact subset of a first countable T_2-space, show that A is closed.

8. Let X be a countably compact space and Y be a first countable T_2-space. Show that a continuous bijection $f : X \to Y$ is a homeomorphism.

9. Let X be a countably compact T_2-space satisfying the first axiom of countability. Show that a sequence in X converges $\Leftrightarrow$ it has a single cluster point.

10. Let X be a first countable space and Y be a countably compact space. If $f : X \to Y$ is a function such that the mapping $X \to X \times Y$, $x \mapsto (x, f(x))$, is closed, show that f is continuous.

11. Prove that a cofinite space X is second countable if and only if X is countable.

12. Show that each of the following spaces is second countable:

    (a) The Hilbert space ℓ_2.

    (b) The set $\mathbb{R}$ with the Smirnov topology (see Exercise 1.4.6).

13. Show that $\mathbb{R}^\omega$ with the uniform metric topology is first countable, but not second countable.

14. Prove that a totally bounded (and therefore a countably compact) metric space is second countable. (For the converse, see Exercise 8.2.22.)

15. Is a locally compact metric space second countable?

16. Prove that every basis of a second countable space contains a countable subfamily which is also a basis.

17. Prove that a second countable, locally compact Hausdorff space has a countable basis consisting of open sets with compact closures.

18. If X is a second countable T_1-space, prove that $|X| \leq c$.

19. Let $\{X_\alpha\}, \alpha \in A$, be a collection of spaces. If all spaces X_α satisfy the first axiom of countability, show that the same holds for their sum $\sum X_\alpha$.

20. Let $\{X_\alpha\}, \alpha \in A$, be a collection of spaces. If the index set A is countable, and each X_α is second countable, prove that $\sum X_\alpha$ is also second countable.

21. Prove that the wedge of a finite family of first countable pointed spaces with closed base points is first countable. Give an example to show that the wedge of an infinite family of first (or second) countable spaces need not be so.

7.2 Separable and Lindelöf Spaces

There are two simple but useful properties of second countable spaces: first, they all contain countable dense sets; and second, open coverings of every one of them contain countable subcoverings. We intend to study here spaces with these properties.

Separability

Definition 7.2.1 A topological space is called *separable* if it has a countable dense subset.

Example 7.2.1 The Euclidean space $\mathbb{R}^n$ is separable, since the set of points $x \in \mathbb{R}^n$ with the coordinates x_i rational numbers is countable and dense.

Example 7.2.2 The space ℓ_2 is separable, since the set of all points (q_i) with q_i's rational numbers and having only a finite number of nonzero terms is countable and dense in ℓ_2.

Example 7.2.3 An uncountable space with the cocountable topology is not separable.

Clearly, every second countable space is separable, for a countable dense set can be obtained by taking one point from each (nonempty) member of a countable basis for the space. Example 5.2.5 shows that the converse is not true. Even a separable, first countable space need not be second countable as shown by the Sorgenfrey line $\mathbb{R}_\ell$. But there is a partial converse.

Proposition 7.2.2 *A separable metric space is second countable.*

Proof Let X be a separable metric space and $A \subseteq X$ be a countable dense set. It is easy to see that each open ball $B(x; r)$ in X contains an open ball $B(a; 1/n)$ containing x, where $a \in A$ and $n > 2/r$. Therefore, the family $\{B(a; 1/n) \mid a \in A, n \in \mathbb{N}\}$ is a countable basis for X.   ◇

Corollary 7.2.3 *A metric space is second countable if and only if it is separable.*

Separability is a topological invariant; in fact, it is preserved by continuous mappings. We restate the latter assertion as a proposition and leave the simple proof to the reader.

Proposition 7.2.4 *If f is a continuous mapping of a separable space X onto a space Y, then Y is separable.*

In particular, we see that a quotient space of a separable space is separable. As for subspaces, we see that an open subspace of a separable space is separable. For, if A is a countable dense subset of a space X and $Y \subseteq X$ is open, then $A \cap Y$ is a countable dense subset of the subspace Y. However, even a closed subspace of a separable space need not be separable.

Example 7.2.4 The Sorgenfrey plane $\mathbb{R}_\ell \times \mathbb{R}_\ell$ is separable, since its subset $\mathbb{Q} \times \mathbb{Q}$ is countable and dense. The antidiagonal subspace $L = \{(x, -x) \mid x \in \mathbb{R}\}$ is closed in $\mathbb{R}_\ell \times \mathbb{R}_\ell$, for the open set $[x, x + r) \times [y, y + r)$, where $y \neq -x$, and $r = -(x + y)/3$ if $y < -x$, does not meet L. Also, L is discrete, since $L \cap ([x, \infty) \times [-x, \infty))$ is exactly the point $(x, -x)$. Being uncountable and discrete, L cannot be separable.

In regard to the product of spaces, we content ourselves with the following, though a stronger statement holds good (see Dugundji [3], p. 175).

Proposition 7.2.5 *Let $X_n, n = 1, 2, \ldots$, be a family of separable spaces. Then $\prod X_n$ is separable.*

Proof For every n, let A_n be a countable dense subset of X_n and choose an (fixed) element $c_n \in A_n$. Then the set

$$S = \{(a_n) \in \prod A_n \mid a_n = c_n \text{ for all but finitely many indices } n\}$$

is clearly countable. We show that it is dense in $\prod X_n$. Consider a basic open subset B of $\prod X_n$. Then there exist finitely many open sets $U_{n_i} \subseteq X_{n_i}, i = 1, \ldots, k$ (say) such that $B = \bigcap_1^k p_{n_i}^{-1}(U_{n_i})$, where $p_{n_i} : \prod X_n \to X_{n_i}$ is the projection. Since A_{n_i} is dense in X_{n_i}, we can find an element $a_{n_i} \in U_{n_i} \cap A_{n_i}$ for every i. Put $x_n = c_n$ for $n \neq n_1, \ldots, n_k$ and $x_{n_i} = a_{n_i}$ for $i = 1, \ldots, k$. Then the point $x = (x_n)$ obviously lies in $B \cap S$, and it follows that every open set in $\prod X_n$ intersects S. Therefore, S is dense in $\prod X_n$.   ◇

Lindelöf Spaces

Theorem 7.2.6 (Lindelöf) *Every open covering of a second countable space X has a countable subcovering.*

Proof Let $\{B_n \mid n = 1, 2, \ldots\}$ be a countable basis for X. Given an open cover $\mathscr{G}$ of X, let M be the set of all those integers n for which B_n is contained in some member of $\mathscr{G}$. Then, for each $m \in M$, choose a member $G_m \in \mathscr{G}$ such that $B_m \subseteq G_m$. Obviously, the family $\{G_m \mid m \in M\}$ is countable. We observe that this subfamily of $\mathscr{G}$ also covers X. For each point $x \in X$, there exists a $G \in \mathscr{G}$ containing x. Since G is open in X, there is a basis element B_n such that $x \in B_n \subseteq G$. Then $n \in M$ and so $x \in G_n$. Thus, the family $\{G_m \mid m \in M\}$ covers X. $\diamond$

A topological space with the above property is referred to as Lindelöf. For the sake of emphasis, we restate this as follows.

Definition 7.2.7 A space X is Lindelöf if each open covering of X admits a countable subcovering.

It is obvious that a compact space is Lindelöf but it need not satisfy other countability axioms; this can be seen from the various examples given in Sect. 5.1. By the above theorem, every second countable space is Lindelöf. However, a Lindelöf space need not be second countable. For example, an uncountable space with the cocountable (or cofinite) topology is clearly Lindelöf but it is not second countable (see Example 5.2.5).

Clearly, a countably compact Lindelöf space is compact. For, given an open covering of a countably compact Lindelöf space, one can first find a countable subcovering, and then reduce it to a finite subcovering. It follows that the conditions of compactness and sequential compactness are equivalent to the B-W property for second countable T_1-spaces.

For metrizable spaces, we have the following.

Proposition 7.2.8 *A Lindelöf metric space is second countable.*

Proof Let X be a Lindelöf metric space. Then, for each integer $n > 0$, there is a countable set $A_n \subseteq X$ such that the open balls $B(a; 1/n), a \in A_n$, cover X. If $G \subseteq X$ is open and $x \in G$, then there exists an open ball $B(x; r) \subseteq G$. For each $n = 1, 2, \ldots$, $x \in B(a_x; 1/n)$ for some point $a_x \in A_n$. Choose an integer n such that $2 < nr$. Then we have $x \in B(a_x; 1/n) \subseteq G$. Therefore, the family $\{B(a; 1/n) \mid a \in A_n, n \in \mathbb{N}\}$ is a countable basis of X, and X is second countable. $\diamond$

It follows that a compact metrizable space satisfies all the countability axioms. Also, the conditions of separability and being Lindelöf coincide with second countability on *metric spaces*, although there is no general relationship between them.

Example 7.2.5 A separable space, which is not Lindelöf. It has been shown in Example 7.2.4 that $\mathbb{R}_\ell \times \mathbb{R}_\ell$ is separable and its subset $L = \{(x, -x) \mid x \in \mathbb{R}\}$ is closed. If $\mathbb{R}_\ell \times \mathbb{R}_\ell$ were Lindelöf, then the open covering which consists of $\mathbb{R}_\ell \times \mathbb{R}_\ell - L$ and the sets $[x, x + r) \times [-x, -x + r), x \in \mathbb{R}$, would have a countable subcovering. But this is impossible, since L is uncountable and each set $[x, x + r) \times [-x, -x + r)$ meets L in exactly one point, viz., $(x, -x)$.

Example 7.2.6 A Lindelöf space, which is not separable. Consider the ordinal space $[0, \Omega]$. Given any open covering $\mathscr{G}$ of this space, choose a member G of $\mathscr{G}$ with $\Omega \in G$. Then there is an ordinal number $\beta < \Omega$ such that $(\beta, \Omega] \subseteq G$. The interval $[0, \beta]$ is countable (see Proposition A.7.7), so it can be covered by countably many members of $\mathscr{G}$. It follows that $\mathscr{G}$ has a countable subcovering, and $[0, \Omega]$ is Lindelöf. On the other hand, if A is a countable subset of $[0, \Omega]$, then $\xi = \sup (A - \{\Omega\}) < \Omega$. Obviously, the open set (ξ, Ω) does not meet A, and therefore A cannot be dense in $[0, \Omega]$.

Note that the ordinal space $[0, \Omega)$ is not Lindelöf, for the open covering $\{[0, \alpha) \mid \alpha < \Omega\}$ fails to have a countable subcovering. It follows that a subspace of a Lindelöf space need not be Lindelöf. However, we have the following.

Proposition 7.2.9 *A closed subspace of a Lindelöf space is Lindelöf.*

Proof Let X be a Lindelöf space and Y be a closed subspace of X. Let $\{H_\alpha\}_{\alpha \in A}$ be any open covering of Y. Then for each index α, there is an open subset G_α of X such that $H_\alpha = Y \cap G_\alpha$. Since Y is closed in X, $X - Y$ is open. So the family $\{X - Y\} \cup \{G_\alpha \mid \alpha \in A\}$ is an open covering of X. Since X is Lindelöf, this open covering has a countable subcovering. Accordingly, we find a countable subset B of A such that $Y \subseteq \bigcup\{G_\beta \mid \beta \in B\}$. It follows that $Y = \bigcup\{H_\beta \mid \beta \in B\}$, and thus $\{H_\beta : \beta \in B\}$ is a countable subcovering of $\{H_\alpha : \alpha \in A\}$.   ◇

The property of being a Lindelöf space is a topological invariant; in fact, we have the following.

Proposition 7.2.10 *The continuous image of a Lindelöf space is Lindelöf.*

Proof Suppose that X is a Lindelöf space and $f : X \to Y$ is a continuous function. Let $\mathscr{U} = \{U_\alpha \mid \alpha \in A\}$ be an open covering of $f(X)$. Then, for each $\alpha \in A$, there exists an open set $G_\alpha \subseteq Y$ such that $U_\alpha = f(X) \cap G_\alpha$. Clearly, the sets $f^{-1}(G_\alpha)$ form an open covering of X. Since X is Lindelöf, we find countably many sets G_{α_n}, $n = 1, 2, \ldots$, in the family $\{G_\alpha\}_{\alpha \in A}$ such that $X = \bigcup_n f^{-1}(G_{\alpha_n})$. This implies that $f(X) = \bigcup_n U_{\alpha_n}$, as desired.   ◇

It follows that Lindelöfness is transmitted to quotient spaces. But unlike other countability properties, the product of even two Lindelöf spaces is not necessarily Lindelöf.

Example 7.2.7 As seen in Example 7.2.5, the Sorgenfrey plane $\mathbb{R}_\ell \times \mathbb{R}_\ell$ is not a Lindelöf space. However, the space $\mathbb{R}_\ell$ satisfies the Lindelöf condition. To see this, consider an open covering $\{G_\alpha \mid \alpha \in A\}$ of $\mathbb{R}_\ell$, and put $E = \bigcup_\alpha G_\alpha^\circ$, where G_α° denotes the interior of G_α in the Euclidean space $\mathbb{R}$. Since $\mathbb{R}$ is second countable, so is E. Because a second countable space is Lindelöf, there is a countable set $B \subseteq A$ such that $\{G_\beta^\circ \mid \beta \in B\}$ covers E. Next, we observe that $F = \mathbb{R} - E$ is countable. Given $x \in F$, there is an index $\alpha \in A$ such that $x \in G_\alpha$. As G_α is open in $\mathbb{R}_\ell$, there exists a rational number r_x such that $[x, r_x) \subseteq G_\alpha$. Obviously, $(x, r_x) \subseteq G_\alpha^\circ \subseteq E$. So $(x, r_x) \cap F = \varnothing$, and $r_x \leq y < r_y$ for every $x < y$ in F. It follows that the mapping $x \to r_x$ is an injection from F into $\mathbb{Q}$, and hence F is countable. Accordingly, there is a countable set $\Gamma \subseteq A$ such that $\{G_\gamma \mid \gamma \in \Gamma\}$ covers F. It is now clear that the family $\{G_\alpha \mid \alpha \in B \cup \Gamma\}$ is a countable subcovering of $\{G_\alpha \mid \alpha \in A\}$, and thus $\mathbb{R}_\ell$ is Lindelöf.

Exercises

1. Show that a discrete space is separable if and only if it is countable.

2. Show that a cofinite space is separable.

3. Find a countable dense subset of the space $\mathbb{R} - \mathbb{Q}$ (of all irrationals).

4. Prove that the space $\mathcal{B}(I)$ of Example 1.1.5 is not separable but its subspace $\mathcal{C}(I)$ is.

5. Prove that the space in Example 1.1.4 is separable.

6. If every closed ball in a metric space X is compact, show that X is separable.

7. Let X be a metric space in which every infinite subset has a limit point. Prove that X is separable.

8. Give an example of a separable metric space that is not countably compact.

9. Prove that a separable metric space X is locally compact if and only if X is the union of a countable family of open subsets U_n such that $\overline{U}_n$ is compact and $\overline{U}_n \subseteq U_{n+1}$ for every n.

10. Is Sorgenfrey line $\mathbb{R}_\ell$ metrizable?

11. Prove that every subspace of a separable metric space is separable.

12. Let X be an uncountable set, and let $x_0 \in X$ be a fixed point. Let $\mathcal{T}$ be the family of sets $G \subseteq X$ such that $G = \varnothing$ or $x_0 \in G$. Show that $(X, \mathcal{T})$ is a separable space which has a non-separable subspace.

13. Let X be a second countable or separable space. Show that every family of pairwise disjoint open subsets of X is countable. (This property of a space is referred to as "the countable chain condition".)

14. (a) Show that the ordered space $I \times I$ in Exercise 1.4.15 is first countable and Lindelöf but not second countable or separable.

   (b) Show that the subspace $(0, 1) \times I$ of the ordered space $I \times I$ is not Lindelöf.

15. Show that $I \times I$ with the television topology is not metrizable.

16. • Show that the ordinal space $[0, \Omega)$ satisfies the first axiom of countability, but no other countability properties.

17. • Show that the ordinal space $[0, \Omega]$ does not satisfy any countability condition except the Lindelöf property. Deduce that it cannot be given a metric consistent with its topology.

18. Let X be an uncountable set, and let $x_0 \in X$ be a fixed point. Let $\mathscr{U}$ be the family of sets $G \subseteq X$ such that $G = X$ or $x_0 \notin G$. Show that $(X, \mathscr{U})$ is a Lindelöf space which has a subspace that is not Lindelöf.

19. Let X be a Lindelöf space. If every sequence in X has a cluster point, show that X is compact.

20. Let X be a Lindelöf space. Show that each uncountable subset of X has a limit point.

21. Show that an uncountable subset Y of a second countable space X contains uncountably many of its limit points. Is the converse also true?

22. If X is compact and Y is Lindelöf, show that $X \times Y$ is Lindelöf.

23. Let $f : X \to Y$ be a proper surjection. If Y is Lindelöf, show that so is X.

24. Let X be a separable T_2-space. Show that (a) $|\mathcal{C}(X)| \leq c$, (b) $|X| \leq 2^c$ and (c) if X is also first countable, then $|X| \leq c$.

Chapter 8
Separation Axioms

In general, the properties of topological spaces are quite different from those of metric spaces, so some additional restrictions are often imposed on the topology of a space in order to bring its properties closer to those of metric spaces. The reader might be aware of the fact that two distinct points or disjoint closed sets in a metric space can be separated by disjoint open sets. There are more than ten such "separation axioms" for topological spaces, traditionally denoted by $T_0, T_1, \ldots$. These specify the degree to which points or closed sets may be separated. The axioms T_0, T_1, and T_2 have already been studied in Chap. 4. In the present chapter, we shall treat the axioms $T_3, T_{3\frac{1}{2}}$, and T_4 only. Sect. 8.1 concerns with the axiom T_3 which stipulates the separation of points and closed sets by disjoint open sets. Sect. 8.2 is devoted to the axiom T_4 which specifies that every pair of disjoint closed sets have disjoint nbds. For spaces satisfying this axiom, we shall prove several important theorems of topology such as Urysohn's Lemma, Tietze Extension Theorem, and Urysohn Metrization Theorem. The other axiom is stronger than T_3-axiom and weaker than T_4-axiom for most of the spaces. This requires separation of points and closed sets by real-valued continuous functions, and its study is the object of Sect. 8.3. It will be seen that the spaces satisfying axioms T_1 and $T_{3\frac{1}{2}}$ can be embedded into compact Hausdorff spaces; so some interesting theorems can be proved for them.

8.1 Regular Spaces

Definition 8.1.1 A space X is T_3 if for each point $x \in X$ and each closed set $F \subseteq X$ with $x \notin F$, there are disjoint open sets U and V such that $x \in U$ and $F \subseteq V$. The space X is called *regular* if it satisfies both T_1 and T_3 axioms.

(Caution: The opposite terminology is used by many authors for these and the other separation properties are discussed later in this chapter.)

Example 8.1.1 A trivial space with more than one point is T_3 (vacuously), but not T_1.

© Springer Nature Singapore Pte Ltd. 2019
T. B. Singh, *Introduction to Topology*,
https://doi.org/10.1007/978-981-13-6954-4_8

Example 8.1.2 An infinite cofinite space is T_1 but not T_3.

Example 8.1.3 The Euclidean space $\mathbb{R}^n$ is regular; more generally, any metric space is regular. For, if F is a closed subset of a metric space X and $x \in X - F$, then there is an open ball $B(x; r) \subseteq X - F$. Clearly, $B(x; r/2)$ and $X - B[x; r]$ are disjoint nbds of x and F, respectively.

Example 8.1.4 The Sorgenfrey line $\mathbb{R}_\ell$ is regular. For, if F is a closed subset of $\mathbb{R}_\ell$ and $x \in \mathbb{R} - F$, then there is a real number $c > x$ such that $[x, c) \subseteq \mathbb{R} - F$. Clearly, $[x, c)$ and its complement in $\mathbb{R}_\ell$ are disjoint nbds of x and F, respectively. Also, given any two reals $x < y$, $[x, y)$ and $[y, y + 1)$ are disjoint nbds of x and y in $\mathbb{R}_\ell$, respectively, and so $\mathbb{R}_\ell$ is Hausdorff.

A regular space is Hausdorff, since points are closed; however, the Hausdorff property is not a consequence of T_3-axiom. Also, a Hausdorff space need not be regular.

Example 8.1.5 In the set $\mathbb{R}$ of real numbers, consider the topology $\mathscr{T}$ generated by the set $\mathbb{Q}$ of rational numbers and the (open) intervals (a, b), $a < b$ reals, as subbasis. This topology is finer than the usual topology for $\mathbb{R}$, so $(\mathbb{R}, \mathscr{T})$ is Hausdorff. The set $\mathbb{R} - \mathbb{Q}$ is obviously closed in $\mathscr{T}$. If $q \in \mathbb{Q}$ and U is a nbd of q, then there is an open interval (a, b) such that $q \in (a, b) \cap \mathbb{Q} \subseteq U$. Let x be an irrational number in (a, b). Then, every nbd of x intersects U so that there is no nbd of $\mathbb{R} - \mathbb{Q}$ which is disjoint from U. Thus $(\mathbb{R}, \mathscr{T})$ is not regular.

This example also shows that a topology finer than a regular topology is not necessarily regular.

By a nbd of a subset A of a space X is meant a subset $N \subseteq X$ such that there is an open set $U \subseteq X$ with $A \subseteq U \subseteq N$. With this terminology, the condition of T_3-axiom can be rephrased as: Each point $x \in X$ and each closed subset F of X not containing x have disjoint nbds. An equivalent formulation is given by

Proposition 8.1.2 *A space X is T_3 if and only if for each point $x \in X$ and each nbd U of x, there exists a nbd V of x such that $\overline{V} \subseteq U$ (that is, the closed nbds of x form a nbd basis at x).*

Proof Suppose that X satisfies the T_3 axiom. Let U be an open nbd of x. Then, there exist disjoint open sets V and W such that $x \in V$ and $X - U \subseteq W$. Since $V \cap W = \emptyset$ and $X - W$ is closed, we have $\overline{V} \subseteq X - W \subseteq U$.

Conversely, let F be a closed subset of X and $x \in X - F$. By our hypothesis, there exists an open set V such that $x \in V \subseteq \overline{V} \subseteq X - F$. Put $U = X - \overline{V}$. Then U is a nbd of F and $U \cap V = \emptyset$.   ◇

By the proof of Theorem 5.1.8, we see that every compact Hausdorff is regular. Moreover, the preceding proposition and Theorem 5.4.2 show that every locally compact Hausdorff space is also regular.

Clearly, regularity is a topological invariant. But, like the Hausdorff property, it is also not preserved under continuous closed or open maps.

Example 8.1.6 Consider the subspace $X = I \times \{0, 1\}$ of $\mathbb{R}^2$ and let $\sim$ be the equivalence relation $(t, 0) \sim (t, 1), t \in [0, 1)$. Clearly, the projection map $\pi: X \to X/\sim$ is open, and the quotient space $X/\sim$ is T_1. Also, the points $x_0 = (1, 0)$ and $x_1 = (1, 1)$ do not have disjoint saturated open nbds in X, so the points $\pi(x_0)$ and $\pi(x_1)$ cannot be separated by open sets. Thus $X/\sim$ fails to satisfy the T_3-axiom.

Example 8.1.7 Let $H = \{(x, y) \in \mathbb{R}^2 \mid y > 0\}$ and $L = \{(x, 0) \mid x \in \mathbb{R}\}$. Let $\mathscr{T}$ be the topology on $Z = H \cup L$ generated by the basis consisting of the open balls contained in H, and the sets $\{(x, 0)\} \cup B((x, r); r)$. Observe that the latter sets are open balls in the upper half plane which are tangent to the x-axis together with the points of tangency (see Fig. 8.1). It is easy to see that $\mathscr{T}$ is finer than the Euclidean subspace topology for Z so that it is T_1. To see that Z is regular, let U be an open nbd of $p \in Z$. If $p \in H$, then there is an open ball $B(p; r) \subseteq U$, and the set $V = B(p; r/2)$ satisfies $p \in V \subseteq \overline{V} \subseteq U$. If $p = (x, 0) \in L$, then there is an open ball $B((x, r); r)$ such that $\{p\} \cup B((x, r); r) \subseteq U$. The set $\{p\} \cup B((x, r/2); r/2) = V$ is a nbd of p with $\overline{V} \subseteq U$. By Proposition 8.1.2, we see that $(Z, \mathscr{T})$ is regular.

Notice that H is open in Z and so are the sets $H \cup \{p\}, p \in L$. Thus L as well as each $L - \{p\}$ is closed in Z. Consequently, every set $F \subseteq L$ is closed, being the intersection of closed sets $L - \{p\}, p \notin F$. In particular, both of the sets $A = \{(q, 0) \mid q \text{ is rational}\}$ and $B = \{(p, 0) \mid p \text{ is irrational}\}$ are closed. Consider the equivalence relation $\sim$ on Z, which identifies A to a point. We show that the quotient space $Z/\sim$ is not regular. Clearly, the quotient map $\pi: Z \to Z/\sim$ is closed, so $\pi(B)$ is a closed subset of $Z/\sim$. Also, $\pi(B)$ does not contain the point $\pi(A)$, since $A \cap B = \varnothing$. We observe that there are no disjoint nbds of $\pi(A)$ and $\pi(B)$ in $Z/\sim$. Assume otherwise. Then there exist disjoint open sets G and H in Z such that $A \subseteq G$ and $B \subseteq H$. Consequently, for each rational q and each irrational p, we have a reals $\delta_q > 0$ and $\delta_p > 0$ such that

$$U_q = \{(q, 0)\} \cup B\left((q, \delta_q); \delta_q\right) \subseteq G \text{ and}$$
$$V_p = \{(p, 0)\} \cup B\left((p, \delta_p); \delta_p\right) \subseteq H.$$

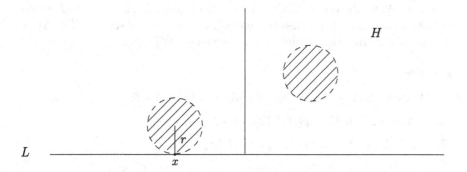

Fig. 8.1 Basic open sets in Z

Now, for each $n = 1, 2, \ldots$, let $S_n = \{(p, 0) \in B \mid \delta_p > 1/n\}$. Then the family of all these sets S_n together with the one-point sets $\{x\}$, $x \in A$, covers L. As a subspace of the Euclidean space $\mathbb{R}^2$, L is complete, and hence int $\left(\overline{S_{n_0}}\right) \neq \varnothing$ for some n_0 (by Theorem 10.3.5). So there exist reals $s < t$ such that $N = \{(x, 0) \mid s < x < t\} \subseteq \overline{S_{n_0}}$. Choose a rational $q \in (s, t)$ and then a real ϵ such that $0 < \epsilon < \min\{q - s, t - q, 2\sqrt{\delta_q/n_0}\}$. Since $(q, 0) \in \overline{S_{n_0}}$, there exists a point $(p, 0) \in S_{n_0}$ such that $|p - q| < \epsilon$. Then $U_q \cap V_p \neq \varnothing$, and therefore $G \cap H \neq \varnothing$, a contradiction.

The property of regularity behaves well under formation of subspaces and products.

Proposition 8.1.3 *Every subspace of a regular space is regular.*

Proof Let X be a regular space and let Y be a subspace of X. Let A be a closed subset of Y and $y \in Y - A$. Then there exists a closed set $F \subseteq X$ such that $A = Y \cap F$. Clearly, $y \notin F$. So, there exist disjoint open sets U and V in X such that $y \in U$ and $F \subseteq V$, since X is regular. Then $Y \cap U$ and $Y \cap V$ are disjoint open nbds of y and A, respectively, in Y. Thus Y satisfies T_3-axiom. Being a subspace of a T_1-space, Y also satisfies T_1-axiom. $\diamond$

Theorem 8.1.4 *Let X_α, $\alpha \in A$, be a family of regular spaces. Then the product space $\prod X_\alpha$ is regular.*

Proof Since each X_α satisfies the T_1-axiom, so does $\prod X_\alpha$. To see that it is a T_3-space, let $x = (x_\alpha) \in \prod X_\alpha$ and O be any nbd of x. Then there is a basic open set $\prod U_\alpha$ in $\prod X_\alpha$ such that $x \in \prod U_\alpha \subseteq O$. Assume that $U_\alpha = X_\alpha$ for all $\alpha \neq \alpha_1, \ldots, \alpha_n$. For each $i = 1, \ldots, n$, there exists a nbd V_{α_i} of x_{α_i} such that $\overline{V}_{\alpha_i} \subseteq U_{\alpha_i}$, since X_{α_i} is regular. Write $V_\alpha = X_\alpha$ for $\alpha \neq \alpha_1, \ldots, \alpha_n$. Then $\prod V_\alpha$ is a nbd of x such that $\prod \overline{V}_\alpha \subseteq \prod U_\alpha \subseteq O$. By Proposition 8.1.2, $\prod X_\alpha$ is T_3, and thus regular. $\diamond$

Note that the converse of the above theorem also holds good. For, if $\prod X_\alpha$ is regular, then each X_α, being homeomorphic to a subspace of $\prod X_\alpha$, is regular.

We conclude this section with the remark that a quotient space $X/\sim$ is T_3 if and only if for each equivalence class $[x]$ and each saturated closed set $F \subseteq X$ such that $F \cap [x] = \varnothing$, there exist disjoint saturated open nbds of $[x]$ and F in X.

Exercises

1. On the set $\{a, b, c\}$, find a topology which is T_3 but not T_2.

2. Is the space X in Example 6.1.12 regular?

3. Let H, L and Z be as in Example 8.1.7. Show:

   (a) The family $\mathscr{B}$ of open balls contained in H, and the sets
       $\{(x, 0)\} \cup [B((x, 0); r) \cap H]$ is a basis for a topology on Z.

   (b) The topology generated by $\mathscr{B}$ is T_2 but not T_3.

4. Let X be a space such that each point of X admits a closed nbd which is T_3. Show that X is T_3.

5. Let A be a closed subspace of the regular space X. Show that (a) for each point $x \notin A$, there are open nbds U of A and V of x such that $\overline{U} \cap \overline{V} = \varnothing$, and (b) A coincides with the intersection of its closed nbds.

6. Let X be a space such that for each $x \in X$ and each subbasic nbd U of x, there exists a nbd V of x with $\overline{V} \subseteq U$. Prove that X satisfies the T_3-axiom.

7. Let $f : X \to Y$ be a proper surjection. If X is regular, show that Y is also regular.

8. • If X is regular and A is closed in X, show that X/A is Hausdorff.

9. Prove that a countably compact, first countable T_2-space is regular.

10. Let X be a regular space and $A \subseteq X$ infinite. Show that there exists a countable family $\{U_n\}$ of open sets such that $A \cap U_n \neq \varnothing$ for all n and $\overline{U}_n \cap \overline{U}_m = \varnothing$ for $n \neq m$.

11. Let X be a regular locally connected space, and $U \subseteq X$ be a connected, open set. If $x, y \in U$, prove that there is a closed connected set $C \subseteq X$ such that $x, y \in C \subseteq U$.

12. • Let A and B be the subsets of the space Z as in Example 8.1.7, and consider the equivalence relation $\sim$ on Z which identifies A and B to points. If $\pi : Z \to Z/\sim$ is the natural projection, show that $\pi \times \pi$ is not a quotient map.

13. • Let X be a regular space. Show:

    (a) The relation $x \sim y$ if every nbd of x contains y is an equivalence relation on X.

    (b) The natural projection $\pi : X \to X/\sim$ is both open and closed.

    (c) $X/\sim$ is regular.

    (d) If Y is T_2 and $\phi : X \to Y$ is continuous, then there exists a continuous map $\psi : X/\sim \to Y$ such that $\phi = \psi\pi$.

14. • (a) If, for an equivalence relation $\sim$ on a regular space X, the quotient map $X \to X/\sim$ is closed, show that $G(\sim)$ is closed in $X \times X$.

    (b) Give an example of an equivalence relation $\sim$ on a regular space X such that $G(\sim)$ is closed in $X \times X$, but the projection $X \to X/\sim$ is not closed.

15. Prove that the sum of a family of T_3-spaces is T_3.

16. Give an example to show that an adjunction space of two regular spaces need not be regular.

17. Let A be a closed nbd retract of a regular space X (that is, A is closed in X, and has a nbd N in X such that there exists a continuous map $r : N \to A$ fixing every element of A). If Y is Hausdorff, prove that $X \cup_f Y$ is Hausdorff for every continuous map $f : A \to Y$.

8.2 Normal Spaces

Definition 8.2.1 A space X is T_4 if each pair of disjoint closed subsets of X have disjoint nbds. X is *normal* if it satisfies both the T_1 and T_4 axioms.

Example 8.2.1 Every compact Hausdorff space is normal, by Proposition 5.1.9.

Example 8.2.2 Every metric space is normal. Let X be a metric space and A, B be disjoint closed subsets of X. For each $a \in A$, there is an open ball $B(a; r_a)$ disjoint from B. Put $U = \bigcup_{a \in A} B(a; r_a/2)$. Then U is a nbd of A. Similarly, construct a nbd V of B, where $V = \bigcup_{b \in B} B(b; r_b/2)$ and $B(b; r_b) \cap A = \varnothing$ for all $b \in B$. If $U \cap V \neq \varnothing$, then there exist points $a \in A$ and $b \in B$ such that $d(a, b) < \max\{r_a, r_b\}$. This forces $a \in B(b; r_b)$ or $b \in B(a; r_a)$, a contradiction in either case. Therefore the nbds U and V are disjoint.

In particular, each Euclidean space $\mathbb{R}^n$ is normal.

Axiom-T_4 is neither stronger nor weaker than axiom-T_3.

Example 8.2.3 The right-hand topology on $\mathbb{R}$, which consists of $\varnothing$, $\mathbb{R}$ and all open right rays $(a, +\infty)$ is T_4, since there are no disjoint closed sets. But it is not T_3, for there are no disjoint open sets.

Example 8.2.4 The regular space Z in Example 8.1.7 is not T_4, since the sets $A = \{(q, 0) \mid q \text{ is rational}\}$ and $B = \{(r, 0) \mid r \text{ is irrational}\}$ are disjoint and closed in Z, but there are no disjoint nbds of A and B in Z.

A normal space is obviously regular but, as seen in the preceding example, a regular space is not normal in general. For Lindelöf spaces, regularity forces normality.

Theorem 8.2.2 *A regular Lindelöf space is normal.*

Proof Let X be a Lindelöf regular space and A, B be disjoint closed sets in X. Then, for each $a \in A$, there exists an open nbd U_a of a such that $\overline{U}_a \subseteq X - B$. Similarly, for each $b \in B$, we find an open nbd V_b of b such that $\overline{V}_b \subseteq X - A$. Since X is Lindelöf, the open covering $\{U_a \mid a \in A\} \cup \{V_b \mid b \in B\} \cup \{X - (A \cup B)\}$ has a countable subcovering. So there is a countable subfamily $\{U_{a_n}\}$ of $\{U_a \mid a \in A\}$ which covers A, and a countable subfamily $\{V_{b_n}\}$ of $\{V_b \mid b \in B\}$ which covers B. For each integer $n \geq 1$, let $G_n = U_{a_n} - \bigcup\{\overline{V}_{b_i} \mid i \leq n\}$, and $H_n = V_{b_n} - \bigcup\{\overline{U}_{a_i} \mid i \leq n\}$. Then the sets G_n and H_n are open for each n and $G_n \cap H_m = \varnothing$ for all n and m. Since $\{U_{a_n}\}$ covers A and no $\overline{V}_{b_i}$ meets A, $\{G_n\}$ is an open cover of A. Similarly, $\{H_n\}$ is an open cover of B. It follows that $\bigcup G_n$ and $\bigcup H_n$ are disjoint open nbds of A and B, respectively, and X is normal. $\diamond$

Corollary 8.2.3 *Every second countable regular space is normal.*

Regarding the invariance properties of normal spaces, we have

Proposition 8.2.4 *Let X be normal space. Then*

(a) *every closed subspace of X is normal, and*
(b) *each closed continuous image of X is normal.*

Proof (a): This follows from the fact that a closed subset of a closed subspace is a closed subset of the space.

(b): Let $f : X \to Y$ be a continuous closed surjection. Then Y is obviously T_1. Suppose that A, B are disjoint closed sets in Y. Then $f^{-1}(A)$ and $f^{-1}(B)$ are disjoint closed subsets of X. So there exist disjoint open sets $U \supseteq f^{-1}(A)$ and $V \supseteq f^{-1}(B)$. Since f is closed, $U' = Y - f(X - U)$ and $V' = Y - f(X - V)$ are open subsets of Y with $A \subseteq U'$ and $B \subseteq V'$. Since $f^{-1}(U') \subseteq U$, $f^{-1}(V') \subseteq V$ and $U \cap V = \emptyset$, we have $U' \cap V' = \emptyset$. This proves (b). $\diamond$

By the preceding proposition, normality is a topological invariant, but its not preserved by continuous (open) functions in general (refer to Example 8.1.6). We also remark that non-closed subspaces of normal spaces may fail to be normal; however, it is not convenient at this stage to give an example supporting the remark. Moreover, the product of two normal spaces need not be normal.

Example 8.2.5 The Sorgenfrey line $\mathbb{R}_\ell$ is normal. To see this, let A, B be disjoint closed subsets of $\mathbb{R}_\ell$. Then for each $a \in A$, there exists $x_a > a$ such that $[a, x_a) \subseteq \mathbb{R} - B$, and for each $b \in B$, there exists $x_b > b$ such that $[b, x_b) \subseteq \mathbb{R} - A$. It is easily checked that $[a, x_a) \cap [b, x_b) = \emptyset$ for all $a \in A$ and $b \in B$. So $U = \bigcup \{[a, x_a) \mid a \in A\}$ and $V = \bigcup \{[b, x_b) \mid b \in B\}$ are disjoint nbds of A and B, respectively. But the product space $\mathbb{R}_\ell^2$ is not normal. For, the antidiagonal $L = \{(x, y) \in \mathbb{R}_\ell^2 \mid x + y = 0\}$ of $\mathbb{R}_\ell^2$ is closed and discrete (see Example 7.2.4), and so a proof similar to Example 8.1.7 can be given to show that the disjoint closed sets $\{(x, y) \in L \mid x, y \in \mathbb{Q}\}$ and $\{(x, y) \in L \mid x, y \in \mathbb{R} - \mathbb{Q}\}$ do not have disjoint nbds in $\mathbb{R}_\ell^2$.

Alternatively, if $\mathbb{R}_\ell^2$ were normal, then for every subset $F \subseteq L$, where L is as above, there would be disjoint open sets $U(F)$ and $V(F)$ such that $F \subseteq U(F)$ and $L - F \subseteq V(F)$. Consequently, for any two subsets $F_1 \not\subseteq F_2$ of L, $U(F_1) \cap V(F_2)$ is a nonempty open subset of $\mathbb{R}_\ell^2$. Since $D = \{(x, y) \mid x, y \in \mathbb{Q}\}$ is dense in $\mathbb{R}_\ell^2$, we have $D \cap U(F_1) \cap V(F_2) \neq \emptyset$. It follows that $D \cap U(F_1)$ and $D \cap U(F_2)$ are distinct subsets of D, for $D \cap U(F_2)$ is disjoint from $D \cap V(F_2)$. Thus, there is an injection from $\mathscr{P}(L)$ (the power set of L) into $\mathscr{P}(D)$, and hence, by Propositions A.8.1, A.8.5 and A.8.6, we obtain the contradiction $c = |L| < |\mathscr{P}(L)| \leq |\mathscr{P}(D)| = c$.

In view of the above fact, the following consequence of the Tychonoff theorem is interesting: *The product of any family of closed unit intervals is compact Hausdorff, and therefore normal.*

As already seen, not every quotient of a normal space is normal (or even T_4). It is clear that the quotient space $X/\sim$ of a space X is T_4 if and only if for any saturated, closed subsets of A and B of X with $A \cap B = \emptyset$, there exist disjoint saturated open nbds of A and B in X. Also, if the saturation of every closed subset of a normal space X under an equivalence relation $\sim$ is closed, then the quotient space $X/\sim$ is normal, by Proposition 8.2.4b. In particular, if X is normal and $A \subseteq X$ is closed, then X/A is normal. Moreover, this property behaves well with the construction of adjunction spaces.

Theorem 8.2.5 *If X and Y are normal spaces and $A \subseteq X$ is closed, then $X \cup_f Y$ is normal for any map $f : A \to Y$.*

Proof Since each of the equivalence classes of $X + Y$ is closed, $X \cup_f Y$ is T_1. To see that it is T_4, let B_1 and B_2 be disjoint closed subsets of $X \cup_f Y$, and let $\pi : X + Y \to X \cup_f Y$ be the natural projection. Then $F_i = \pi^{-1}(B_i) \cap Y, i = 1, 2$, are disjoint closed sets in Y. As Y is normal, there exist open sets $G_i \subseteq Y$ such that $F_i \subseteq G_i$ for $i = 1, 2$, and $\overline{G}_1 \cap \overline{G}_2 = \varnothing$ (see Exercise 3). Since $\pi | Y$ is a closed injection, the sets $C_i = B_i \cup \pi\left(\overline{G}_i\right), i = 1, 2$, are disjoint and closed in $X \cup_f Y$. Consequently, $\pi^{-1}(C_1) \cap X$ and $\pi^{-1}(C_2) \cap X$ are disjoint closed subsets of X. By the normality of X, there exist disjoint open sets O_1 and O_2 in X such that $\pi^{-1}(C_i) \cap X \subseteq O_i$ for $i = 1, 2$. Put $U_i = \pi\left((O_i - A) \cup G_i\right)$. Then $B_i \subseteq U_i$ and $U_1 \cap U_2 = \varnothing$. We show that U_1 and U_2 are open. Clearly, $\pi^{-1}(U_i) \cap Y = G_i$ and $\pi^{-1}(U_i) \cap X = (O_i - A) \cup f^{-1}(G_i)$. By the continuity of f, both $f^{-1}(G_1)$ and $f^{-1}(G_2)$ are open in A. So there exist open sets V_1 and V_2 in X such that $f^{-1}(G_i) = V_i \cap A$. As $f^{-1}(G_i) \subseteq \pi^{-1}(\pi(G_i)) \cap X \subseteq O_i$, we have

$$(O_i - A) \cup f^{-1}(G_i) = (O_i - A) \cup (O_i \cap V_i \cap A) =$$
$$O_i \cap ((X - A) \cup (V_i \cap A)) = O_i \cap ((X - A) \cup V_i),$$

which is open. This completes the proof. $\diamond$

We now come to the first major result of the book, the proof of which consists of a different idea not encountered so far. It guarantees the existence of nonconstant continuous real-valued functions on normal spaces—a purely topological assumption. This fact is used to prove many important theorems for normal spaces; we shall content ourselves with just two of them—the Urysohn metrization theorem and the Tietze extension theorem. These theorems justify the use of the adjective "normal" for this class of the spaces despite the property not behaving well with the operations of forming subspaces and products.

A dyadic rational number is a number of the form $m/2^n$, where m and $n > 0$ are integers. The set of all dyadic rational numbers is dense in the real line $\mathbb{R}$, for if $r < r'$ are two real numbers, then we can find integers m and $n > 0$ such that $2^{-n} < r' - r$ and $m - 1 < 2^n r < m$.

Theorem 8.2.6 (Urysohn Lemma) *If X is a normal space, then for each pair of nonempty disjoint closed subsets A and B of X, there is a continuous function $f : X \to I$ such that $f(A) = \{0\}$ and $f(B) = \{1\}$.*

Proof Let D be the set of dyadic rational numbers of I. For each $t \in D$, we associate an open nbd U_t of A such that

(a) $B \cap U_t = \emptyset$, and
(b) $\overline{U}_t \subseteq U_{t'}$ for $t < t'$.

Set $U_1 = X - B$. Then U_1 is an open nbd of A. Since X is T_4, there exists an open set $U_0 \subseteq X$ such that $A \subseteq U_0 \subseteq \overline{U}_0 \subseteq U_1$. For the same reason, we find an open set $U_{1/2}$ such that $\overline{U}_0 \subseteq U_{1/2} \subseteq \overline{U}_{1/2} \subseteq U_1$. By induction, suppose that for some integer $n \geq 2$ the open sets $U_{m/2^n}$ satisfying conditions (a) and (b) have already been defined. Then, using the normality of X again, we can find open sets $U_{m/2^{n+1}}$ $(m = 1, 3, 5, \ldots, 2^{n+1} - 1)$ such that $\overline{U}_{(m-1)/2^{n+1}} \subseteq U_{m/2^{n+1}} \subseteq \overline{U}_{m/2^{n+1}} \subseteq U_{(m+1)/2^{n+1}}$. This completes the induction argument, and we conclude that there exist open sets U_t, $t \in D$, meeting conditions (a) and (b) hold.

Now, define a function $f : X \to I$ by

$$f(x) = \begin{cases} \inf \{t \mid x \in U_t\} & \text{for } x \in U_1, \text{ and} \\ 1 & \text{for } x \in X - U_1. \end{cases}$$

It is evident that $f(x) = 0$ if $x \in A$, and $f(x) = 1$ if $x \in B$. To establish the continuity of f, we note that the intervals $[0, r)$ and $(r, 1]$, $0 < r < 1$, constitute a subbase of I. Therefore it suffices to prove that the sets $f^{-1}([0, r))$ and $f^{-1}((r, 1])$ are open for every $0 < r < 1$. Clearly, $f^{-1}([0, r)) = \bigcup_{t < r} U_t$, and $f^{-1}((r, 1]) = X - f^{-1}([0, r])$. We further observe that $f^{-1}([0, r]) = \bigcap_{t > r} U_t = \bigcap_{t > r} \overline{U}_t$. To see the first equality, assume that $f(x) \leq r$. If $t \in D$ and $r < t$, then $f(x) < t$, so there an $s \in D$ such that $s < t$ and $x \in U_s$. By condition (b), $x \in U_t$. Conversely, suppose that $x \in U_t$ for all $t > r$ and $t \in D$. If $r < f(x)$, then there is a $t \in D$ such that $r < t < f(x)$. By our assumption, $x \in U_t$, which implies that $f(x) \leq t$, a contradiction. It follows that $f(x) \leq r$ and the equality $f^{-1}([0, r]) = \bigcap_{t > r} U_t$ holds. For the second equality, the inclusion $\bigcap_{t > r} U_t \subseteq \bigcap_{t > r} \overline{U}_t$ is obvious. To see the reverse inclusion, assume that $x \notin U_t$ for some $t > r$. Then there is an $s \in D$ such that $t > s > r$ and, by (b), $x \notin \overline{U}_s$. Hence $\bigcap_{t > r} \overline{U}_t \subseteq \bigcap_{t > r} U_t$, and the second equality holds. Thus $f^{-1}((r, 1]) = \bigcup_{t > r} (X - \overline{U}_t)$ is also open, and this completes the proof. ◇

Remark 8.2.7 (a) The proof of the above theorem does not use axiom T_1; accordingly, the theorem is true for any T_4-space.
(b) A continuous function $f : X \to I$ such that $f(A) = 0$ and $f(B) = 1$ is referred to as a Urysohn function for the pair A, B. Notice that a Urysohn function f for a pair A, B obviously satisfies $A \subseteq f^{-1}(0)$; the theorem does not assert that $A = f^{-1}(0)$. Thus, f may take the value 0 outside of A.
(c) If each pair of disjoint closed subsets of a space X admits a Urysohn function, then it is clear that X is T_4. So, the converse of Urysohn Lemma holds good for T_1-spaces.

(d) The unit interval I can be replaced by any closed interval $[a, b]$. In fact, the composition of a Urysohn function for the pair A, B with the homeomorphism $t \to a + (b - a)t$ between I and $[a, b]$ is a continuous function from X into $[a, b]$ which maps A into a and B into b.

As a first important application of the Urysohn Lemma, we see a sufficient condition for metrizability. Recall that a topological space X is metrizable if there exists a metric on the set X that induces the topology of X.

Theorem 8.2.8 (Urysohn Metrization Theorem) *Every regular second countable space is metrizable.*

Proof Let X be a regular space with a countable basis $\mathscr{B}$. Since a space homeomorphic to a subspace of a metric space is metrizable, it suffices to prove that X can be embedded into a metric space. Each point $x \in X$ belongs to some member B of $\mathscr{B}$ and, by regularity of X, we find another member B' of $\mathscr{B}$ such that $x \in B' \subseteq \overline{B'} \subseteq B$. So the collection $\mathscr{C} = \left\{ (B, B') \mid B, B' \in \mathscr{B} \text{ and } \overline{B'} \subseteq B \right\}$ is nonempty and countable. We index the pairs (B, B') in $\mathscr{C}$ by positive integers. By Corollary 8.2.3, X is normal, and therefore, for each integer $n > 0$, there is a Urysohn function $f_n : X \to I$ such that $f_n(x) = 0$ if $x \in \overline{B'_n}$, and $f_n(x) = 1$ if $x \notin B_n$. For $x \in X$, write $\phi(x) = (f_1(x), f_2(x)/2, f_3(x)/3, \ldots)$. Then $\phi(x) \in \ell_2$, the Hilbert space. Thus, there is a function $\phi : X \to \ell_2$. We assert that ϕ is an embedding. First, if $x \neq y$ in X, then there is a pair (B_n, B'_n) in $\mathscr{C}$ such that $x \in B'_n$ and $y \notin B_n$, by regularity of X. So $f_n(x) = 0$ and $f_n(y) = 1$. This implies that $\phi(x) \neq \phi(y)$, and ϕ is injective. Next, to see the continuity of ϕ, let $x_0 \in X$ be a fixed point and $\epsilon > 0$ be given. Since the series $\sum n^{-2}$ converges, we can choose an integer $m > 0$ such that $\sum_{n > m} n^{-2} < \epsilon^2/2$. By the continuity of f_n, there is a nbd V_n of x_0 such that $\left| f_n(x) - f_n(x_0) \right| < n\epsilon/\sqrt{2m}$ for all $x \in V_n$. Then, $U = \bigcap_1^m V_n$ is a nbd of x_0 such that

$$\sum_1^m n^{-2} \left| f_n(x) - f_n(x_0) \right|^2 < \epsilon^2/2$$

for every $x \in U$. It follows that $\| \phi(x) - \phi(x_0) \| < \epsilon$ for all $x \in U$, and ϕ is continuous at x_0. Finally, we show that the function $\psi : \phi(X) \to X$, the inverse of ϕ, is continuous. Given a point $y_0 \in \phi(X)$ and a nbd N of $\psi(y_0) = x_0$, we find a pair $\left(B_{n_0}, B'_{n_0} \right)$ in $\mathscr{C}$ such that $x_0 \in B'_{n_0}$ and $B_{n_0} \subseteq N$. Then, for $\epsilon = 1/n_0$ and $\phi(x) = y$, $\| y - y_0 \| < \epsilon \Rightarrow \sum n^{-2} \left| f_n(x) - f_n(x_0) \right|^2 < \epsilon^2 \Rightarrow \left| f_{n_0}(x) - f_{n_0}(x_0) \right| < 1$. Since f_{n_0} vanishes on B'_{n_0}, we have $\left| f_{n_0}(x) \right| < 1$, which forces $x \in B_{n_0}$. Thus, if $y \in \phi(X)$ and $\| y - y_0 \| < \epsilon$, then $\psi(y) \in N$. This establishes the continuity of ψ at y_0, and our assertion follows. $\diamond$

Since a metric space is certainly regular, the Urysohn Metrization theorem can be restated as

Theorem 8.2.9 *A second countable space is metrizable if and only if it is regular.*

The last theorem is a characterization of second countable metric spaces. As there are metric spaces which are not separable and therefore not second countable, the Urysohn theorem is not a characterization of general metric spaces. In Chap. 9, we shall see a complete characterization of metric spaces, separable or not.

By the proof of Theorem 8.2.8, we immediately conclude the following.

Theorem 8.2.10 (Urysohn Embedding Theorem) *Every regular second countable space can be embedded in the Hilbert space ℓ_2.*

We now turn to another fundamental problem in topology: Whether or not a continuous function defined on a subspace of a topological space admits a continuous extension to the whole space. This is referred to as the "extension problem". For example, the Urysohn lemma implies that the function $g \colon A \cup B \to I$ defined by $g(A) = 0$, $g(B) = 1$ has an extension over X, if A and B are closed subsets of the normal space X. On the other hand, the identity function on the unit circle $\mathbb{S}^1$ cannot be extended to a continuous function on the unit disc $\mathbb{D}^2$. It is difficult to justify it at this stage, but we shall do so in Chap. 14.

Theorem 8.2.11 (Tietze Extension Theorem) *If A is a closed subset of the normal space X, then every continuous function $f \colon A \to \mathbb{R}$ extends to a continuous function $g \colon X \to \mathbb{R}$. Moreover, g can be chosen so that $\inf_{x \in X} g(x) = \inf_{a \in A} f(a)$ and $\sup_{x \in X} g(x) = \sup_{a \in A} f(a)$.*

Proof To begin with, we shall assume that f is bounded, and let $\alpha = \inf_{a \in A} f(a)$ and $\beta = \sup_{a \in A} f(a)$. Then f maps A into $[\alpha, \beta]$. If f is constant, the theorem is obviously true. So we may assume that $\alpha < \beta$. Then the mapping $h \colon x \to (2x - \beta - \alpha)/(\beta - \alpha)$ is a homeomorphism between $[\alpha, \beta]$ and $[-1, 1]$. It is clear that if $g \colon X \to [-1, 1]$ is a continuous extension of hf, then $h^{-1}g \colon X \to [\alpha, \beta]$ is a continuous extension of f. So we can assume further that $\alpha = -1$ and $\beta = 1$. Now, define two subsets E_1 and F_1 of A by $E_1 = f^{-1}([-1, -1/3])$ and $F_1 = f^{-1}([1/3, 1])$. Since A is closed in X, E_1 and F_1 are nonempty, disjoint, closed subsets of X. By Theorem 8.2.6, there is a continuous function $g_1 \colon X \to [-1/3, 1/3]$ which takes the value $-1/3$ on E_1, and $1/3$ on F_1. Observe that $|f(x) - g_1(x)| \leq 2/3$ for $x \in A$, and $|g_1(x)| \leq 1/3$ for $x \in X$. Assume, by induction, that we have already constructed continuous functions $g_i \colon X \to \mathbb{R}$, $i = 1, 2, \ldots, n$, such that

$$\left| f(x) - \sum_1^n g_i(x) \right| \leq (2/3)^n \quad \text{if } x \in A, \text{ and}$$
$$|g_i(x)| \leq 2^{i-1}/3^i \quad \text{if } x \in X. \tag{$*$}$$

Then, to construct g_{n+1}, we repeat the above process with $f - \sum_1^n g_i$ in place of f. Specifically, we consider the closed sets

$$E_{n+1} = \left\{ x \in A \mid f(x) - \sum_1^n g_i(x) \leq -2^n/3^{n+1} \right\}, \text{ and}$$
$$F_{n+1} = \left\{ x \in A \mid f(x) - \sum_1^n g_i(x) \geq 2^n/3^{n+1} \right\},$$

and apply Theorem 8.2.6 to find a continuous function

$$g_{n+1} \colon X \longrightarrow \left[-2^n/3^{n+1}, 2^n/3^{n+1}\right]$$

such that $g_{n+1}(E_{n+1}) = -2^n/3^{n+1}$ and $g_{n+1}(F_{n+1}) = 2^n/3^{n+1}$. Clearly, $\left|f(x) - \sum_1^{n+1} g_i(x)\right| \le (2/3)^{n+1}$ for $x \in A$, and $|g_{n+1}(x)| \le 2^n/3^{n+1}$ for $x \in X$.

We thus have a sequence of functions $g_n \colon X \to \mathbb{R}$, $n = 1, 2, \ldots$, satisfying $(*)$. Since the series $\sum 2^{n-1}/3^n$ converges to 1, we see, by the second inequality in $(*)$, that the series $\sum g_n(x)$ converges to a number in $[-1, 1]$ for every $x \in X$. We may therefore define a function $g \colon X \to \mathbb{R}$ by $g(x) = \sum g_n(x)$. Obviously, $|g(x)| \le 1$ for all $x \in X$ and, by the first inequality in $(*)$, $g(x) = f(x)$ for every $x \in A$. It remains to show that g is continuous. Let $x_0 \in X$ be an arbitrary but fixed point. Then, given $\epsilon > 0$, we can find an integer m such that $\sum_{i>m} 2^{i-1}/3^i < \epsilon/4$. Since the function g_i is continuous, there exists an open nbd U_i of x_0 such that $|g_i(x) - g_i(x_0)| < \epsilon/2m$ for $x \in U_i$. If $x \in U_1 \cap \cdots \cap U_m$, then we have (by the second inequality in $(*)$)

$$|g(x) - g(x_0)| \le \sum_1^m |g_i(x) - g_i(x_0)| + 2\sum_{i>m} 2^{i-1}/3^i < \epsilon,$$

and so g is continuous at x_0.

Now, suppose that f is not bounded. We choose an appropriate homeomorphism h

(a) $(-\infty, \infty) \approx (-1, 1)$ in case f is unbounded in both directions,

(b) $[\alpha, \infty) \approx [-1, 1)$ in case f is bounded below by α,

(c) $(-\infty, \beta] \approx (-1, 1]$ in case f is bounded above by β,

and consider the composition hf. This is bounded by $-1, 1$ and has a continuous extension $\phi \colon X \to [-1, 1]$. Put $B = \{x \in X \mid |\phi(x)| = 1\}$ in the case (a), $B = \{x \in X \mid \phi(x) = 1\}$ in the case (b), and $B = \{x \in X \mid \phi(x) = -1\}$ in the case (c). Then B is closed in X and $B \cap A = \varnothing$. By Theorem 8.2.6, there is a Urysohn function $\psi \colon X \to [0, 1]$ such that $\psi(B) = 0$ and $\psi(A) = 1$. We define $g \colon X \to \mathbb{R}$ by $g(x) = \phi(x)\psi(x)$. Then g is a continuous extension of hf and maps X into $(-1, 1)$ in case (a), into $[-1, 1)$ in case (b) and into $(-1, 1]$ in case (c). It is now immediate that $h^{-1}g$ is the desired extension of f. $\diamond$

Remark 8.2.12 (a) The theorem fails if we omit the assumption that the subspace A is closed. For example, the continuous function $x \to \log(x/1 - x)$ defined on $(0, 1)$ does not extend to a continuous function on $[0, 1]$ because any continuous function on $[0, 1]$ must be bounded.

(b) We have proved the Tietze theorem by invoking the Urysohn lemma. Conversely, the Urysohn lemma can be derived from the Tietze theorem. For, given disjoint closed subsets A, B of X, the function $f \colon A \cup B \to \mathbb{R}$ defined by $f(a) = 0$ for all $a \in A$, and $f(b) = 1$ for all $b \in B$ is continuous. Then an extension g of f is a Urysohn function for the pair A and B.

(c) It is clear from (a) and Remark 8.2.7c that a T_1-space X is normal if each continuous real-valued function on a closed subset of X extends continuously to X.

(d) The theorem can be generalized to maps into $\mathbb{R}^n$. If A is a closed subset of the normal space X, then every continuous function $A \to \mathbb{R}^n$ extends to continuous function $X \to \mathbb{R}^n$. To see this, one needs to apply the above theorem to the coordinate functions of the given map $A \to \mathbb{R}^n$. The claim remains true if one takes the cube I^n instead of $\mathbb{R}^n$.

Corollary 8.2.13 *Let A be a closed subset of a normal space X, and let f be a continuous function of A into the n-sphere $\mathbb{S}^n$, $n \geq 0$. Then there is an open set $U \subseteq X$ such that $A \subseteq U$, and f has a continuous extension $U \to \mathbb{S}^n$.*

Proof By the proof of Example 2.2.6, there is a continuous function $h \colon \mathbb{S}^n \to \partial I^{n+1}$. So the composition $A \xrightarrow{f} \mathbb{S}^n \xrightarrow{h} I^{n+1}$ is continuous. If $p_i \colon I^{n+1} \to I$ is the projection map onto the ith factor, $1 \leq i \leq n+1$, then each composition $p_i h f$ can be extended to a continuous mapping $g_i \colon X \to I$, by the Tietze theorem. Define $g \colon X \to I^{n+1}$ so that $p_i g = g_i$ for every i. Then g is a continuous extension of hf. Now, let z denote the center of I^{n+1}. Then, by Lemma 5.3.4, the ray from z through each point $y \neq z$ meets ∂I^{n+1} in exactly one point $k(y)$ (say). The function $k \colon I^{n+1} - \{z\} \to \partial I^{n+1}$ taking y into $k(y)$ is clearly continuous and fixes the points in ∂I^{n+1}. Set $U = g^{-1}\left(I^{n+1} - \{z\}\right)$. Then U is open in X and $A \subseteq U = X - g^{-1}(z)$. It is obvious that g maps U into $I^{n+1} - \{z\}$, and the composition $kg \colon U \to \partial I^{n+1}$ is defined and continuous. If $x \in A$, then $kg(x) = khf(x) = hf(x)$, so $h^{-1}kg(x) = f(x)$. Thus $h^{-1}kg \colon U \to \mathbb{S}^n$ is a desired extension of f. $\diamond$

Exercises

1. On the set $\{a, b, c\}$, find a topology which is T_4 but not T_3.

2. Let I be the unit interval with the Euclidean subspace topology, and let $J \subseteq I$ be the set of all irrationals. Consider the topology $\mathscr{T}$ generated by the subbasis consisting of subsets of J and the open subsets of I. Prove that $(I, \mathscr{T})$ is normal.

3. • (a) Prove that a T_1-space is normal $\Leftrightarrow$ for each closed $A \subseteq X$ and for every open $U \supseteq A$, there exists an open set $V \subseteq X$ such that $A \subseteq V \subseteq \overline{V} \subseteq U$.

    (b) If A, B are disjoint closed subsets of a normal space X, show that there are open sets $U \supseteq A$ and $V \supseteq B$ such that $\overline{U} \cap \overline{V} = \varnothing$.

    (c) Prove (b) for finitely many closed sets $A_1, \ldots, A_n$ with $\bigcap_1^n A_i = \varnothing$.

4. Let X be a T_3-space, K a compact subset of X and U be an open nbd of K in X. Show that there exists an open set $V \subseteq X$ such that $K \subseteq V \subseteq \overline{V} \subseteq U$. Deduce that a compact T_3-space is T_4.

5. In the space $\mathbb{R}_\ell^2$, consider the sets $A = \{(x, -x) \mid x \in \mathbb{Q}\}$, and $B = \{(x, -x) \mid x \in \mathbb{R} - \mathbb{Q}\}$. Show that A and B are disjoint closed subsets of $\mathbb{R}_\ell^2$ which fail to have disjoint nbds.

6. Show that the space Z in Example 8.1.7 is separable, but not Lindelöf. (Thus, a regular separable space need not be normal.)

7. Prove that the ordinal space $[0, \Omega)$ is normal. Generalize this to an ordered space with a well-ordering.

8. Give an example of a proper map $f : X \to Y$ such that Y is normal, but X is not.

9. Prove that the sum and the wedge sum of a family of normal spaces are normal.

10. Suppose that a space Y is a coherent union of a countable family of closed subsets X_n. If each X_n is normal, show that Y is normal.

11. Let D be a dense subset of the nonnegative reals. Suppose that for each $r \in D$, there is an open subset U_r of a space X such that $r < s \Rightarrow \overline{U}_r \subseteq U_s$, and $\bigcup_{r \in D} U_r = X$. Show that the function $f : X \to \mathbb{R}$ defined by $f(x) = \inf \{r \mid x \in U_r\}$ is continuous.

12. Show that the condition of closedness on the sets A and B in the Urysohn lemma is essential.

13. Does there exist a countable connected regular space?

14. Let K be a compact subset of a locally compact Hausdorff space X and U be an open set in X with $K \subseteq U$. Show that there exists a continuous function $f : X \to I$ such that $f(x) = 0$ for $x \notin U$ and $f(x) = 1$ for $x \in K$.

15. Prove that a T_1-space X is normal if and only if for each finite covering $\{U_1, \ldots, U_n\}$ of X by open sets, there exist continuous functions $f_1, \ldots, f_n$ of X into I such that $f_i(x) = 0$ for $x \notin U_i$ and $\sum f_i(x) = 1$ for all $x \in X$.

16. (a) Let X be a topological space. If $f : X \to \mathbb{R}$ is a continuous function, show that $A = f^{-1}(0)$ is a closed G_δ-set.

    (b) If X is normal and A is a closed G_δ-set in X, prove that there exists a continuous real-valued function f such that $A = f^{-1}(0)$.

17. Let X be a normal space. If A and B are disjoint closed G_δ-sets in X, show that there exists a continuous map $f : X \to I$ such that $A = f^{-1}(0)$ and $B = f^{-1}(1)$.

18. Suppose that X is a compact space. If there exists a continuous function $f : X \times X \to \mathbb{R}$ such that $f(x, y) = 0 \Leftrightarrow x = y$, prove that the diagonal Δ in $X \times X$ is a G_δ-set, and hence deduce that X is second countable. (Thus X is metrizable.)

19. Is a second countable Hausdorff space metrizable?

20. Prove that every separable metric space is homeomorphic to a subspace of a Fréchet space $\mathbb{R}^\omega$ (refer to Exercise 2.2.19).

21. Show that the product space I^I is separable, normal but not first countable (and hence is not metrizable).

22. Prove that a metrizable space has a totally bounded metric $\Leftrightarrow$ it is second countable.

23. Prove that a countably compact Hausdorff space is metrizable if and only if it is second countable.

24. Prove that the continuous image of a compact metric space in a Hausdorff space is metrizable.

25. Prove that every infinite subset of a pseudocompact normal space has a limit point.

26. Let X be a metrizable space. Prove that the following conditions are equivalent:

   (a) X is compact.

   (b) X is bounded in every metric that gives the topology of X.

   (c) X is pseudocompact.

27. (a) If a Hausdorff space X is the union of two compact metrizable spaces, show that X is metrizable. (This is known as the *addition theorem* for compacta.)

   (b) Give an example of a non-metrizable space that is the union of two metrizable subsets.

28. • Show that the one-point compactification of a second countable, locally compact Hausdorff space is metrizable. (This implies that a second countable, locally compact Hausdorff space is metrizable.)

29. A space X is called T_5 if for every pair of separated sets A, B (i.e., with $\overline{A} \cap B = \varnothing = A \cap \overline{B}$) in X, there exist disjoint open sets containing them. X is said to be *completely normal* if it is both T_1 and T_5.

   Prove: (a) A metric space is completely normal.

   (b) $\mathbb{R}_\ell$ is completely normal.

   (c) $\mathbb{R}_\ell^2$ is not completely normal.

   (d) Complete normality is hereditary.

   (e) X is completely normal if and only if every subspace of X is normal.

   (f) A regular second countable space is completely normal.

   (g) An ordered space with the least upper bound property is completely normal.

   (h) The product space I^I is not completely normal.

30. A normal space in which each closed set is G_δ is called *perfectly normal*.

   (a) Prove that a metric space is perfectly normal.

   (b) Show that a perfectly normal space is completely normal.

   (c) Give an example of a completely normal space which is not perfectly normal.

8.3 Completely Regular Spaces

Let A and B be a disjoint pair of subsets of the space X. The previous section is concerned with the question of whether A and B can be separated by open sets: Do there exist disjoint open nbds U, V of A, B, respectively? The present section is concerned with the question whether A and B can be separated by a continuous function: Does there exist a Urysohn function f for the pair A, B? The latter condition implies the former, since we can take $U = f^{-1}[0, 1/2)$, $V = f^{-1}(1/2, 1]$. The Urysohn lemma says that in a normal space we can separate each pair of disjoint closed sets by a continuous function, and thus the two notions are equivalent. Naturally, one would be interested to know if an analogous statement holds good in a regular space, where points can be separated from closed sets by open sets. The answer is no, because there is a regular space, due to E. Hewitt, on which every continuous real-valued function is constant. This suggests a new separation condition intermediate between regularity and normality.

Definition 8.3.1 A space X is $T_{3\frac{1}{2}}$ if for each closed set $F \subseteq X$ and each point $x \in X - F$, there exists a continuous function $f : X \to I$ such that $f(F) = 0$ and $f(x) = 1$. A space which satisfies both T_1 and $T_{3\frac{1}{2}}$ axioms is called *completely regular* or *Tychonoff*.

A function $f : X \to I$ satisfying the condition of the preceding definition is said to separate F and x, and is also referred to as a Urysohn function for F and x.

Example 8.3.1 The Euclidean space $\mathbb{R}^n$ is completely regular. More generally, every metric space X is completely regular. For, if F is a closed subset of X and $p \in X - F$, then $g : X \to \mathbb{R}$ defined by $g(x) = \text{dist}(x, F)$ is continuous. If $r = g(p) > 0$, then the mapping $f : x \mapsto |\sin(g(x)\pi/2r)|$ is a Urysohn function for F and p.

It is obvious that a completely regular space is regular. Since points are closed in normal spaces, it follows from the Urysohn lemma that a normal space is completely regular. There are, however, completely regular spaces which fail to be normal.

Example 8.3.2 The space Z in Example 8.1.7 is completely regular. To see this, consider a closed subset $F \subset Z$ and a point p in $Z - F$. Let U be a basic nbd of p contained in $Z - F$. If $p \in H$ (the open upper half plane), then U is an open disc B centered at p, and if $p \in L$ (the x-axis), then $U = B \cup \{p\}$, where B is an open disc tangent to L at p. We define a function $f : Z \to \mathbb{R}$ by sending each straight line segment from p to a point on the boundary of B linearly onto I, and $z \notin U$ to 1. More specifically, if $U = B(p; r)$, we put $f(z) = |z - p|/r$ for $z \in U$, and if $U = B \cup \{p\}$, we put

$$f(z) = \begin{cases} 0 & \text{for } z = p, \\ |z - p|^2/2ry & \text{for } z \in B, \end{cases}$$

where y is the ordinate of z and r is the radius of B. To see the continuity of f in the first case, we observe that the function $z \to |z - p|/r$ is defined on ∂U and assumes value 1. Therefore the Gluing lemma applies, and f is continuous. Similarly, the continuity of f in the latter case is established, once we check that it is continuous at $p = (x, 0)$. But this is clear because the nbd $B((x, \delta); \delta) \cup \{p\}$ of p is mapped by f into $[0, \epsilon)$, where $\delta = r\epsilon$. It follows that f is a Urysohn function for F and p, and Z is completely regular. We have already seen (in Example 8.2.4) that it is not normal.

Proposition 8.3.2 *Every subspace of a completely regular space is completely regular.*

Proof Suppose that Y is a subspace of a completely regular space X. Let A be a closed subset of Y and $y \notin A$. We have $A = Y \cap F$ for some closed subset F of X. Since X is completely regular and $y \notin F$, there is a Urysohn function $f : X \to I$ for F and y. Then the restriction of f to Y separates A and y. $\diamond$

Corollary 8.3.3 *Every locally compact Hausdorff space is completely regular.*

Proof Let X be a locally compact Hausdorff space and consider its one-point compactification X^*. Being a compact Hausdorff space, X^* is normal and hence completely regular. Since X is a subspace of X^*, it must be completely regular, by the preceding proposition. $\diamond$

The complete regularity also behaves nicely under the formation of products.

Theorem 8.3.4 *The product of a family of completely regular spaces is completely regular.*

Proof Let X be the product of a family of completely regular spaces $X_\alpha, \alpha \in A$. Since each X_α is T_1, so is X. Given a closed set $F \subset X$ and a point $x = (x_\alpha)$ in $X - F$, we choose a basic open nbd $\prod U_\alpha$ of x that does not meet F. Then $U_\alpha = X_\alpha$ for all but finitely many indices α, say, $\alpha = \alpha_1, \ldots, \alpha_n$. Since X_{α_i} is completely regular, there is a continuous function $f_i : X_{\alpha_i} \to I$, for each $i = 1, \ldots, n$, such that $f_i(x_{\alpha_i}) = 1$ and $f_i(X_{\alpha_i} - U_{\alpha_i}) = 0$. The composition $X \xrightarrow{p_{\alpha_i}} X_{\alpha_i} \xrightarrow{f_i} I$ takes x into 1 and vanishes outside $p_{\alpha_i}^{-1}(U_{\alpha_i})$, where the mapping p_{α_i} is natural projection. We define $g : X \to I$ by $g(y) = \min \{f_i p_{\alpha_i}(y) \mid i = 1, \ldots, n\}$ for every $y \in X$. Then g is continuous and $g(x) = 1$, and $g = 0$ throughout $F \subseteq X - \prod U_\alpha = \bigcup_{i=1}^{n} (X - p_{\alpha_i}^{-1}(U_{\alpha_i}))$. $\diamond$

The converse of the above theorem is also true. To see this, suppose that $f : X \to Y$ is an embedding and Y is completely regular. Then, given a closed set $F \subset X$ and a point $x \in X - F$, there exists a continuous function $g : f(X) \to I$ separating the closed set $f(F)$ and the point $f(x)$, since $f(X)$ is completely regular. It is now clear that the composition $gf : X \to I$ separates x and F, and X is completely regular. Since each factor of a product space can be embedded in it, the assertion follows.

The above argument also shows that complete regularity is a topological invariant. But, this property is not preserved by even continuous closed or open maps in general.

Example 8.3.3 The quotient space of the completely regular space Z in Example 8.3.2 obtained by identifying the closed set $A = \{(x, 0) \mid x \in \mathbb{Q}\}$ to a point is not even regular because the point $[A]$ and the closed set $F = \{(x, 0) \mid x \in \mathbb{R} - \mathbb{Q}\}$ cannot be separated by open sets. Hence Z/A is not completely regular, although the natural map $Z \to Z/A$ is closed.

There is an analogue of the Urysohn Embedding Theorem for completely regular spaces without even the condition of second countability. We recall that for an indexing set A, the cartesian product of A copies of the unit interval I is denoted by I^A and, with the product topology, it is referred to as a *cube*. We have the following

Theorem 8.3.5 (Tychonoff Embedding Theorem) *A space is completely regular if and only if it can be embedded in a cube.*

Proof Let X be a completely regular space and Φ denote the family of all continuous functions $X \to I$. Consider the function $e\colon X \to I^\Phi$ defined by $e(x)_f = f(x)$, $f \in \Phi$. We show that e is an embedding. If $p_f\colon I^\Phi \to I$ is the fth projection, then, by the definition of the mapping e, we have $p_f \circ e = f$ for all $f \in \Phi$. Hence e is continuous. Next, if $x \neq y$ are two points of X, then there exists a $f \in \Phi$ such that $f(x) = 0$ and $f(y) = 1$, for X satisfies T_1-axiom. So $e(x) \neq e(y)$, and e is injective. It remains to prove that if $U \subseteq X$ is open, then $e(U)$ is open in the subspace $e(X)$ of I^Φ. To see this, let $x \in U$ be arbitrary. Since X is T_3, there exists a $f \in \Phi$ such that $f(X - U) = 0$ and $f(x) = 1$. Then $p_f^{-1}(0, 1]$ is open in I^Φ, so $G = e(X) \cap p_f^{-1}(0, 1]$ is open in $e(X)$. Clearly, $e(x) \in G \subseteq e(U)$, and therefore $e(U)$ is open in $e(X)$.

For the converse, we note that a cube is completely regular, by Theorem 8.3.4, since the unit interval I is completely regular. So, by Proposition 8.3.2, it is clear that any space X, which can be embedded in a cube, is completely regular.   ◇

Since a cube is normal, it follows that a space is completely regular if and only if it is homeomorphic to a subspace of a normal space. Furthermore, using the above theorem, we give an apparently shorter proof of the Urysohn Metrization Theorem.

Theorem 8.3.6 *A completely regular, second countable space X is metrizable.*

Proof Let $\mathscr{B}$ be a countable basis for X, and let $\mathscr{C}$ be the family of all pairs (B, B') of members of $\mathscr{B}$ such that $B \supseteq \overline{B'}$. Clearly, $\mathscr{C}$ is countable. For each pair (B, B') in $\mathscr{C}$, we select a continuous function $f\colon X \to I$ such that $f \equiv 0$ on $X - B$ and $f \equiv 1$ on $\overline{B'}$, provided such a function exists. Since $\mathscr{C}$ is countable, so is the family Φ of the functions f. Therefore the cube I^Φ is metrizable, by Theorem 2.2.11. We show that X can be embedded in I^Φ, and hence it is metrizable. By the proof of Tychonoff Embedding theorem, it suffices to prove that if $F \subseteq X$ is closed and $x \in X - F$, then there exists a function $f \in \Phi$ such that $f(F) = 0$ and $f(x) = 1$. Since $\mathscr{B}$ is a base of X, there is a $B \in \mathscr{B}$ such that $x \in B \subseteq X - F$. Then there is a continuous function $g\colon X \to I$ such that $g \equiv 0$ on $X - B$ and $g(x) = 1$, for X is completely regular. The function $h\colon I \to I$ given by $h(t) = 2t$ for $0 \leq t \leq 1/2$ and $h(t) = 1$ for $1/2 \leq t \leq 1$ is continuous, so the composition $hg = f\colon X \to I$ is continuous.

Obviously, $f \equiv 1$ on the open set $g^{-1}((3/4, 1])$ and $f \equiv 0$ on $X - B$. Choose a basic open nbd B' of x contained in $g^{-1}((3/4, 1])$. Then $f \equiv 1$ on $\overline{B'}$, so $\overline{B'} \subseteq B$. Thus we have $f \in \Phi$, where $f(F) = 0$ and $f(x) = 1$, as desired. ◇

Stone–Čech Compactification

In general, a *compactification* of a space X is a compact space $\alpha(X)$ together with an embedding $\eta: X \to \alpha(X)$ such that $\eta(X)$ is dense in $\alpha(X)$. In Sect. 5.4, we have already discussed one method of compactifying a space, viz., the construction of one-point compactification. This construction is somewhat unsatisfactory because it fails to admit continuous extensions. For example, the continuous function $x \xrightarrow{f} \sin(1/x)$ defined on $(0, 1]$ cannot be extended continuously over $[0, 1]$, the one-point compactification of $(0, 1]$. However, it is possible to find a compact space K which contains a homeomorphic copy X' of $(0, 1]$ and admits a continuous extension of f carried over to X' via the homeomorphism $X' \approx (0, 1]$. Let X' be the graph of f. Then $(0, 1] \approx X'$. Since X' is contained in the product $[0, 1] \times [-1, 1]$, $K = \overline{X'}$ is compact. Obviously, the function $K \to \mathbb{R}$, $(x, y) \mapsto y$, is an extension of the composition $X' \approx (0, 1] \xrightarrow{f} \mathbb{R}$.

In the year 1937, M.H. Stone and E. Čech each published important papers which provided independent proofs of the existence of a compactification of a space X with the property that every bounded real-valued continuous function on X extends continuously over it. This compactification of X is called the *Stone–Čech compactification*, and usually denoted by $\beta(X)$. Moreover, since a compact Hausdorff space is completely regular and the latter property is hereditary, only completely regular spaces X can have such a Hausdorff compactifications.

Theorem 8.3.7 (Stone–Čech) *For every completely regular space X, there exists a compact Hausdorff space $\beta(X)$ such that*

(a) *X is homeomorphic to a dense subspace X' of $\beta(X)$, and*

(b) *every bounded real-valued continuous function on X' can be extended uniquely to a continuous function $\beta(X) \to \mathbb{R}$.*

Proof Let Φ be the family of all continuous functions $f: X \to I$ (the unit interval). Then, by the proof of Theorem 8.3.5, the mapping $e: X \to I^\Phi$, $x \mapsto (f(x))$, is an embedding. Let X' denote the range of e and write $\beta(X) = \overline{X'}$ (the closure of X' in I^Φ). Then the subspace $\beta(X) \subseteq I^\Phi$ compact and Hausdorff, since I^Φ is so. Obviously, X is homeomorphic to the subspace X' of $\beta(X)$, and X' is dense in $\beta(X)$.

To establish the last statement, assume that $g: X' \to \mathbb{R}$ is a bounded continuous function. We choose appropriate positive numbers b and c so that $0 \leq b(ge(x) + c) \leq 1$ holds for every $x \in X$. Then $b(ge + c) = f$ for some $f \in \Phi$. By the definition of e, we have $p_f e(x) = f(x)$ for all $x \in X$, where p_f is the fth projection map $I^\Phi \to I$. Accordingly, we define $F: \beta(X) \to I$ to be the restriction of p_f to $\beta(X)$. Then, for $x \in X$, $Fe(x) = f(x)$, while $ge(x) = b^{-1}f(x) - c$. It follows that $b^{-1}F - c = G$ is a continuous extension of g over $\beta(X)$. The map G is unique because X' is dense in $\beta(X)$. ◇

Remark 8.3.8 (a) We usually identify the space X with the subspace $X' \subseteq \beta(X)$, and consider X as a subspace of $\beta(X)$.

(b) The proof of the preceding theorem also establishes that if X is a compact T_2-space, then $\beta(X) \approx X$.

The next theorem shows that not only a real-valued bounded continuous function on X extends continuously over $\beta(X)$, but also any continuous function of X into a compact Hausdorff space does so.

Theorem 8.3.9 (Stone) *Let f be a continuous function of a completely regular space X into a compact Hausdorff space Y. Then there exists a unique continuous extension of f over $\beta(X)$.*

Proof Let Φ be the family of all continuous functions of Y into $\mathbb{R}$. For each $\phi \in \Phi$, denote the range of ϕ by R_ϕ. Note that R_ϕ is a closed bounded subset of $\mathbb{R}$. By Theorem 8.3.7, there exists a continuous map $F_\phi \colon \beta(X) \to \mathbb{R}$ such that $F_\phi(x) = (\phi f)(x)$ for every $x \in X$. Since $\beta(X) = \overline{X}$ and R_ϕ is closed, it follows that F_ϕ maps $\beta(X)$ into R_ϕ. Consequently, there is a continuous function G from $\beta(X)$ into the product space $\prod R_\phi$ given by $p_\psi(G(x)) = F_\psi(x)$ for all $x \in \beta(X)$, where the mapping $p_\psi \colon \prod R_\phi \to R_\psi$ is projection. Let $i \colon X \to \beta(X)$ be the inclusion map and let $j \colon Y \to \prod R_\phi$ be the function defined by $p_\phi(j(y)) = \phi(y)$. Clearly, j is continuous and we have $p_\psi \circ j \circ f = \psi \circ f = F_\psi \circ i = p_\psi \circ G \circ i$ for every ψ, that is, the following diagram of topological spaces and continuous functions commutes for all ψ:

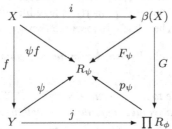

Therefore we deduce that $G \circ i = j \circ f$, and so

$$G\,(\beta(X)) \subseteq \overline{(Gi)(X)} = \overline{(jf)(X)} \subseteq \overline{j(Y)}.$$

Since Y is compact and $\prod R_\phi$ is Hausdorff, j is a closed mapping. In particular, $j(Y)$ is closed in $\prod R_\phi$, so we have $G\,(\beta(X)) \subseteq j(Y)$. Further, if $y \neq y'$, then there exists a $\phi \in \Phi$ such that $\phi(y) = 0$ and $\phi(y') = 1$, since Y is normal. It follows that j is injective and thus a closed embedding. Consequently, the mapping $j^{-1} \colon j(Y) \to Y$ is defined and continuous. Denote the composition $j^{-1} \circ G$ by H. Then $H \circ i = f$, so H is a continuous extension of f. Since X is dense in $\beta(X)$ and Y is a T_2-space, Corollary 4.4.3 shows that H is the unique. $\diamond$

Corollary 8.3.10 *Any Hausdorff compactification $\alpha(X)$ of a completely regular space X such that every continuous function of X into a compact space has an* *extension over it is homeomorphic to* $\beta(X)$*; indeed, there is a homeomorphism* $\alpha(X) \approx \beta(X)$ *which leaves points of X fixed.*

Proof Let $i: X \to \beta(X)$ and $j: X \to \alpha(X)$ be the inclusions. By our hypothesis, there exists a continuous map $\tilde{i}: \alpha(X) \to \beta(X)$ such that $\tilde{i} \circ j = i$. And, by Theorem 8.3.9, there exists a continuous map $\tilde{j}: \beta(X) \to \alpha(X)$ such that $\tilde{j} \circ i = j$. So $\tilde{i} \circ \tilde{j} \circ i = i$, thus $\tilde{i} \circ \tilde{j}$ and the identity map on $\beta(X)$ agree on X. Since X is dense in $\beta(X)$, $\tilde{i} \circ \tilde{j}$ is the identity map on $\beta(X)$. Similarly, $\tilde{j} \circ \tilde{i}$ is the identity map on $\alpha(X)$, and therefore $\tilde{j}$ is a homeomorphism with $\tilde{i} = \tilde{j}^{-1}$. Also, it is immediate that $\tilde{j}$ fixes the points of X. ◇

From the last two results, it is clear that the Stone–Čech compactification $\beta(X)$ of a completely regular space X is essentially unique. We also see that $\beta(X)$ is characterized by the property that every bounded real-valued continuous function on X extends continuously over it.

Example 8.3.4 The ordinal space $[0, \Omega]$ is the Stone–Čech compactification of the subspace $[0, \Omega)$. We know that $[0, \Omega)$ is dense in the space $[0, \Omega]$, since Ω is a limit point of $[0, \Omega)$. By Example 5.1.9, $[0, \Omega]$ is compact, so it is a Hausdorff compactification of $[0, \Omega)$. To prove that it is Stone–Čech compactification, it suffices to show that every bounded continuous function $f: [0, \Omega) \to \mathbb{R}$ can be extended continuously over $[0, \Omega]$. With this end in view, we first observe that for each integer $n > 0$, there exists an $x_n < \Omega$ such that $|f(x_n) - f(y)| < 1/n$ for all $y \geq x_n$ in $[0, \Omega)$. Assume, on the contrary, that there is an integer $m > 0$ such that for each ordinal number $x < \Omega$, there exists $y \in (x, \Omega)$ with $|f(y) - f(x)| \geq 1/m$. Let y_1 be the least element of $(0, \Omega)$ such that $|f(y_1) - f(0)| \geq 1/m$. Next, let y_2 be the least element of (y_1, Ω) such that $|f(y_2) - f(y_1)| \geq 1/m$. Continuing in this way, we obtain a sequence of ordinals $0 = y_0 < y_1 < y_2 < \cdots$ such that $|f(y_n) - f(y_{n-1})| \geq 1/m$. If y is the supremum of $\{y_0, y_1, \ldots\}$, then $0 < y < \Omega$, and each basic open nbd $(x, y]$ of y in $[0, \Omega)$ contains all but finitely many y_n. Consequently, f fails to map $(x, y]$ into the open nbd $(f(y) - 1/3m, f(y) + 1/3m)$ of $f(y)$. This contradicts the continuity of f at y, and hence our claim. Now, for each positive integer n, choose an ordinal number $x_n < \Omega$ satisfying the above condition. Let $b < \Omega$ be an upper bound of $\{x_1, x_2, \ldots\}$ (such a b exists, since $\sup x_n < \Omega$). Then, for any $x \geq b$, we have

$$|f(x) - f(b)| \leq |f(x) - f(x_n)| + |f(x_n) - f(b)| < 2/n$$

for all n. Therefore $f(x) = f(b)$ for all $x \geq b$ in $[0, \Omega)$. Clearly, a desired continuous extension F of f can be obtained by defining $F(\Omega) = f(b)$ and $F(x) = f(x)$ for $x < \Omega$.

In view of the fact that, even for such a simple space as $\mathbb{N}$, $\beta(N)$ is fairly complicated (see Exercise 17), the above example is rather exceptional.

As another consequence of Theorem 8.3.9, we have

Corollary 8.3.11 *Any Hausdorff compactification of X is a continuous image of* $\beta(X)$ *under a mapping which leaves points of X fixed.*

Proof Suppose that $\alpha(X)$ is a Hausdorff compactification of X. By Theorem 8.3.9, there is a continuous map $\tilde{j}: \beta(X) \to \alpha(X)$ extending the inclusion map $j: X \to \alpha(X)$. Clearly, the image of $\tilde{j}$ is a closed set containing X. Since X is dense in $\alpha(X)$, we have $\alpha(X) = \operatorname{im}(\tilde{j})$, that is, $\tilde{j}$ is surjective. $\diamond$

If $\alpha_1(X)$ and $\alpha_2(X)$ are two compactifications of a space X, we define $\alpha_2(X) \preceq \alpha_1(X)$, if there exists a continuous mapping F of $\alpha_1(X)$ onto $\alpha_2(X)$ such that $F(x) = x$ for all $x \in X$. Then $\preceq$ is a partial ordering on the collection of Hausdorff compactifications of X. From the above corollary, it follows that $\beta(X)$ is a maximal element in the collection of Hausdorff compactifications of X, while one-point compactification of a non-compact space is a minimal element.

We close this section with the following theorem which characterizes the topology of a completely regular space.

Theorem 8.3.12 *Let X and Y be completely regular, first countable spaces. Then* $X \approx Y \Leftrightarrow \beta(X) \approx \beta(Y)$.

Proof Suppose first that $h: X \to Y$ is a homeomorphism, and let $i: X \to \beta(X)$ and $j: Y \to \beta(Y)$ be the inclusions. By Theorem 8.3.9, there exist unique continuous functions $\phi: \beta(X) \to \beta(Y)$ and $\psi: \beta(Y) \to \beta(X)$ such that $\phi i = jh$ and $\psi j = ih^{-1}$. Thus we have $\psi\phi(x) = x$ for all $x \in X$, and $\phi\psi(y) = y$ for all $y \in Y$. Since $\overline{X} = \beta(X)$ and $\overline{Y} = \beta(Y)$, we deduce that $\psi\phi = 1_X$ and $\phi\psi = 1_Y$, and so ϕ is a homeomorphism with ψ as its inverse.

Conversely, suppose that $\phi: \beta(X) \to \beta(Y)$ is a homeomorphism. We show that ϕ maps X onto Y. Clearly, this follows from

$$\phi(\beta(X) - X) \cap Y = \varnothing = \phi(X) \cap (\beta(Y) - Y).$$

To this end, we prove that there is no countable open basis at any point of $\beta(X) - X$ or $\beta(Y) - Y$. Assume otherwise, and let $\{U_n\}$ be a countable open basis at a point $p \in \beta(X) - X$. By regularity, we may assume that $\overline{U}_{n+1} \subseteq U_n$ for every n. Since X is dense in $\beta(X)$, each U_n contains infinitely many points of X. We construct two sequences $\langle s_n \rangle$ and $\langle t_n \rangle$ of distinct points of X such that $s_n, t_n \in U_n$ and $s_n \neq t_m$ for all n and m. Clearly, no point of X belongs to $\{p\} = \bigcap U_n = \bigcap \overline{U}_{n+1}$. So, for each integer $i > 0$, there is an integer n_i such that $s_i \notin \overline{U}_{n_i}$. Note that $\overline{U}_{n_i}$ contains all but finitely many s_n's and t_n's. So we can find an open nbd V_i of s_i in X such that $V_i \subseteq U_i - \overline{U}_{n_i}$ and V_i does not contain any t_j and $s_j \neq s_i$. Since X is completely regular, there exists a continuous map $f_i: X \to I$ such that $f_i(s_i) = 1$ and $f_i(x) = 0$ for all $x \notin V_i$. We define a function $g: X \to \mathbb{R}$ by setting $g(x) = \sup_i f_i(x)$. We assert that g is continuous. It is obvious that $g^{-1}(a, \infty) = \bigcup_i f_i^{-1}(a, \infty)$, which is open for all $a \in \mathbb{R}$. To see that $g^{-1}(-\infty, a)$ is also open, consider a point x with $0 \leq g(x) < a$. We have an integer k such that $x \notin \overline{U}_k$. As the function f_i vanishes on $X - \overline{U}_k$ for all $i \geq k$, $W = (X - \overline{U}_k) \cap (\bigcap_{i=1}^{k-1} f_i^{-1}(-\infty, a))$ is a nbd of x. It is easily checked that g maps W into $(-\infty, a)$, and our assertion follows. Now, we have $g(s_n) = 1$ and $g(t_n) = 0$ for all n. Since $\lim s_n = p = \lim t_n$, g cannot have a continuous extension over $\beta(X)$, a contradiction. Therefore there is no countable open basis in $\beta(X)$ at p. A similar argument shows that $\beta(Y)$ also does not have a countable open basis at any point of $\beta(Y) - Y$. $\diamond$

Exercises

1. Show that $\mathbb{R}_\ell^2$ is completely regular.

2. If a space X satisfies the T_3 and T_4 axioms, show that it satisfies $T_{3\frac{1}{2}}$.

3. Let X be a completely regular space and $F \subseteq X$ be closed. Show that for each point $x \in X - F$, there is a continuous function $f : X \to I$ such that $f \equiv 0$ on F and $f \equiv 1$ on a nbd of x.

4. Let X be completely regular, and V be an open nbd of $x \in X$. Prove that a necessary and sufficient condition that there be a continuous function $f : X \to I$ such that $f^{-1}(1) = x$, $f(X - V) = 0$, is that $\{x\}$ be a G_δ-set.

5. Show that a space X is $T_{3\frac{1}{2}}$ if and only if its topology is induced by all real-valued bounded continuous functions on it.

6. Prove that a T_1-space X is completely regular if and only if the topology of X is generated by the cozero sets $X - f^{-1}(0)$ of continuous functions $f : X \to \mathbb{R}$ (as a base).

7. Let X be a completely regular space. Let F be a closed subset and K a compact subset of X with $K \cap F = \varnothing$. Show that there exists a continuous function $f : X \to I$ such that $f(x) = 0$ for all $x \in K$ and $f(x) = 1$ for all $x \in F$.

8. Let $f : X \to Y$ be a proper, open surjection. If X is completely regular, show that Y is also completely regular.

9. Show that a continuous open image of completely regular space need not be completely regular.

10. Let X be a space, and consider the relation $x \sim y$ if and only if $\{x\}$ and $\{y\}$ have the same closure (equivalently, x and y have the same nbd base). Show that $\sim$ is an equivalence relation on X, and if X satisfies the T_i-axiom, $i = 3\frac{1}{2}$ or 4, so does $X/\sim$ (cf. Exercise 8.1.13).

11. Give an example to show that the adjunction space of two completely regular spaces need not be completely regular. (cf. Theorem 8.2.5).

12. If X is a connected, completely regular space having more than one point, show that every nonempty open subset of X is uncountable.

13. Justify the following:

    (a) $[0, 1]$ is not $\beta((0, 1])$.

    (b) $[-1, +1]$ is not $\beta((-1, 1))$.

    (c) $\mathbb{S}^1$ is not $\beta(\mathbb{R}^1)$.

14. Let X be a completely regular space, Y a compact Hausdorff space, and let f be a homeomorphism of X into Y. Show that the Stone–Čech extension $F : \beta(X) \to Y$ sends $\beta(X) - X$ into $Y - f(X)$.

15. Prove that a completely regular space X is connected $\Leftrightarrow \beta(X)$ is connected.

16. Let X be a discrete space. Prove that the closure of every open subset of $\beta(X)$ is open and hence deduce that $\beta(X)$ is totally disconnected.

17. ● Show that a sequence in $\mathbb{N}$ converges in $\beta(\mathbb{N})$ if and only if it converges in $\mathbb{N}$. Conclude that $\beta(\mathbb{N})$ is not second countable or metrizable. (The sequence $\langle 1, 2, \ldots \rangle$ in $\beta(\mathbb{N})$ has no convergent subsequence, although it has a convergent subnet. The space $\beta(\mathbb{N})$ is one of the most widely studied topological spaces. The interested reader may see the text by Walker [15].)

18. If a sequence $\langle x_n \rangle$ in a normal space X converges to some point y in $\beta(X)$, show that $y \in X$.

19. Suppose that the Stone–Čech compactification of a completely regular X is metrizable. Show that X is a compact metric space.

Chapter 9
Paracompactness and Metrizability

Paracompact Hausdorff spaces are very close to metrizable spaces, and have proved quite useful in differential geometry and topology. Section 9.1 concerns with the study of these spaces. It will be seen that every metric space is paracompact Hausdorff. On the other hand, a locally metrizable paracompact Hausdorff space is metrizable. In fact, there are several theorems that guarantee the metrizability of a topological space X. In the previous chapter, we have seen such a theorem, viz., the Urysohn Metrization Theorem. In Sect. 9.2, we treat the most important one, it gives a complete characterization of metric spaces.

9.1 Paracompact Spaces

The concept of paracompactness was introduced in 1944 by J. Diedonné as a generalization of compactness. It is formulated using the following terminology.

Definition 9.1.1 Let $\{U_\alpha \mid \alpha \in A\}$ and $\{V_\beta \mid \beta \in B\}$ be two coverings of a space X. We say that $\{U_\alpha\}$ is a *refinement of* $\{V_\beta\}$ if for each $\alpha \in A$, there is a $\beta \in B$ such that $U_\alpha \subseteq V_\beta$. In this case, we also say that $\{U_\alpha\}$ *refines* $\{V_\beta\}$. If each member of $\{U_\alpha\}$ is open (resp. closed), we call $\{U_\alpha\}$ an open (resp. closed) refinement of $\{V_\beta\}$.

Recall that a family $\{U_\alpha \mid \alpha \in A\}$ of subsets of a space X is locally finite if each point $x \in X$ has a nbd which meets at most finitely many U_α,s (Definition 2.1.8).

Definition 9.1.2 A space X is called *paracompact* if each open covering of X has a locally finite open refinement.

A discrete space is paracompact, since the open covering by singleton sets is locally finite and refines every open covering of the space. It is also obvious that a finite covering is locally finite; so every compact space is paracompact. Also, every

© Springer Nature Singapore Pte Ltd. 2019
T. B. Singh, *Introduction to Topology*,
https://doi.org/10.1007/978-981-13-6954-4_9

metric space is paracompact; however, the proof of this fact is not straightforward and is deferred for a while (Theorem 9.1.7).

We turn now to the study of certain reformulations of the condition for paracompactness. We shall see later that every paracompact Hausdorff space is regular (see the proof of Theorem 9.1.9 below), and thus satisfies T_3-Axiom. Interestingly, for T_3-spaces, there are several reformulations of Definition 9.1.2; we will treat here a few of them.

Theorem 9.1.3 *In a T_3-space X, the following conditions are equivalent:*

(a) *X is paracompact.*

(b) *Each open covering of X has a locally finite refinement (consisting of sets not necessarily either open or closed).*

(c) *Each open covering of X has a locally finite closed refinement.*

Proof (a) $\Rightarrow$ (b): Obvious.

(b) $\Rightarrow$ (c): Let $\mathscr{U}$ be an open covering of X. Since X satisfies the T_3-axiom, there exists an open covering $\mathscr{V}$ of X such that the closure of each member of $\mathscr{V}$ is contained in some member of $\mathscr{U}$. By our hypothesis, $\mathscr{V}$ has a locally finite refinement $\mathscr{W}$, say. Then the family of closed sets $\{\overline{W} \mid W \in \mathscr{W}\}$ is also locally finite and refines $\mathscr{U}$.

(c) $\Rightarrow$ (a): Suppose that X is a space with the property that every open covering of X has a locally finite closed refinement. Let $\mathscr{U}$ be an open covering of X, and let $\mathscr{F} = \{F_\alpha \mid \alpha \in A\}$ be a locally finite closed refinement of $\mathscr{U}$. Then for each $x \in X$, there exists an open nbd V_x of x such that $\{\alpha \in A \mid V_x \cap F_\alpha \neq \varnothing\}$ is finite. Next, find a locally finite closed refinement $\{E_\beta \mid \beta \in B\}$ of the open cover $\{V_x \mid x \in X\}$ of X, and set $W_\alpha = X - \bigcup \{E_\beta \mid E_\beta \cap F_\alpha = \varnothing\}$ for every $\alpha \in A$. Since the family $\{E_\beta \mid \beta \in B\}$ is locally finite, each W_α is open. It is also clear that $F_\alpha \subseteq W_\alpha$ for every α, and hence X is covered by the family $\{W_\alpha \mid \alpha \in A\}$. We observe that it is locally finite, too. Let $x \in X$ be arbitrary. Then there exists a nbd G of x and a finite subset $\Gamma \subseteq B$ such that $G \cap E_\beta = \varnothing$ for all $\beta \notin \Gamma$. So $G \subseteq \bigcup_{\gamma \in \Gamma} E_\gamma$. Since each E_γ is contained in some V_x, the set $\{\alpha \in A \mid E_\gamma \cap F_\alpha \neq \varnothing\}$ is finite for every $\gamma \in \Gamma$. Moreover, by the definition of W_α, $E_\gamma \cap W_\alpha \neq \varnothing \Leftrightarrow E_\gamma \cap F_\alpha \neq \varnothing$. It follows that $\{\alpha \in A \mid G \cap W_\alpha \neq \varnothing\}$ is finite, and hence our assertion. Now, for each index α, choose a member U_α of $\mathscr{U}$ such that $F_\alpha \subseteq U_\alpha$. Then $\{U_\alpha \cap W_\alpha \mid \alpha \in A\}$ is a locally finite open refinement of $\mathscr{U}$, and X is paracompact. $\diamond$

We apply the formulation 9.1.3(b) to obtain one more characterization of paracompactness, which is expressed briefly in the following terminology.

Definition 9.1.4 A family $\{U_\alpha \mid \alpha \in A\}$ of subsets of a space X is *σ-locally finite* if $A = \bigcup_{n \in \mathbb{N}} A_n$ and for each n, the family $\{U_\alpha \mid \alpha \in A_n\}$ is locally finite.

The following specification of a σ-locally finite family is quite useful. Suppose that $\{U_\alpha \mid \alpha \in A\}$ is a σ-locally finite family of subsets of a space X. If we put $U_{n,\alpha} = U_\alpha$ for $\alpha \in A_n$, and $U_{n,\alpha} = \varnothing$ for $\alpha \notin A_n$, then we obtain a family $\{U_{n,\alpha} \mid n \in \mathbb{N} \text{ and } \alpha \in A\}$ such that for each fixed $n \in \mathbb{N}$, the family $\{U_{n,\alpha} \mid \alpha \in A\}$

is locally finite. Note that $\bigcup_\alpha U_\alpha = \bigcup_{n,\alpha} U_{n,\alpha}$. Notice that a locally finite family is σ-locally finite, and so is a countable family.

Theorem 9.1.5 *A T_3-space X is paracompact if and only if each open covering of X has a σ-locally finite open refinement.*

Proof By definition, each open covering of a paracompact space has a locally finite open refinement, and thus the necessity of the condition is obvious. For sufficiency, suppose that every open covering of a T_3-space X has a σ-locally finite open refinement, and let $\mathscr{U}$ be any open covering of X. Then there is an open refinement $\mathscr{V} = \{V_{n,\alpha} \mid n \in \mathbb{N} \text{ and } \alpha \in A\}$ of $\mathscr{U}$ such that for each fixed n, the family $\{V_{n,\alpha} \mid \alpha \in A\}$ is locally finite. For each n, put $V_n = \bigcup_\alpha V_{n,\alpha}$ and set $W_1 = V_1$ and $W_n = V_n - \bigcup_{i<n} V_i$ for $n > 1$. We show that the family $\{W_n \cap V_{n,\alpha} \mid n \in \mathbb{N} \text{ and } \alpha \in A\}$ is a locally finite refinement of $\mathscr{U}$. Given $x \in X$, there is obviously a least integer $n(x)$ such that $x \in V_{n(x)}$. Then $x \in W_{n(x)} \cap V_{n(x),\beta}$ for some $\beta \in A$. Also, it is clear that $V_{n(x),\beta}$ does not intersect W_n for $n > n(x)$. For $i \leq n(x)$, there is an open nbd G_i of x such that $\{\alpha \in A \mid G_i \cap V_{i,\alpha} \neq \varnothing\}$ is finite, since the family $\{V_{i,\alpha} \mid \alpha \in A\}$ is locally finite. Then $\bigcap_{i=1}^{n(x)} \left(G_i \cap V_{n(x),\beta}\right)$ is a nbd of x, which meets only finitely many sets $W_n \cap V_{n,\alpha}$. Thus the family $\{W_n \cap V_{n,\alpha} \mid n \in \mathbb{N} \text{ and } \alpha \in A\}$ is a locally finite refinement of $\mathscr{U}$, and hence X is paracompact, by Theorem 9.1.3. $\diamond$

As an immediate consequence of the above theorem, we have

Corollary 9.1.6 *Every Lindelöf T_3-space is paracompact.*

Proof Let $\mathscr{U}$ be an open covering of a Lindelöf T_3-space X. Then there exists a countable subcovering $\{U_n \mid n \in \mathbb{N}\}$ of $\mathscr{U}$. Obviously, this is an open refinement of $\mathscr{U}$ and decomposes into countably many locally finite families, each consisting of just one set U_n. $\diamond$

Example 9.1.1 The Sorgenfrey line $\mathbb{R}_\ell$ is paracompact, for it is Lindelöf (by Example 7.2.7) and regular (by Example 8.1.4).

Example 9.1.2 The Euclidean space $\mathbb{R}^n$ is paracompact, being a second countable regular space.

More generally, we have the following result, due to A. H. Stone.

Theorem 9.1.7 *Every metric space is paracompact.*

Proof Let (X, d) be a metric space and $\mathscr{U} = \{U_\alpha \mid \alpha \in A\}$ be an open covering of X. Then, for each integer $n \in \mathbb{N}$ and each $\alpha \in A$, let $E_{n,\alpha} = \{x \in U_\alpha \mid \text{dist}(x, X - U_\alpha) \geq 1/n\}$ (see Fig. 9.1). Note that some $E_{n,\alpha}$ may be empty. Now, by the well-ordering principle, we assume that the indexing set A is well ordered by an ordering $\prec$, and put $F_{n,\alpha} = E_{n,\alpha} - \bigcup_{\alpha'<\alpha} U_{\alpha'}$. We show that the family $\mathscr{V}$ of the open sets $V_{n,\alpha} = \bigcup\{B(x; 1/3n) \mid x \in F_{n,\alpha}\}, n \in \mathbb{N} \text{ and } \alpha \in A$, is a σ-locally finite refinement of $\mathscr{U}$. First, observe that $\mathscr{V}$ is a covering of X. For, given $x \in X$, there is an index $\alpha \in A$ such that $x \in U_\alpha$ and $x \notin U_{\alpha'}$ for every $\alpha' \prec \alpha$ (by the well-ordering of A). Then we find an

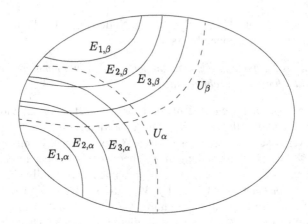

Fig. 9.1 Proof of Theorem 9.1.7

integer $n > 0$ such that $x \in E_{n,\alpha}$; this is possible, for dist$(x, X - U_\alpha) > 0$. Clearly, $x \in F_{n,\alpha} \subseteq V_{n,\alpha}$, and $\mathcal{V}$ covers X. Next, we show that $V_{n,\alpha} \subseteq U_\alpha$ for all n. By the definition of $V_{n,\alpha}$, for each $x \in V_{n,\alpha}$, there exists $y \in F_{n,\alpha}$ such that $d(x, y) < 1/3n$. Since dist$(y, X - U_\alpha) \geq 1/3n$, we must have $x \in U_\alpha$. Thus $\mathcal{V}$ is an open refinement of $\mathcal{U}$. It remains to show that for a fixed n, the family $\{V_{n,\alpha} \mid \alpha \in A\}$ is locally finite. Consider an open ball $B(x; r)$ in X, and assume that it contains points $y \in V_{n,\alpha}$ and $z \in V_{n,\beta}$, where $\alpha \neq \beta$. Then there exist $p \in F_{n,\alpha}$ and $q \in F_{n,\beta}$ such that $d(y, p) < 1/3n$ and $d(z, q) < 1/3n$. So $d(p, q) \leq d(p, y) + d(y, z) + d(z, q) < d(y, z) + 2/3n$. We have either $\beta \prec \alpha$ or $\alpha \prec \beta$; accordingly, either $F_{n,\alpha} \subseteq X - U_\beta$ or $F_{n,\beta} \subseteq X - U_\alpha$. Therefore $d(p, q) \geq 1/n$ in either case, and we find that $d(y, z) > 1/3n$. On the other hand, we have $d(y, z) \leq d(y, x) + d(x, z) < 2r$, and the two inequalities are clearly inconsistent for an $r \leq 1/6n$. Therefore the open ball $B(x; r)$, where $r \leq 1/6n$, can meet at most one of the sets $V_{n,\alpha}$, $\alpha \in A$, as desired. $\diamond$

It is easily seen that paracompactness is a topological invariant. In fact, we shall see later that this property is preserved by continuous closed maps, but, unlike compactness, it is not preserved by continuous maps.

Moving on to the behavior of paracompactness under the formation of subspaces, we see that, like compactness, this property is not hereditary, but inherited by closed subspaces.

Example 9.1.3 The ordinal space ordinal space $[0, \Omega]$ is paracompact, by Example 8.3.4, but its subspace $[0, \Omega)$ is not. To see this, consider the open covering $\mathcal{G} = \{[0, x) \mid 0 < x < \Omega\}$ of $[0, \Omega)$ and assume that there is a locally finite open refinement $\{U_j \mid j \in J\}$ of $\mathcal{G}$. Since the sets $(y, x]$, $0 < x < \Omega$, together with $\{0\}$ form a basis of $[0, \Omega)$, for each nonzero ordinal number $x < \Omega$, we can choose an ordinal number $f(x) < x$ such that $(f(x), x] \subseteq U_j$ for some j. And, define $f(0) =$

0. By our assumption, there exists a finite set $K \subseteq J$ (depending on x) such that $x \notin U_j$ for $j \notin K$. Then we find an ordinal number $\alpha < \Omega$ such that $U_k \subseteq [0, \alpha)$ for every $k \in K$. Clearly, $x \leq f(z)$ for all $z \geq \alpha$, and the set $\{y \mid f(z) \geq x$ for all $z \geq y\}$ is nonempty. Denote its least element by $g(x)$. Thus we obtain a function g from $[0, \Omega)$ to itself. By the principle of recursive definition (Theorem A.5.6), there exists a function $h: [0, \omega) \to [0, \Omega)$ such that $h(n+1) = g(h(n))$, where $h(0) = 1$. Then $s = \sup_{n<\omega} h(n)$ exists and $s < \Omega$. Obviously, $h(n) < s + 1$ for every $n < \omega$. Hence, by the definition of $h(n)$, there is an ordinal $y_n \leq s$ such that $f(z) \geq h(n)$ for all $z \geq y_n$. In particular, $f(s) \geq h(n)$ for every $n < \omega$, so we have $f(s) \geq s$. This contradicts the choice of $f(s)$. Therefore the open covering $\mathscr{G}$ of $[0, \Omega)$ does not have a locally finite open refinement, and the assertion follows.

Theorem 9.1.8 *A closed subspace of a paracompact space is paracompact.*

Proof Let F be a closed set in a paracompact space X, and let $\mathscr{V} = \{V_\alpha \mid \alpha \in A\}$ be an open covering of the subspace F. For each $\alpha \in A$, there exists an open set $U_\alpha \subseteq X$ such that $V_\alpha = F \cap U_\alpha$. Since X is paracompact, the open covering $\{U_\alpha \mid \alpha \in A\} \cup \{X - F\}$ has a locally finite open refinement $\{W_\lambda \mid \lambda \in \Lambda\}$, say. Then the family $\{F \cap W_\lambda \mid \lambda \in \Lambda\}$ is obviously a locally finite open refinement of $\mathscr{V}$. $\diamond$

Paracompact spaces have another interesting property in common with compact spaces, given by the following.

Theorem 9.1.9 *A paracompact Hausdorff space X is normal.*

Proof We first show that X is regular. Let F be a closed subset of X and $x_0 \in X - F$. Then for each $x \in F$, there exist disjoint open sets U_x and V_x such that $x_0 \in U_x$ and $x \in V_x$. Since X is paracompact, the open covering $\{V_x \mid x \in F\} \cup \{X - F\}$ has a locally finite open refinement $\{W_\alpha \mid \alpha \in A\}$, say. Let $B = \{\alpha \in A \mid W_\alpha \subseteq V_x$ for some $x \in F\}$. Then for each $\alpha \in B$, there is an $x \in F$ such that $U_x \cap W_\alpha = \varnothing$. So $x_0 \notin \overline{W}_\alpha$. Put $U = X - \bigcup_{\alpha \in B} \overline{W}_\alpha$ and $V = \bigcup_{\alpha \in B} W_\alpha$. Then U is open, by Proposition 2.1.9, and $x_0 \in U$. Moreover, V is open and disjoint from U. It is also clear that each W_α, which intersects F, is contained in V_x for some $x \in F$. Therefore $F \subseteq V$, and we see that X is regular.

Now, let E and F be two disjoint closed subsets of X. For each $x \in E$, find disjoint open nbds U_x of x and V_x of F, as above. Obviously, the sets U_x, $x \in E$, together with $X - E$ cover X. Let $\{G_\lambda \mid \lambda \in \Lambda\}$ be a locally finite open refinement of this open covering of X, and put

$$M = \{\lambda \in \Lambda \mid G_\lambda \subseteq U_x \text{ for some } x \in E\}.$$

Then, for each $\lambda \in M$, there exists an $x \in E$ such that $G_\lambda \cap V_x = \varnothing$, so $F \cap \overline{G}_\lambda = \varnothing$. Put $G = \bigcup_{\lambda \in M} G_\lambda$ and $H = X - \bigcup_{\lambda \in M} \overline{G}_\lambda$. Then G and H are disjoint open nbds of E and F, respectively. $\diamond$

From the proof of the preceding theorem, it is clear that *a paracompact T_3-space is T_4*.

The following result is obviously a generalization of Theorem 9.1.8.

Proposition 9.1.10 *Every F_σ-set in a paracompact Hausdorff space is paracompact. (Recall that a subset A of a space X is F_σ if it is the union of at most countably many closed sets.)*

Proof Let Y be an F_σ-set in a paracompact space X, and suppose that $Y = \bigcup_n F_n$, where each F_n is closed in X. Let $\mathscr{V} = \{V_\alpha \mid \alpha \in A\}$ be an open covering of the subspace Y. Then for each $\alpha \in A$, we find an open set $U_\alpha \subseteq X$ such that $V_\alpha = Y \cap U_\alpha$. Clearly, for each fixed n, $\{U_\alpha \mid \alpha \in A\} \cup \{X - F_n\}$ is an open covering of X. Therefore it has a locally finite open refinement $\{W_{n,\lambda} \mid \lambda \in \Lambda_n\}$. For each n, let

$$\mathscr{G}_n = \{Y \cap W_{n,\lambda} \mid F_n \cap W_{n,\lambda} \neq \varnothing\}.$$

Then each family $\mathscr{G}_n$ is locally finite and a covering of F_n by open subsets of Y. Therefore the family $\bigcup_n \mathscr{G}_n$ is an open covering of Y. Note that if $W_{n,\lambda} \neq \varnothing$, then $W_{n,\lambda} \subseteq U_\alpha$ for some α. So $\bigcup_n \mathscr{G}_n$ refines $\mathscr{V}$. Since X is regular (by Theorem 9.1.9), the subspace $Y \subseteq X$ is also so. Thus Theorem 9.1.5 applies to Y, and we conclude that it is paracompact.   ◇

We remark that there is no analogue of this theorem for compact spaces.

Coming to the behavior of paracompactness under the formation of products, we note that even the product of two paracompact spaces need not be paracompact. For example, the product of the paracompact Hausdorff space $\mathbb{R}_\ell$ with itself is not paracompact, since it is not normal (see Example 8.2.5). In this regard, the following result is quite useful.

Proposition 9.1.11 *The product of a paracompact space and a compact space is paracompact.*

Proof Let X be a paracompact space and Y be a compact space. Consider an open covering $\mathscr{U}$ of $X \times Y$, and let $x \in X$ be fixed. Then, for each $y \in Y$, choose a member U in $\mathscr{U}$ with $(x, y) \in U$. Now, find open nbds G_y of x and H_y of y such that $(x, y) \in G_y \times H_y \subseteq U$. Clearly, the open covering $\{H_y \mid y \in Y\}$ of the compact space Y has a finite subcovering. Thus, there exist finitely many points $y_1, \ldots, y_{n_x}$ (say) in Y such that $Y = \bigcup_{j=1}^{n_x} H_{y_j}$. Put $G_x = \bigcap_{j=1}^{n_x} G_{y_j}$. Then each $G_x \times H_{y_j}$ is contained in some member of $\mathscr{U}$. Since X is paracompact, the open covering $\{G_x \mid x \in X\}$ has a locally finite open refinement $\{V_\lambda\}_{\lambda \in \Lambda}$, say. For each $\lambda \in \Lambda$, choose a point $\xi(\lambda) \in X$ such that $V_\lambda \subseteq G_{\xi(\lambda)}$ and write $H_{j,\lambda} = H_{y_j}$ for $1 \leq j \leq n_{\xi(\lambda)}$. Then $\{V_\lambda \times H_{j,\lambda} \mid j = 1, \ldots, n_{\xi(\lambda)} \text{ and } \lambda \in \Lambda\}$ is a locally finite open refinement of $\mathscr{U}$, and $X \times Y$ is paracompact.   ◇

We now turn to an important result about open coverings of paracompact spaces, known as "*the shrinking lemma.*"

Lemma 9.1.12 *Let X be a paracompact Hausdorff space, and let $\{U_\alpha\}_{\alpha \in A}$ be an open covering of X. Then there is a locally finite open refinement $\{V_\alpha\}_{\alpha \in A}$ of $\{U_\alpha\}_{\alpha \in A}$ such that $\overline{V}_\alpha \subseteq U_\alpha$ for every $\alpha \in A$.*

Proof Since X is paracompact and Hausdorff, it is regular. So there is an open covering $\{W_\beta\}_{\beta \in B}$ of X such that each $\overline{W}_\beta$ is contained in some U_α. Then we find a locally finite open refinement $\{G_\gamma\}_{\gamma \in \Gamma}$ of $\{W_\beta\}_\beta \in B$. Obviously, each $\overline{G}_\gamma \subseteq U_\alpha$ for some $\alpha \in A$. Now, for each $\gamma \in \Gamma$, choose $f(\gamma) \in A$ such that $\overline{G}_\gamma \subseteq U_{f(\gamma)}$. Thus we obtain a function $f : \Gamma \to A$. Put $\Gamma_\alpha = f^{-1}(\alpha)$ for every $\alpha \in A$. Note that some Γ_α may be empty and $\Gamma_\alpha \cap \Gamma_{\alpha'} = \varnothing$ for $\alpha \neq \alpha'$. Put $V_\alpha = \bigcup_{\gamma \in \Gamma_\alpha} G_\gamma$. Then each V_α is an open subset (possibly empty) of X and $V_\alpha \subseteq U_\alpha$. In fact, we have $\overline{V}_\alpha = \bigcup_{\gamma \in \Gamma_\alpha} \overline{G}_\gamma \subseteq U_\alpha$ for all $\alpha \in A$, since the family $\{G_\gamma\}_{\gamma \in \Gamma}$ is locally finite. Moreover, it is clear that a subset $N \subseteq X$ intersects V_α if and only if $N \cap G_\gamma \neq \varnothing$ for some $\gamma \in \Gamma_\alpha$. As $\Gamma_\alpha \cap \Gamma_{\alpha'} = \varnothing$ for $\alpha \neq \alpha'$ and the family $\{G_\gamma\}_{\gamma \in \Gamma}$ is locally finite, we see that the family $\{V_\alpha\}_{\alpha \in A}$ is also locally finite. Thus $\{V_\alpha\}_{\alpha \in A}$ is a desired refinement of $\{U_\alpha\}_{\alpha \in A}$. $\diamond$

As a first application of the preceding lemma, we prove

Theorem 9.1.13 *Let $f : X \to Y$ be a continuous closed surjection. If X be a paracompact Hausdorff space, then Y is also paracompact.*

Proof By Theorem 9.1.9 and Proposition 8.2.4, we see that Y is normal, and thus regular. Therefore it suffices, by Theorem 9.1.5, to show that every open covering of Y has a σ-locally finite open refinement. Let $\{U_\alpha \mid \alpha \in A\}$ be an open covering of Y. Then $\{f^{-1}(U_\alpha)\}_{\alpha \in A}$ is an open covering of X. Consider a well-ordering $\preceq$ on A. For each $n \in \mathbb{N}$, we construct by induction a locally finite open covering $\{V_{n,\alpha} \mid \alpha \in A\}$ of X such that

(a) $f(\overline{V}_{n,\alpha}) \subseteq U_\alpha$ for every $\alpha \in A$, and
(b) for $n > 1$, $f(\overline{V}_{n,\beta}) \cap f(\overline{V}_{n-1,\alpha}) = \varnothing$ if $\alpha \prec \beta$.

By Lemma 9.1.12, the open covering $\{f^{-1}(U_\alpha)\}_{\alpha \in A}$ of X has a locally finite open refinement $\{V_\alpha\}_{\alpha \in A}$ such that $\overline{V}_\alpha \subseteq f^{-1}(U_\alpha)$ for every $\alpha \in A$. Put $V_{1,\alpha} = V_\alpha$ for every α and assume that, for an $n \geq 1$, the family $\{V_{n,\alpha}\}_{\alpha \in A}$ has been defined. Then, for every $\beta \in A$, the set $W_{n,\beta} = U_\beta - \bigcup_{\alpha \prec \beta} f(\overline{V}_{n,\alpha})$ is open, since f is closed and $\{\overline{V}_{n,\alpha}\}_{\alpha \in A}$ is locally finite. We observe that $\{W_{n,\alpha}\}_{\alpha \in A}$ covers Y. Let $y \in Y$ be arbitrary. By the well-ordering of A, there exists a first element β_0 in A such that $y \in U_{\beta_0}$. Then $y \notin f(\overline{V}_{n,\alpha})$ for every $\alpha \prec \beta_0$, since $\overline{V}_{n,\alpha} \subseteq f^{-1}(U_\alpha)$. Accordingly, $y \in W_{n,\beta_0}$, and hence the assertion. It follows that $\{f^{-1}(W_{n,\alpha})\}_{\alpha \in A}$ is an open covering of X. We invoke the above lemma again to find a locally finite open refinement $\{V_{n+1,\alpha}\}_{\alpha \in A}$ of $\{f^{-1}(W_{n,\alpha})\}_{\alpha \in A}$ such that $\overline{V}_{n+1,\alpha} \subseteq f^{-1}(W_{n,\alpha}) \subseteq f^{-1}(U_\alpha)$ for every $\alpha \in A$. Clearly, $f(\overline{V}_{n+1,\beta}) \cap f(\overline{V}_{n,\alpha}) = \varnothing$ for $\alpha \prec \beta$, since $f(\overline{V}_{n+1,\beta}) \subseteq W_{n,\beta}$ which does not meet $f(\overline{V}_{n,\alpha})$ when $\alpha \prec \beta$. Thus both conditions (a) and (b) are satisfied by the sets $V_{n+1,\alpha}$, $\alpha \in A$. Hence, by induction (or rather A.5.6), $V_{n,\alpha}$ is defined for all $n \in \mathbb{N}$ and $\alpha \in A$.

Next, for every $n \in \mathbb{N}$ and $\alpha \in A$, define $G_{n,\alpha} = Y - \bigcup_{\beta \neq \alpha} f\left(\overline{V}_{n,\beta}\right)$. Clearly, each $G_{n,\alpha}$ is open, $G_{n,\alpha} \subseteq f\left(\overline{V}_{n,\alpha}\right) \subseteq U_\alpha$ and, if $\alpha \neq \beta$, $G_{n,\alpha} \cap G_{n,\beta} = \varnothing$. Moreover, $Y = \bigcup_{n,\alpha} G_{n,\alpha}$. To see the last assertion, suppose $y \in Y$. Then, for each $n \in \mathbb{N}$, there is an $\alpha \in A$ such that $y \in f\left(\overline{V}_{n,\alpha}\right)$. Denote the first element of the set $\{\alpha \in A \mid y \in f\left(\overline{V}_{n,\alpha}\right)\}$ by λ_n, and let λ_m be the first element of the set $\{\lambda_n \mid n \in \mathbb{N}\}$. Then, by the definition of λ_m, $y \notin f\left(\overline{V}_{m+1,\beta}\right)$ for every $\beta \prec \lambda_m$. Also, if $\lambda_m \prec \beta$, then $y \notin f\left(\overline{V}_{m+1,\beta}\right)$, by (b). Therefore $y \in G_{m+1,\lambda_m}$ and $\{G_{n,\alpha} \mid n \in \mathbb{N}, \ \alpha \in A\}$ is a covering of Y. Writing $G_n = \bigcup_\alpha G_{n,\alpha}$, we obviously have $Y = \bigcup_n G_n$. So $\{f^{-1}(G_n) \mid n \in \mathbb{N}\}$ is an open covering of X. Then, by Lemma 9.1.12 again, we obtain a locally finite open refinement $\{H_n\}_{n \in \mathbb{N}}$ of $\{f^{-1}(G_n)\}_{n \in \mathbb{N}}$ such that $f\left(\overline{H}_n\right) \subseteq G_n$ for every n. Since Y is normal, there exists an open set $O_n \subseteq Y$ such that $f\left(\overline{H}_n\right) \subseteq O_n \subseteq \overline{O}_n \subseteq G_n$. Then, for each integer n, we construct an open set $L_n = O_n - \bigcup_{k<n} f\left(\overline{H}_k\right)$. Clearly, given $y \in Y$, if n is the least integer such that $y \in O_n$, then $y \in L_n \cap G_n$. It follows that the family $\{L_n \cap G_{n,\alpha} \mid n \in \mathbb{N}$ and $\alpha \in A\}$ is an open refinement of $\{U_\alpha\}_{\alpha \in A}$. We assert that it is also σ-locally finite. Notice that, for each integer n, the family $\{G_{n,\alpha} \mid \alpha \in A\} \cup \{Y - \overline{O}_n\}$ is an open covering of Y. Since, $G_{n,\alpha} \cap G_{n,\beta} = \varnothing$ for $\alpha \neq \beta$ and $G_{n,\alpha} \cap (Y - \overline{O}_n) = \varnothing$ for all α, the assertion follows. This completes the proof. $\diamond$

Notice that while proving the above theorem, we have shown that the family $\{L_n \cap G_{n,\alpha} \mid \alpha \in A\}$ actually satisfies a stronger condition rather than it being simply locally finite. This leads to the following

Definition 9.1.14 A family $\mathscr{F}$ of subsets of a topological space X is called *discrete* if each point of X has a nbd which meets at most one member of $\mathscr{F}$. The family $\mathscr{F}$ is called σ-*discrete* if it can be decomposed as the union of countably many discrete families of subsets of X.

With this terminology, we obtain yet another characterization of paracompactness.

Theorem 9.1.15 *A regular space X is paracompact $\Leftrightarrow$ every open covering of X has a σ-discrete open refinement.*

Proof The sufficiency of the condition is immediate from Theorem 9.1.5, since a discrete family is obviously locally finite. The proof of the necessary part is very similar to that of Theorem 9.1.13, and is left to the reader. $\diamond$

Next, we move on perhaps to the most important property of paracompact Hausdorff spaces. This requires the knowledge of the following terminologies.

Definition 9.1.16 Let X be a space. For a continuous map $f : X \to \mathbb{R}$, the *support* of f is defined to be the closure of the set $\{x \in X \mid f(x) \neq 0\}$, and denoted by $\operatorname{supp} f$.

Observe that if x lies outside the support of f, then f vanishes on a nbd of x, and conversely.

Definition 9.1.17 A *partition of unity on* X is a family $\{f_\alpha \mid \alpha \in A\}$ of continuous maps $f_\alpha : X \to I$ such that

(a) the family $\{\operatorname{supp} f_\alpha\}_{\alpha \in A}$ is a locally finite covering of X, and
(b) $\sum_\alpha f_\alpha(x) = 1$ for each $x \in X$.

If $\{U_\alpha\}_{\alpha \in A}$ is an open covering of X, we say that a partition of unity $\{f_\alpha \mid \alpha \in A\}$ is *subordinate* to (or *dominated* by) $\{U_\alpha\}_{\alpha \in A}$ if the support of each f_α is contained in the corresponding U_α.

Theorem 9.1.18 *Every open covering of a paracompact Hausdorff space has a partition of unity subordinate to it.*

Proof Let X be a paracompact Hausdorff space and $\{U_\alpha\}_{\alpha \in A}$ be an open covering of X. By Lemma 9.1.12, we find a locally finite open refinement $\{V_\alpha\}_{\alpha \in A}$ of $\{U_\alpha\}_{\alpha \in A}$ such that $\overline{V}_\alpha \subseteq U_\alpha$ for every $\alpha \in A$. We further shrink $\{V_\alpha\}_{\alpha \in A}$ to get a locally finite open refinement $\{W_\alpha\}_{\alpha \in A}$ of $\{V_\alpha\}_{\alpha \in A}$ such that $\overline{W}_\alpha \subseteq V_\alpha$ for every α. Then, by Urysohn's lemma, there exists a continuous function $f_\alpha : X \to I$ such that $f_\alpha\left(\overline{W}_\alpha\right) = 1$ and $f_\alpha(X - V_\alpha) = 0$. Obviously, the set

$$S_\alpha = \{x \in X \mid f_\alpha(x) > 0\} \subseteq V_\alpha,$$

and so $\operatorname{supp} f_\alpha \subseteq \overline{V}_\alpha \subseteq U_\alpha$. Since $\{V_\alpha\}_{\alpha \in A}$ is locally finite, so is $\{S_\alpha\}_{\alpha \in A}$. It follows that for each $x \in X$, $f(x) = \sum_\alpha f_\alpha(x)$ is a real number, and we obtain a function $f : X \to \mathbb{R}$. We prove that f is continuous. Suppose that $N(x)$ is an open nbd of x in X which meets only finitely many sets $S_{\alpha_1}, \ldots, S_{\alpha_n}$, say. Then, for every $y \in N(x)$, $f(y) = \sum_1^n f_{\alpha_i}(y)$ and f is continuous on $N(x)$. Since $\{N(x) \mid x \in X\}$ covers X, f is continuous. We also note that $f(x) \geq 1$ for every $x \in X$ because there is an $\alpha \in A$ such that $f_\alpha(x) = 1$. Now, for each $\alpha \in A$, we define a continuous function $g_\alpha : X \to I$ by setting $g_\alpha(x) = f_\alpha(x)/f(x)$, $x \in X$. It is easily seen that the family $\{g_\alpha \mid \alpha \in A\}$ is a partition of unity subordinate to $\{U_\alpha\}_{\alpha \in A}$. ◇

We remark that the converse of the preceding theorem (due to C. H. Dowker) is also true. In view of Exercise 10, the proof of the preceding theorem also applies to T_4-spaces, and we see that a partition of unity subordinate to a locally finite open covering of a T_4-space always exists.

We conclude this section with an application of the preceding theorem to "manifolds" which are the main object of study in Algebraic Topology and Differential Topology. A topological n-*manifold* (without boundary) is a second countable Hausdorff space X such that each point of X has an open nbd which is homeomorphic to an open subset of the Euclidean space $\mathbb{R}^n$.

Clearly, a manifold is regular and hence paracompact, by Corollary 9.1.6. The Euclidean space $\mathbb{R}^n$, any of its open subspaces and the n-sphere $\mathbb{S}^n$ are some of the examples of n-manifolds. Generally, manifolds are studied with some "differentiable structures" on them, and it can be proven that any such n-manifold imbeds in $\mathbb{R}^{2n+1}$. However, we rest content here with the following.

Theorem 9.1.19 *A compact n-manifold X ($n \geq 1$) can be embedded in an Euclidean space $\mathbb{R}^m$.*

Proof Since X is compact, there is a finite open covering $\mathcal{U} = \{U_1, \ldots, U_k\}$ of X such that, for each i, there exists a homeomorphism $\phi_i : U_i \to V_i$, where V_i is an open subset of the Euclidean space $\mathbb{R}^n$. Let $\{f_1, \ldots, f_k\}$ be a partition of unity subordinate to $\mathcal{U}$. Put $m = nk + k$ and define $\eta : X \to \mathbb{R}^m$ by

$$\eta(x) = (f_1(x)\phi_1(x), \ldots, f_k(x)\phi_k(x), f_1(x), \ldots, f_k(x)),$$

where we take $\phi_i(x) = 0$ for $x \notin U_i$. Since $\operatorname{supp} f_i \subseteq U_i$, we have $f_i \phi_i(x) = 0$ for all $x \notin \operatorname{supp} f_i$. So $x \mapsto f_i(x)\phi_i(x)$ is continuous for every i, and hence η is continuous. Next, we observe that η is injective. Suppose that $\eta(x) = \eta(y)$ for $x, y \in X$. Then $f_i(x) = f_i(y)$ for every i. Since $\sum_i f_i(x) = 1$, we have $f_i(x) \neq 0$ for some i. So $x, y \in \operatorname{supp} f_i \subseteq U_i$, and the equation $f_i(x)\phi_i(x) = f_i(y)\phi_i(y)$ implies that $x = y$, for ϕ_i is injective. Finally, since X is compact and $\mathbb{R}^m$ is Hausdorff, we see that η is a closed embedding.   ◇

Exercises

1. Prove that a space X is paracompact if and only if each open covering $\{U_\alpha\}_{\alpha \in A}$ of X has a locally finite open refinement $\{V_\alpha\}_{\alpha \in A}$ such that $V_\alpha \subseteq U_\alpha$ for each $\alpha \in A$.

2. If each open subset of a paracompact space is paracompact, show that every subspace is paracompact.

3. Let $\mathscr{F}$ be a locally finite family of closed subsets of a regular space X. If each member of $\mathscr{F}$ is a paracompact subspace of X, show that the union of sets in $\mathscr{F}$ is a paracompact.

4. Prove that a second countable locally compact Hausdorff space is paracompact.

5. Show that a locally compact Hausdorff space that is a countable union of compact sets is paracompact.

6. Suppose that a regular space Y is the union of countably many compact sets. If X is a paracompact Hausdorff space, show that $X \times Y$ is paracompact.

7. Let $p : X \to Y$ be a proper map. If Y is paracompact, show that X is also paracompact.

8. Give an example of a normal space which is not paracompact.

9. Prove that a space X is $T_4 \Leftrightarrow$ each finite open covering of X has a locally finite closed refinement.

10. ● Prove that a space X is $T_4 \Leftrightarrow$ each point-finite open covering $\{U_\alpha \mid \alpha \in A\}$ of X has an open refinement $\{V_\alpha \mid \alpha \in A\}$ such that $\overline{V_\alpha} \subseteq U_\alpha$ for every α. (This is known as *the shrinking lemma for normal spaces*.)

9.2 A Metrization Theorem

The following theorem was proved independently by J. Nagata (1950) and Yu M. Smirnov (1951).

Theorem 9.2.1 (Nagata–Smirnov Metrization Theorem) *A space is metrizable if and only if it is regular and has a basis that is the union of at most countably many locally finite families of open sets.*

To prove this, we begin with a generalization of the concept of metric spaces.

A *pseudo-metric* on a set X is a function $d: X \times X \to \mathbb{R}$ such that the following conditions are satisfied for all $x, y, z \in X$:

(a) $d(x, x) = 0$,
(b) $d(x, y) = d(y, x)$, and
(c) $d(x, z) \leq d(x, y) + d(y, z)$.

If d is a pseudo-metric on a set X, then we have $d(x, y) \geq 0$ for all x, y in X. The pair (X, d) is called a pseudo-metric space. The value $d(x, y)$ on a pair of points $x, y \in X$ is called the *distance* between x and y. As in the case of metric spaces, we have the notions of "an open ball," "an open set," "diameter of a set," etc., in a pseudo-metric space. Notice that a pseudo-metric space satisfying the T_0-axiom is actually a metric space. A topological space X is called *pseudo-metrizable* if there is a pseudo-metric d on X such that the topology induced by d is the same as that of X.

Proposition 9.2.2 *Let $(X_n, d_n), n = 1, 2, \ldots,$ be a sequence of pseudo-metric spaces. Then the product space $P = \prod X_n$ is pseudo-metrizable.*

Proof If d is a pseudo-metric on a set X, then $d_1(x, y) = \min\{1, d(x, y)\}$ is also a pseudo-metric on X, and the topology induced by d_1 is the topology of (X, d). So we may assume that $\text{diam}(X_n) \leq 1$ for every n. Then the function $\delta: P \times P \to \mathbb{R}$, defined by

$$\delta((x_n), (y_n)) = \sum 2^{-n} d_n(x_n, y_n),$$

is easily checked to be a pseudo-metric on P. We show that δ induces the product topology of P. Let $O \subseteq P$ be a subbasic open set of the product topology. Then there exists an integer $i > 0$ and an open set $U \subseteq X_i$ such that $O = \{x \in P \mid x_i \in U\}$. If $x \in O$, then we can find a real $\epsilon > 0$ such that $B_{d_i}(x_i; \epsilon) \subseteq U$. Obviously, $\delta(x, y) < 2^{-i}\epsilon \Rightarrow y_i \in U$ so that $B_\delta(x; 2^{-i}\epsilon) \subseteq O$. Thus O is open relative to the metric δ, and the product topology is weaker than the metric topology for P. To see the converse, we observe that for each $x \in P$ and any real number $r > 0$, the open ball $B_\delta(x; r)$ contains a nbd of x in the product topology. Let n be so large that $r^{-1} < 2^n$. Then

$$V = B_{d_1}(x_1; r/2) \times \cdots \times B_{d_{n+1}}(x_{n+1}; r/2) \times \prod_{n+2}^{\infty} X_i$$

is a nbd of x in the product topology. For any $y \in V$, we have

$$\delta(x, y) = \sum_1^{n+1} 2^{-i} d_i(x_i, y_i) + \sum_{n+2}^{\infty} 2^{-i} d_i(x_i, y_i)$$
$$< \sum_1^{n+1} 2^{-i-1} r + \sum_{n+2}^{\infty} 2^{-i}$$
$$< r/2 + r/2 = r.$$

So $y \in B_\delta(x; r)$ and $V \subseteq B_\delta(x; r)$. It follows that the topology induced by δ on P is weaker than the product topology. $\diamond$

Next, let Y_α, $\alpha \in A$, be a family of spaces and let Φ be a family of functions $f_\alpha : X \to Y_\alpha$, each defined on a fixed space X. If for each pair of points $x \neq x'$ in X, there is an $\alpha \in A$ such that $f_\alpha(x) \neq f_\alpha(x')$, then we say that Φ *separates points* of X. Also, recall that the family Φ separates points and closed sets in X if for each closed set $F \subset X$ and each point $x \in X - F$, there is an $\alpha \in A$ such that $f_\alpha(x) \notin \overline{f_\alpha(F)}$ (refer to §7.6). Moreover, there is a function $e : X \to \prod Y_\alpha$ defined by $e(x)_\alpha = f_\alpha(x)$, $x \in X$. This is referred to as the *evaluation map* induced by Φ.

With the above notations and terminology, we have

Proposition 9.2.3 (a) *If each f_α is continuous, then e is continuous.*

  (b) *If Φ separates points, then e is an injection, and conversely.*

  (c) *If Φ separates points and closed sets, then e is an open map of X onto $e(X)$.*

Proof (a): Clearly, the composition of e with the projection $p_\beta : \prod Y_\beta \to Y_\beta$ is f_β. So, by Theorem 2.2.10, it is immediate that e is continuous if f_β is continuous for every $\beta \in A$.

  (b): This is obvious.

  (c): Let $U \subseteq X$ be open, and $x \in U$. Then, by our hypothesis, there is an index $\beta \in A$ such that $f_\beta(x)$ does not belong to $\overline{f_\beta(X - U)}$. If $O = Y_\beta - \overline{f_\beta(X - U)}$, then $p_\beta^{-1}(O)$ is a subbasic open subset of $\prod Y_\alpha$ containing $e(x)$, where p_β is the projection map $\prod Y_\alpha \to Y_\beta$. So $G = e(X) \cap p_\beta^{-1}(O)$ is an open nbd of $e(x)$ in subspace $e(X)$. It is evident that $G \subseteq e(U)$, and hence $e(U)$ is open in $e(X)$. $\diamond$

Now, we can prove the Nagata–Smirnov theorem.

Proof of Theorem 9.2.1: If X is a metrizable space, then it is regular and paracompact, by Theorem 9.1.7. So, for each integer $n > 0$, the covering of X by open balls $B(x; 1/n)$, $x \in X$, has a locally finite open refinement $\mathscr{V}_n$, say. We observe that $\mathscr{B} = \bigcup_n \mathscr{V}_n$ is a base for X. Let $U \subseteq X$ be a nbd of x. Then there exists a sufficiently large integer $n > 0$ so that $B(x; 1/n) \subseteq U$. Now, we find a $V \in \mathscr{V}_{2n}$ with $x \in V$. Since $V \subseteq B(y; 1/2n)$ for some $y \in X$, we have $V \subseteq B(x; 1/n)$ and the assertion follows.

Conversely, suppose that $\mathscr{B}$ is a σ-locally finite base of a regular space X, and let $\mathscr{B} = \bigcup_{n=1}^{\infty} \mathscr{B}_n$, where each $\mathscr{B}_n$ is locally finite. We first observe that X is normal. Suppose that E and F are disjoint nonempty closed subsets of X. For each n, consider the subfamily

$$\mathscr{C}_n = \left\{ B \in \mathscr{B}_n \mid \overline{B} \subseteq X - F \text{ and } B \cap E \neq \varnothing \right\}$$

of $\mathscr{B}_n$ and put $U_n = \bigcup_{B \in \mathscr{C}_n} B$. Note that some $\mathscr{C}_n$ and hence U_n may be empty. Since $\mathscr{B}_n$ is locally finite, $\overline{U}_n = \bigcup_{B \in \mathscr{C}_n} \overline{B} \subseteq X - F$. Also, by the regularity of X, we have $E \subseteq \bigcup_n U_n$. Similarly, we have a countable family of open sets $V_n \subset X$ such that $\overline{V}_n \subseteq X - E$ and $F \subseteq \bigcup_n V_n$. Now, for each n, put $G_n = U_n - \bigcup \{\overline{V}_i \mid i \le n\}$, and $H_n = V_n - \bigcup \{\overline{U}_i \mid i \le n\}$. Clearly, G_n and H_n are open in X and $G_n \cap H_m = \varnothing$ for all n and m. Since the sets U_n cover E and no $\overline{V}_m$ meets E, $\{G_n\}_{n \in \mathbb{N}}$ is an open covering of E. Similarly, $\{H_n\}_{n \in \mathbb{N}}$ is an open covering of F. Thus $\bigcup G_n$ and $\bigcup H_n$ are disjoint open nbds of E and F, respectively, and X is normal.

Next, we construct a pseudo-metrizable space. Notice that, for each fixed integer $n > 0$ and any $V \subseteq X$, the family $\{\overline{B} \subseteq V \mid B \in \mathscr{B}_n\}$ is locally finite, since $\mathscr{B}_n$ is locally finite. Therefore $F(V) = \bigcup \{\overline{B} \mid B \in \mathscr{B}_n \text{ and } \overline{B} \subseteq V\}$ is closed in X, by Proposition 2.1.9. Thus, for an open subset $V \subseteq X$, we have disjoint closed sets $X - V$ and $F(V)$. Since X is normal, there exists a continuous function $\phi_{n,V} : X \to I$ such that $\phi_{n,V}(X - V) = 0$ and $\phi_{n,V}(F(V)) = 1$ if $F(V) \ne \varnothing$. If $F(V) = \varnothing$, we take $\phi_{n,V}(F(V))$ to be the constant map at 0. Now, for any integer $m > 0$ and an ordered pair (x, y) of points in X, the sum $d_{m,n}(x, y) = \sum_{V \in \mathscr{B}_m} |\phi_{n,V}(x) - \phi_{n,V}(y)|$ is finite, since each point of X belongs to at most finitely many members of $\mathscr{B}_m$. It is easily checked that $d_{m,n}$ is a pseudo-metric on X. Thus, for each pair (m, n) of positive integers, we have the pseudo-metric space $(X, d_{m,n})$, denoted by $X_{m,n}$, and so is the product space $P = \prod_{m,n \in \mathbb{N}} X_{m,n}$, by Proposition 9.2.2.

Finally, we show that X can be embedded into P and is, therefore, metrizable. To this end, for each pair (m, n), $m, n \in \mathbb{N}$, we construct a continuous function $f_{m,n} : X \to X_{m,n}$ such that the family $\{f_{m,n} \mid m, n \in \mathbb{N}\}$ satisfies the conditions of Proposition 9.2.3. Define $f_{m,n}(x) = x$ for all $x \in X$. To see the continuity of $f_{m,n}$, consider a point $x_0 \in X$ and a real $\epsilon > 0$. Then find a nbd U_0 of x_0 in X which meets at most finitely many members of $\mathscr{B}_m$. If U_0 does not meet any member of $\mathscr{B}_m$, then $d_{m,n}(x_0, x) = 0$ for all $x \in U_0$, and we are through. So assume that U_0 meets the sets $V_1, \dots, V_k$ in $\mathscr{B}_m$ and is disjoint from the other members. Now, by the continuity of the functions ϕ_{n,V_i}, $1 \le i \le k$, there exists an nbd U_i of x_0 in X such that $|\phi_{n,V_i}(x_0) - \phi_{n,V_i}(x)| < \epsilon/k$ for every $x \in U_i$. Then $W = \bigcap_0^k U_i$ is a nbd of x_0 and $d_{m,n}(x_0, x) < \epsilon$ for all $x \in W$. This implies that $f_{m,n}$ is continuous at x_0. To see other conditions, note that the points in X are closed, so we only need to show that the family $\{f_{m,n} \mid m, n \in \mathbb{N}\}$ separates points and closed sets in X. Suppose that $E \subseteq X$ is closed and $x \in X - E$. Then there exists an integer $m \in \mathbb{N}$ and a set $V \in \mathscr{B}_m$ such that $x \in V \subseteq X - E$. Since X is regular, we find a basic open set $W \in \mathscr{B}$ such that $x \in W \subseteq \overline{W} \subseteq V$. If $W \in \mathscr{B}_n$, then $\phi_{n,V}(x) = 1$ and $\phi_{n,V}(E) = 0$, for $x \in F(V)$ and $E \subseteq X - V$. Accordingly, $d_{m,n}(x, y) \ge |\phi_{n,V}(x) - \phi_{n,V}(y)| = 1$ for all $y \in E$, and it follows that x_0 does not belong to the closure of E in $X_{m,n}$. Thus the family $\{f_{m,n} \mid m, n \in \mathbb{N}\}$ separates points from closed sets in X, and we conclude that the evaluation map $e : X \to P$, $x \mapsto (x)$, is an embedding. $\diamond$

The proof of the sufficiency of the condition in Nagata–Smirnov theorem shows the validity of the following:

Theorem 9.2.4 *If a normal space X has a σ-locally finite base, then X can be embedded into a pseudo-metrizable space, and therefore X is metrizable.*

We end this section with another metrization theorem proved independently at the same time by R. H. Bing (1951).

Theorem 9.2.5 (Bing Metrization Theorem) *A space X is metrizable if and only if it is regular and has a σ-discrete base.*

Proof The sufficiency of the condition is immediate from Theorem 9.2.1, since a σ-discrete family is σ-locally finite.

To see the necessity, we consider the covering of X by the open balls $B(x; 1/n)$, $x \in X$ and $n > 0$ a fixed integer. Then, by the proof of Theorem 9.1.7, there is a discrete refinement $\mathscr{V}_n$ of this open cover. As in the proof of Theorem 9.2.1, it is easily seen that $\bigcup_n \mathscr{V}_n$ is a basis for X.   ◇

Exercises

1. A space X is called *locally metrizable* if each point of X has a nbd which is metrizable in the relative topology.
   Show that a compact Hausdorff space X is metrizable if it is locally metrizable. Generalize this to paracompact Hausdorff spaces.

2. If a completely regular space X is the union of a locally finite family of closed, metrizable subspaces, show that X is metrizable.

Chapter 10
Completeness

Unlike most other concepts, the notion of completeness is not a topological invariant. But it is closely related to some important topological properties. Our considerations here are limited to some useful theorems in topology which find frequent applications in analysis. We study complete spaces in Sect. 10.1, and treat the completion of incomplete metric spaces in Sect. 10.2. Baire spaces and Baire category theorem are discussed in Sect. 10.3.

10.1 Complete Spaces

We begin by recalling a familiar concept from analysis.

Definition 10.1.1 Let X be a metric space with metric d. A sequence $\langle x_n \rangle$ in X is called a *Cauchy sequence* if for every real number $\epsilon > 0$, there exists an integer k (depending on ϵ) such that $d(x_m, x_n) < \epsilon$ whenever $m, n > k$.

It is easy to see that every convergent sequence in a metric space is a Cauchy sequence. But the converse is not true in general; for example, the Cauchy sequence $\langle 1/n \rangle$ in the space $(0, \infty)$ with the Euclidean metric fails to converge.

Definition 10.1.2 A metric space X is called *complete* if every Cauchy sequence in X is convergent.

It should be noted that completeness is a property of metrics; it is not a topological invariant.

Example 10.1.1 It is well known that the real line $\mathbb{R}$ is complete in the usual metric (we shall also see this shortly). Consider another metric d' given by $d'(x, y) = |x/(1 + |x|) - y/(1 + |y|)|$. The sequence $\langle n \rangle$ in the space $(\mathbb{R}, d')$ is a Cauchy sequence and does not converge to any point of $\mathbb{R}$. Thus the space $(\mathbb{R}, d')$ is not complete, although the associated topology is Euclidean.

© Springer Nature Singapore Pte Ltd. 2019
T. B. Singh, *Introduction to Topology*,
https://doi.org/10.1007/978-981-13-6954-4_10

Accordingly, we introduce

Definition 10.1.3 A metrizable space X is called *topologically complete* if X has a metric d such that the metric space (X, d) is complete, and d metrizes the topology of X.

The following result leads to some interesting examples of complete spaces.

Theorem 10.1.4 *A compact metrizable space X is complete in every metric which induces its topology.*

Proof By Exercise 1 (below), it suffices to show that $\langle x_n \rangle$ has a convergent subsequence, and this is immediate from Theorem 5.2.7. There is also a direct simple proof of this theorem. Let $\langle x_n \rangle$ be a Cauchy sequence in X relative to a metric which induces the topology of X. If $\langle x_n \rangle$ does not have a convergent subsequence, then no point of X can be a cluster point of $\langle x_n \rangle$. Therefore, each $x \in X$ has an open nbd U_x which contains at most finitely many terms of $\langle x_n \rangle$. Since X is compact, we find finitely many such open sets, say $U_{x_1}, \ldots, U_{x_n}$, covering X. This implies that x_n is defined only for finitely many indices n, a contradiction. Therefore $\langle x_n \rangle$ has a convergent subsequence, proving the theorem.   ◇

It should be noted that there are many complete metric spaces which are not compact.

Example 10.1.2 The real line $\mathbb{R}$ is complete. If $\langle x_n \rangle$ is a Cauchy sequence in $\mathbb{R}$, then the set $\{x_n\}$ is bounded, and therefore it is contained in a closed interval K, say. Since K is compact, $\langle x_n \rangle$ converges to a point of $K \subseteq \mathbb{R}$.

Example 10.1.3 The Euclidean space $\mathbb{R}^k$ is complete. Let $\langle x^{(n)} \rangle$ be a Cauchy sequence in $\mathbb{R}^k$. If $x_i^{(n)}$ denote the ith coordinate of $x^{(n)}$, then $\left| x_i^{(n)} - x_i^{(m)} \right| \leq \left\| x^{(n)} - x^{(m)} \right\|$ for every $i = 1, \ldots, k$. So $\langle x_i^{(n)} \rangle$ is a Cauchy sequence in $\mathbb{R}$. Because $\mathbb{R}$ is complete, there is a real number x_i, say, such that $\lim_{n \to \infty} x_i^{(n)} = x_i$. Clearly, the point $x = (x_i)$ in $\mathbb{R}^k$ is the limit of the sequence $\langle x^{(n)} \rangle$.

Example 10.1.4 The Hilbert space ℓ_2 is complete. Let $\langle \phi(n) \rangle$ be a Cauchy sequence in ℓ_2. Given $\epsilon > 0$, there exists a positive integer k such that $m, n > k \Rightarrow d(\phi(m), \phi(n)) < \epsilon$, where d is the metric on ℓ_2. If $\phi(n)_i$ denote the ith coordinate of $\phi(n)$, then $\left| \phi(m)_i - \phi(n)_i \right| < \epsilon$, for $m, n > k$, and it follows that $\langle \phi(n)_i \rangle$ is a Cauchy sequence in $\mathbb{R}$. By Example 10.1.2, $\phi(n)_i \to x_i$ for some $x_i \in \mathbb{R}$. Put $x = (x_i)$. We observe that $x \in \ell_2$, and $\phi \to x$. For a fixed integer $j > 0$ and integers $m, n > k$, we have $\sum_{i=1}^{j} (\phi(m)_i - \phi(n)_i)^2 < \epsilon^2$. Letting $m \to \infty$, we get $\sum_{i=1}^{j} (x_i - \phi(n)_i)^2 \leq \epsilon^2$. This is true for every integer j, so we have $\sum (x_i - \phi(n)_i)^2 \leq \epsilon^2$. Also, by using the inequality $2ab \leq a^2 + b^2$, which holds for any two reals a, b, we have $x_i^2 \leq 2(x_i - \phi(n)_i)^2 + 2(\phi(n)_i)^2$ for all $n > k$. This implies that $\sum x_i^2 < \infty$ and $d(x, \phi(n)) \leq \epsilon$ for $n > k$.

Example 10.1.5 Let X be a set. Then, the metric space $\mathcal{B}(X)$ of all bounded functions $X \to \mathbb{R}$ with the supremum metric d^* (refer to Example 1.1.5) is complete. To see this, let $\langle f_n \rangle$ be a Cauchy sequence in $\mathcal{B}(X)$. Then, for each $x \in X$, $\langle f_n(x) \rangle$ is a Cauchy sequence in $\mathbb{R}$, since $\left| f_n(x) - f_m(x) \right| \leq d^*(f_n, f_m)$. Therefore $\langle f_n(x) \rangle$ converges to a real number $g(x)$, say. Thus we have a function $g \colon X \to \mathbb{R}$ defined by $g(x) = \lim_{n \to \infty} f_n(x)$ for all $x \in X$. We show that g is bounded, and $f_n \to g$. Given $\epsilon > 0$, there exists an integer k such that $d^*(f_m, f_n) < \epsilon$ for $m, n > k$. Consequently, $\left| f_n(x) - f_m(x) \right| < \epsilon$ for all $x \in X$, and $m, n > k$. While keeping n and x fixed, let $m \to \infty$. Then we have $\left| f_n(x) - g(x) \right| \leq \epsilon$ for all $n > k$, since $f_m(x) \to g(x)$ as $m \to \infty$. In particular, this implies that there is an integer n such that $\left| f_n(x) - g(x) \right| \leq 1$ for every $x \in X$. Since f_n is bounded, we deduce that g is also bounded. Thus $g \in \mathcal{B}(X)$, and $d^*(f_n, g) \leq \epsilon$ for all $n > k$. So $f_n \to g$, and the assertion follows.

As seen from the above examples, the converse of Theorem 10.1.4 does not hold good. But, in totally bounded metric spaces, the concepts of completeness and compactness are equivalent. Since a compact metric space is also totally bounded, we have

Theorem 10.1.5 *A metric space is compact if and only if it is totally bounded and complete.*

Proof Clearly, we need to prove the sufficiency of the conditions. Let (X, d) be a complete and totally bounded metric space. Then, by Theorem 5.3.3, it is enough to prove that X is sequentially compact. Let $\langle x_n \rangle$ be a sequence in X. Since X is a union of finitely many open balls each of radius $1/2$, there exists an open ball of radius $1/2$ which contains infinitely many terms of $\langle x_n \rangle$. So we have a monotonically increasing sequence ϕ_1 of natural numbers such that $d\left(x_{\phi_1(n)}, x_{\phi_1(m)} \right) < 1$ for all n, m. Again, by the total boundedness of X, there is an open ball of radius $1/4$ which contains infinitely many terms of the subsequence $\langle x_{\phi_1(n)} \rangle$. So we have a subsequence ϕ_2 of ϕ_1 such that $d\left(x_{\phi_2(n)}, x_{\phi_1(m)} \right) < 1/2$ for all n, m. Continuing in this way, we obtain a subsequence ϕ_k of ϕ_{k-1} for every integer $k > 1$ such that $d\left(x_{\phi_k(n)}, x_{\phi_k(m)} \right) < 1/k$ for all n, m. We observe that the diagonal sequence $\langle x_{\phi_n(n)} \rangle$ is Cauchy subsequence of $\langle x_n \rangle$. If $\epsilon > 0$, then find an integer $N > 0$ such that $\epsilon < 1/N$. Clearly, for $m, n > N$, we have $d\left(x_{\phi_m(m)}, x_{\phi_n(n)} \right) < \epsilon$, and $\langle x_{\phi_n(n)} \rangle$ is a Cauchy sequence. Since X is complete, $\langle x_{\phi_n(n)} \rangle$ converges in X. Thus X is sequentially compact, completing the proof. $\diamond$

Analogous to the characterization of compactness in terms of closed sets, we have the following.

Theorem 10.1.6 (G. Cantor) *A metric space (X, d) is complete if and only if the intersection of every family $\mathscr{F}$ of closed subsets of X with the finite intersection property and $\inf_{F \in \mathscr{F}} \operatorname{diam}(F) = 0$ contains exactly one point.*

Proof $\Rightarrow$: Suppose that X is complete and let $\mathscr{F}$ be a family of closed subsets of X such that the intersection of every finite subfamily of $\mathscr{F}$ is nonempty, and $\inf_{F \in \mathscr{F}} \operatorname{diam}(F) = 0$. Then, for each integer $n \geq 1$, there is a member of $\mathscr{F}$, say

F_n, such that diam $(F_n) < 1/n$. Since $\bigcap_1^n F_i \neq \varnothing$, we choose a point x_n which lies in each F_i, $1 \leq i \leq n$. Then $\langle x_n \rangle$ is a Cauchy sequence in X. Since X is a complete metric space, $\langle x_n \rangle$ converges to a point $x \in X$. We claim that $x \in \bigcap_{F \in \mathscr{F}} F$. If $x \notin F$ for some $F \in \mathscr{F}$, then there exists an open ball $B(x; r) \subseteq X - F$. So, we can find an integer n sufficiently large so that $1/n < r/2$ and $x_n \in B(x; r/2)$. Then $F_n \subseteq B(x; r)$, which forces $F \cap F_1 \cap \cdots \cap F_n = \varnothing$, a contradiction. Hence our claim. To see the uniqueness of x, assume that there is another point $y \in \bigcap_{F \in \mathscr{F}} F$. Then we have $\mathrm{diam}(F) \geq d(y, x) > 0$ for all $F \in \mathscr{F}$, contrary to our hypothesis.

$\Leftarrow$: Let $\langle x_n \rangle$ be a Cauchy sequence in X. For each n, let $E_n = \{x_m \mid m \geq n\}$. Then diam $(E_n) \to 0$, so the family $\{\overline{E}_n\}$ satisfies our hypothesis. Consequently, there exists a unique point $x \in \bigcap \overline{E}_n$. We assert that $x_n \to x$. Given $\epsilon > 0$, there exists an integer n_ϵ such that diam $(\overline{E}_n) < \epsilon$, for all $n \geq n_\epsilon$. As $x \in \overline{E}_n$, we have $d(x_n, x) < \epsilon$ for all $n \geq n_\epsilon$, and hence our assertion. $\diamond$

We remark that the condition $\inf_{F \in \mathscr{F}} \mathrm{diam}(F) = 0$ in the preceding theorem is essential: The family $F_n = [n, \infty)$, $n \in \mathbb{N}$, in the real line $\mathbb{R}$ has an empty intersection.

The next theorem generalizes the Bolzano–Weierstrass theorem for the real line and gives another characterization of complete metric spaces.

Theorem 10.1.7 *A metric space X is complete if and only if every infinite totally bounded subset of X has a limit point in X.*

Proof Suppose that X is complete, and let $A \subseteq X$ be infinite totally bounded. Choose a sequence $\langle x_n \rangle$ of distinct points from A. Since A is totally bounded, it is covered by a finite number of open balls, each of radius $1/2$. Accordingly, at least one of these balls, say B_1, contains an infinite number of terms of $\langle x_n \rangle$. Thus there exists an infinite set $M_1 \subseteq \mathbb{N}$ such that $x_n \in B_1$ for all $n \in M_1$. Let $n_1 \in M_1$ be the least integer. Again, consider a finite covering of X by the open balls each of radius $1/4$. Then some member of this covering, say B_2, contains x_n for infinitely many indices $n \in M_1$. Accordingly, there is an infinite set $M_2 \subseteq M_1$ such that $x_n \in B_2$ for all $n \in M_2$. Let $n_2 \in M_2$ be the least integer greater than n_1. Then $x_{n_2} \in B_1 \cap B_2$. Proceeding by induction, we obtain a subsequence $\langle x_{n_k} \rangle$ of $\langle x_n \rangle$, and a sequence of open balls B_k of radii $1/2k$ such that $x_{n_k} \in \bigcap_{j=1}^k B_j$. The sequence $\langle x_{n_k} \rangle$ satisfies the Cauchy condition, since the points x_{n_k} and x_{n_l} lie in B_j for $k, l > j$, and $\mathrm{diam}(B_j) \leq 1/j$. Since X is complete, the sequence $\langle x_{n_k} \rangle$ converges to a point $x \in X$. It is clear that x is a limit point of A because $\langle x_{n_k} \rangle$ is composed of distinct points of A.

To prove the converse, suppose that every infinite totally bounded subset of X has a limit point. Let $\langle x_n \rangle$ be Cauchy sequence in X. Then the range $A = \{x_n \mid n = 1, 2, \ldots\}$ of the sequence $\langle x_n \rangle$ is totally bounded. If A is finite, then there exists an integer m such that $x_n = x_m$ for all $n \geq m$ and $\langle x_n \rangle$ converges to x_m. In the other case, A has a limit point $x \in X$. Consequently, we find a subsequence $\langle x_{n_k} \rangle$ of $\langle x_n \rangle$, which converges to x. Since $\langle x_n \rangle$ is a Cauchy sequence, $x_n \to x$ and we see that X is complete. $\diamond$

Turning to subspaces of a complete metric space, we point out that every subspace is not necessarily complete, for example, the subspace $\mathbb{Q}$ of $\mathbb{R}$ is not complete, although $\mathbb{R}$ is. However, we have the following analogues of Theorems 5.1.7 and 5.1.8.

Theorem 10.1.8 *Every closed subspace of a complete metric space is complete; conversely, a complete subspace of a metric space is closed.*

Proof Let X be a complete metric space, and $Y \subseteq X$ be closed. If $\langle y_n \rangle$ is a Cauchy sequence in Y (with the induced metric), then there is a point $x \in X$ such that $y_n \to x$. Therefore $x \in \overline{Y} = Y$, and Y is complete.

The simple proof of the converse is left to the reader.   ◇

Although, open subspaces of a complete metric space in general are not complete, they are *topologically* complete. In fact, we have the following more general result.

Theorem 10.1.9 *If X is a complete metric space, then every G_δ-set in X is topologically complete.*

Proof Let (X, d) be a complete metric space and $Y \subset X$ be a G_δ-set. Suppose that $Y = \bigcap_1^\infty U_n$, where each U_n is open in X. Obviously, we may assume that $U_n \supseteq U_{n+1}$ for all n. Note that if $x \in U_n$, then $\mathrm{dist}(x, X - U_n) > 0$. Put $\phi_n(x) = 1/\mathrm{dist}(x, X - U_n)$ and then, for $x, y \in U_n$, define

$$\psi_n(x, y) = \left|\phi_n(x) - \phi_n(y)\right| / \left(1 + \left|\phi_n(x) - \phi_n(y)\right|\right).$$

Now, consider the function $\rho : Y \times Y \to \mathbb{R}$ given by

$$\rho(x, y) = d(x, y) + \sum_1^\infty 2^{-n} \psi_n(x, y).$$

Clearly, the series $\sum_1^\infty 2^{-n} \psi_n(x, y)$ is convergent, and thus ρ is well defined. Since

$$\left|\phi_n(x) - \phi_n(z)\right| \leq \left|\phi_n(x) - \phi_n(y)\right| + \left|\phi_n(y) - \phi_n(z)\right|,$$

we have $\psi_n(x, z) \leq \psi_n(x, y) + \psi_n(y, z)$, and it is easily checked that ρ is a metric on Y.

Next, we show that the metric space (Y, ρ) is complete. Let $\langle y_k \rangle$ be a ρ-Cauchy sequence in Y. Then it is clearly a d-Cauchy sequence. Since X is d-complete, $y_k \to x$ for some $x \in X$. We assert that $x \in Y$. If $x \notin Y$, then there exists an integer $n_0 > 0$ such that $x \notin U_n$ for all $n > n_0$. Let k be a fixed positive integer and $n > n_0$. Then

$$\mathrm{dist}(y_{k+j}, X - U_n) \leq d(x, y_{k+j}) \to 0 \text{ as } j \to \infty,$$

since $y_{k+j} \to x$. Consequently, $\psi_n(y_k, y_{k+j}) \to 1$ and it follows that $\lim_{j \to \infty} \rho(y_k, y_{k+j}) \geq \sum_{n>n_0}^\infty 2^{-n}$. This contradicts our assumption about the sequence $\langle y_k \rangle$, and hence our assertion. Now, we observe that $y_k \to x$ in the metric

ρ. Given $\epsilon > 0$, choose an integer $n_0 > 0$ so that $\sum_{n>n_0}^{\infty} 2^{-n} < \epsilon/3$. Since $y_k \to x$ in the metric d and the distance function $x \to \text{dist}(x, X - U_n)$ is continuous, we can find an integer $k_0 > 0$ such that $d(y_k, x) < \epsilon/3$ and $\psi_n(x, y_k) < \epsilon/3n_0$ for all $k > k_0$ and $n = 1, \ldots, n_0$. Then, for $k > k_0$, we have $\rho(x, y_k) < \epsilon$, and $y_k \to x$. Thus (Y, ρ) is complete.

Finally, notice that a sequence in Y converges in the metric ρ if and only if it converges in the metric d, so a subset $F \subseteq Y$ is d-closed if and only if it is ρ-closed. Thus ρ is equivalent to d_Y, and this completes the proof.   ◇

An alternative proof of this theorem is graded in Exercise 16.

By the preceding theorem, we see that the subspace of irrational numbers is topologically complete, for $\mathbb{R} - \mathbb{Q} = \bigcap_{q \in \mathbb{Q}}(\mathbb{R} - \{q\})$ and each $\{q\}$ is closed in $\mathbb{R}$. On the other hand, the subspace $\mathbb{Q} \subset \mathbb{R}$ is not topologically complete, as we shall see later on (Example 10.3.4).

We conclude this section by showing the invariance of topological completeness under countable products.

Theorem 10.1.10 *Let* (X_i, d_i), $i = 1, 2, \ldots$, *be a countable family of complete metric spaces. Then the product space* $\prod X_i$ *is topologically complete.*

Proof Choose a sequence $\langle \lambda_i \rangle$ of positive real numbers such that $\lambda_i \to 0$. For each i, define a new metric $\bar{d}_i$ on X_i by $\bar{d}_i(x_i, y_i) = \min\{\lambda_i, d_i(x_i, y_i)\}$. Then X_i is also complete in the metric $\bar{d}_i$, and the metric ρ on $\prod X_i$ given by $\rho((x_i), (y_i)) = \sup_i \bar{d}_i(x_i, y_i)$ metrizes the product topology (ref. Theorem 2.2.11). We assert that $\prod X_i$ is complete in the metric ρ. Let $\langle x^n \rangle$ be a Cauchy sequence in $\prod X_i$. If x_i^n denotes the ith coordinate of x^n, then each coordinate sequence $\langle x_i^n \rangle$ is a Cauchy sequence in X_i. So there exists a point $x_i \in X_i$ such that $x_i^n \to x_i$ in the metric $\bar{d}_i$. By Theorem 4.2.7, we see that the sequence $\langle x^n \rangle$ converges to the point (x_i) in $\prod X_i$.   ◇

Exercises

1. • Prove that a Cauchy sequence in a metric space X is convergent if it has a convergent subsequence.

2. • Let (X, d) be a metric space and d_1 be the standard bounded metric corresponding to d : $d_1(x, y) = \min\{1, d(x, y)\}$. Show that

   (a) d_1 is a metric equivalent to d.

   (b) d_1 does not alter Cauchy sequences in X.

   (c) X is d-complete $\Leftrightarrow$ X is d_1-complete.

3. Prove that every discrete space is topologically complete.

4. Prove that $d(x, y) = |(1 - x)^{-1} - (1 - y)^{-1}|$ defines a metric on the interval $[0, 1)$ and with this metric it is complete.

5. Let z be an irrational number, and define a metric d on the set $\mathbb{Q}$ of all rational numbers by $d(x, y) = \left|(x - z)^{-1} - (y - z)^{-1}\right|$. What are the Cauchy sequences in $\mathbb{Q}$?

6. Is the sequence $\langle f_n \rangle$ defined by $f_n(x) = x^n$ a Cauchy sequence in the space $\mathcal{B}(I)$ with the supremum metric?

7. For each integer $n > 0$, define $f_n : I \to \mathbb{R}$ by

$$
f_n(t) = \begin{cases} 1 & \text{for } 0 \le t \le 1/3, \\ 1 + 3^{n-1} - 3^n t & \text{for } 1/3 < t \le \left(1 + 3^{n-1}\right)/3^n, \\ 0 & \text{for } \left(1 + 3^{n-1}\right)/3^n < t \le 1. \end{cases}
$$

Show that $\langle f_n \rangle$ is a Cauchy sequence in the metric space in Example 1.1.4 which fails to converge.

8. Prove that the Fréchet space $\mathbb{R}^\omega$ is complete.

9. Show that a complete subspace of a metric space X is closed.

10. If every closed and bounded subset of a metric space X is compact, show that X is complete.

11. Let X be a metric space, and suppose that there is a real $r > 0$ such that $\overline{B(x; r)}$ is compact for every $x \in X$. Prove that X is complete.

12. Prove that a metric space X is totally bounded if and only if every sequence in X has a Cauchy subsequence.

13. Let X be a complete metric space. Show that a subset $A \subseteq X$ has compact closure if and only if A is totally bounded.

14. Let X be a complete metric space. If $\{F_n\}$ is a sequence of closed and bounded subsets of X such that $F_n \supseteq F_{n+1}$ and diam $(F_n) \to 0$, show that $\bigcap F_n$ consists of exactly one point. Also, prove the converse.

15. Give an example to show that the image of a complete space under a uniformly continuous map is not necessarily complete.

16. • In a metric space X, prove:

   (a) The graph of the function $x \mapsto 1/\text{dist}(x, X - U)$ of an open set $U \subset X$ is closed in $X \times \mathbb{R}$.

   (b) If X is complete, and $U \subseteq X$ is open, then U is topologically complete.

   (c) If $U_1, U_2, \ldots$ is a sequence of subspaces of X, then the image of the embedding $x \to (x)$ of $\bigcap U_n$ into $\prod U_n$ is closed.

   (d) If X is complete, then every G_δ-set in X is topologically complete.

17. Let (X, d) be a complete metric space, and $f : X \to X$ be a function for which there exists a real $\alpha < 1$ such that $d\left(f(x), f(y)\right) \le \alpha d(x, y) \, \forall \, x, y \in X$. Prove:

(a) f is continuous.

(b) For every $x \in X$, the sequence $\langle f^n(x) \rangle$ is a Cauchy sequence in X.

(c) If $x_0 = \lim f^n(x)$ (which exists), then $f(x_0) = x_0$. (The point x_0 is called a fixed point of f.)

10.2 Completion

In this section, we will see that an incomplete metric space can be always enlarged to become complete. To make this statement precise, we recall that an *isometry* (or an *isometric embedding*) of a metric space (X, d_X) into a metric space (Y, d_Y) is a distance-preserving map $f : X \to Y$, that is, one which satisfies $d_Y\left(f(x), f(x')\right) = d_X(x, x')$ for all $x, x' \in X$. If $f : X \to Y$ is a surjective isometry, then we say that X and Y are *isometric*. It is easily verified that an isometry is an embedding (topological), and $X \approx Y$ if X and Y are isometric. With the above terminology, we have

Theorem 10.2.1 *Any metric space X can be isometrically embedded into a complete metric space.*

Proof Let $C^*(X)$ be the space of all bounded continuous real-valued functions on X with the supremum metric d^*. If $\langle f_n \rangle$ is a Cauchy sequence in $C^*(X)$, then there exists a bounded function $g : X \to \mathbb{R}$ such that $f_n \to g$ uniformly on X (refer to Example 10.1.5). Since f_n is continuous for all n, g is continuous. Thus the sequence $\langle f_n \rangle$ converges to the function g in $C^*(X)$, and $C^*(X)$ is complete. Let us fix a point $x_0 \in X$ and, for each $x \in X$, consider the function $f_x : X \to \mathbb{R}$ defined by $f_x(p) = d(p, x) - d(p, x_0)$ for all $p \in X$, where d is the metric on X. Since d is continuous, so is f_x. Moreover, by the triangle inequality, $\left| f_x(p) \right| \leq d(x, x_0)$ for all $p \in X$, and therefore $f_x \in C^*(X)$. We show that the mapping $x \mapsto f_x$ is an isometric embedding of X into $C^*(X)$. For $x, y \in X$, we obviously have

$$d^*\left(f_x, f_y\right) = \sup_{p \in X}\left\{\left| f_x(p) - f_y(p) \right|\right\} = \sup_{p \in X}\left\{\left| d(p, x) - d(p, y) \right|\right\}.$$

Since $\left| d(p, x) - d(p, y) \right| \leq d(x, y)$ for every $p \in X$ and the equality is attained for $p = y$, $d^*\left(f_x, f_y\right) = d(x, y)$, and the assertion follows. $\diamond$

By Theorem 10.1.8, the closure of the image of the isometry $x \mapsto f_x$ in $C^*(X)$ is also complete. Thus there is a complete metric space $\widehat{X}$ which contains a dense subset isometric to X. We call $\widehat{X}$ a *completion* of X. Of course, this definition of $\widehat{X}$ depends on the choice of the fixed point x_0; however, we shall see in a moment that any two completions of X are isometric. So $\widehat{X}$ is unique up to an isometry, and we may call $\widehat{X}$ *the* completion of X.

Now, we go on to prove our claim. To this end, we first establish the following extension theorem for uniformly continuous functions into complete metric spaces.

Theorem 10.2.2 *Let A be a dense subset of a metric space X, and Y be a complete metric space. Then a uniformly continuous function $f : A \to Y$ has a unique uniformly continuous extension to X.*

Proof We define a function $g : X \to Y$ as follows: For $x \in A$, we put $g(x) = f(x)$. If $x \in X - A$, then there is a sequence $\langle a_n \rangle$ in A which converges to x. Since f is uniformly continuous, it is easily checked that $\langle f(a_n) \rangle$ is a Cauchy sequence in Y. So $\langle f(a_n) \rangle$ converges to a point of Y, since Y is complete. Put $g(x) = \lim f(a_n)$. We observe that $g(x)$ is independent of the sequence $\langle a_n \rangle$ used in its definition. Let $\langle b_n \rangle$ be another sequence in A such that $b_n \to x$. Then $d_X(a_n, b_n) \to 0$ which implies that $d_Y(f(a_n), f(b_n)) \to 0$. Therefore $f(b_n) \to g(x)$. Notice that g extends the function f to X, by its definition.

Next, we show that g is uniformly continuous on X. Given $\epsilon > 0$, choose $\delta > 0$ (by the uniform continuity of f) such that $d_Y(f(a), f(b)) < \epsilon$ whenever $d_X(a, b) < \delta$. Let $x, x' \in X$ satisfy $d_X(x, x') < \delta/3$. Find sequences $\langle a_n \rangle$ and $\langle a'_n \rangle$ in A such that $a_n \to x$ and $a'_n \to x'$. (If $x \in A$, we may take $a_n = x$ for every n.) Then, there exists an integer n_0 such that $d_X(a_n, x) < \delta/3$ and $d_X(a'_n, x') < \delta/3$ for all $n > n_0$. Now, for $n > n_0$, $d_X(a_n, a'_n) < \delta \Rightarrow d_Y(f(a_n), f(a'_n)) < \epsilon$. Because $f(a_n) \to g(x)$ and $f(a'_n) \to g(x')$, we obtain $d_Y(g(x), g(x')) = \lim d_Y(f(a_n), f(a'_n)) \leq \epsilon$ (verify). This shows that g is uniformly continuous.

Finally, it is immediate from Corollary 4.4.3 that g is unique. ◇

We remark that the hypotheses about f and Y in the preceding theorem are essential. This can be seen by considering the functions $\mathbb{R} - \{0\} \to \mathbb{R}$, $x \mapsto x/|x|$, and the identity map on the subspace $\mathbb{Q} \subset \mathbb{R}$.

Corollary 10.2.3 *Suppose that X and Y are dense subsets of the complete metric spaces $\widehat{X}$ and $\widehat{Y}$, respectively. If X and Y are isometric, then so are $\widehat{X}$ and $\widehat{Y}$.*

Proof Let f be an isometry of X onto Y and let $i : X \to \widehat{X}$ and $j : Y \to \widehat{Y}$ be the inclusion maps. Then it is obvious that f is uniformly continuous, and so is the composition jf. By Theorem 10.2.2, there exists a uniformly continuous function $g : \widehat{X} \to \widehat{Y}$ extending jf. Similarly, we have a uniformly continuous function $h : \widehat{Y} \to \widehat{X}$ extending if^{-1}. Thus, we have the following commutative diagram of topological spaces and continuous maps:

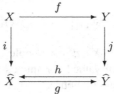

It is clear that the composition hg is an extension of i. Since the identity map on $\widehat{X}$ is also an extension of i, we have $hg = 1_{\widehat{X}}$, by uniqueness. Similarly, $gh = 1_{\widehat{Y}}$, and g is a homeomorphism. Moreover, since $j \circ f$ is an isometry and each point of $\widehat{X}$ is a limit point of a sequence in X, it is easily seen that g is an isometry. ◇

Clearly, the uniqueness property of completion of a metric space is a particular case of the above corollary. We remark that there is a different method of constructing the completion, which is an emulation of the Cantor process for constructing the real numbers from the rationals by means of Cauchy sequences. This construction is outlined in Exercise 5 below.

Two metrics d_1 and d_2 on a set X are called *uniformly equivalent* if given an $\epsilon > 0$, there exists a $\delta > 0$ such that $d_1(x, x') < \delta \Rightarrow d_2(x, x') < \epsilon$ and $d_2(x, x') < \delta \Rightarrow d_1(x, x') < \epsilon$ for all $x, x' \in X$. It follows that d_1 and d_2 are uniformly equivalent if and only if the identity map $i_X : (X, d_1) \rightarrow (X, d_2)$ is a uniform homeomorphism in the following sense:

A homeomorphism $f : X \rightarrow Y$ between metric spaces is a *uniform homeomorphism* if both f and f^{-1} are uniformly continuous.

A surjective isometry is a uniform homeomorphism, but the converse is not true. A property which is preserved under uniform homeomorphisms is called a uniform property; for example, completeness is a uniform property.

Corollary 10.2.4 *Let X and Y be complete metric spaces and $A \subseteq X$ and $B \subseteq Y$ be dense. Then each uniform homeomorphism $f : A \rightarrow B$ can be extended to a uniform homeomorphism $g : X \rightarrow Y$.*

The simple proof is left as an exercise.

Exercises

1. • Let (X, d) be a metric space. Suppose that $\langle x_n \rangle$ and $\langle x'_n \rangle$ are two sequences in X converging to x and x', respectively. Show that $d(x_n, x'_n) \rightarrow d(x, x')$.

2. • Let $f : X \rightarrow Y$ be a uniformly continuous function. If $\langle x_n \rangle$ is a Cauchy sequence in X, show that $\langle f(x_n) \rangle$ is a Cauchy sequence in Y. Give an example of a continuous map which transforms Cauchy sequences into Cauchy sequences, but fails to be uniformly continuous.

3. • Let X be a topological space, and $C^*(X)$ be the set of all real-valued bounded continuous functions on X with the sup metric

$$d^*(f, g) = \sup \left\{ \left| f(x) - g(x) \right| : x \in X \right\}.$$

Show that the metric space $\left(C^*(X), d^* \right)$ is complete. (Notice that if X is compact, then $C^*(X)$ consists of all real-valued continuous functions on X; in particular, $C(I)$ is complete in the supremum metric.)

4. Prove that the completion of a metric space X is separable if and only if X is separable.

5. • Let (X, d) be a metric space, and S be the set of all Cauchy sequences in X. Consider the relation $\sim$ on S defined by $\langle x_n \rangle \sim \langle y_n \rangle \Leftrightarrow \lim d(x_n, y_n) = 0$. Prove:

(a) $\sim$ is an equivalence relation.

(b) If $\langle x_n \rangle$ and $\langle y_n \rangle$ are Cauchy sequences in X, then $d(x_n, y_n)$ is a Cauchy sequence in $\mathbb{R}$, and hence converges.

(c) If $\hat{x}$ and $\hat{y}$ are equivalence classes of $\langle x_n \rangle$ and $\langle y_n \rangle$, respectively, then $\hat{d}(\hat{x}, \hat{y}) = \lim d(x_n, y_n)$ is independent of the choice of these representations and defines a metric on $\widehat{X} = S/\sim$.

(d) The metric space $(\widehat{X}, \hat{d})$ is complete.

(e) If $f(x)$ denotes the equivalence class of the constant sequence $\langle x \rangle, x \in X$, then $\hat{d}(f(x), f(y)) = d(x, y)$. Thus the mapping $x \mapsto f(x)$ is an isometry of X into $\widehat{X}$.

(f) $f(X)$ is dense in $\widehat{X}$, and $f(X) = \widehat{X}$ when X is complete.

6. (a) Give an example of a uniform homeomorphism which is not an isometry.

   (b) Show that the homeomorphism $x \to x^3$ of $\mathbb{R}$ onto itself is not a uniform homeomorphism, although it preserves Cauchy sequences in $\mathbb{R}$.

7. (a) Show that the Euclidean metric on $\mathbb{R}^n$ and the two metrics described in Exercise 1.1.4 are all uniformly equivalent.

   (b) Find a metric on $\mathbb{R}^n$ which is equivalent but not uniformly equivalent to the Euclidean metric.

8. (a) If d is a metric on X, show that it is uniformly equivalent to the metric $d' = d/(1 + d)$.

   (b) Prove that (i) boundedness is a metric but not a uniform property, and (ii) total boundedness is a uniform property, but not a topological property.

10.3 Baire Spaces

In this section, we prove one of the most important theorems in topology, which has extensive applications in analysis.

Theorem 10.3.1 *Let (X, d) be a complete metric space. Then the intersection of any countable family of open dense subsets of X is nonempty; in fact, it is dense in X.*

Proof Let $\{U_n\}_{n \in \mathbb{N}}$ be a countable collection of open sets in X such that $\overline{U}_n = X$ for every n. We show that $\bigcap (G \cap U_n) \neq \emptyset$ for each nonempty open set $G \subseteq X$. Since U_1 is dense, $G \cap U_1 \neq \emptyset$. Because $G \cap U_1$ is open, there is an open ball B_1 of radius $r_1 \leq 1$ and centered at $x_1 \in G \cap U_1$ such that $\overline{B}_1$ is contained in $G \cap U_1$. Since U_2 is dense, there is a point $x_2 \in B_1 \cap U_2$. So we can find an open ball B_2 of radius

$r_2 \leq 1/2$ and centered at x_2 such that $\overline{B}_2 \subseteq U_2$. For $r_2 < r_1 - d(x_1, x_2)$, we have $\overline{B}_2 \subseteq B_1$. Proceeding inductively, we obtain a sequence of open balls $B_n \subseteq U_n$ such that $\mathrm{diam}(B_n) \leq 2/n$ and $\overline{B}_{n+1} \subseteq B_n \cap U_{n+1}$. Let x_n denote the center of B_n. Then $\langle x_n \rangle$ is a Cauchy sequence in X, since $x_m \in B_n$ for all $m \geq n$, and $\mathrm{diam}(B_n) \to 0$. By the completeness of X, $x_n \to x$ for some $x \in X$. It is clear that $x \in \overline{B}_n$ for every n. Since $\overline{B}_1 \subseteq G$ and $\overline{B}_n \subseteq U_n$ for every n, we have $x \in \bigcap \overline{B}_n \subseteq \bigcap (G \cap U_n)$. $\diamond$

The preceding theorem can be also established for locally compact Hausdorff spaces. This property of locally compact Hausdorff spaces makes them interesting to analysts.

Theorem 10.3.2 *The intersection of any countable family of open dense sets in a locally compact Hausdorff space X is dense.*

Proof Let $\{U_n\}_{n \in \mathbb{N}}$ be a countable family of open sets, each U_n being dense in X. Let O be any nonempty open subset of X. We show that $\bigcap (U_n \cap O) \neq \varnothing$. Since $X = \overline{U}_1$, $O \cap U_1 \neq \varnothing$. By Theorem 5.4.2, there exists a nonempty open set V_1 such that $\overline{V}_1$ is compact and $\overline{V}_1 \subseteq O \cap U_1$. For the same reason, we find a nonempty open set V_2 such that $\overline{V}_2$ is compact and $\overline{V}_2 \subseteq V_1 \cap U_2$. Proceeding by induction, we obtain for each positive integer n, a nonempty open set V_n such that $\overline{V}_n$ is compact and $\overline{V}_n \subseteq V_{n-1} \cap U_n$. Obviously, the sets $\overline{V}_n$ are closed in $\overline{V}_1$, and the family $\left\{ \overline{V}_n \right\}_{n \in \mathbb{N}}$ has the finite intersection property. Hence $\bigcap \overline{V}_n \neq \varnothing$, since $\overline{V}_1$ is compact. As $\overline{V}_n \subseteq U_n$ for every n and $\overline{V}_1 \subseteq O$, we have $\bigcap \overline{V}_n \subseteq \bigcap (U_n \cap O)$, and the assertion follows. $\diamond$

We remark that the hypotheses "open" and "countable" in the above theorems are essential, for the set $\mathbb{Q}$ and its complement are dense in $\mathbb{R}$, and the family $\{\mathbb{R} - \{x\} \mid x \in \mathbb{R}\}$ has empty intersection, although all its members are open and dense.

Definition 10.3.3 A space X in which the intersection of each countable family of dense open sets is dense is referred to as a *Baire space*.

With this terminology, the above theorems can be unified as: Each locally compact Hausdorff space and each topologically complete space are Baire spaces.

A more standard formulation of Baire spaces is given by using the following terminology.

Definition 10.3.4 A subset A of a space X is called *nowhere dense* if its closure has an empty interior (i.e., $\mathrm{int}\left(\overline{A}\right) = \varnothing$).

Clearly, A is nowhere dense in $X \Leftrightarrow$ no nonempty open subset of X is contained in $\overline{A} \Leftrightarrow X - \overline{A}$ is dense.

Example 10.3.1 The set $\mathbb{Z}$ of integers is nowhere dense in the real line $\mathbb{R}$.

Example 10.3.2 The Cantor set C is nowhere dense in $\mathbb{R}$ (refer to Example 1.3.12). Clearly, every open interval $(a, b) \subset I$ contains an interval of the form $\left(\frac{1+3k}{3^n}, \frac{2+3k}{3^n}\right)$ for n large enough so that $4 < 3^n(b - a)$. By definition, C is disjoint from all such intervals, and therefore $\text{int}(C) = \varnothing$. Also, C is closed, and so it is nowhere dense in $\mathbb{R}$.

Theorem 10.3.5 (Baire Category Theorem) *If X is a Baire space, then it is not a union of countable number of nowhere dense sets.*

Proof Let $\{E_n\}_{n \in \mathbb{N}}$ be a countable family of nowhere dense subsets of a Baire space X. Then $U_n = X - \overline{E}_n$ is an open dense set for every n, and so there is a point $x \in \bigcap U_n$. This implies that $x \notin E_n$ for all n. $\diamond$

From the preceding theorem, it is immediate that a Baire space cannot be decomposed into countably many closed sets, each of which has no interior.

Originally, R. Baire expressed Theorem 10.3.1 using the following terminology. A subset E of a space X is said to be of the *first category* if it is the union of a countable family of nowhere dense sets. A set which is not of the first category is said to be of the *second category* (or *nonmeager*). In these terms, the Baire Category Theorem can be rephrased as: *A Baire space space is a set of the second category.* But, the other way round, a second category space is not necessarily a Baire space.

Example 10.3.3 Let $X = \mathbb{Q} \cup \{p\}$, where p is an irrational number, and consider the topology for X generated by the open subsets of the subspace $\mathbb{Q} \subset \mathbb{R}$ and $\{p\}$. Clearly, X is a set of the second category, but it is not a Baire space. For, the intersection of open and dense sets $U_q = X - \{q\}$, $q \in \mathbb{Q}$, is not dense in X.

Example 10.3.4 In the real line $\mathbb{R}$, the set $\mathbb{Q}$ of rationals is of the first category, and its complement is of the second category. For, the family $\{\{r\} \mid r \in \mathbb{Q}\}$ is a countable closed covering of $\mathbb{Q}$, and no $\{r\}$ is an open set. If $\mathbb{R} - \mathbb{Q}$ were also of the first category, then the complete space $\mathbb{R}$ would be a countable union of nowhere dense sets, and hence $\mathbb{R}$ is not a Baire space, by Theorem 10.3.5, This is contrary to Theorem 10.3.1.

It follows that $\mathbb{Q}$ is not topologically complete, for otherwise, it would be of the second category. We also notice that a Hausdorff space of the second category cannot be countable and perfect.

We conclude this section with an application of the Baire Category Theorem.

Theorem 10.3.6 *Let X be a complete metric space, and let $\mathscr{F}$ be a family of real-valued continuous functions on X with the property that for each $x \in X$, there is a number M_x such that $|f(x)| \leq M_x$ for all $f \in \mathscr{F}$. Then there exists a nonempty open set $U \subseteq X$ and a number M such that $|f(x)| \leq M$ for all $f \in \mathscr{F}$ and all $x \in U$.*

Proof For each $f \in \mathscr{F}$ and each integer $n > 0$, let $E_{n,f} = \{x \in X \mid |f(x)| \le n\}$. Then $E_{n,f}$ is closed, for it is the inverse image of $[-n, n]$ under f. Set $E_n = \bigcap \{E_{n,f} \mid f \in \mathscr{F}\}$. Then E_n is closed and $X = \bigcup E_n$. By Theorem 10.3.5, some set E_n is not nowhere dense. Since E_n is closed, it must contain a nonempty open set U, say. Then, for every $x \in U$, $|f(x)| \le n$ for all $f \in \mathscr{F}$.   ◇

The above theorem is known as the *uniform boundedness principle*. For another application of Baire's Theorem, we ask the reader to prove Exercise 24 below. It goes to show the existence of a continuous real-valued function on I which has no derivative at any point.

Exercises

1. Prove, without using Theorem 10.3.2, that a compact Hausdorff space is a Baire space.

2. Show that the Sorgenfrey line $\mathbb{R}_\ell$ is a Baire space.

3. Prove that the set $\mathbb{Q}$ of rational numbers is not a G_δ-set in $\mathbb{R}$.

4. Give an example of an uncountable nowhere dense set.

5. Let X be a space and $A \subseteq X$. Show that A is nowhere dense $\Leftrightarrow A \subseteq \overline{(X - \overline{A})}$.

6. Prove that a closed set is nowhere dense $\Leftrightarrow$ its complement is everywhere dense. Is this true for an arbitrary set?

7. Show that the boundary of a closed (or open) set is nowhere dense. Is this true for an arbitrary set?

8. Prove that the union of two nowhere dense sets is nowhere dense.

9. Prove: (a) Every subset of a set of the first category is of the first category.

   (b) A countable union of sets of the first category is a set of the first category.

10. Let X be a Baire space and $A \subset X$ be a set of the first category. Show that (a) A has no interior, and (b) $X - A$ is a set of the second category and it is dense in X.

11. Give an example to show that the complement of a second category set in a Baire space need not be a first category set.

12. Show that a space X is a set of the second category $\Leftrightarrow$ the intersection of every countable family of open dense sets in X is nonempty.

13. Prove that the intersection of countably many open dense sets in a second category space is a set of the second category.

14. Prove that an open (or closed) subset of a topologically complete space is of the second category.

15. Prove that an open subset of a Baire space is also a Baire space.

16. Give an example of a second category set which contains an open subset that is a set of the first category.

17. Prove that the notion of category is not topological.

18. If X is a complete metric space which has no isolated points, prove that X is uncountable.

19. Prove that a locally compact Hausdorff space cannot be countable and perfect both.

20. Suppose that each point of a space X has a nbd that is a Baire space. Show that X is a Baire space.

21. Show that Baire spaces are invariant under continuous open surjections.

22. Prove that the set of points of discontinuity of a real-valued function on a space is an F_σ-set. Deduce that there is no function from $\mathbb{R}$ to itself that is continuous precisely on the rationals.

23. Construct a function that is continuous at each irrational number and is discontinuous at each rational number.

24. • Consider the metric space $C(I)$ with the sup metric d^*. For each positive integer n, let F_n denote the set of all $f \in C(I)$ for which there exists $x \in I$ such that $\left| \frac{f(x+\delta)-f(x)}{\delta} \right| \le n$ for all $0 < \delta$ with $x + \delta \in I$. Prove:

    (a) If a sequence $\langle f_p \rangle$ in F_n converges to $f \in C(I)$, then $f \in F_n$. (So F_n is closed.)

    (b) Given $f \in F_n$ and $r > 0$, there exists a piecewise linear function g such that $d^*(f, g) < r/2$.

    (c) There exists a "sawtoothed" function h within $r/2$ of g such that the absolute value of the gradient of each line segment of h is greater than n. (Thus F_n is nowhere dense.)

    (d) A function in the complement $\bigcup F_n$ is not differentiable anywhere.

Chapter 11
Function Spaces

In this chapter, we shall learn various methods of introducing topologies on a collection of functions of a space X into another space Y. The reader might have studied the notions of pointwise convergence and uniform convergence in Analysis. Section 11.1 concerns with generalizations of these notions in the realm of topological spaces and two topologies for a family of functions $X \to Y$, one describing the pointwise convergence and the other describing the uniform convergence. The next section is devoted to the "compact-open topology", which is predominantly preferred for function spaces. In Sect. 11.3, we confine ourselves to families of functions of a topological space into a metric space. We shall discuss here "the topology of compact convergence" and prove a theorem for compactness of a function space in this topology, known as Arzela–Ascoli theorem.

11.1 Topology of Pointwise Convergence

Let (X, d_X) and (Y, d_Y) be metric spaces. We say that a sequence of functions $f_n : X \to Y$ converges

(a) *simply* (or *pointwise*) to a function $f : X \to Y$ if for each $x \in X$ and each $\epsilon > 0$, there is an integer $n_0 > 0$ (depending on x and ϵ) such that $d_Y(f_n(x), f(x)) < \epsilon$ for all $n \geq n_0$.

(b) *uniformly* to a function $f : X \to Y$ if for each $\epsilon > 0$, there is an integer $n_0 > 0$ (depending on ϵ) such that $d_Y(f_n(x), f(x)) < \epsilon$ for all $n \geq n_0$, and for every $x \in X$.

(c) *continuously* to a function $f : X \to Y$ if for each $x \in X$ and each sequence $\langle x_n \rangle$ in X converging to x, the sequence $\langle f_n(x_n) \rangle$ converges to $f(x)$ in Y.

Note that the metric of X does not play any role in the first two types of convergence.

© Springer Nature Singapore Pte Ltd. 2019
T. B. Singh, *Introduction to Topology*,
https://doi.org/10.1007/978-981-13-6954-4_11

Given nonempty sets X and Y, let $Y_x = Y$ for every $x \in X$. Then, by definition, the cartesian product $\prod_{x \in X} Y_x$, also denoted by Y^X, is the set of all functions $f : X \to Y$. So, if Y is a topological space, then Y^X can be given the product topology. This topology for Y^X will be usually denoted by $\mathscr{T}_p$. Recall that if X is finite and $X = \{x_1, \ldots, x_n\}$, then we also have another definition of the cartesian product of the family $\{Y_{x_i} \mid 1 \le i \le n\}$, namely $Y_{x_1} \times \cdots \times Y_{x_n}$. Although, the topologies for $\prod_{x_i} Y_{x_i}$ and $Y_{x_1} \times \cdots \times Y_{x_n}$ are defined in different ways, there is a homeomorphism $\eta : \prod_{x_i} Y_{x_i} \to Y_{x_1} \times \cdots \times Y_{x_n}$, $f \mapsto (f(x_i))$, where $f(x_i)$ is the ith coordinate of $(f(x))$ (see Sect. 2.2). So the two spaces can be identified. By Theorem 4.2.7, a net $\langle f_\alpha \rangle$ in $(Y^X, \mathscr{T}_p)$ converges to a function $f \in Y^X$ if and only if $f_\alpha(x) \to f(x)$ in Y for every $x \in X$. This kind of convergence for nets (or sequences) of functions is called the *pointwise convergence*. Because of this, the topology $\mathscr{T}_p$ for Y^X is appropriately called the *topology of pointwise convergence* or simply the *pointwise topology*.

For each $x \in X$, the projection mapping $Y^X \to Y$, $f \mapsto f(x)$, is referred to as the *evaluation mapping at* x, and denoted by e_x. A subbasis for the topology $\mathscr{T}_p$ on Y^X consists of the sets

$$e_x^{-1}(U) = \{f : X \to Y \mid f(x) \in U\},$$

where $x \in X$ and $U \subseteq Y$ is open (Fig. 11.1). Accordingly, $\mathscr{T}_p$ is also referred to as the *point-open topology*.

From our knowledge acquired in the preceding chapters, we see that the space $(Y^X, \mathscr{T}_p)$ is compact, connected or T_i $(i = 1, 2, 3, 3\frac{1}{2})$ if Y is so. As normality and

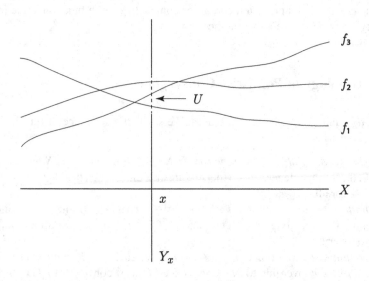

Fig. 11.1 Three elements of $e_x^{-1}(U)$

the countability axioms are not product invariant, these properties may not transfer from Y to $(Y^X, \mathscr{T}_p)$. From Theorem 3.4.4 (resp. Theorem 5.4.6), it follows that $(Y^X, \mathscr{T}_p)$ is locally connected (resp. locally compact) if and only if Y is locally connected and connected (resp. locally compact and compact), in which case Y^X is also connected (resp. compact). Note that if X is finite, then only the condition of local connectedness (resp. local compactness) on Y is needed for $(Y^X, \mathscr{T}_p)$ to be locally connected (resp. locally compact).

It is quite common to be concerned with a subset Φ of Y^X. A natural way to topologize Φ would be to give it the relative topology induced from the topology $\mathscr{T}_p$ for Y^X. The subspace $\Phi \subseteq (Y^X, \mathscr{T}_p)$ is denoted by Φ_p, and its topology is referred to by any of the four adjectives used to describe the topology $\mathscr{T}_p$. Clearly, the pointwise topology on Φ is generated by the subbase consisting of sets $\{f \in \Phi \mid f(x) \in U\}$, where U are open subsets of Y and x varies over X.

In functional analysis, it is often important to know whether a function space Φ_p is compact. From the Tychonoff theorem, it is evident that Φ_p is compact if Y is a compact Hausdorff space and Φ is closed in $(Y^X, \mathscr{T}_p)$. A more general characterization of compact subspaces of Y^X in the pointwise topology is given by

Theorem 11.1.1 *Let X be a set and Y be a Hausdorff space. A subspace Φ of $(Y^X, \mathscr{T}_p)$ is compact if and only if Φ is closed in Y^X, and each set $\{f(x) \mid f \in \Phi\}$, $x \in X$, has compact closure in Y.*

The simple proof is omitted.

Note that Hausdorffness of Y is not needed for "if" part of the theorem.

A useful fact about the pointwise topology for a subset Φ of Y^X, where Y is Hausdorff, is that Φ with any topology strictly finer than this one is not compact. For, if $(\Phi, \mathscr{T})$ is compact and $\mathscr{T}$ finer than the pointwise topology, then the identity function $(\Phi, \mathscr{T}) \to \Phi_p$ is a homeomorphism, since Φ_p is Hausdorff. So $\mathscr{T}$ coincides with the pointwise topology. This simple remark provides the standard method of checking the compactness of a function space $(\Phi, \mathscr{T})$. One needs to prove that the space Φ_p is compact, and a net in Φ converges to f in the topology $\mathscr{T}$ if it converges to $f \in \Phi$ in the pointwise topology.

Topology of Uniform Convergence

Let Y be a metrizable space. By Corollary 1.4.9, there exists a bounded metric d on Y which induces the topology of Y. Clearly, for any set X and any two functions $f, g \colon X \to Y$, $\sup \{d(f(x), g(x)) \mid x \in X\}$ is a real number. We leave it to the reader to verify that d^* defined by

$$d^*(f, g) = \sup \{d(f(x), g(x)) \mid x \in X\}$$

is a metric on Y^X (cf. Exercise 1.1.7).

Definition 11.1.2 The metric d^* on Y^X is called the *sup metric* or the *uniform metric* induced by d, and the topology it induces for Y^X is called the *topology of uniform convergence*.

Note that, for any real $\epsilon > 0$, $d^*(f, g) \leq \epsilon \Leftrightarrow d\left(f(x), g(x)\right) \leq \epsilon$ for all $x \in X$. Therefore a sequence $\langle f_n \rangle$ in the metric space $\left(Y^X, d^*\right)$ converges to $f : X \to Y$ if and only if, for each $\epsilon > 0$, there exists a positive integer $n_0(\epsilon)$ such that $n \geq n_0$ implies that $d\left(f_n(x), f(x)\right) \leq \epsilon$ for all $x \in X$. Thus $f_n \to f$ in $\left(Y^X, d^*\right)$ if and only if $\langle f_n \rangle$ converges uniformly to f. This justifies the use of the adjective "uniform" with the metric d^*. It should be emphasized that the topology of uniform convergence for Y^X generally depends not only on the topology of Y, but also on the particular bounded metric used in Y, as is shown by the following.

Example 11.1.1 Let $X = Y = \mathbb{R}$, and d_1 and d_2 be the metrics on Y defined by

$$d_1(y, y') = \min\left\{1, |y - y'|\right\}, \quad \text{and}$$
$$d_2(y, y') = \left|y/(1 + |y|) - y'/(1 + |y'|)\right|.$$

Both the metrics d_1 and d_2 are bounded and generate the usual topology for Y. But the induced metrics d_1^* and d_2^* on Y^X are not equivalent. To see this, consider the functions $f_n : X \to Y$ for each $n = 1, 2, \ldots$, given by

$$f_n(x) = \begin{cases} x & \text{if } x \leq n, \text{ and} \\ n & \text{if } x > n. \end{cases}$$

Let $f : X \to Y$ be the identity map. Then we have $d_1^*(f_n, f) = 1$, and $d_2^*(f_n, f) = 1/(1 + n)$. Accordingly, $f_n \to f$ in the metric d_2^*, but fails to do so in the metric d_1^*. It follows that the metrics d_1^* and d_2^* generate different topology for Y^X.

However, if Y is a compact metric space, then the topology of uniform convergence for Y^X is independent of the metric used to give the topology of Y. To see this, let d_1, d_2 be two metrics on Y, each generating its topology. Then, by compactness of Y, both metrics d_1 and d_2 are bounded. Also, given $\epsilon > 0$, there exists $\delta_1 > 0$ such that $d_1(y, y') < \delta_1 \Rightarrow d_2(y, y') < \epsilon$ for all $y, y' \in Y$. So, for f, g in Y^X, $d_1^*(f, g) < \delta_1 \Rightarrow d_1\left(f(x), g(x)\right) \leq \delta_1$ for all $x \in X \Rightarrow d_2\left(f(x), g(x)\right) \leq \epsilon$ for all $x \in X \Rightarrow d_2^*(f, g) \leq \epsilon$. By symmetry, we find a real $\delta_2 > 0$ such that $d_2^*(f, g) < \delta_2 \Rightarrow d_1^*(f, g) \leq \epsilon$. Hence $d_1^* \sim d_2^*$.

For notational convenience, a subspace Φ of the metric space $\left(Y^X, d^*\right)$ will be usually denoted by Φ_u, even if (Y, d) is not compact. We note that each evaluation map $e_x : \left(Y^X, d^*\right) \to (Y, d)$, $f \mapsto f(x)$, is uniformly continuous. This is immediate from the definition of d^*. Because the pointwise topology for Y^X is the coarsest topology such that all evaluation mappings e_x are continuous, *the topology of uniform convergence is finer than the topology of pointwise convergence*; in other words, every uniformly convergent sequence in Y^X is simply convergent. We remark that, in general, the two topologies for Y^X disagree (refer to Exercise 10).

We now assume that X also has a topology, and denote the set of all continuous maps $X \to Y$ by $C(X, Y)$. When considered as a subspace of the metric space Y^X with the sup metric, the following proposition describes the main properties of this set.

Proposition 11.1.3 *Let X be a topological space and (Y, d) be a bounded metric space. Then $C(X, Y)$ is closed in the metric space $\left(Y^X, d^*\right)$.*

Proof Let $\langle f_n \rangle$ be a sequence in $C(X, Y)$, and suppose that $f_n \to f$ in the metric space Y^X. We need to show that f is continuous. Let $x_0 \in X$ be arbitrary but fixed. Given $\epsilon > 0$, there is an integer $n_0(\epsilon)$ such that $d\left(f_{n_0}(x), f(x)\right) < \epsilon/3$ for every $x \in X$. We have $d\left(f(x), f(x_0)\right) \leq d\left(f(x), f_{n_0}(x)\right) + d\left(f_{n_0}(x), f_{n_0}(x_0)\right) + d\left(f_{n_0}(x_0), f(x_0)\right)$. Since f_{n_0} is continuous, we can find an open nbd G of x_0 such that $f_{n_0}(G) \subseteq B\left(f_{n_0}(x_0); \epsilon/3\right)$. Then $d\left(f(x), f(x_0)\right) < \epsilon$ for all $x \in G$, and f is continuous at x_0. Thus $f \in C(X, Y)$, and the proposition follows.

Observe that the proof of the preceding proposition also establishes the classical result: *The uniform limit of continuous functions is continuous.*

It should be, however, noted that the pointwise limit of continuous functions need not be continuous, that is, $C(X, Y)$ is not always closed in the topology of pointwise convergence for Y^X.

Example 11.1.2 Let $X = I$ and $Y = \mathbb{R}$. For each positive integer n, let $f_n : I \to \mathbb{R}$ be the function defined by $f_n(t) = t^n$, $t \in I$. Then the sequence $\langle f_n \rangle$ converges to f in the pointwise topology on Y^X, where f is given by $f(t) = 0$ if $t \neq 1$, and $f(1) = 1$. Notice that each f_n is continuous, while f is discontinuous.

Proposition 11.1.4 *Let X be a topological space and (Y, d) be a bounded, complete metric space. Then the metric space $\left(Y^X, d^*\right)$ is complete, and so is its subspace $C(X, Y)$.*

Proof Let $\langle f_n \rangle$ be a Cauchy sequence in Y^X. Given $\epsilon > 0$, there is an integer $n_0(\epsilon)$ such that $d^*\left(f_n, f_m\right) < \epsilon$ whenever $n, m \geq n_0(\epsilon)$. So for every $x \in X$ and $n, m \geq n_0$, we have $d\left(f_n(x), f_m(x)\right) \leq d^*\left(f_n, f_m\right) < \epsilon$. This shows that $\langle f_n(x) \rangle$ is a Cauchy sequence in Y for each $x \in X$. Since Y is complete, there is a point in Y, say $f(x)$, such that $f_n(x) \to f(x)$. If $n \geq n_0$, then

$$d\left(f_n(x), f(x)\right) = d\left(f_n(x), \lim_{m \to \infty} f_m(x)\right)$$
$$= \lim_{m \to \infty} d\left(f_n(x), f_m(x)\right) \leq \epsilon$$

for every $x \in X$. Thus $d^*\left(f_n, f\right) \leq \epsilon$ for all $n \geq n_0$, and $f_n \to f$ in Y^X. By Proposition 11.1.3, $C(X, Y)$ is closed in Y^X, and then Theorem 10.1.8 shows the completeness of $C(X, Y)$. $\diamond$

Note that if an unbounded metric d is chosen to generate the topology of Y, then $d^*(f, g)$ is still a real number for every pair of d-bounded functions $f, g \colon X \to Y$. In fact, taking a fixed point $x_0 \in X$, we have

$$d\left(f(x), g(x)\right) \leq d\left(f(x), f(x_0)\right) + d\left(f(x_0), g(x_0)\right) + d\left(g(x_0), g(x)\right)$$
$$\leq \operatorname{diam}\left(f(Y)\right) + d\left(f(x_0), g(x_0)\right) + \operatorname{diam}\left(g(Y)\right)$$

for every $x \in X$. This implies that $d^*(f, g) < \infty$. It is now easily verified that d^* is a metric for the set $\mathcal{B}(X, Y)$ of all d-bounded functions of X into Y (refer to Exercise 1.1.7). Consequently, we see that the foregoing discussion can be carried on with $\mathcal{B}(X, Y)$ and its subspace $\mathcal{C}^*(X, Y)$ of all d-bounded continuous functions $X \to Y$ (Exercise 15). Observe that if X is a compact space, then the metric d^* is defined on the set $\mathcal{C}(X, Y)$, since every continuous function from X to Y is bounded in this case. Later, we will see that *the topology on $\mathcal{C}(X, Y)$ induced by d^* is independent of the metric d on Y* (see Theorem 11.2.5).

Exercises

1. Let $X = Y = \mathbb{R}$ and $f : X \to Y$ be a function. Given a real $\epsilon > 0$, and a finite set $F \subseteq X$, define

$$B(f, \epsilon, F) = \left\{g \in Y^X \mid |g(x) - f(x)| < \epsilon \text{ for all } x \in F\right\}.$$

   Show that the sets $B(f, \epsilon, F)$, as F ranges over all finite subsets of X, and ϵ ranges over all positive reals, form a nbd basis at f in the pointwise topology on Y^X.

2. Suppose that $f \in \Phi \subseteq Y^X$, where X is a set and Y is a space. Given an integer $n > 0$, let $x_i \in X$ and $U_i \subseteq Y$ be a nbd of $f(x_i), i = 1, \ldots, n$. Prove that the set $\{g \in \Phi \mid g(x_i) \in U_i \text{ for each } i = 1, \ldots, n\}$ is a nbd of f in Φ_p, and the family of all such sets is a nbd basis at f.

3. Let X be a completely regular space. Prove that $\mathcal{C}(X)$ is dense in the space $\mathbb{R}^X$ with pointwise topology.

4. If Y is Hausdorff (regular, completely regular), prove that $\mathcal{C}(X, Y)_p$ is Hausdorff (regular, completely regular) for any space X.

5. Which of the following is compact in the pointwise topology?

   (a) $\{f : I \to I \mid f(0) = 0\}$,   (b) $\{f \in \mathcal{C}(I, I) \mid f(0) = 0\}$.

6. Is $\mathbb{R}^I$ separable in the topology of pointwise convergence?

7. (a) Is the sequence $\langle f_n \rangle$ given by $f_n(t) = t^n, t \in I$, convergent in the topology of uniform convergence for I^I?
   (b) Are the topology of pointwise convergence and the topology of uniform convergence for I^I identical?

8. ● Suppose that a sequence $\langle f_n \rangle$ of continuous functions of a topological space X into a metric space Y converges uniformly to f. Prove that $f_n(x_n) \to f(x)$ whenever $x_n \to x$ in X.

9. (a) For each integer $n = 1, 2 \ldots$, define a function $f_n : I \to \mathbb{R}$ by $f_n(x) = nx(1-x)^n$. Show that the sequence $\langle f_n \rangle$ converges pointwise to the constant function at 0, but fails to converge uniformly.

   (b) Do as in (a) with the sequence $\langle g_n \rangle$, where $g_n(x) = nx(1-x^2)^n$ for all $x \in I$.

10. ● Let d be a bounded metric in $\mathbb{R}$. Prove that the topology of uniform convergence for $\mathbb{R}^{\mathbb{R}}$ defined by d is strictly finer than the topology of pointwise convergence.

11. Compare the topology of pointwise convergence for $\mathcal{C}(I)$ with the topology of uniform convergence.

12. Discuss the compactness and separablity of the following spaces in the topology of uniform convergence:

   (a) I^I, (b) $\mathcal{C}(I, I)$, (c) $\mathcal{C}(I, \mathbb{R})$.

13. Suppose that X_i and Y_i, $i = 1, 2$, are topological spaces such that $X_1 \approx X_2$ and $Y_1 \approx Y_2$. If Y_1 and Y_2 are compact and metrizable, show that $\mathcal{C}(X_1, Y_1)_u \approx \mathcal{C}(X_2, Y_2)_u$.

14. If a Cauchy sequence $\langle f_n \rangle$ in the metric space $\mathcal{C}(I)_u$ converges pointwise to f, show that $\langle f_n \rangle$ converges uniformly to f.

15. ● Let X be a topological space and (Y, d) be a complete metric space.

   (a) Let Φ be a set of bounded functions $X \to Y$. If $d_1 = \min\{1, d\}$, show that the sup metric $d_1^* = \min\{1, d^*\}$ on Φ.

   (b) Show that both the set $\mathcal{B}(X, Y)$ of all bounded functions $X \to Y$ and the set $\mathcal{C}^*(X, Y)$ of all bounded continuous functions $X \to Y$ are closed in the metric d_1^* on Y^X.

   (c) Show that the spaces $\mathcal{B}(X, Y)$ and $\mathcal{C}^*(X, Y)$, both, are complete in metric d^*.

16. Let $Y = \mathbb{R} - \{0\}$ have the standard bounded metric. Discuss the completeness of the followings in the uniform metric:
   (a) $Y^{\mathbb{R}}$, (b) $\mathcal{C}(\mathbb{R}, Y)$, and (c) $\{f : \mathbb{R} \to Y \mid |f| \geq 1\}$.

11.2 Compact-Open Topology

Given a set Φ of functions of one space into another, we have studied two topologies in the previous section, namely, the pointwise topology and the uniform metric topology (when the range space of the functions in Φ is metrizable). The first one is too small for many purposes though it inherits many properties of the range space and is easy to handle, while the second one is quite large. This remark can be seen from the following examples.

Example 11.2.1 Let $X = \mathbb{R}$ and $Y = I$, the unit interval. For each finite set $A \subset X$, consider the characteristic function $f_A \colon X \to Y$ (that is, $f_A(x) = 1$ if $x \in A$, and $f_A(x) = 0$ otherwise). The family $\mathscr{A}$ of the sets A is directed by the inclusion, and therefore we have a net $\{f_A, \ A \in \mathscr{A}\}$ in Y^X. Let $c \colon X \to Y$ be the constant function at 1, and let U be any open set in Y with $1 \in U$. Then, for each $x \in X$, $e_x^{-1}(U)$ is a subbasic open set in $\left(Y^X\right)_p$ and contains c. Because each singleton $\{x\}$ is a member of $\mathscr{A}$, the net $\langle f_A \rangle$ is eventually in $e_x^{-1}(U)$, for $f_A(x) = 1 \in U$ if $\{x\} \subseteq A$. Hence $f_A \to c$ in the pointwise topology on Y^X. This convergence is, however, contrary to our intuition, since no term of the net $\langle f_A \rangle$ seems close to c.

Example 11.2.2 Consider the space $C^* (\mathbb{R})$ of all real-valued bounded continuous functions on $\mathbb{R}$ with the uniform metric topology. For each positive integer n, define the function $f_n \colon \mathbb{R} \to \mathbb{R}$ by

$$
f_n(x) = \begin{cases} 0 & \text{for } |x| \le n, \\ 1 & \text{for } |x| > n + 2 \text{ and} \\ (|x| - n)/2 & \text{for } n \le |x| \le n + 2. \end{cases}
$$

The sequence $\langle f_n \rangle$ clearly fails to converge in the space $C^* (\mathbb{R})$; however, it converges to the constant function $f(x) = 0$ on every bounded interval of $\mathbb{R}$ (Fig. 11.2).

One would like to see a topology for $C^* (X)$ in which sequences such as in Example 11.2.2 are convergent. With this end in view, we examine the pointwise topology further. Suppose that X and Y are topological spaces and a set $\Phi \subseteq Y^X$ is given the pointwise topology. Then the evaluation mapping $e_x \colon f \mapsto f(x)$ from Φ to Y is continuous for each $x \in X$, since the inverse image of an open set $U \subseteq Y$ under e_x is a subbasic open set of Φ_p. In fact, the pointwise topology on Φ is the smallest topology for which each evaluation map e_x is continuous. Moreover, if Φ consists of continuous functions, then it is obvious that the natural function $e \colon \Phi_p \times X \to Y$, defined by $e(f, x) = f(x)$, is separately continuous in each of its variables. However,

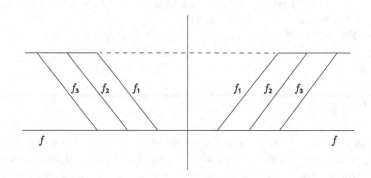

Fig. 11.2 First three terms of sequence $\langle f_n \rangle$ in Example 11.2.2

e may not be jointly continuous in both the variables, f and x, if $\Phi_p \times X$ is given the product topology. This is shown by the following.

Example 11.2.3 Consider the sequence of functions $f_n : I \to \mathbb{R}$ given by $f_n(x) = nxe^{-nx^2}$. Clearly, $f_n \to 0$ in $C(I, \mathbb{R})_p$, and $f_n(1/n) \to 1$. If the function $e : C(I, \mathbb{R})_p \times I \to \mathbb{R}$, $(f, x) \mapsto f(x)$, were continuous, then the sequence $\langle f_n(1/n) \rangle$ would have converged to 0. Hence e is not continuous.

The natural function $e : \Phi \times X \to Y$, $(f, x) \mapsto f(x)$, plays an important role in the study of function spaces, and e is referred to as the *evaluation map* of Φ. Clearly, its continuity requires that the functions in Φ be continuous. Accordingly, our main interest centers around the set $C(X, Y)$ of all continuous functions $X \to Y$.

Definition 11.2.1 A topology for $\Phi \subseteq C(X, Y)$ is called *admissible* (or *jointly continuous*) if the evaluation map $e : \Phi \times X \to Y$, $(f, x) \mapsto f(x)$, is continuous.

Because the pointwise topology is the smallest topology on Φ which makes the evaluation map $e : \Phi \times X \to Y$ separately continuous, an admissible topology for Φ must be finer than this topology. At the other extreme, given any two spaces X and Y, the discrete topology on $\Phi \subseteq C(X, Y)$ is admissible. For, if $(f, x) \in e^{-1}(U)$, where $U \subseteq Y$, then there is an open nbd O of x such that $f(O) \subseteq U$, since f is continuous. Now, the set $\{f\} \times O$ is an open nbd of $(f, x) \in C(X, Y) \times X$, and is mapped by the evaluation map e into U. Therefore $e^{-1}(U)$ is a nbd of each of its points, and hence open in Φ. This proves our assertion. So it is natural to seek the smallest admissible topology, if it exists. However, there is no such smallest topology, in general. In fact, R. Arens (1946) has shown that X must be locally compact Hausdorff if there is an admissible topology on $C(X, I)$, which is coarsest. Thus, for example, there is no coarsest admissible topology on $C(\mathbb{R}^\omega, I)$.

As observed in the previous section, the metric topology for a set $\Phi \subseteq Y^X$ depends on the metric in Y, and the pointwise topology for Φ depends only on it and the topology of Y. The topology of X does not play any role in the definitions of these topologies and the results mentioned or proved in §1. We shall now discuss a method of topologizing Φ in which the topology of domain space plays a significant role. This topology for Φ, viz. *the compact-open topology*, is fairly satisfactory and of particular importance in Analysis.

Let Φ be a set of functions from X to Y. For each pair of sets $A \subseteq X$, and $B \subseteq Y$, define

$$(A, B) = \{f \in \Phi \mid f(A) \subseteq B\}.$$

With this notation, the subbase generating the pointwise topology $\mathcal{T}_p$ on Φ is the family of the sets (x, U), where $x \in X$ and $U \subseteq Y$ is open. It is easily observed that the family

$$\{(F, U) \mid F \subseteq X \text{ finite, and } U \subseteq Y \text{ open}\}$$

also generates the topology $\mathcal{T}_p$. But, if we use compact sets in place of the one-point sets used in the defining subbase for $\mathcal{T}_p$, we obtain a new topology.

Definition 11.2.2 Let Φ be a set of functions from a space X to a space Y. The *compact-open* topology on Φ is the topology generated by the subbasis

$$\big\{ (K, U) \mid K \subseteq X \text{ compact, and } U \subseteq Y \text{ open} \big\}.$$

It is also referred to as the *k-topology* for Φ, and will be generally denoted by $\mathscr{T}_{co}$.

It is easy to verify that the compact-open topology for Φ coincides with the relative topology it inherits from the space $\left(Y^X, \mathscr{T}_{co} \right)$ as a subspace. The space Φ with this topology will be usually denoted by Φ_{co}.

Notice that if Y is indiscrete, then the compact-open topology on Y^X is also indiscrete for all spaces X. And, if X is discrete, then the compact-open topology and the pointwise topology for Y^X are identical for all Y, because the only compact sets in X are the finite sets. However, the two topologies are distinct, in general (refer to Exercise 3). The compact-open topology $\mathscr{T}_{co}$, of course, is finer than the pointwise topology $\mathscr{T}_p$. This is immediate from the fact that the defining subbase for the topology $\mathscr{T}_{co}$ contains a subbasis for the topology $\mathscr{T}_p$, since each one-point subset of X is compact.

The following proposition shows that the topology $\mathscr{T}_p$ on $\mathcal{C}(X, Y)$ is not admissible if $\mathscr{T}_{co}$ is strictly finer than $\mathscr{T}_p$.

Proposition 11.2.3 *Let Φ be a set of continuous functions from X to Y. The compact-open topology for Φ is coarser than any admissible topology.*

Proof Let Φ_a denote the space Φ with an admissible topology $\mathscr{T}_a$. It suffices to prove that each subbasis member (K, U) of Φ_{co} is open in Φ_a. Let $e: \Phi \times X \to Y$ be the evaluation map $(f, x) \mapsto f(x)$. If $f \in (K, U)$, then $e^{-1}(U)$ contains (f, x) for every $x \in K$. By our hypothesis, $e^{-1}(U)$ is open in the product space $\Phi_a \times X$. So there exist open sets $V_x \subseteq X$ and $G_x \subseteq \Phi_a$ such that $(f, x) \in G_x \times V_x \subseteq e^{-1}(U)$. Since K is compact, we find finitely many points $x_1, \ldots, x_n$, say, in K such that $K \subseteq \bigcup_1^n V_{x_i}$. Put $G = \bigcap_1^n G_{x_i}$. Then G is obviously a nbd of f in Φ_a. If $g \in G$, and $x \in V_{x_i}$, then $(g, x) \in G_{x_i} \times V_{x_i} \subseteq e^{-1}(U)$. So $g(x) \in U$, and it follows that $G \subseteq (K, U)$. Thus (K, U) is a nbd of f in the topology $\mathscr{T}_a$, and we see that (K, U) is open in Φ_a. $\diamond$

On the other hand, the topology of uniform convergence for $\mathcal{C}(X, Y)$ is always admissible.

Proposition 11.2.4 *Let X be a topological space and (Y, d) be a bounded metric space. Then the topology of $\mathcal{C}(X, Y)_u$ is admissible.*

Proof Let $e: \mathcal{C}(X, Y)_u \times X \to Y$ be the evaluation map. To see the continuity of e, consider a point $x_0 \in X$, and a continuous function $f: X \to Y$. Given $\epsilon > 0$, let U be the open ball in $\mathcal{C}(X, Y)_u$ with radius ϵ and center f. Clearly, $V = f^{-1}(B(f(x_0); \epsilon))$ is a nbd of x_0 in X, for f is continuous. Thus $U \times V$ is an open nbd of (f, x_0) in $\mathcal{C}(X, Y)_u \times X$. Since

$$d\left(g(x), f(x_0)\right) \le d\left(g(x), f(x)\right) + d\left(f(x), f(x_0)\right) < 2\epsilon$$

for every $g \in U$ and $x \in V$, we see that the function e maps $U \times V$ into the open ball $B(f(x_0); 2\epsilon)$. This implies that e is continuous at the point (f, x_0), and the proposition follows. $\diamond$

We note that the above proposition remains valid for the space $C^*(X, Y)_u$ of all bounded continuous functions of a topological space X into a metric space Y if the metric in Y is unbounded. From the last two results, it follows that the compact-open topology for $C^*(X, Y)$ is coarser than topology of uniform convergence. Moreover, the two topologies coincide when X is compact, as the following theorem shows.

Theorem 11.2.5 *Let X be a compact space and (Y, d) be a metric space. Then the topology of uniform convergence determined by d for $C(X, Y)$ is identical with the compact-open topology.*

Proof Since X is compact, the family $C^*(X, Y)$ coincides with $C(X, Y)$ and, as seen above, the topology of $C(X, Y)_u$ is finer than the topology of $C(X, Y)_{co}$. For an alternative argument, consider a member (K, U) of the subbase for $C(X, Y)_{co}$, and let $f \in (K, U)$. Then $K \subseteq X$ is compact, $U \subseteq Y$ is open and $f(K) \subseteq U$. So, for each $y \in f(K)$, we can find a real $r_y > 0$ such that $B(y; 2r_y) \subseteq U$. By compactness of $f(K)$, there exist finitely many points, say, $y_1, \ldots, y_n$ in $f(K)$ such that $f(K) \subseteq \bigcup_1^n B(y_i; r_{y_i})$. Put $\epsilon = \min\{r_{y_1}, \ldots, r_{y_n}\}$. Then $\epsilon > 0$, and $B(f; \epsilon) \subseteq (K, U)$. For, if $d^*(g, f) < \epsilon$ and $x \in K$, then there is an index i such that $d(y_i, f(x)) < r_{y_i}$. So $d(g(x), y_i) < 2r_{y_i} \Rightarrow g(x) \in U$. Since $x \in K$ is arbitrary, we have $g \in (K, U)$. Thus $B(f; \epsilon) \subseteq (K, U)$ and (K, U) is a nbd of f in $C(X, Y)_u$. It follows that (K, U) is open in $C(X, Y)_u$, and hence the assertion.

To see the converse, consider an open ball $B(f; \epsilon)$ in the metric space $C(X, Y)_u$. Choose a real δ such that $0 < 3\delta < \epsilon$. By compactness of X, we find finitely many points $x_1, \ldots, x_n$, say, in X such that $f(X) \subseteq \bigcup_1^n B(f(x_i); \delta)$. For each $i = 1, \ldots, n$, let $K_i = f^{-1}\left(\overline{B(f(x_i); \delta)}\right)$ and $V_i = B(f(x_i); 2\delta)$. Then each K_i is compact, being a closed subset of the compact space X, and each V_i is open in Y. Obviously, $f(K_i) \subseteq \overline{B(f(x_i); \delta)} \subseteq V_i$ so that $f \in (K_i, V_i)$ for every i. Thus $\bigcap_1^n (K_i, V_i)$ is an open nbd of f in $C(X, Y)_{co}$. We assert that it is contained in $B(f; \epsilon)$. Suppose that $g \in (K_i, V_i)$ for every $i = 1, \ldots, n$. If $x \in X$, then $f(x)$ belongs to some $B(f(x_i); \delta)$. Accordingly, $x \in K_i$ and $g(x) \in V_i$. So $d(g(x), f(x)) \le d(g(x), f(x_i)) + d(f(x_i), f(x)) < 3\delta$, which implies that $d^*(g, f) < \epsilon$. Thus $g \in B(f; \epsilon)$, and hence our assertion. Then Proposition 1.4.6 shows that $\mathcal{T}_{co}$ is finer than $\mathcal{T}_{d^*}$. $\diamond$

From the preceding theorem, it is immediate that if X is compact and Y is metrizable, then *the topology of uniform convergence on $C(X, Y)$ is independent of the metric d on Y.* Furthermore, the compact-open topology for $C(X, Y)$ is metrizable and, by Proposition 11.2.4, admissible. The last statement holds good even for locally compact Hausdorff spaces X, as we see now.

Proposition 11.2.6 *Let Φ be a set of continuous functions of a space X into a space Y. If X is locally compact Hausdorff, then the compact-open topology on Φ is the coarsest admissible topology.*

Proof Suppose that X is locally compact Hausdorff. In view of Proposition 11.2.3, we simply need to prove that the evaluation mapping $e: \Phi_{co} \times X \to Y$ is continuous. Let $f \in \Phi$ and $x \in X$ be arbitrary, and let $V \subseteq Y$ be an open set containing $f(x)$. Then $f^{-1}(V)$ is an open nbd of x, by the continuity of f. Since X is a locally compact Hausdorff space, there exists a compact nbd K of x such that $K \subseteq f^{-1}(V)$. Clearly, $f \in (K, V)$, and e maps $(K, V) \times K^{\circ}$ into V. Since (K, V) is open in Φ_{co} and K° is a nbd of x, we see that $e: \Phi_{co} \times X \to Y$ is continuous at (f, x). $\diamond$

We remark that the condition of local compactness on X in the preceding proposition is essential.

Example 11.2.4 Recall that the subspace $\mathbb{Q} \subset \mathbb{R}$ of the rational numbers is not locally compact (see Example 5.4.4). Let $c: \mathbb{Q} \to I$ be the constant map $x \mapsto 1$, and let $q \in \mathbb{Q}$ be a fixed point. We show that the evaluation map $e: \mathcal{C}(\mathbb{Q}, I)_{co} \times \mathbb{Q} \to I$, $(f, x) \mapsto f(x)$, is not continuous at (c, q). If $B = \bigcap_1^n (K_i, U_i)$ is a basic nbd of c in $\mathcal{C}(\mathbb{Q}, I)_{co}$, then each K_i is compact and $1 = c(K_i) \subseteq U_i$ for every i. Obviously, $F = \bigcup_1^n K_i$ is compact, and hence closed in $\mathbb{Q}$. If a nbd N of q in $\mathbb{Q}$ is contained in F, then its closure $\overline{N}$ in $\mathbb{Q}$ is compact, and so $\overline{N}$ would be closed in $\mathbb{R}$. But, this is obviously false. Therefore we can find a point $x \in N - F$. Since $\mathbb{Q}$ is completely regular, there exists a continuous function $f: \mathbb{Q} \to I$ such that $f(x) = 0$ and $f(F) = 1$. Then $f(K_i) = 1 \in U_i$ for every i, so $f \in B$. We deduce that e fails to map $B \times N$ into $(0, 1]$. Since $(0, 1]$ is a nbd of $c(q)$ and the sets $B \times N$ form a local base at (c, q), it follows that e is not continuous at this point. Thus the compact-open topology for $\mathcal{C}(\mathbb{Q}, I)$ is not admissible.

The above example also shows that Proposition 11.2.6 cannot be generalized to k-spaces.

Next, we discuss the concept of continuous convergence and its relationship with the notion of the compact-open topology.

Definition 11.2.7 Let X and Y be topological spaces. We say that a net $\{f_\nu, \nu \in N\}$ in $\mathcal{C}(X, Y)$ *converges continuously* to $f \in \mathcal{C}(X, Y)$ if for each $x \in X$ and each net $\{x_\nu, \nu \in N\}$ in X converging to x, the net $\langle f_\nu(x_\nu)\rangle$ converges to $f(x)$ in Y.

For example, if a sequence $\langle f_n\rangle$ of continuous functions of a topological space X into a metric space Y converges uniformly to a function f, then it converges continuously to f (Exercise 11.1.8).

We observe that the condition of continuous convergence is equivalent to the following: For each $x \in X$ and each open nbd $O_{f(x)}$ of $f(x)$ in Y, there exist an open nbd U_x of x in X and an index $\mu \in N$ such that $f_\nu(U_x) \subseteq O_{f(x)}$ for all $\nu \succeq \mu$. To see this, suppose that there is a point $x \in X$ and an open nbd $O_{f(x)}$ of $f(x)$ such that for each open nbd U_x of x and each $\nu \in N$, there exist an index $\theta(U_x, \nu) \succeq \nu$ and a point $x_{\theta(U_x, \nu)}$ in U_x with $f_{\theta(U_x, \nu)}\left(x_{\theta(U_x, \nu)}\right) \notin O_{f(x)}$. Direct the set $\mathscr{U}_x \times N$ by

$$(U_x, \nu) \preceq (V_x, \mu) \iff U_x \supseteq V_x \text{ and } \nu \preceq \mu,$$

where $\mathscr{U}_x$ is the set of all open nbds of x in X. It is obvious that (X, ν) is a member of $\mathscr{U}_x \times N$ for all $\nu \in N$, and $\theta(U_x, \mu) \succeq \nu$ for all $(U_x, \mu) \succeq (X, \nu)$. Accordingly, $\langle x_{\theta(U_x, \nu)} \rangle$ is a subnet of $\langle x_\nu \rangle$. Clearly, the net $\langle x_{\theta(U_x, \nu)} \rangle$ converges to x while $\langle f_{\theta(U_x, \nu)}\left(x_{\theta(U_x, \nu)}\right)\rangle$ fails to converge to $f(x)$.

Although the concept of continuous convergence does not require the topology of the relevant function space, we see that a sequence of continuous functions $f_n: X \to Y$ which converges to $f \in C(X, Y)$ in the compact-open topology also converges continuously to f. For, if $x_n \to x$ in X, then $f(x_n) \to f(x)$, by the continuity of f. So, for an open nbd U of $f(x)$ in Y, there exists an integer $n_0 > 0$ such that $f(x_n) \in U$ for every $n \geq n_0$. The subset $K = \{x\} \cup \{x_n \mid n \geq n_0\}$ of X is compact, and (K, U) contains f. By our hypothesis, there is an integer n_1 such that $f_n \in (K, U)$ for every $n \geq n_1$. Now, for $n \geq \max\{n_0, n_1\}$, we have $f_n(x_n) \in U$, which implies that $f_n(x_n) \to f(x)$ in Y. This proves the direct part of the following

Proposition 11.2.8 *Let X be a first countable space and Y be any space. Then a sequence of continuous functions $f_n: X \to Y$ converges to f relative to the compact-open topology on $C(X, Y)$ if and only if $\langle f_n \rangle$ converges continuously to f.*

Proof To see the converse, suppose that $\langle f_n \rangle$ fails to converge to f relative to the compact-open topology for $C(X, Y)$. Then there exist a compact subset $K \subseteq X$, an open subset $U \subseteq Y$ such that $f(K) \subseteq U$ and $\langle f_n \rangle$ is frequently in the complement of (K, U) in $C(X, Y)$. Consequently, we find a sequence of integers $n_1 < n_2 < n_3 < \cdots$ such that $f_{n_i}(K) \nsubseteq U$ for every i. For each integer i, put $g_i = f_{n_i}$ and choose a point $x_i \in K$ such that $g_i(x_i) \notin U$. Since K is compact, the sequence $\langle x_i \rangle$ has a cluster point, say, x_0 in K. Since X is first countable, the sequence $\langle x_i \rangle$ has a subsequence $\langle x_{i_k} \rangle$ which converges to x_0. If $\langle f_n \rangle$ converges continuously to f, then $\langle g_{i_k}\left(x_{i_k}\right)\rangle$ would converge to $f(x_0)$. But this is clearly not true, and hence the result. $\diamond$

In particular, if both X and $C(X, Y)_{co}$ satisfy the axiom of first countability, then we see that the compact-open topology on $C(X, Y)$ is admissible, and the topology of $C(X, Y)$ is determined by continuous convergence. In fact, the compact-open topology is determined by continuous convergence whenever it is admissible; this is an immediate consequence of Proposition 11.2.3 and the following

Proposition 11.2.9 *In the smallest admissible topology for $C(X, Y)$, if it exists, continuous convergence of a net is equivalent to its convergence.*

Proof Let $\mathscr{T}$ be an admissible topology for $C(X, Y)$. If a net $\langle f_\nu \rangle$ in the space $C(X, Y)$ converges to $f \in C(X, Y)$, and a net $\langle x_\nu \rangle$ in X converges to $x \in X$, then $(f_\nu, x_\nu) \to (f, x)$ in the product space $C(X, Y) \times X$. Since the evaluation map

$e: C(X, Y) \times X \to Y$ is continuous, we find that $f_\nu(x_\nu) \to f(x)$, and thus $\langle f_\nu \rangle$ converges continuously to f.

Conversely, suppose that a net $\{f_\nu, \nu \in N\}$ in $C(X, Y)$ converges continuously to a continuous function $f: X \to Y$. Put $T_\nu = \{f_\mu \mid \nu \preceq \mu\}$ and consider the family

$$\mathscr{U} = \{U \subseteq C(X, Y) \mid f \in U \Rightarrow T_\nu \subseteq U \text{ for some } \nu \in N\}.$$

It is easily seen that $\mathscr{U}$ is a topology for $C(X, Y)$, and $f_\nu \to f$ in this topology. We claim that $\mathscr{U}$ is admissible: Let $(g, x) \in C(X, Y) \times X$ and let $O_{g(x)}$ be an open nbd of $g(x)$ in Y. By the continuity of g, there exists a nbd V_x of x in X such that $g(V_x) \subseteq O_{g(x)}$. If $g \neq f$, then $\{g\} \in \mathscr{U}$ and $e(\{g\} \times V_x) \subseteq O_{g(x)}$. So e is continuous at (g, x). If $g = f$, then there exist a nbd W_x of x and an index $\nu_0 \in N$ such that $f_\nu(W_x) \subseteq O_{f(x)}$ for all $\nu \succeq \nu_0$, since $f_\nu \to f$ continuously. Write $G_x = V_x \cap W_x$ and $U = \{f\} \cup T_{\nu_0}$. Then G_x is a nbd of x in X and we have $f \in U \in \mathscr{U}$. Moreover, it is obvious that $e(U \times G_x) \subseteq O_{f(x)}$. This proves the continuity of e at (f, x), and hence our claim. Now, if $\mathscr{T}$ is the smallest admissible topology for $C(X, Y)$, then $\mathscr{T}$ is coarser than $\mathscr{U}$, and therefore $f_\nu \to f$ with respect to $\mathscr{T}$, for this holds good in the topology $\mathscr{U}$. $\diamond$

We now turn to separation properties of compact-open topology.

Theorem 11.2.10 *Let X and Y be topological spaces. Then*

(a) $\left(Y^X, \mathscr{T}_{co}\right)$ *satisfies T_i-axiom for $i = 0, 1, 2$ if Y does so.*

(b) $C(X, Y)_{co}$ *satisfies T_i-axiom for $i = 0, 1, 2, 3, 3\frac{1}{2}$ if Y does so.*

Proof (a): Suppose that Y is a T_2-space. Let $f, g: X \to Y$ be distinct functions. Then there is an $x \in X$ such that $f(x) \neq g(x)$. Since Y is T_2, there exist disjoint open sets U and V in X containing $f(x)$ and $g(x)$, respectively. Clearly, (x, U) and (x, V) are disjoint nbds of f and g, respectively.

Similarly, the remaining cases ($i = 0, 1$) can be proved.

(b): Since the properties T_0, T_1 and T_2 are hereditary, the statements in these cases follow from (a).

Case $i = 3$: Suppose that Y is a T_3-space. Let $f: X \to Y$ be a continuous function, and let (K, U) be a subbasic nbd of f. Then $f(K)$ is compact and contained in U. Since Y is T_3, for each $y \in f(K)$, there exists an open set $G_y \subseteq Y$ such that $y \in G_y \subseteq \overline{G_y} \subseteq U$. By compactness of $f(K)$, we find finitely many points $y_1, \ldots, y_n$ in $f(K)$ such that $f(K) \subseteq \bigcup_1^n G_{y_i} = V$, say. We assert that $\overline{(K, V)} \subseteq (K, U)$. Let $g: X \to Y$ be a continuous function such that $g \notin (K, U)$. Then, for some $x \in K$, $g(x) \notin U$. Obviously, we have $\overline{V} = \bigcup_1^n \overline{G_{y_i}} \subseteq U$ so that $g \in (x, Y - \overline{V})$. Since $(x, Y - \overline{V})$ is open in $C(X, Y)_{co}$ and disjoint from (K, V), $g \notin \overline{(K, V)}$, and hence the assertion. Thus we have $f \in (K, V) \subseteq \overline{(K, V)} \subseteq (K, U)$, where (K, V) is a subbasic open set in $C(X, Y)_{co}$. Now, if $\mathscr{N}$ is any nbd of f in $C(X, Y)_{co}$, then there exist finitely many compact sets $K_1, \ldots, K_m$ in X, and open sets $U_1, \ldots, U_m$ in Y such that

$f \in \bigcap_1^m (K_i, U_i) \subseteq \mathcal{N}$. As above, we also find open sets $V_1, \ldots, V_m$ in Y such that $f \in (K_i, V_i) \subseteq \overline{(K_i, V_i)} \subseteq (K_i, U_i)$ for every $i = 1, \ldots, m$. Then $f \in \bigcap_1^m (K_i, V_i) \subseteq \bigcap_1^m \overline{(K_i, V_i)} \subseteq \mathcal{N}$, and thus the axiom T_3 is satisfied by $\mathcal{C}(X, Y)_{co}$.

Case $i = 3\frac{1}{2}$: Let Y be a $T_{3\frac{1}{2}}$-space, and let f_0 be a continuous function $X \to Y$. We first show the existence of a Urysohn function for the pair $\{f_0\}$ and the complement of a subbasic open set (K, U) containing f_0 in the space $\mathcal{C}(X, Y)_{co}$. By the definition of compact-open topology, $K \subseteq X$ is compact and $U \subseteq Y$ is open. Since Y is $T_{3\frac{1}{2}}$ and $f_0(K) \subseteq U$, we have, for each $y \in f_0(K)$, a continuous function $\phi_y : Y \to I$ such that $\phi_y(y) = 0$ and $\phi_y(Y - U) = 1$. Choose a real $0 < r < 1$ and put $V_y = \phi_y^{-1}[0, r)$. Then V_y is open and the family $\{V_y \mid y \in f_0(K)\}$ covers the compact set $f_0(K)$. So there exist finitely many points $y_1, \ldots, y_n$ in $f_0(K)$ such that $f_0(K) \subseteq \bigcup_1^n V_{y_i}$. Let $\xi = \min \{\phi_{y_1}, \ldots, \phi_{y_n}\}$. Then $\xi : Y \to I$ is continuous and $\xi(y) < r$ for every $y \in f_0(K)$, and $\xi(Y - U) = 1$. Consider the function $\mu = \max \{\xi, \chi\}$, where χ denotes the constant map $Y \to I$ at r. Obviously, μ is continuous and maps Y into $[r, 1]$. Also, $\mu f_0(K) = r$ and $\mu(Y - U) = 1$. Since K is compact, for any continuous map $f : X \to Y$, the real-valued function $\mu f | K$ attains its maximum. Consequently, there exists a point $x_0 \in K$ (depending upon f) such that $\mu f(x_0) = \sup \{\mu f(x) \mid x \in K\}$. Define a mapping $\zeta : \mathcal{C}(X, Y) \to [r, 1]$ by $\zeta(f) = \mu f(x_0)$. Clearly, the number $\zeta(f)$ is independent of the choice of point x_0 in K, and $\zeta(f_0) = r$, and $\zeta(f) = 1$ if $f \notin (K, U)$. To verify the continuity of ζ at f, let $\epsilon > 0$ be given. By the continuity of μ, there exists an open nbd W_1 of $f(x_0) = y_0$ in Y such that $|\mu(y) - \mu(y_0)| < \epsilon$ for every $y \in W_1$. Put $\zeta(f) = \mu(y_0) = s$ and let $W_2 = \mu^{-1}[r, s + \epsilon)$ if $s + \epsilon \leq 1$, and $W_2 = Y$ otherwise. Then $\mathcal{G} = (x_0, W_1) \cap (K, W_2)$ is an open subset of $\mathcal{C}(X, Y)_{co}$. For $x \in K$, we have $r \leq \mu(f(x)) \leq \mu(f(x_0)) = s < s + \epsilon$, so $f \in \mathcal{G}$. If $g \in \mathcal{G}$, then $g(x_0) \in W_1$. Therefore $\mu g(x_0) > s - \epsilon$, and it follows that $\zeta(g) = \sup \{(\mu g)(x) \mid x \in K\} > s - \epsilon$, for $x_0 \in K$. Since $g(K) \subseteq W_2$, we have $\mu g(x) < s + \epsilon$ for all $x \in K$; so $\zeta(g) < s + \epsilon$. Thus $|\zeta(g) - s| < \epsilon$, and ζ is continuous at f. The composition of ζ with the linear homeomorphism $[r, 1] \approx [0, 1]$ yields a continuous function $\psi : \mathcal{C}(X, Y)_{co} \to I$ such that $\psi(f_0) = 0$ and $\psi(f) = 1$ for $f \notin (K, U)$. Now, let $\mathcal{N}$ be any open nbd of f_0 in $\mathcal{C}(X, Y)_{co}$. Then there exist finitely many compact sets $K_1, \ldots, K_n$ in X and open sets $U_1, \ldots, U_n$ in Y such that $f_0 \in \bigcap_1^n (K_i, U_i) \subseteq \mathcal{N}$. For each $i = 1, \ldots, n$, we find, as above, a continuous function $\psi_i : \mathcal{C}(X, Y)_{co} \to I$ such that $\psi_i(f_0) = 0$, and $\psi_i(f) = 1$ for all $f \notin (K_i, U_i)$. Then the function $\psi = \max \{\psi_1, \ldots, \psi_n\}$ is continuous and assumes value 0 at $\psi(f_0)$ and 1 at $\psi(f)$ for all $f \notin \mathcal{N}$. $\diamond$

The converse of the preceding theorem is also valid, since the property of being a T_i-space is hereditary for every $i = 0, 1, 2, 3, 3\frac{1}{2}$ and Y can be embedded in $\mathcal{C}(X, Y)_{co}$. We prove the last assertion as

Proposition 11.2.11 *The mapping* $\lambda : Y \to \mathcal{C}(X, Y)_{co}, y \mapsto \lambda(y)$, *where* $\lambda(y)(x) = y$ *for all* $x \in X$, *is an embedding. Moreover, if* Y *is Hausdorff, then* $\lambda(Y)$ *is closed in* $\mathcal{C}(X, Y)_{co}$.

Proof The continuity of λ is clear, for if $\mathcal{G} = (K, U)$ is a subbasic open set in $\mathcal{C}(X, Y)_{co}$, then $\lambda^{-1}(\mathcal{G}) = U$ is open in Y. Moreover, if $U \subseteq Y$ is open, then we have $\lambda(U) = \lambda(Y) \cap (x, U)$ for any $x \in X$. As (x, U) is open in $\mathcal{C}(X, Y)_{co}$, we find that $\lambda(U)$ is open in the subspace $\text{im}(\lambda)$ of $\mathcal{C}(X, Y)_{co}$. Obviously, λ is injective, and thus it is an embedding.

To prove the last statement, suppose that Y is T_2. We show that the complement $\mathcal{O}$ of $\lambda(Y)$ in $\mathcal{C}(X, Y)_{co}$ is open. If $f \notin \lambda(Y)$, then there exist two distinct points x_1, x_2 in X such that $f(x_1) \neq f(x_2)$. Since Y is T_2, we find disjoint open nbds U_i of $f(x_i)$, $i = 1, 2$, in Y. Set $\mathcal{G} = (x_1, U_1) \cap (x_2, U_2)$. Then $\mathcal{G}$ is open in $\mathcal{C}(X, Y)_{co}$ and contains f. If $g \in \mathcal{G}$, then $g(x_1) \neq g(x_2)$, for $U_1 \cap U_2 = \varnothing$. So $g \notin \lambda(Y)$, and we have $\mathcal{G} \subseteq \mathcal{O}$. It follows that $\mathcal{O}$ is a nbd of f. Since $f \in \mathcal{O}$ is arbitrary, it is open. $\diamond$

The map λ in the preceding proposition is referred to as the natural injection of Y into $\mathcal{C}(X, Y)_{co}$.

As regards normality of the space $\mathcal{C}(X, Y)$ in the compact-open topology, we recall that its topology coincides with the pointwise topology in the case X is a discrete space. Since normality is not preserved even under the formation of finite products, there is no hope that the space $\mathcal{C}(X, Y)_{co}$ would be normal. A similar inference can also be made about any of the countability properties. However, for some nice spaces X, the compact-open topology on $\mathcal{C}(X, Y)$ is second countable whenever Y is. Specifically, we prove

Theorem 11.2.12 *Let X be a second countable locally compact Hausdorff space. Then, for any second countable space Y, $\mathcal{C}(X, Y)_{co}$ is second countable.*

Proof Suppose that both X and Y satisfy the second axiom of countability. Let $\mathscr{A}$ and $\mathscr{B}$ be countable bases for X and Y, respectively, and consider the family $\Sigma = \{(\overline{A}, B) \mid A \in \mathscr{A}, B \in \mathscr{B} \text{ and } \overline{A} \text{ is compact}\}$. Obviously, it is a countable family of open sets in $\mathcal{C}(X, Y)_{co}$. If X is also locally compact Hausdorff, then we show that Σ is a subbase for the space $\mathcal{C}(X, Y)_{co}$. To this end, it suffices to prove that each subbasic open subset (K, U) of $\mathcal{C}(X, Y)_{co}$ is open in the topology generated by Σ. Suppose that $f \in (K, U)$. Then, for each $x \in K$, we find a $B_x \in \mathscr{B}$ such that $f(x) \in B_x \subseteq U$. Since X is locally compact Hausdorff, there exists an open set $V_x \subseteq X$ such that $x \in V_x \subseteq \overline{V}_x \subseteq f^{-1}(B_x)$ and $\overline{V}_x$ is compact. Then we find an $A_x \in \mathscr{A}$ such that $x \in A_x \subseteq V_x$, for $\mathscr{A}$ is a basis of X. Obviously, $\overline{A}_x \subseteq \overline{V}_x \subseteq f^{-1}(B_x)$, so $\overline{A}_x$ is compact and we have $f \in (\overline{A}_x, B_x) \in \Sigma$. Since K is compact there are finitely many points $x_1, \ldots, x_n$ in K such that $K \subseteq \bigcup_1^n A_{x_i}$. Then, for each $i = 1, \ldots, n$, we have $f(\overline{A}_{x_i}) \subseteq B_{x_i} \subseteq U$, and therefore

$$f \in \bigcap_1^n (\overline{A}_{x_i}, B_{x_i}) \subseteq \left(\bigcup_i \overline{A}_{x_i}, \bigcup_i B_{x_i}\right) \subseteq (K, U).$$

It follows that (K, U) is a nbd of f in the topology generated by Σ, and thus we conclude that Σ is a subbasis of $\mathcal{C}(X, Y)_{co}$. It is now clear that $\mathcal{C}(X, Y)_{co}$ is second countable, since Σ is countable. $\diamond$

It should be noted that the conditions of local compactness on X and second countability on Y in the preceding theorem are essential. The necessity of the condition on X is displayed by the space $\mathcal{C}(\mathbb{Q}, I)_{co}$ (which is not second countable), and that on Y is evident from the fact that it can be embedded into $\mathcal{C}(X, Y)_{co}$.

As we have already seen, the compact-open topology for $\mathcal{C}(X, Y)$ is metrizable if X is compact and Y is a metric space. By Proposition 11.1.4 and Exercise 10.1.2, it follows that such a function space is topologically complete whenever Y is.

The compactness of $\mathcal{C}(X, Y)$ in the compact-open topology generally requires some strong condition to be satisfied, which shall be studied in detail in the next section.

We now discuss the most useful feature of the compact-open topology. Suppose that X, Y and Z are three spaces. Given a function $f : X \times Y \to Z$, for each fixed $x \in X$, we have a function $f_x : Y \to Z$ defined by $f_x(y) = f(x, y)$. If f is continuous, then each f_x is continuous, since $Y \approx \{x\} \times Y$ and $f|(\{x\} \times Y)$ is continuous. Thus a continuous function $f : X \times Y \to Z$ determines a one-parameter family of continuous functions $f_x : Y \to Z$ indexed by X. So we can define a function $\hat{f} : X \to \mathcal{C}(Y, Z)$ by setting $\hat{f}(x) = f_x$, $x \in X$. The other way around, given a function $\phi : X \to \mathcal{C}(Y, Z)$, we can define a function $f : X \times Y \to Z$ by setting $f(x, y) = \phi(x)(y)$. Then we have $\hat{f} = \phi$ so that each function $f_x : Y \to Z$ defined by f is continuous. We refer to the function $\hat{f}$ as the *associate* of f, and vice versa. For example, the identity map on $\mathcal{C}(X, Y)$ and the evaluation map $e : \mathcal{C}(X, Y) \times X \to Y$ are associates of each other. The relationship between the continuity of f and $\hat{f}$ is given by the following.

Theorem 11.2.13 *If a function $f : X \times Y \to Z$ is continuous, then the associated function $\hat{f} : X \to \mathcal{C}(Y, Z)$, $x \mapsto f_x$, is also continuous in the compact-open topology on $\mathcal{C}(Y, Z)$. Conversely, the function associated with a continuous function $\phi : X \to \mathcal{C}(Y, Z)$ is continuous, provided the topology of $\mathcal{C}(Y, Z)$ is admissible.*

Proof For the direct part, it is clearly enough to prove that the inverse image of each subbasic open set $(K, U) \subseteq \mathcal{C}(Y, Z)_{co}$ under $\hat{f}$ is open in X. Let x be a point in $\hat{f}^{-1}(K, U)$. Then $f_x \in (K, U)$ so that $f(x, y) \in U$ for every $y \in K$. This implies that $x \times K \subseteq f^{-1}(U)$. Since f is continuous, and $U \subseteq Z$ is open, $f^{-1}(U)$ is open in $X \times Y$. By Lemma 5.1.18, there exists a nbd N of x such that $N \times K \subseteq f^{-1}(U)$. Clearly, this implies that $\hat{f}$ maps N into (K, U). Consequently, $\hat{f}^{-1}(K, U)$ is a nbd of x, and the desired result follows.

To prove the converse, we note that f is the composition

$$X \times Y \xrightarrow{\phi \times 1} \mathcal{C}(Y, Z) \times Y \xrightarrow{e} Z,$$

where 1 denotes the identity map on Y and e is the evaluation map. Hence it is clear that f would be continuous if both of the functions e and ϕ are continuous. $\diamond$

This theorem describes an important feature of the compact-open topology and is often used to establish the continuity of a function $\phi : X \to \mathcal{C}(Y, Z)_{co}$ by showing that the associated function $e \circ (\phi \times 1)$ is continuous on $X \times Y$. Of course, there are also some interesting applications of its direct part. As an illustration, we give an *alternative proof of Theorem* 6.2.8.

Let $f: X \to Y$ be an identification, and suppose that Z is a locally compact Hausdorff space. To show that $f \times 1: X \times Z \to Y \times Z$ is an identification, consider a space T and a function $g: Y \times Z \to T$ such that the composition

$$X \times Z \xrightarrow{f \times 1} Y \times Z \xrightarrow{g} T$$

is continuous. Put $h = g \circ (f \times 1)$. Then, by the preceding theorem, the associated function $\hat{h}: X \to C(Z, T)_{co}$ is continuous. Now, given $y \in Y$, there is an $x \in X$ with $f(x) = y$. We have $g(y, z) = h(x, z) = \hat{h}(x)(z)$ for every $z \in Z$. This shows that $\hat{h}(x)(z)$ is independent of the choice of x in $f^{-1}(y)$. So we can define a function $\theta: Y \to C(Z, T)_{co}$ by $\theta(y) = \hat{h}(x)$, where $f(x) = y$. It is obvious that the composite $\theta f = \hat{h}$, which is continuous. Since f is an identification, we see by Theorem 6.2.6 that θ is continuous. Since Z is locally compact Hausdorff, the evaluation map $e: C(Z, T)_{co} \times Z \to T$ is continuous. Therefore the composition $e \circ (\theta \times 1) = g$ is continuous, and by the converse part of this theorem, we conclude that $f \times 1$ is an identification.

By Theorem 11.2.13, we have a function

$$\alpha: C(X \times Y, Z) \to C(X, C(Y, Z)_{co})$$

defined by $\alpha(f) = \hat{f}$. This is referred to as the *association* function. Clearly, α is injective. For, if f and g are distinct members of $C(X \times Y, Z)$, then there is a point $(x, y) \in X \times Y$ such that $f(x, y) \neq g(x, y)$. As seen above, if the topology of $C(Y, Z)$ is admissible, then α is surjective, too. In particular, for locally compact Hausdorff spaces Y, the association function is a bijection between $C(X \times Y, Z)$ and $(X, C(Y, Z)_{co})$. Interestingly, this function becomes a homeomorphism when its domain and range are given the compact-open topology. More precisely, we have

Theorem 11.2.14 (The Exponential Law) *Let X and Y be Hausdorff spaces. Then, for any topological space Z, the association map*

$$\alpha: C(X \times Y, Z)_{co} \to C(X, C(Y, Z)_{co})_{co},$$

$f \mapsto \hat{f}$, *is an embedding. If, in addition, Y is locally compact, then α is a homeomorphism.*

Before proving this result, we find some convenient subbases for the compact-open topology on the function spaces involved therein. To this end, the following relationships between subbasic open sets come in useful:

$$\bigcap_\alpha (K_\alpha, V) = \left(\bigcup_\alpha K_\alpha, V\right),$$
$$\bigcap_\alpha (K, V_\alpha) = \left(K, \bigcap_\alpha V_\alpha\right) \text{ and }$$
$$\bigcap_\alpha (K_\alpha, V_\alpha) \subseteq \left(\bigcup_\alpha K_\alpha, \bigcup_\alpha V_\alpha\right).$$

Lemma 11.2.15 *Let X be a Hausdorff space and Y be any topological space. If $\mathscr{S}$ is a subbase for the topology of Y, then the family $\{(K, V) \mid K \subseteq X$ compact and $V \in \mathscr{S}\}$ is a subbase for the compact-open topology on $\mathcal{C}(X, Y)$.*

Proof Let Σ be the family of all sets $(K, V) \subseteq \mathcal{C}(X, Y)$, where $K \subseteq X$ is compact, and $V \in \mathscr{S}$, the subbasis for Y. It suffices to prove that each subbasic open set (K, U) of $\mathcal{C}(X, Y)_{co}$ is open in the topology generated by Σ. Let $f \in (K, U)$ be arbitrary. Then $f : X \to Y$ is a continuous function such that $f(K) \subseteq U$. If $\mathscr{B}$ is the base for Y determined by the subbasis $\mathscr{S}$, then U is the union of a subfamily $\{W_\alpha\} \subseteq \mathscr{B}$. Since $f(K) \subseteq U$, the family $\{f^{-1}(W_\alpha)\}$ covers K. Clearly, K is regular. Hence, if $x \in K \cap f^{-1}(W_\alpha)$, then we have an open set H_x in K such that $x \in H_x \subseteq \overline{H}_x \subseteq K \cap f^{-1}(W_\alpha)$. Since K is compact, its open covering $\{H_x \mid x \in K\}$ has a finite subcovering $\{H_{x_j} \mid j = 1, \dots, m\}$, say. Now, for each j, we choose a set W_{α_j} in the family $\{W_\alpha\}$ such that $\overline{H}_{x_j} \subseteq f^{-1}(W_{\alpha_j})$. Then $f \in (\overline{H}_{x_j}, W_{\alpha_j})$, where $\overline{H}_{x_j}$ is compact, being a closed subset of K. By definition, each member of $\mathscr{B}$ is a finite intersection of sets in $\mathscr{S}$. So, for each j, we have finitely many sets $V_{j1}, \dots, V_{jn_j}$ in $\mathscr{S}$ such that $W_{\alpha_j} = \bigcap_{i=1}^{n_j} V_{ji}$. Then $f \in (\overline{H}_{x_j}, \bigcap_{i=1}^{n_j} V_{ji}) = \bigcap_{i=1}^{n_j} (\overline{H}_{x_j}, V_{ji})$. Obviously, this is true for every $j = 1, \dots, m$, so we have

$$f \in \bigcap_{j=1}^{m} \left[\bigcap_{i=1}^{n_j} (\overline{H}_{x_j}, V_{ji})\right] = \bigcap_j (\overline{H}_{x_j}, W_{\alpha_j})$$
$$\subseteq \left(\bigcup_j \overline{H}_{x_j}, \bigcup_j W_{\alpha_j}\right) \subseteq (K, U).$$

It follows that (K, U) is open in the topology generated by the subbase Σ, as desired.   ◇

Lemma 11.2.16 *Let X and Y be Hausdorff spaces. Then, for any space Z, the sets $(A \times B, W)$, where $A \subseteq X$, $B \subseteq Y$ are compact and $W \subseteq Z$ is open, form a subbase for the compact-open topology on $\mathcal{C}(X \times Y, Z)$.*

Proof Let (K, W) be a subbasic open set in $\mathcal{C}(X \times Y, Z)_{co}$, and let the function $f \in (K, W)$ be arbitrary. Then $f : X \times Y \to Z$ is continuous with $f(K) \subseteq W$. So $f^{-1}(W)$ is a nbd of K. Let $p_X : X \times Y \to X$, $p_Y : X \times Y \to Y$ be the projection maps, and put $E = p_X(K)$ and $F = p_Y(K)$. Then $E \subseteq X$ and $F \subseteq Y$ are compact, since K is so. Also, we have $K \subseteq E \times F$. Clearly, $E \times F$ is compact Hausdorff, and hence regular. So, for each $k \in K$, we can find an open nbd N_k of k in $E \times F$ such that $\overline{N}_k \subseteq (E \times F) \cap f^{-1}(W)$. Notice that the closures of N_k in $E \times F$ and $X \times Y$ are identical. Moreover, we may assume that $N_k = U_k \times V_k$, where U_k is open in E, and V_k is open in F. Then we have $f \in (\overline{N}_k, W) = (\overline{U}_k \times \overline{V}_k, W)$. Since K is compact, there exist finitely many points $k_1, \dots, k_n$ in K such that $K \subseteq \bigcup_1^n N_{k_i} \subseteq \bigcup_1^n \overline{N}_{k_i}$. Consequently, we have

$$f \in \bigcap_1^n \left(\overline{U}_{k_i} \times \overline{V}_{k_i}, W \right) = \bigcap_1^n \left(\overline{N}_{k_i}, W \right) = \left(\bigcup_1^n \overline{N}_{k_i}, W \right) \subseteq (K, W).$$

Since E and F are compact, all $\overline{U}_k$'s and $\overline{V}_k$'s are compact, and we see that (K, W) is open in the topology generated by the sets of the desired form. Because the compact-open topology for $\mathcal{C}(X \times Y, Z)$ is generated by the sets (K, W), the lemma follows. $\diamond$

Proof of the Exponential Law. Let Φ and Ψ denote the spaces $\mathcal{C}(X \times Y, Z)$ and $\mathcal{C}(X, \mathcal{C}(Y, Z)_{co})$, respectively, with the compact-open topology. We have already seen that $\alpha \colon \Phi \to \Psi$ is injective. By Lemmas 11.2.15 and 11.2.16, we see that the topology of Ψ is generated by the sets $(A, (B, W))$, and that of Φ by the sets $(A \times B, W)$, where $A \subseteq X$, $B \subseteq Y$ are compact, and $W \subseteq Z$ is open. It is also clear that

$$\begin{aligned} f \in (A \times B, W) &\Leftrightarrow f_x(y) \in W \text{ for all } x \in A \text{ and all } y \in B \\ &\Leftrightarrow f_x \in (B, W) \text{ for all } x \in A \\ &\Leftrightarrow \alpha(f) = \hat{f} \in (A, (B, W)). \end{aligned}$$

Therefore we have $\alpha^{-1}(A, (B, W)) = (A \times B, W)$. Since the sets $(A \times B, W)$ are open in Φ, we deduce that α is continuous. Moreover, the image of $(A \times B, W)$ under α is $(A, (B, W)) \cap \alpha(\Phi)$, and thus α carries the members of a subbase for Φ to open sets in $\alpha(\Phi)$. Since α is injective, it preserves the intersections, and therefore $\alpha \colon \Phi \to \alpha(\Phi)$ is open.

For the last assertion, we note that the compact-open topology on $\mathcal{C}(Y, Z)$ is admissible if Y is locally compact and Hausdorff, by Proposition 11.2.6. So, given a function $\psi \in \Psi$, the function $f = e \circ (\psi \times 1_Y)$ is in Φ and $\alpha(f) = \psi$, obviously. $\diamond$

If the notation Y^X is used to denote the space $\mathcal{C}(X, Y)$ with the compact-open topology, then the above theorem can be expressed as $Z^{X \times Y} \approx \left(Z^Y \right)^X$, justifying its name. We close this section with the remark that the Exponential Law holds good under some other conditions, too (e.g., see Exercises 16 and 24). This theorem is highly appreciated by the mathematicians working in the field of Functional Analysis (duality theory) or Algebraic Topology (homotopy theory).

Exercises

1. Show that the family $\{(F, U) \mid F \subseteq X \text{ finite, and } U \subseteq Y \text{ open}\}$ generates the pointwise topology on Y^X. (For this reason, the pointwise topology for Y^X is also referred to as the *finite-open topology*.)

2. If X is a discrete space with n elements, show that $\mathcal{C}(X, Y)_{co}$ is homeomorphic to the product space $Y \times \cdots \times Y$ (n-factors).

3. ● Prove that the compact-open topology for $\mathcal{C}(\mathbb{R})$ is different from the pointwise topology as well as the topology of uniform convergence defined by the standard bounded metric.

4. Prove that the function $\psi \colon \mathbb{R}^{n+1} \to C(\mathbb{R})_{co}$ defined by $\psi(a_0, \ldots, a_n) = a_0 + a_1 t + \cdots + a_n t^n$ is continuous.

5. (a) Prove that $\overline{(A, B)} \subseteq (A, \overline{B})$ in the compact-open topology on Y^X, and hence deduce that $C(I, I)$ is closed in $C(I)_{co}$.

   (b) Is $C(I)$ closed in the space $\mathbb{R}^I$ with the compact-open topology?

6. Let F be a closed subset of a space Y and let X be a locally compact Hausdorff space. Prove that $\{(f, x) \mid f(x) \in F\}$ is closed in $C(X, Y)_{co} \times X$.

7. Let X be a compact Hausdorff space, and let $G \subseteq X$ be open. Prove:

   (a) If $F \subseteq Y$ is closed, then $\{f \mid f^{-1}(F) \subseteq G\}$ is open in $C(X, Y)_{co}$.
   (b) $\{(f, y) \mid f^{-1}(y) \subseteq G\}$ is open in $C(X, Y)_{co} \times Y$, if Y is Hausdorff.

8. Let A be a closed subspace of a space X, and let y_0 be an element of a space Y. Let $\Phi = \{f \in C(X, Y) \mid f(X - A) = \{y_0\}\}$. If A is locally compact Hausdorff, prove that the evaluation map $e \colon \Phi_{co} \times X \to Y$ is continuous.

9. Give an example of a locally compact, second countable space X and a second countable space Y such that $C(X, Y)_p$ is not second countable. (This shows that the analogue of Theorem 11.2.12 does not hold in pointwise topology.)

10. Let X be a second countable compact Hausdorff space, and let Y be a second countable space. Prove that $C(X, Y)_{co}$ is metrizable $\Leftrightarrow Y$ is regular.

11. For $f, g \in C(\mathbb{R})$ and each integer $n > 0$, denote the supremum of $\{|f(x) - g(x)| \colon |x| \le n\}$ by c_n. Show that ρ defined by $\rho(f, g) = \sum_1^\infty \frac{c_n}{2^n(1+c_n)}$ is a metric on $C(\mathbb{R})$ and it induces the compact-open topology.

12. Let X be a discrete space of all positive integers, and let Y be the discrete two-point space. Show that the compact-open topology for $C(X, Y)$ satisfies the second axiom of countability. What is the uniform metric topology for it? Are the two topologies identical?

13. Let Z be a subspace of a space Y. Show that $C(X, Z)_{co}$ is homeomorphic to a subspace of $C(X, Y)_{co}$ and the mapping $C(Y, X)_{co} \to C(Z, X)_{co}$, $f \mapsto f|Z$, is continuous for every space X.

14. Let Φ be a set of continuous functions $X \to Y$, and suppose that $\mathcal{T}$ is a topology on Φ such that if a net in Φ converges to $f \in \Phi$ in the topology $\mathcal{T}$, then it converges continuously to f. Show that $\mathcal{T}$ is admissible.

15. Let Y and Z be spaces, and suppose that $C(Y, Z)$ is given a topology. If the association function $C(X \times Y, Z) \to C(X, C(Y, Z))$ is surjective for every space X, show that the topology on $C(Y, Z)$ is admissible.

16. • Let X and Y be first countable spaces. Prove:

   (a) If $\psi \colon X \to C(Y, Z)_{co}$ is continuous for a space Z, then the associated function $\hat{\psi} \colon X \times Y \to Z$, $(x, y) \mapsto \psi(x)(y)$, is continuous.

(b) The association map $C(X \times Y, Z) \longrightarrow C(X, C(Y, Z)_{co})$, $f \mapsto \hat{f}$, is a homeomorphism in the compact-open topology on both function spaces if X and Y are also Hausdorff.

17. Show that $C(\mathbb{Q}, I)_{co}$ is not second countable, where $\mathbb{Q}$ is the subspace of the rationals.

18. Let X and Z be Hausdorff spaces, and let Y be a locally compact space. Show that the function $\theta \colon C(Y, Z)_{co} \times C(X, Y)_{co} \to C(X, Z)_{co}$, defined by $\theta(g, f) = gf$, is continuous.

19. Given three spaces X, Y, Z, and a continuous function $\phi \colon Y \to Z$, define the function $\phi_* \colon C(X, Y)_{co} \to C(X, Z)_{co}$ by $\phi_*(f) = \phi f$. Prove:

   (a) ϕ_* is continuous.

   (b) If $\psi \colon Z \to T$ is another continuous function, then $(\psi\phi)_* = \psi_* \circ \phi_*$.

   (c) If ϕ is an embedding, then so is ϕ_*.

20. Given three spaces X, Y, Z, and a continuous function $\phi \colon X \to Y$, define the function $\phi^* \colon C(Y, Z)_{co} \to C(X, Z)_{co}$ by $\phi^*(f) = f\phi$. Prove:

   (a) ϕ^* is continuous.

   (b) If $\psi \colon T \to X$ is another continuous function, then $(\phi\psi)^* = \psi^* \circ \phi^*$.

   (c) If ϕ is surjective, then ϕ^* is injective for any space Z.

21. If ϕ is a proper map of a Hausdorff space X onto a locally compact Hausdorff space Y, show that ϕ^* is an embedding.

22. Let X be a compact T_2 space and Y be any space. If $\pi \colon X \to X/A$ is the quotient map, where $A \subseteq X$ closed, show that the image of $\pi^* \colon C(X/A, Y)_{co} \to C(X, Y)_{co}$ is the subspace consisting of continuous functions $f \colon X \to Y$ such that $f(A)$ is a singleton set.

23. Prove: (a) $C(X_1 + X_2, Y)_{co} \approx C(X_1, Y)_{co} \times C(X_2, Y)_{co}$ for all spaces X_1, X_2 and Y.

   (b) If X is a Hausdorff space, then, for every space Y_1 and Y_2, $C(X, Y_1 \times Y_2)_{co} \approx C(X, Y_1)_{co} \times C(X, Y_2)_{co}$.

   (c) If X_1 and X_2 are Hausdorff, then the mapping $(f_1, f_2) \mapsto f_1 \times f_2$ of the product space $C(X_1, Y_1)_{co} \times C(X_2, Y_2)_{co}$ into $C(X_1 \times X_2, Y_1 \times Y_2)_{co}$ is an embedding.

24. • Given three spaces X, Y, and Z, let $\psi \colon X \to C(Y, Z)_{co}$ be a continuous map and $\hat{\psi} \colon X \times Y \to Z$ be its associate. Prove:

   (a) $\hat{\psi} | (X \times K)$ is continuous for every compact set $K \subseteq Y$.

   (b) If $X \times Y$ is a k-space, then the association map $C(X \times Y, Z) \longrightarrow C(X, C(Y, Z)_{co})$ is a homeomorphism in the compact-open topology on both spaces.

11.3 Topology of Compact Convergence

Throughout this section, we assume that Y is a metric space and study another useful topology for Y^X. It will be seen that this topology on $\mathcal{C}(X, Y)$ is independent of the metric on Y; in fact, it coincides with the compact-open topology.

Definition 11.3.1 Let X be a set and (Y, d) be a bounded metric space. Given a family $\mathscr{A}$ of sets $A \subseteq X$, consider the sup metric for each set Y^A determined by d, and let γ_A be the function $Y^X \to Y^A$ defined by $\gamma_A(f) = f|A$. The topology on Y^X induced by the family of functions $\{\gamma_A \mid A \in \mathscr{A}\}$ is called the *topology of uniform convergence on members of $\mathscr{A}$*, and will be denoted by $\mathscr{T}(d^*|\mathscr{A})$.

It is useful to know a neighborhood basis at a point f of Y^X in the topology $\mathscr{T}(d^*|\mathscr{A})$. Since the open balls about a point in a metric space form a neighborhood basis at the point, it is clear from the definition of $\mathscr{T}(d^*|\mathscr{A})$ that a nbd N of f in this topology contains a finite intersection $\bigcap_A \gamma_A^{-1}(B(f|A; \epsilon_A))$, where $B(f|A; \epsilon_A)$ is the open ball of radius ϵ_A centered at $f|A$ in the metric space (Y^A, d^*). For each real number $\epsilon > 0$ and each set $A \in \mathscr{A}$, put

$$B(f, \epsilon, A) = \{g \in Y^X \mid d(g(x), f(x)) < \epsilon \text{ for all } x \in A\}.$$

Then $B(f, \epsilon, A)$ is a nbd of f in Y^X, being the inverse image of the open ball $B(f|A; \epsilon) = \{h \in Y^A \mid d(h(x), f(x)) < \epsilon \text{ for all } x \in A\}$ under the continuous map $\gamma_A: Y^X \to Y^A$, and N contains a finite intersection of such nbds of f. Hence, we see that the family of all finite intersections of sets $B(f, \epsilon, A)$, where ϵ varies over $\mathbb{R}_+$, the set of positive real numbers, and A varies over the family $\mathscr{A}$, is a nbd basis at f. Furthermore, if $\mathscr{A}$ is *closed under formation of finite unions* (e.g., when $\mathscr{A}$ is the family of all compact subsets of X), then the family $\{B(f, \epsilon, A) \mid \epsilon \in \mathbb{R}_+ \text{ and } A \in \mathscr{A}\}$ itself is a nbd base at f.

Definition 11.3.2 Let $A \subseteq X$ be sets, and (Y, d) be a metric space. A net $\langle f_\nu \rangle$ in Y^X is said to *converge to $g: X \to Y$ uniformly on A* if for each $\epsilon > 0$, there is an index μ, depending on ϵ and A, such that $\mu \preceq \nu \Rightarrow d(f_\nu(x), g(x)) < \epsilon$ for all $x \in A$.

Let X be a set and (Y, d) be a bounded metric space. If a net $\langle f_\nu \rangle$ in Y^X converges to a function g uniformly on a set $A \subseteq X$, then it is obvious that the net $\langle f_\nu|A \rangle$ converges to $g|A$ in the metric topology on Y^A determined by the metric d. Conversely, suppose that $\mathscr{A}$ is a covering of X and let $\langle f_\nu \rangle$ be a net in Y^X such that the net $\langle f_\nu|A \rangle$ converges to g_A in the metric space Y^A for every $A \in \mathscr{A}$. Since $\mathscr{A}$ covers X, each point $x \in X$ belongs to some set $A \in \mathscr{A}$. Write $g(x) = g_A(x)$, if $x \in A$. Then $g(x) = \lim f_\nu(x)$, and so it is well defined. It is now trivial to see that $\langle f_\nu \rangle$ converges to g uniformly on each $A \in \mathscr{A}$.

The topology of uniform convergence on members of a family of subsets of a set X derives its name from the following.

Proposition 11.3.3 *Let X be a set and (Y, d) be a bounded metric space. Suppose that $\mathscr{A}$ is a covering of X, and Y^X is given the topology of uniform convergence on members of $\mathscr{A}$. Then, a net $\langle f_\nu \rangle$ in the space Y^X is convergent if and only if the net $\langle f_\nu | A \rangle$ is convergent in the space Y^A with the metric topology determined by d, for every $A \in \mathscr{A}$.*

Proof Suppose that $f_\nu \to g$ in the space Y^X. By the definition of the topology on Y^X, the function $\gamma_A : Y^X \to (Y^A)_u, f \mapsto f | A$, is continuous for every $A \in \mathscr{A}$. Therefore $f_\nu | A \to g | A$ in $(Y^A)_u$ for all $A \in \mathscr{A}$.

Conversely, suppose that, for every $A \in \mathscr{A}$, $f_\nu | A \to g_A$ in the metric space (Y^A, d_A^*), where d_A^* is the metric on Y^A determined by d. Then there exists a function $g : X \to Y$ such that $g | A = g_A$. We assert that $f_\nu \to g$ in the space Y^X. To see this, consider a nbd of g of the form $B(g, \epsilon, A)$. By our hypothesis, there exists an index μ such that $\mu \preceq \nu \Rightarrow d_A^* (f_\nu | A, g | A) < \epsilon \Rightarrow d (f_\nu(x), g(x)) < \epsilon$ for all $x \in A$. So $f_\nu \in B(g, \epsilon, A)$ for all $\nu \succeq \mu$. Since the family of finite intersections of the sets $B(g, \epsilon, A)$ is a nbd basis at g, we see that $f_\nu \to g$ relative to the topology of Y^X. $\diamond$

Now, we introduce the topology of main concern here.

Definition 11.3.4 Let X be a topological space and (Y, d) be a bounded metric space. The topology of uniform convergence on the compact subsets of X for Y^X is called *the topology of compact convergence* (or *the topology of uniform convergence on compacta*.)

We will denote the topology of compact convergence for Y^X by $\mathscr{T}_c$, and the subspace $\mathcal{C}(X, Y) \subseteq (Y^X, \mathscr{T}_c)$ by $\mathcal{C}(X, Y)_c$. If $\mathscr{T}_u$ denote the topology of uniform convergence for Y^X, then we have $\mathscr{T}_u \supseteq \mathscr{T}_c \supseteq \mathscr{T}_p$. This is an immediate consequence of the following theorem, since X is clearly covered by the family of all its compact subsets.

Theorem 11.3.5 (a) *The topology of uniform convergence for Y^X is finer than the topology of uniform convergence on members of $\mathscr{A}$, and coincides with it if $X \in \mathscr{A}$.*

(b) *If the family $\mathscr{A}$ covers X, then the topology of pointwise convergence for Y^X is coarser than the topology of uniform convergence on members of $\mathscr{A}$.*

The theorem follows directly by invoking definitions of the relevant topologies.

Notice that the topologies $\mathscr{T}_u$ and $\mathscr{T}_c$ are identical when X is compact, and $\mathscr{T}_c$ and $\mathscr{T}_p$ are identical when X is discrete. But, in general, the three topologies are distinct from each other, as we see below.

Example 11.3.1 For each $n = 1, 2, \ldots,$ define $f_n : I \to \mathbb{R}$ by setting $f_n(t) = t^n (1 - t^n)$. The sequence $\langle f_n \rangle$ in $\mathcal{C}(I)$ clearly converges to the constant function c at 0 in the pointwise topology. But, given n, $f_n(t) = 1/4$ at $t = 2^{-1/n}$ and therefore it fails to converge uniformly to c. This shows that the topology of compact convergence on $\mathcal{C}(I)$ is strictly finer than the topology of pointwise convergence.

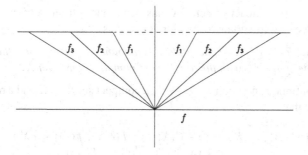

Fig. 11.3 First three terms of $\langle f_n \rangle$ in Example 11.3.2 and its pointwise limit

Example 11.3.2 Let $C^*(\mathbb{R})$ be the set of all real-valued bounded continuous functions. For each $n = 1, 2, \ldots$, define a function $f_n : \mathbb{R} \to \mathbb{R}$ by $f_n(x) = |x|/n$ for $|x| \leq n$, and 1 otherwise (Fig. 11.3). It is clear that the sequence $\langle f_n \rangle$ converges to the constant function $f(x) = 0$ in the pointwise topology for $C^*(\mathbb{R})$. But it does not converge to f in the topology of uniform convergence, since for each integer $n > 0$, $|f_n(x) - f(x)| = 1$ for $x > n$. If $\mathscr{A}$ is a covering of $\mathbb{R}$ consisting of bounded sets, then we show that the sequence $\langle f_n \rangle$ converges to f in the topology of uniform convergence on members of $\mathscr{A}$. Consider a subbasic nbd $B(f, \epsilon, A)$ of f. Since A is bounded, there is a real number δ such that $|x| \leq \delta$ for every $x \in A$. Now choose an integer $n_0 > \delta/\epsilon$. Then, for $n > n_0$, we have $|f_n(x) - f(x)| = |x|/n < \epsilon$ for all $x \in A$. This implies that $f_n \in B(f, \epsilon, A)$ for all $n > n_0$, and hence $f_n \to f$. In particular, $\langle f_n \rangle$ converges to f in the topology of compact convergence, and we deduce that the topology of uniform convergence for $C^*(\mathbb{R})$ is strictly finer than the topology of compact convergence.

It is known from Real Analysis that a sequence of continuous functions $f_n : \mathbb{R} \to \mathbb{R}$, $n = 1, 2, \ldots$, has a continuous limit function, if it converges uniformly on every compact subset of $\mathbb{R}$. This result is also true for nets of continuous functions of a k-space into a metric space. To this end, we first prove

Proposition 11.3.6 *Let X be a topological space and (Y, d) be a metric space. If a net $\langle f_\nu \rangle$ in $C(X, Y)$ converges to a function $g : X \to Y$ uniformly on a set $A \subseteq X$, then $g|A$ is continuous.*

Proof Let $x_0 \in A$ be arbitrary. Given $\epsilon > 0$, there is an index μ such that $d\left(f_\mu(x), g(x)\right) < \epsilon/3$ for all $x \in A$. Since f_μ is continuous, $G_x = f_\mu^{-1}\left(B(f_\mu(x); \epsilon/3)\right)$ is open in X. Thus $A \cap G_{x_0}$ is a nbd of x_0 in A and, for every $x \in A \cap G_{x_0}$, we have $d\left(g(x), g(x_0)\right) \leq d\left(g(x), f_\mu(x)\right) + d\left(f_\mu(x), f_\mu(x_0)\right) + d\left(f_\mu(x_0), g(x_0)\right) < \epsilon$. This implies that $g|A$ is continuous at x_0. $\diamond$

From Theorem 5.4.14 and the preceding proposition, it follows immediately that a net of continuous functions of a k-space X into a metric space Y, which converges uniformly on every compact subset of X, has a continuous limit. A trivial, but useful, reformulation of the foregoing discussion is given as

Corollary 11.3.7 *If X is a k-space, and Y is a bounded metric space, then $C(X, Y)$ is closed in the space Y^X with the topology of compact convergence.*

For any continuous function $f : X \to Y$, a compact set $K \subseteq X$ and a real number $\epsilon > 0$, we see that

$$C(X, Y) \cap B(f, \epsilon, K) = \{g \in C(X, Y) \mid d(g(x), f(x)) < \epsilon \ \forall \ x \in K\}$$
$$= C(X, Y) \cap \gamma_K^{-1}(B(f|K; \epsilon)),$$

and thus it is an open nbd of f in the subspace $C(X, Y)_c$ of $(Y^X, \mathcal{T}_c)$. For convenience, we will denote this set, too, by $B(f, \epsilon, K)$. Then the family $\{B(f, \epsilon, K) \mid f \in C(X, Y), \epsilon \in \mathbb{R}_+ \text{ and } K \in \mathcal{K}\}$ is a base of $C(X, Y)_c$, since the collection $\mathcal{K}$ of all compact sets $K \subseteq X$ is closed under formation of finite unions. It is also easily seen that the topology of $C(X, Y)_c$ is induced by the functions $C(X, Y) \to C(K, Y)_u$, $f \mapsto f|K$, where K varies over $\mathcal{K}$ (cf. Exercise 6.6.7). We further note that the topology of $C(X, Y)_c$ is independent of the metric on Y, since the topology of $C(K, Y)_u$ is so. In fact, we have

Theorem 11.3.8 *Let X be a topological space and Y be a metric space. Then the topology of compact convergence and the compact-open topology for $C(X, Y)$ are identical.*

Proof Assume first that (K, U) is a member of the subbase for the compact-open topology for $C(X, Y)$ and let $f \in (K, U)$ be arbitrary. We adapt the first part of the proof of Theorem 11.2.5 to obtain a real $\epsilon > 0$ such that the open ball $B(f(x); \epsilon) \subseteq U$ for every $x \in K$. Then, by definition, every member of the open subset $B(f, \epsilon, K)$ of $C(X, Y)_c$ maps K into U and (K, U) contains the nbd $B(f, \epsilon, K)$ of f. This shows that (K, U) is a union of open sets in $C(X, Y)_c$, and hence open in this space. Since (K, U) is an arbitrary member of a subase of $C(X, Y)_{co}$, we conclude that its topology is coarser than that of $C(X, Y)_c$.

To see the converse, let $f : X \to Y$ be a continuous map. Since the sets $B(f, \epsilon, K)$, where $\epsilon > 0$ is a real number and $K \subseteq X$ is compact, form a nbd base at f in the topology of compact convergence, it suffices to prove that each set $B(f, \epsilon, K)$ is a nbd of f in the compact-open topology. To this end, we make a slight modification in the second part of the proof of Theorem 11.2.5 to find finitely many closed subsets K_i of K and open sets $U_i \subseteq Y$ such that $f \in \bigcap_i (K_i, U_i) \subseteq B(f, \epsilon, K)$.   ◇

An alternative proof of the above theorem is graded in Exercise 5. As an immediate consequence of this result, we see that the topology of compact convergence for $C(X, Y)$ does not depend upon the metric of Y, as the same is true of the compact-open topology. Moreover, we prove

Corollary 11.3.9 *Let X be a locally compact Hausdorff space, and let Y be a metric space. Then continuous convergence in $C(X, Y)$ is equivalent to uniform convergence on compacta.*

Proof This is evident from Propositions 11.2.6 and 11.2.9 and the preceding theorem. ◇

To characterize compact subsets of a function space $C(X, Y)$ in the topology compact convergence, we need one more notion which we study now.

Definition 11.3.10 Let X be a topological space and Y be a metric space. A subset $\Phi \subseteq Y^X$ is *equicontinuous at a point* $x \in X$ if for each $\epsilon > 0$, there is a nbd U of x such that $f(U) \subseteq B(f(x); \epsilon)$ for every $f \in \Phi$. The family Φ is *equicontinuous* if it is equicontinuous at every point of X.

Obviously, Φ is equicontinuous at x if and only if

$$\bigcap \{f^{-1}(B(f(x); \epsilon)) \mid f \in \Phi\}$$

is a nbd of x for each $\epsilon > 0$. It is also clear that the members of an equicontinuous family Φ are continuous. Note that equicontinuity depends on the metric used in Y.

Example 11.3.3 Any finite family $\Phi \subseteq C(X, Y)$ is equicontinuous.

Example 11.3.4 For each real $c > 0$, the family

$$\Phi = \{f \in C(I) \mid |f'(t)| \leq c \text{ for all } t \in (0, 1)\}$$

is equicontinuous on I, since $|f(t) - f(t')| \leq |t - t'|c$, by the mean value theorem.

Example 11.3.5 For each $n = 1, 2, \ldots$, define $f_n : I \to \mathbb{R}$ by

$$f_n(x) = x^2 / [x^2 + (1 - nx)^2].$$

The family $\{f_n\}$ is obviously not equicontinuous at $0 \in I$, since $f_n(1/n) = 1$ for every n.

Lemma 11.3.11 *Let Φ be an equicontinuous family of functions of a topological space X into a metric space (Y, d). Then*

(a) *the topology of pointwise convergence for Φ coincides with the topology of compact convergence, and*
(b) *the closure of Φ in the space Y^X with the pointwise topology is equicontinuous.*

Proof (a): We have already seen that the topology of compact convergence is finer than the topology of pointwise convergence (ref. Theorem 11.3.5). We prove the reverse by showing that if a net $\langle f_\nu \rangle$ in Φ converges to f in the pointwise topology on Φ, then it also converges to f in the topology of compact convergence. Clearly, the sets

$$B(f, \epsilon, K) = \{g \in \Phi \mid d(g(x), f(x)) < \epsilon \text{ for every } x \in K\},$$

where $K \subseteq X$ is compact, and $\epsilon > 0$ a real, form a nbd basis at f in the subspace Φ_c of $C(X, Y)_c$. Since Φ is equicontinuous, each point $x \in X$ has an open nbd U_x such that $g(U_x) \subseteq B(g(x); \epsilon/4)$ for all $g \in \Phi$. So we have $d(f_\nu(u), f_\nu(x) \le \epsilon/4)$ for all ν and all $u \in U_x$. Also, by our assumption, $f_\nu(x) \to f(x)$ in Y. Therefore there exists an index μ_x such that $d(f_\nu(x), f(x)) \le \epsilon/4$ for all $\nu \succeq \mu_x$, and thus we have

$$d(f_\nu(u), f(x)) \le d(f_\nu(u), f_\nu(x)) + d(f_\nu(x), f(x)) < \epsilon/2$$

for all $u \in U_x$ and $\nu \succeq \mu_x$. This implies that $d(f(u), f(x)) = \lim d(f_\nu(u), f(x)) \le \epsilon/2$ for every $u \in U_x$. By compactness of K, there exist finitely many points $x_1, \ldots, x_n$ in K such that $K \subseteq \bigcup_1^n U_{x_i}$. Then choosing an index μ_0 such that $\mu_0 \succeq \mu_{x_i}$ for every i, we find that

$$d(f_\nu(u), f(u)) \le d(f_\nu(u), f(x_i)) + d(f(x_i), f(u)) < \epsilon$$

for all $\nu \succeq \mu_0$ and all $u \in U_{x_i}$. It follows that $f_\nu \in B(f, \epsilon, K)$ for every $\nu \succeq \mu_0$, and hence $f_\nu \to f$ in the topology of compact convergence.

(b): Let $\overline{\Phi}$ denote the closure of Φ in $(Y^X, \mathscr{T}_p)$ and $\epsilon > 0$ be given. Then, for an arbitrary but fixed $x_0 \in X$, there is a nbd U of x_0 such that $f(U) \subseteq B(f(x_0); \epsilon/3)$ for every $f \in \Phi$, since Φ is equicontinuous at x_0. If $g \in \overline{\Phi}$, then there is a net $\langle f_\nu \rangle$ in Φ such that $f_\nu(x) \to g(x)$ for every $x \in X$. So there exists an index μ_{x_0} such that $d(f_\nu(x_0), g(x_0)) < \epsilon/3$ for all $\nu \succeq \mu_{x_0}$. Consequently, for each $x \in U$ and $\nu \succeq \mu_{x_0}$, we have

$$d(f_\nu(x), g(x_0)) \le d(f_\nu(x), f_\nu(x_0)) + d(f_\nu(x_0), g(x_0)) < 2\epsilon/3.$$

Hence $d(g(x), g(x_0)) = \lim d(f_\nu(x), g(x_0)) \le 2\epsilon/3 < \epsilon$ for every $x \in U$, and we have $g(U) \subseteq B(g(x_0); \epsilon)$, as desired. $\diamondsuit$

Now, we come to the main result of this section.

Theorem 11.3.12 (Arzela–Ascoli) *Let X be a locally compact hausdorff space, and let (Y, d) be a metric space. If Φ is a compact subset of $C(X, Y)_c$, then it is closed and equicontinuous, and $\overline{e_x(\Phi)}$ is compact for every $x \in X$. The converse holds for every space X.*

Proof Suppose that Φ is a compact subset of $C(X, Y)_c$. Since Y is Hausdorff and the topology of compact convergence is finer than the pointwise topology, we see that $C(X, Y)_c$ is Hausdorff and each evaluation map $e_x : C(X, Y)_c \to Y$, $f \mapsto f(x)$, is continuous. Hence Φ is closed in $C(X, Y)_c$, and the sets $e_x(\Phi) \subseteq Y$ are compact so that these are also closed. Consequently, $\overline{e_x(\Phi)} = e_x(\Phi)$ is compact for every x. Thus, it remains to prove that Φ is equicontinuous. Let $x_0 \in X$ and $\epsilon > 0$ be arbitrary. By

Theorem 11.3.8 and Proposition 11.2.6, the evaluation map $e\colon \mathcal{C}(X, Y)_c \times X \to Y$ is continuous. Therefore, for each $f \in \Phi$, there exist open sets $O_f \subseteq \mathcal{C}(X, Y)_c$ and $U_f \subseteq X$ such that $f \in O_f$, $x_0 \in U_f$ and $g\left(U_f\right) \subseteq B\left(f(x_0); \epsilon/2\right)$ for every $g \in O_f$. Since $d\left(g(x), g(x_0)\right) \leq d\left(g(x), f(x_0)\right) + d\left(f(x_0), g(x_0)\right)$, we have $g\left(U_f\right) \subseteq B\left(g(x_0); \epsilon\right)$ for every $g \in O_f$. Clearly, the open sets O_f, $f \in \Phi$, cover the compact set Φ. So we find finitely of them, say $O_{f_1}, \ldots, O_{f_n}$, covering Φ. Let $U_{f_1}, \ldots, U_{f_n}$ be the corresponding nbds of x_0 and put $U = \bigcap_1^n U_{f_i}$. Then, U is a nbd of x_0 and $f(U) \subseteq B\left(f(x_0); \epsilon\right)$ for every $f \in \Phi$. Thus Φ is equicontinuous at x_0, as desired.

Conversely, suppose that Φ is a family of functions $X \to Y$ with the given properties. Since $\prod_x \overline{e_x(\Phi)}$ is closed in Y^X in the pointwise topology $\mathcal{T}_p$ and $\Phi \subseteq \prod_x \overline{e_x(\Phi)}$, we have $\overline{\Phi} \subseteq \prod_x \overline{e_x(\Phi)}$, where $\overline{\Phi}$ denotes the closure of Φ in the space $\left(Y^X, \mathcal{T}_p\right)$. By the Tychonoff theorem, $\prod_x \overline{e_x(\Phi)}$ is compact, for each $\overline{e_x(\Phi)}$ is compact. Being a closed subset of a compact space, $\overline{\Phi}$ is a compact subspace of $\prod_x \overline{e_x(\Phi)}$. Moreover, by Lemma 11.3.11, $\overline{\Phi} \subseteq \mathcal{C}(X, Y)$ and the relative topologies induced from the spaces $\mathcal{C}(X, Y)_c$ and $\prod_x \overline{e_x(\Phi)}$ on $\overline{\Phi}$ coincide. Therefore $\overline{\Phi}$ is also compact as a subspace of $\mathcal{C}(X, Y)_c$. Since Φ is closed in $\mathcal{C}(X, Y)_c$, it is certainly closed in the subspace $\overline{\Phi} \subseteq \mathcal{C}(X, Y)_c$, and we conclude that Φ is compact subset of $\mathcal{C}(X, Y)_c$. $\diamond$

The foregoing theorem generalizes to k-spaces; in addition, the requirement of equicontinuity on Φ can also be weakened in terms of the following notion.

Definition 11.3.13 Let X be a topological space and (Y, d) be a metric space. A family $\Phi \subseteq \mathcal{C}(X, Y)$ is called *equicontinuous on a set* $K \subseteq X$ if the family $\{f|K : f \in \Phi\}$ is equicontinuous at each point of K.

Clearly, a family of continuous functions $X \to Y$ that is equicontinuous at every point of $K \subseteq X$ is equicontinuous on K, but the converse is not true, in general. With the above terminology, we have

Theorem 11.3.14 (Arzela–Ascoli) *Let X be a k-space and Y be a metric space. Then a subspace Φ of $\mathcal{C}(X, Y)_c$ is compact if and only if it is closed, equicontinuous on each compact subset of X, and $\overline{e_x(\Phi)}$ is compact for every $x \in X$.*

Proof The proof of necessity of the conditions goes through as in Theorem 11.3.12 because if $\Phi \subseteq \mathcal{C}(X, Y)_c$ is compact, then $\{f|K : f \in \Phi\}$ is a compact subset of $\mathcal{C}(K, Y)_u$ for every compact subset $K \subseteq X$, since the restriction map $\gamma_K\colon \mathcal{C}(X, Y)_c \to \mathcal{C}(K, Y)_u, f \mapsto f|K$, is continuous. And, the evaluation map $\mathcal{C}(K, Y)_u \times K \to Y$ is continuous, by Proposition 11.2.4.

For sufficiency of the conditions, we first show that every function in $\overline{\Phi}$, the closure of Φ in the space Y^X with the pointwise topology, is continuous. Let K be a compact subset of X. It is obvious that the restriction map $\gamma_K\colon Y^X \to Y^K, f \mapsto f|K$, is continuous in the pointwise topology on both spaces, and hence $\gamma_K\left(\overline{\Phi}\right)$ is contained in $\overline{\gamma_K(\Phi)}$, the closure of $\gamma_K(\Phi)$ in $(Y^K)_p$. By our hypothesis, $\gamma_K(\Phi) = \{f|K : f \in \Phi\}$ is equicontinuous, and then, by Lemma 11.3.11, $\overline{\gamma_K(\Phi)}$ is also equicontinuous. It follows that $\gamma_K\left(\overline{\Phi}\right)$, being a subfamily of an equicontinuous family of functions $K \to Y$, is equicontinuous at every point of K; in particular, $g|K$ is continuous for all $g \in \overline{\Phi}$. Since X is a k-space, we see that each $g \in \overline{\Phi}$ is continuous on X, and thus $\overline{\Phi} \subseteq \mathcal{C}(X, Y)$.

Next, we show that the topology of compact convergence for $\overline{\Phi}$ agrees with the pointwise topology. By Theorem 11.3.5, it is clear that the pointwise topology for $\overline{\Phi}$ is coarser than the topology of compact convergence. To see the converse, recall that the sets

$$B\left(f, \epsilon, K\right) = \gamma_K^{-1}\left(B\left(f|K; \epsilon\right)\right) \cap \mathcal{C}(X, Y),$$

where $f \in \mathcal{C}(X, Y)$, $\epsilon \in \mathbb{R}_+$ and $K \subseteq X$ are compact sets, form a base for the topology of $\mathcal{C}(X, Y)_c$. Therefore the topology of $\overline{\Phi}_c$ is generated by the sets $\gamma_K^{-1}\left(B\left(f|K; \epsilon\right)\right) \cap \overline{\Phi}$, and hence it suffices to show that these sets are open in the pointwise topology for $\overline{\Phi}$. By Theorems 11.2.5 and 11.3.8, we see that $B\left(f|K; \epsilon\right)$ is open in $\mathcal{C}(K, Y)$ in the topology of compact convergence. As seen above, $\gamma_K\left(\overline{\Phi}\right)$ is equicontinuous, and so, by Lemma 11.3.11, $B\left(f|K; \epsilon\right) \cap \gamma_K\left(\overline{\Phi}\right)$ is open in $\gamma_K\left(\overline{\Phi}\right)$ in the pointwise topology. Then the continuity of the map $\overline{\Phi}_p \to \gamma_K\left(\overline{\Phi}\right)_p, g \mapsto g|K$, implies that $\gamma_K^{-1}\left(B\left(f|K; \epsilon\right)\right) \cap \overline{\Phi}$ is open $\overline{\Phi}_p$, as desired.

Now, as in Theorem 11.3.12, we see that $\overline{\Phi}_c$ is compact. Also, it is clear that Φ is closed in $\overline{\Phi}_c$, and hence Φ is a compact subset of $\mathcal{C}(X, Y)_c$. $\diamondsuit$

Remark 11.3.15 (a) The form of Arzela–Ascoli theorem given here is quite general; we will develop in Exercise 15 (below) the classical form of the theorem.

(b) The concept of uniform convergence and the results discussed in this section can be generalized to a class of spaces which lie between metric spaces and topological spaces. These spaces are referred to as *the uniform spaces*. A study of this notion is outside the scope of this book; the interested reader may refer to James [5], Kelley [6] or Willard [16].

Exercises

1. Does the sequence in Example 11.3.5 converge to some function in the pointwise topology or the topology compact convergence for $\mathcal{C}(I)$?

2. (a) For each integer $n = 1, 2 \ldots$, define a function $f_n : \mathbb{R} \to \mathbb{R}$ by $f_n(x) = (n+1)x/n$. Prove that the sequence $\langle f_n \rangle$ converges uniformly on compacta, but fails to converge uniformly.

    (b) Do as in (a) with the functions $f_n(x) = x/n$.

    (c) What about the sequence $\langle f_n \rangle$, where $f_n(x) = n \sin(x/n)$?

3. Let X denote the subspace $(-1, 1)$ of $\mathbb{R}$, and for each integer $n = 1, 2 \ldots$, define $s_n : X \to \mathbb{R}$ by $s_n(x) = \sum_1^n ix^i$. Show that the sequence $\langle s_n \rangle$ converges to a continuous function uniformly on every compact subset of X, but convergence is not uniform.

4. Let X be a space and $\langle f_\nu \rangle$ be a monotonically decreasing (or increasing) net of continuous functions $f_\nu : X \to \mathbb{R}$. If $\langle f_\nu \rangle$ converges pointwise to a continuous function $f : X \to \mathbb{R}$, show that $\langle f_\nu \rangle$ converges to f uniformly on compacta.

5. • Let X be a topological space and Y be a metric space. Without using Theorem 11.3.8, prove that a net $\langle f_\nu \rangle$ in $C(X, Y)$ converges to an $f \in C(X, Y)$ relative to the compact-open topology if and only if $\langle f_\nu \rangle$ converges to f uniformly on every compact subset of X.

6. Let X be a first countable space, Y a metric space, and let $f : X \to Y$ be a continuous map. Show that a sequence $\langle f_n \rangle$ in $C(X, Y)$ converges continuously to f if and only if it converges uniformly to f on every compact subset of X.

7. Let X be a topological space and (Y, d) be a bounded metric space. Let $\mathscr{A}$ be a family of subsets of X, and suppose that $X = \bigcup \{A \mid A \in \mathscr{A}\}$. Prove:

   (a) A net $\langle f_\nu \rangle$ in Y^X converges in the topology $\mathscr{T}\left(d^* \mid \mathscr{A}\right) \Leftrightarrow \langle f_\nu \rangle$ converges pointwise and $\langle f_\nu \mid A \rangle$ is a Cauchy net in $\left(Y^A\right)_u$ for every $A \in \mathscr{A}$.

   (b) If Φ is the family of all functions $f : X \to Y$ such that $f \mid A : A \to Y$ is continuous for each $A \in \mathscr{A}$, then Φ is closed in the space Y^X with the topology $\mathscr{T}\left(d^* \mid \mathscr{A}\right)$.

8. Let Φ be a family of functions from a space X to another space Y, and let $\mathscr{K}$ be the family of all compact subsets of X. Prove:

   (a) Any topology on Φ which makes the evaluation map $e_K : \Phi \times K \to Y$, $e_K(f, x) = f(x)$, continuous for every $K \in \mathscr{K}$, is finer than the compact-open topology.

   (b) If X is T_2 or T_3, and each function in Φ is continuous on every $K \in \mathscr{K}$, then the evaluation map $e_K : \Phi_{co} \times K \to Y$ is continuous for every $K \in \mathscr{K}$.

9. Let X be a topological space and (Y, d) be a bounded metric space. Let $\mathscr{A}$ be a family of subsets of X, and suppose that $X = \bigcup \{A \mid A \in \mathscr{A}\}$. Prove:

   (a) If Φ is a family of functions $f : X \to Y$ such that $f \mid A : A \to Y$ is continuous for each $A \in \mathscr{A}$ and has the relative topology induced by $\mathscr{T}\left(d^* \mid \mathscr{A}\right)$, then the evaluation map $e_A : \Phi \times A \to Y$ is continuous for every $A \in \mathscr{A}$.

   (b) If the family $\mathscr{A}$ consists of all compact subsets of X, then the relative topology for Φ is finer than the compact-open topology.

10. Prove that the pointwise topology on an equicontinuous family of functions of a space X into a metric space Y is admissible.

11. Let X be a space and Y be a metric space. If a sequence $\langle f_n \rangle$ in $C(X, Y)$ converges uniformly, show that the family $\{f_n\}$ is equicontinuous.

12. Let X be a space and Y a metric space. Suppose that an equicontinuous sequence of functions $f_n \in C(X, Y)$ converges pointwise to $f : X \to Y$. Prove that f is continuous and $\langle f_n \rangle$ converges uniformly to f on every compact subset of X.

13. Let X be a space and (Y, d) be a compact metric space. If $\Phi \subseteq C(X, Y)_u$ is totally bounded, show that it is equicontinuous.

14. Let X be a compact space and (Y, d) be a compact metric space. If a set $\Phi \subseteq C(X, Y)$ is equicontinuous, show that Φ is totally bounded under d^*. Deduce that the closure of Φ in the compact-open topology on $C(X, Y)$ is compact.

15. • Let (X, d) be a compact metric space, and $\Phi \subseteq C(X)$. Prove:

(a) If Φ is an equicontinuous family, then, for each real $\epsilon > 0$, there exists a real $\delta > 0$ such that $d(x', x'') < \delta \Rightarrow |f(x') - f(x'')| < \epsilon$ for all $f \in \Phi$.

(b) If Φ is equicontinuous and pointwise bounded (i.e., $\{f(x) \mid f \in \Phi\}$ is bounded for every $x \in X$), then it is uniformly bounded (i.e., there exists a real M such that $|f(x)| \leq M$ for all $f \in \Phi$ and all $x \in X$).

(c) (Ascoli theorem) In case (b), the closure of Φ in $C(X)_u$ is compact.

Chapter 12
Topological Groups

"Topological Groups" are mathematical structures in which the notions of group and topology are blended. These partly geometric objects form a rich territory of interesting examples in topology and geometry, due to the presence of the two basic interrelated mathematical structures in one and the same set. The first section concerns with the basic properties and examples of topological groups. Here, we also deal with their separation properties. The notion of subgroups of a topological group is studied in the second section. In the third section, we treat quotient groups and isomorphisms theorems for topological groups. The notions of the decomposition of a topological group into direct product and semidirect product of its subgroups are discussed in the last section.

12.1 Basic Properties

The abstract notion of a topological group was introduced by O. Schreier (1925) and F. Leja (1927), although particular topological groups can be traced to Klein's programme (1872) of study of geometries through transformation groups associated with them, and to the work of Lie (1873) on "continuous groups (known nowadays as Lie Groups)."

Definition 12.1.1 A *topological group* is a topological space G with a group structure such that the group operations

$$\mu: G \times G \to G, \ (x, y) \mapsto xy, \quad \text{and}$$
$$\iota: G \to G, \ x \mapsto x^{-1},$$

(in the multiplicative notation) are continuous, where $G \times G$ is given the product topology.

© Springer Nature Singapore Pte Ltd. 2019
T. B. Singh, *Introduction to Topology*,
https://doi.org/10.1007/978-981-13-6954-4_12

If G is a topological group and A, $B \subseteq G$, then AB denotes the set $\{ab \mid a \in A,$ and $b \in B\}$, and A^{-1} denotes the set $\{a^{-1} \mid a \in A\}$. Obviously, AB is the direct image of $A \times B$ under the multiplication mapping μ, and A^{-1} is the image of A under the inversion mapping ι. It is easily seen that the conditions of continuity of the mappings μ and ι is equivalent to the continuity of the single mapping $G \times G \to G$, $(x, y) \mapsto xy^{-1}$. Accordingly, if x and y are two elements of G, then for each nbd W of xy^{-1}, there exist nbds U of x and V of y such that $UV^{-1} \subseteq W$.

Example 12.1.1 A group with the discrete topology is a topological group.

Example 12.1.2 The real line $\mathbb{R}$ with the addition of real numbers as the group operation is a topological group. More generally, the Euclidean space $\mathbb{R}^n$ is a topological group, the group multiplication being the usual addition. In particular, the complex plane $\mathbb{C}$ is a topological group, the group multiplication being the addition of complex numbers.

Example 12.1.3 The punctured real line $\mathbb{R}_0 = \mathbb{R} - \{0\}$ with the relative topology and the multiplication of reals as a group operation is a topological group. In fact, we have already seen (in Sect. 2.1) that the function $\mathbb{R} \times \mathbb{R} \to \mathbb{R}$, $(x, y) \mapsto xy$ is continuous. As the range of the restriction of this function to $\mathbb{R}_0 \times \mathbb{R}_0$ is $\mathbb{R}_0$, it is clear that the multiplication mapping $\mu : \mathbb{R}_0 \times \mathbb{R}_0 \to \mathbb{R}_0$ is continuous. To check the continuity of the inversion mapping $\iota : \mathbb{R}_0 \to \mathbb{R}_0$, $x \mapsto x^{-1}$, notice that the intersection of an open interval (a, b) with $\mathbb{R}_0$ is (a, b) or $(a, 0) \cup (0, b)$. Clearly, the inverse image of $(a, b) \subset \mathbb{R}_0$ under the mapping ι is (b^{-1}, a^{-1}) when $a \neq 0 \neq b$, and those of $(a, 0)$ and $(0, b)$ are $(-\infty, a^{-1})$ and (b^{-1}, ∞), respectively. Thus, the inverse image of every member of a basis of $\mathbb{R}_0$ under ι is open in $\mathbb{R}_0$, and hence ι is continuous.

Similarly, we see that $\mathbb{C}_0 = \mathbb{C} - \{0\}$ and $\mathbb{H}_0 = \mathbb{H} - \{0\}$ are topological groups with the multiplications and the relative topologies induced from the complex line $\mathbb{C}$ and the quaternionic line $\mathbb{H}$, respectively.

Example 12.1.4 The group $\mathbb{Z}$ of integers with the cofinite topology is not a topological group because addition is not continuous, although it is continuous in each variable separately.

General Linear Groups

Consider the set $GL\,(n, \mathbb{R})$ of all invertible real $n \times n$ matrices; it consists of all nonsingular matrices, that is,

$$GL\,(n, \mathbb{R}) = \{A \in \mathscr{M}\,(n, \mathbb{R}) \mid \det(A) \neq 0\},$$

where $\det(A)$ denotes the determinant of A. This is obviously a group under the matrix multiplication. We find a topology for $GL\,(n, \mathbb{R})$ which turns it into a topological group. To this end, consider the set $\mathscr{M}\,(n, \mathbb{R})$ of all real $n \times n$ matrices. Each matrix (a_{ij}) in $\mathscr{M}\,(n, \mathbb{R})$ determines by stringing out its rows a unique ordered n^2-tuple

$$(a_{11}, \ldots, a_{1n}, a_{21}, \ldots, a_{n1}, \ldots, a_{nn})$$

of real numbers, and there is a bijective mapping of $\mathscr{M}(n, \mathbb{R})$ onto $\mathbb{R}^{n^2}$. This mapping induces a topology on $\mathscr{M}(n, \mathbb{R})$ so that it is homeomorphic to the Euclidean space $\mathbb{R}^{n^2}$. With this topology for $\mathscr{M}(n, \mathbb{R})$, we have continuous maps $p_{ij} : \mathscr{M}(n, \mathbb{R}) \to \mathbb{R}$, $(a_{ij}) \mapsto a_{ij}$, for all $1 \leq i, j \leq n$. By Theorem 2.2.4, we see that a function of a space into $\mathscr{M}(n, \mathbb{R})$ is continuous if and only if its composition with each mapping p_{ij} is continuous. Now, if $A = (a_{ij})$ and $B = (b_{ij})$ are in $\mathscr{M}(n, \mathbb{R})$, then $p_{ij}(AB) = \sum_{k=1}^{n} a_{ik}b_{kj}$. Since the multiplication and the addition of real numbers are continuous operations, it is clear that the composition of the matrix multiplication

$$\mu : \mathscr{M}(n, \mathbb{R}) \times \mathscr{M}(n, \mathbb{R}) \longrightarrow \mathscr{M}(n, \mathbb{R})$$

with each p_{ij} is continuous. Hence, μ is continuous, and we deduce that the group multiplication in the subspace $GL(n, \mathbb{R}) \subset \mathscr{M}(n, \mathbb{R})$ is continuous. For the continuity of the inversion function $A \mapsto A^{-1}$ on $GL(n, \mathbb{R})$, we note that $p_{ij}(A^{-1})$ is $A_{ji}/\det(A)$, where A_{ji} is the (ji)th cofactor of A. Because $\det(A)$ is a sum of products (with appropriate signs) of entries in A, the function $\det(A)$ of A is continuous on $\mathscr{M}(n, \mathbb{R})$. A similar argument shows that the function $\mathscr{M}(n, \mathbb{R}) \to \mathbb{R}$, $A \mapsto A_{ij}$, is also continuous. Since the determinant function does not vanish on $GL(n, \mathbb{R})$, we deduce that the function

$$GL(n, \mathbb{R}) \to \mathbb{R}, \quad A \mapsto p_{ij}(A^{-1})$$

is continuous for every i and j. Therefore, the function $A \mapsto A^{-1}$ on $GL(n, \mathbb{R})$ is continuous, and thus $GL(n, \mathbb{R})$ is a topological group. It is called the *general linear group* over the real numbers.

In the same way, there are *general linear groups* $GL(n, \mathbb{C})$ and $GL(n, \mathbb{H})$ of invertible $n \times n$ matrices over the field $\mathbb{C}$ (of complex numbers) and the skew field $\mathbb{H}$ (of quaternions), respectively. The group $GL(n, \mathbb{C})$ is given the relative topology induced from the complex n^2-space $\mathbb{C}^{n^2}$, and the group $GL(n, \mathbb{H})$ is given the relative topology induced from the quaternionic n^2-space $\mathbb{H}^{n^2}$.

It is well known that $GL(n, \mathbb{R})$ is the complement of the inverse image of $\{0\}$ under the continuous map $A \mapsto \det(A)$, so this is an open subset of $\mathscr{M}(n, \mathbb{R}) \approx \mathbb{R}^{n^2}$. For a similar reason, $GL(n, \mathbb{C})$ is open in $\mathbb{C}^{n^2}$. The group $GL(n, \mathbb{H})$ is also open in $\mathbb{H}^{n^2}$, although the argument in this case is harder, since the determinant function is not defined for quaternionic matrices. Therefore, we find that all these general linear groups are locally compact and Hausdorff. However, $GL(n, \mathbb{R})$ is not compact or connected, for its image under the above continuous function is $\mathbb{R}_0$, which is not compact or connected.

We now return to the general discussion. Let G be a topological group and $g \in G$ be a fixed element. Then, the function $\rho_g : G \to G$, $x \mapsto xg$ is continuous, since it is the composition of the continuous function $G \to G \times G$, $x \mapsto (x, g)$ with the multiplication function $\mu: G \times G \to G$. This is referred to as the *right translation* by the element g. Clearly, $\rho_h \circ \rho_g = \rho_{gh}$ for every $g, h \in G$, and therefore ρ_g is a homeomorphism with $\rho_{g^{-1}}$ as its inverse. Similarly, the *left translation* $\lambda_g: x \mapsto gx$ is a homeomorphism. It is immediate that the conjugation (inner automorphism) determined by g is also a homeomorphism. Notice that if $x, y \in G$, then the homeomorphism $\rho_{x^{-1}y}$ carries x into y. So G is a *homogeneous space* by which it is meant that for every pair of points $x, y \in G$, there exists a homeomorphism $h: G \to G$ such that $h(x) = y$. This fact can be used to show that a particular topological space, e.g., the unit interval I, cannot be made into a topological group under any multiplication.

If A is closed (or open) in G, then all its translates $Ag = \{ag \mid a \in A\}$ and $gA = \{ga \mid a \in A\}$, $g \in G$, are clearly closed (resp. open), and so also is A^{-1}. For every $A, B \subseteq G$, we have $AB = \bigcup_{b \in B} Ab = \bigcup_{a \in A} aB$. Hence, the products AB and BA are open in G whenever A or B is open in G. On the other hand, the product of two closed subsets need not be closed, even if they are subgroups. For example, let $A = \mathbb{Z}$, the subgroup of integers in the real line $\mathbb{R}$, and $B = \{n\sqrt{2} \mid n \in \mathbb{Z}\}$. The product AB is not a closed set (cf. Example 12.3.4). However, we will soon see that if one of the two closed sets is compact, then the product is necessarily closed.

If U is an open nbd of the identity element e of a topological group G, then Ux is obviously an open nbd of $x \in G$. Conversely, if O is an open nbd of $x \in G$, then Ox^{-1} is an open nbd of e in G. Therefore, a subset $H \subseteq G$ is open if and only if for each $x \in H$ there is a nbd U of e such that $Ux \subseteq H$. In other words, H is open if and only if Hx^{-1} is a nbd of e for every $x \in H$. It is now evident that if $\mathcal{V}_e$ is a nbd basis at the identity element $e \in G$, then $\{Vx \mid V \in \mathcal{V}_e\}$ is a nbd basis at the point x of G (so also is $\{xV \mid V \in \mathcal{V}_e\}$). Therefore, the topology of G is completely determined by the nbd basis $\mathcal{V}_e$.

Proposition 12.1.2 *Let G be a topological group with the identity element e. Then, a nbd basis $\mathcal{V}_e$ at e has the following properties:*

(a) *Each $V \in \mathcal{V}_e$ is nonempty.*
(b) *If $V_1, V_2 \in \mathcal{V}_e$, then there is a $V_3 \in \mathcal{V}_e$ such that $V_3 \subseteq V_1 \cap V_2$.*
(c) *If $V \in \mathcal{V}_e$, then there is a $W \in \mathcal{V}_e$ such that $WW^{-1} \subseteq V$.*
(d) *For every $V \in \mathcal{V}_e$ and $x \in G$, there is a $W \in \mathcal{V}_e$ such that $xWx^{-1} \subseteq V$.*

Proof (a) and (b) are readily seen by the definition of nbd basis.

To establish (c), consider the function $\phi: G \times G \to G$ defined by $\phi(x, y) = xy^{-1}$. Since ϕ is continuous, given $V \in \mathcal{V}_e$, there exist nbds M and N of e such that $\phi(m, n) = mn^{-1} \in V$ for all $m \in M$ and $n \in N$. Now, we have a set $W \in \mathcal{V}_e$ such that $W \subseteq M \cap N$. Then, for all $x, y \in W$, we have $xy^{-1} \in V$.

(d) is easily seen from the continuity of the function $y \to xyx^{-1}$ on G.   ◇

Interestingly, a nonempty family of subsets of a group G having the above properties determines a topology for G so that it becomes a topological group.

Theorem 12.1.3 *Let G be an abstract group with the identity element e, and $\mathscr{V}_e$ be a nonempty family of subsets of G having the properties (a)–(d) described in Proposition 12.1.2. Then, there is a unique topology on G which turns it into a topological group with $\mathscr{V}_e$ as a local base at e.*

Proof For each $x \in G$, put $\mathscr{B}_x = \{xV \mid V \in \mathscr{V}_e\}$. Then, $\mathscr{B}_x$ is nonempty, for $\mathscr{V}_e$ is so. Since each member of $\mathscr{V}_e$ is nonempty, $e \in V$ for every $V \in \mathscr{V}_e$, by (c). So $x \in \mathscr{B}_x$. Also, by (b), for every two members B_1 and B_2 of $\mathscr{B}_x$, we find a $B_3 \in \mathscr{B}_x$ such that $B_3 \subseteq B_1 \cap B_2$. Next, we observe that, for each $V \in \mathscr{V}_e$, there is a $U \in \mathscr{V}_e$ such that $U^2 \subseteq V$. By (c), there exists a $W \in \mathscr{V}_e$ such that $WW^{-1} \subseteq V$. Then, again by (c), we find a $U \in \mathscr{V}_e$ such that $UU^{-1} \subseteq W$. Since $e \in U$, we obviously have both $U \subseteq W$ and $U^{-1} \subseteq W$. So $U^2 \subseteq WW^{-1} \subseteq V$, and it follows that $yU \subseteq xU^2 \subseteq xV$ for every $y \in xU$. Thus, the collection $\{\mathscr{B}_x \mid x \in X\}$ satisfies the hypotheses of Theorem 1.4.11, and there is a unique topology $\mathscr{T}$ for G in which $\mathscr{B}_x$ is a local basis at x.

Moreover, $(G, \mathscr{T})$ is a topological group. To see this, consider the mapping $\phi \colon G \times G \to G$ defined by $\phi(x, y) = xy^{-1}$. Clearly, given a nbd N of xy^{-1} in G, there exists a $V \in \mathscr{V}_e$ such that $xy^{-1}V \subseteq N$. By (d), we find a $W \in \mathscr{V}_e$ such that $yWy^{-1} \subseteq V$, and then, by (c), there is a $U \in \mathscr{V}_e$ such that $UU^{-1} \subseteq W$. So $(xU)(yU)^{-1} \subseteq xWy^{-1} \subseteq xy^{-1}V$, and we conclude that ϕ is continuous at (x, y).   ◇

A subset A of a topological group G is called *symmetric* if $A = A^{-1}$. It is clear that $A \cap A^{-1}$ is symmetric for every $A \subseteq G$, and so is AA^{-1}. Also, the intersection of symmetric sets is symmetric. If U is a nbd of the identity e of G, then so is U^{-1}. Obviously, U contains the nbd $V = U \cap U^{-1}$ of e, which is symmetric. It follows that *the symmetric nbds of e form a local base at e, and this local base at e completely describes the topology of G.*

As indicated earlier, we can now prove that the product of a compact subset and a closed subset of a topological group is closed.

Proposition 12.1.4 *Let G be a topological group. If $A \subseteq G$ is closed, and $B \subseteq G$ is compact, then AB is closed in G.*

Proof Suppose that A is closed in G, and B is a compact subset of G. Let $x \in G - AB$ be arbitrary. Then, the identity element e of G does not belong to $x^{-1}AB$. Since A is closed in G, so is $x^{-1}Ab$ for every $b \in B$. Therefore, by Proposition 12.1.2, there exists a nbd W_b of e such that $W_b W_b^{-1} \subseteq G - x^{-1}Ab$. Clearly, B is covered by the open sets bW_b, $b \in B$. Since B is compact, we find finitely many points $b_1, \ldots, b_n$ in B such that $B \subseteq \bigcup_1^n b_i W_{b_i}$. Set $W = \bigcap_1^n W_{b_i}$. Then, xW is obviously a nbd of x. We claim that $xW \subseteq (G - AB)$. Assume otherwise, and let $xw = ab$ for some $w \in W$, $a \in A$ and $b \in B$. Obviously, $b \in b_i W_i$ for some i. Then, there exists a $w_i \in W_{b_i}$ such that $xw = ab_i w_i$, and so $ww_i^{-1} = x^{-1}ab$. This contradicts the definition of W_{b_i}, and hence our claim. It follows that $G - AB$ is a nbd of each of its points, and therefore open.   ◇

Separation Properties

It is generally required that the identity element e of a topological group G be closed. From the homogeneity of G, it follows that this is equivalent to the requirement that $\{x\}$ be closed for every $x \in G$. Thus, a topological group with this property is a T_1-space. We will see here that a topological group satisfying the separation condition T_0 is actually completely regular.

Lemma 12.1.5 *Let G be a topological group with the identity e. If U is a nbd of e in G, then there is a symmetric nbd V of e such that $VV \subseteq U$.*

Proof By the continuity of multiplication $\mu: G \times G \to G$, there exist nbds W_1 and W_2 of e such that $W_1 \times W_2 \subseteq \mu^{-1}(U)$. So U contains the nbd $W = W_1 \cap W_2$ of e, and also the product WW. It is now clear that the set $V = W \cap W^{-1}$ satisfies the requirements of the lemma.   ◇

Proposition 12.1.6 (a) *Every topological group G satisfies T_3-Axiom.*
   (b) *If G is T_0, then it is T_1, and hence regular.*

Proof (a): Let G be a topological group with the identity element e. By Proposition 8.1.2, it suffices to show that, for each nbd U of an element x in G, there exists a nbd V of x such that $\overline{V} \subseteq U$. By the homogeneity of G, it suffices to consider the nbds of e. Let U be a nbd of e. Since $ee^{-1} = e$, there exits a nbd V of e such that $VV^{-1} \subseteq U$. We observe that $\overline{V} \subseteq U$. Let $x \in \overline{V}$ be arbitrary. Then, every nbd of x intersects V. So $xV \cap V \neq \varnothing$, and we can find points $v, v' \in V$ such that $xv = v'$. This implies that $x = v'v^{-1} \in VV^{-1} \subseteq U$, as claimed.

   (b): Clearly, it suffices prove that $\{e\}$ is closed in G. Let $x \neq e$ be any point of G. Since G satisfies the T_0-axiom, $G - \{e\}$ is a nbd of x or $G - \{x\}$ is a nbd of e. In the latter case, $G - \{e\}$ is a nbd of x^{-1}, since the left translation by x^{-1} maps $G - \{x\}$ onto $G - \{e\}$. So the image of $G - \{e\}$ under the inversion homeomorphism ι is a nbd of x. Obviously, ι maps $G - \{e\}$ into itself, so $G - \{e\}$ is a nbd of x. Therefore $G - \{e\}$ is open, as desired.   ◇

   By the preceding proposition, it follows that a topological group satisfying T_0-axiom is Hausdorff, for a regular space is so. This can also be deduced directly from T_1-ness of G. In fact, if $x, y \in G$, and $x \neq y$, then $xy^{-1} \neq e$. By Lemma 12.1.5, there exists a symmetric nbd V of e such that $VV \subseteq G - \{xy^{-1}\}$. Then, Vx and Vy are nbds of x and y, respectively. Moreover, if $vx = v'y$, then $xy^{-1} = v^{-1}v' \in V^{-1}V$, a contradiction. So $Vx \cap Vy = \varnothing$, and G is Hausdorff.

   The proof of the following theorem reminds us of the construction of the map in the Urysohn Lemma.

Theorem 12.1.7 *A topological group satisfying the T_0-axiom is completely regular.*

Proof Let G be a topological group with the identity element e. Since the left trans-lation $\lambda_g : x \mapsto gx$ is a homeomorphism of G onto itself for every $g \in G$, it suffices to prove the existence of a continuous function $f : G \to I$ such that $f(e) = 0$ and $f(F) = 1$ for each closed set $F \subset G$ not containing e. We first associate an open nbd U_r of e with each dyadic rational $r = k/2^n$ between 0 and 1 such that

$$U_{k/2^n} U_{1/2^n} \subseteq U_{(k+1)/2^n} \text{ for all } k < 2^n \text{ and}$$
$$U_r \subseteq U_{r'} \text{ for } r < r' \leq 1.$$

We set $V_0 = G - F$, and find a symmetric open nbd V_1 of e such that $V_1^2 \subseteq V_0$. Then, find a symmetric open nbd V_2 of e such that $V_2^2 \subseteq V_1$, and so on. Thus, we obtain a sequence of symmetric open nbds V_n of e such that $V_{n+1}^2 \subseteq V_n$. Obviously, for each positive integer k, there exists a unique integer $m \geq 0$ such that $2^m \leq k < 2^{m+1}$. So we can write $k = 2^m a_1 + 2^{m-1} a_2 + \cdots + a_{m+1}$, where the a_i are uniquely determined integers 0 or 1. Then, for $n = 0, 1, 2, \ldots$, and $1 \leq k \leq 2^n$, we define $U_1 = V_0$ and $U_{k/2^n} = V_{n-m}^{a_1} \cdots V_n^{a_{m+1}}$, where $V_i^1 = V_i$ and $V_i^0 = \{e\}$. Observe that $U_{1/2^n} = V_n$, and $U_{2k/2^n} = U_{k/2^{n-1}}$, so each U_r depends on the dyadic rational r and not on the particular representation as $k/2^n$. We show that the family $\{U_r\}$ has the desired proprties. To check the inclusion $U_{k/2^n} U_{1/2^n} \subseteq U_{(k+1)/2^n}$, we apply induction on n. If $n = 1$, then $k = 1$ and $U_{1/2} U_{1/2} = V_1 V_1 \subseteq V_0 = U_1$. Now, assume that $n > 1$ and $U_{k/2^{n-1}} U_{1/2^{n-1}} \subseteq U_{(k+1)/2^{n-1}}$. Then, we have

$$
\begin{aligned}
U_{2k/2^n} U_{1/2^n} &= U_{k/2^{n-1}} U_{1/2^n} \\
&= V_{n-m-1}^{a_1} \cdots V_{n-1}^{a_{m+1}} V_n^1 \\
&= U_{(2k+1)/2^n} \qquad \text{and}
\end{aligned}
$$

$$
\begin{aligned}
U_{(2k+1)/2^n} U_{1/2^n} &= U_{k/2^{n-1}} U_{1/2^n} U_{1/2^n} \\
&= U_{k/2^{n-1}} V_n^2 \subseteq U_{k/2^{n-1}} U_{1/2^{n-1}} \\
&\subseteq U_{(k+1)/2^{n-1}} \quad \text{(by our assumption)} \\
&= U_{(2k+2)/2^n}.
\end{aligned}
$$

This completes the inductive argument, and we have the inclusion $U_{k/2^n} U_{1/2^n} \subseteq U_{(k+1)/2^n}$ for all n and all $k < 2^n$.

To see that $U_r \subseteq U_{r'}$ for $r < r'$, suppose that $r = k/2^n$, $r' = k'/2^{n'}$ and write $k = 2^m a_1 + 2^{m-1} a_2 + \cdots + a_{m+1}$ and $k' = 2^{m'} a_1' + 2^{m'-1} a_2' + \cdots + a_{m'+1}'$, where the a_i and the a_j' are 0 or 1, and $a_1 = 1 = a_1'$. If $n - m > n' - m'$, then we have

$$
\begin{aligned}
U_r = V_{n-m}^{a_1} \cdots V_n^{a_{m+1}} &\subseteq V_{n-m} \cdots V_n \\
&\subseteq V_{n-m} \cdots V_n V_n \subseteq V_{n-m} \cdots V_{n-1} V_{n-1} \\
&\subseteq \cdots \subseteq V_{n-m-1} \subseteq V_{n'-m'} \\
&\subseteq V_{n'-m'}^{a_1'} \cdots V_{n'}^{a_{m'+1}'} = U_{r'}.
\end{aligned}
$$

In the case $n - m = n' - m'$, we find the least integer l such that $a_l \neq a'_l$. Then, $a_l = 0$ and $a'_l = 1$. Put $W = V_{n-m}^{a_1} \cdots V_{n-m+l-2}^{a_{l-1}}$. As $V_{n-m+l-1}^{a_l} = \{e\}$, we have

$$U_r = W V_{n-m+l}^{a_{l+1}} \cdots V_n^{a_{m+1}} \subseteq W V_{n-m+l} \cdots V_n$$

$$\subseteq W V_{n-m+l} \cdots V_n V_n \subseteq W V_{n-m+l} \cdots V_{n-1} V_{n-1}$$

$$\subseteq \cdots \subseteq W V_{n-m+l-1} = V_{n'-m'}^{a'_1} \cdots V_{n'-m'+l-1}^{a'_l}$$

$$\subseteq V_{n'-m'}^{a'_1} \cdots V_{n'}^{a'_{m'+1}} = U_{r'},$$

as desired.

Now, we define $f : G \to I$ by

$$f(x) = \begin{cases} 0 & \text{if } x \in U_r \text{ for every } r, \\ 1 & \text{if } x \notin U_1 \quad \text{and} \\ \sup\{r \,|\, x \notin U_r\} & \text{otherwise.} \end{cases}$$

It is obvious that $f(F) = 1$, and $f(e) = 0$, since $e \in U_r$ for all r. To see the continuity of f, let $x \in G$ and $\epsilon > 0$ be given. Find a positive integer n such that $2^{-n} < \epsilon$. If $f(x) = 0$, then each $U_{1/2^m}$ is a nbd of x, and $|f(x) - f(y)| \leq 1/2^m$ for every $y \in U_{1/2^m}$. Consequently, we have $|f(x) - f(y)| < \epsilon$ for all $y \in U_{1/2^m}$ and $m > n$. If $0 < f(x) < 1$, then there exists a positive integer m such that $m > n$ and $2^{-m} < \min\{f(x), 1 - f(x)\}$. Also, there is an integer k such that $1 < k < 2^m$ and $x \in U_{k/2^m} - U_{(k-1)/2^m}$, for $x \in U_{(2^m-1)/2^m} - U_{1/2^m}$. Then, $(k-1)/2^m \leq f(x) \leq k/2^m$. Note that $xU_{1/2^m}$ is an open nbd of x. We verify that $|f(x) - f(y)| < \epsilon$ for all $y \in xU_{1/2^m}$. If $y \in xU_{1/2^m}$, then $y \in U_{k/2^m} U_{1/2^m} \subseteq U_{(k+1)/2^m}$. Since $x^{-1}y \in U_{1/2^m}$ which is symmetric, we have $y^{-1}x \in U_{1/2^m}$, and so $x \in yU_{1/2^m}$. If $y \in U_{(k-2)/2^m}$, then we would have x in $U_{(k-1)/2^m}$, a contradiction. Accordingly, $y \notin U_{(k-2)/2^m}$ and $(k-2)/2^m \leq f(y) \leq (k+1)/2^m$. It follows that $|f(x) - f(y)| \leq 2/2^m \leq 1/2^n < \epsilon$. Finally, suppose that $f(x) = 1$. Then, choose an integer $m > n$, and consider the nbd $xU_{1/2^m}$ of x. Let $y \in xU_{1/2^m}$ be arbitrary. Then, $x \in yU_{1/2^m}$, since $U_{1/2^m}$ is symmetric. If $y \in U_{k/2^m}$ and $k/2^m \leq 1 - 2/2^m$, then we have $x \in U_{(k+1)/2^m}$ which implies that $f(x) \leq (k+1)/2^m < 1$, a contradiction. Therefore, $y \notin U_{k/2^m}$ for $k/2^m \leq 1 - 2/2^m$, and we have $1 \geq f(y) \geq 1 - 2/2^m$. This implies that $|f(x) - f(y)| \leq 2/2^m \leq 1/2^n < \epsilon$, completing the proof. $\diamond$

Exercises

1. Verify that the set $\mathcal{M}(n, \mathbb{R})$ (resp. $\mathcal{M}(n, \mathbb{C})$) of all $n \times n$ matrices over $\mathbb{R}$ (resp. $\mathbb{C}$) under addition with the usual topology is a topological group.

2. • Prove that the set $C(I)$ of all real-valued continuous functions on I with the topology induced by the supremum metric

$$d^*(f, g) = \sup\{|f(t) - g(t)| : 0 \leq t \leq 1\}$$

is a topological group under the pointwise addition.

3. Let X be a compact Hausdorff space. Show that $\mathcal{C}(X)_{co}$ is a topological group under pointwise addition (that is, $(f + g)(x) = f(x) + g(x)$ and $(-f)(x) = -f(x)$).

4. • Prove that the group $\mathbb{R}^{\mathbb{R}}$ with pointwise addition forms a topological group in the topology of uniform convergence defined by a bounded metric uniformly equivalent to the usual metric in $\mathbb{R}$.

5. • Let (X, d) be a compact metric space. Show that the group Homeo(X) topologized by the metric $d^*(f, g) = \sup_{x \in X} d(f(x), g(x))$ is a topological group.

6. Let G' be a homeomorphic copy of a topological group G. Show that G' can be given a topological group structure.

7. Prove that the subspace $(\mathbb{R} \times \{0\}) \cup (\{0\} \times \mathbb{R})$ of $\mathbb{R}^2$ cannot be given a topological group structure.

8. Let p be a prime and $\mathcal{V}_0$ be a family of all subsets $V \subseteq \mathbb{Z}$ such that for some integer $n > 0$, $kp^n \in V$ for all $k \in \mathbb{Z}$. Show that $\mathcal{V}_0$ is a nbd basis at 0 relative to a topology which makes the group $(\mathbb{Z}, +)$ a topological group. (This is referred to as the p-adic topology for $\mathbb{Z}$.) Prove that it is totally disconnected.

9. Let p be a fixed prime. For each $k \in \mathbb{Z}$, define sets
$$U_k = \{mp^k/n \mid m, n \in \mathbb{Z} \text{ and } (p, n) = 1\}.$$
Show that there is a topology on the group $\mathbb{Q}$, which makes it a topological group with the family $\{U_k\}$ as a local basis at 0. (This topology for $\mathbb{Q}$ is referred to as the p-adic topology for $\mathbb{Q}$.) Is it totally disconnected?

10. Let G be a topological group with the identity e and U be a nbd of e. Show that for each integer $n > 0$, there is a symmetric nbd V of e such that $V^n \subseteq U$, where $V^n = V \cdots V$ (n factors).

11. Let $x_1, \ldots, x_n$ be n elements of a topological group G, and let $y = x_1^{k_1} \cdots x_n^{k_n}$, where $k_1, \ldots, k_n$ are integers. If W is a nbd of y, show that there exist nbds U_i of x_i, $1 \le i \le n$, such that $U_1^{k_1} \ldots U_n^{k_n} \subseteq W$.

12. Let G be a topological group and U be an nbd of g in G. Show that there exists a symmetric nbd V of the identity element e of G such that $VgV^{-1} \subseteq U$.

13. In the topological group $\mathbb{R}^2$, what is the product AB, where $A = \{(0, y) \mid y \in \mathbb{R}\}$ and $B = \{(x, 1/x) \mid x > 0 \text{ real}\}$? Is AB closed in $\mathbb{R}^2$?

14. Let G be a topological group, and A, B be compact subsets of G. Show that AB is compact.

15. Prove that a topological group G is Hausdorff $\Leftrightarrow$ the intersection of all nbds of e is $\{e\}$.

16. Prove that a Lindelöf (or separable) first countable topological group is second countable.

17. Let G be a group, and suppose that there is a locally compact Hausdorff topology on G such that the group multiplication $\mu: G \times G \to G$ is continuous. Prove:

    (a) For each compact $K \subseteq G$ and open $U \subseteq G$, $\{g \mid gK \subseteq U\}$ is open in G.

    (b) If G is also locally connected, then it is a topological group.

18. • Let $F = \mathbb{R}$, $\mathbb{C}$ or $\mathbb{H}$. A map $f: F^n \to F^n$ is called an *affine map* if there exists a linear map $\lambda: F^n \to F^n$ and an element $v \in F^n$ such that $f(x) = \lambda(x) + v$ for all $x \in F^n$.

    Prove: (a) The set $Aff_n(F)$ of all invertible affine maps of F^n form a group under the composition of mappings.

    (b) There is a canonical bijection $Aff_n(F) \longleftrightarrow GL(n, F) \times F^n$, so the group $Aff_n(F)$ receives a topology from the product space $GL(n, F) \times F^n$.

    (c) The group $Aff_n(F)$ with the above topology is a topological group.

    (d) Is $Aff_1(\mathbb{R})$ abelian?

12.2   Subgroups

Let G be a topological group and $H \subset G$ be a subgroup. If H is given the relative topology, then it becomes a topological group. This follows from the fact that if $f: X \to Y$ is continuous, and $A \subseteq X$ and $f(A) \subseteq Y$ are given the relative topologies, then the map $g: A \to f(A)$ defined by f is also continuous. A subgroup of a topological group is always assumed to be given the relative topology when it itself is being considered as a topological group.

Example 12.2.1 Let F be one of the fields $\mathbb{R}$, $\mathbb{C}$, and $\mathbb{H}$. The unit sphere

$$S(F) = \{x \in F : \|x\| = 1\}$$

in F forms a subgroup of the topological group $F_0 = F - \{0\}$ under the multiplication in F. Note that $S(F) \approx \mathbb{S}^0$, $\mathbb{S}^1$ or $\mathbb{S}^3$, according as $F = \mathbb{R}$, $\mathbb{C}$ or $\mathbb{H}$. It is an interesting fact that these are the only spheres which admit a topological group structure.

Example 12.2.2 Recall that a real matrix A is orthogonal if $A^t A = I$, the identity matrix. The set $O(n)$ of all orthogonal matrices forms a subgroup of $GL(n, \mathbb{R})$, and is called the *orthogonal group*. We observe that it is a closed and bounded subset of $\mathscr{M}(n, \mathbb{R}) \approx \mathbb{R}^{n^2}$. If $A \in O(n)$ and $A = (a_{ij})$, then we have $\sum_k a_{ik}a_{jk} = \delta_{ij}$ for $1 \le i, j \le n$. For every pair of indices i and j, the functions $\phi_{ij}: \mathscr{M}(n, \mathbb{R}) \to \mathbb{R}$, $A \mapsto \sum_k a_{ik}a_{jk}$, are clearly continuous, and $O(n)$ is the intersection of the sets $\phi_{ii}^{-1}(1)$, and $\phi_{ij}^{-1}(0)$, $i \ne j$. Therefore $O(n)$, being a finite intersection of closed subsets of $\mathscr{M}(n, \mathbb{R})$, is closed. The boundedness of $O(n)$ follows from the fact that each row of $A \in O(n)$ has unit length. Hence, $O(n)$ is a compact subgroup of $GL(n, \mathbb{R})$.

Example 12.2.3 A complex matrix A is unitary if $\overline{A}^t A = I$ the identity matrix. The set $U(n)$ of all unitary matrices forms a subgroup of $GL(n, \mathbb{C})$, and is called the *unitary group*.

Similarly, there is a subgroup of $GL(n, \mathbb{H})$ consisting of quaternionic matrices A such that $\overline{A}^t A = I$, where $\overline{A}^t = (b_{ij})$, b_{ij} being the quaternionic conjugate of jith entry of A. It is called the *symplectic group*, and denoted by $Sp(n)$

As in Example 12.2.2, we see that both $U(n)$ and $Sp(n)$ are compact groups.

Observe that $O(1) \approx \mathbb{S}^0$, $U(1) \approx \mathbb{S}^1$, and $Sp(1) \approx \mathbb{S}^3$ as topological spaces. To see the geometric interpretations of the topological groups $O(n)$, $U(n)$, and $Sp(n)$ in general, recall that a subset $S \subseteq F^n$, where $F = \mathbb{R}, \mathbb{C}$ or $\mathbb{H}$, is *orthonormal* if $\langle x, y \rangle = 0$ for every pair of distinct points x and y in S, and $\|x\| = 1$ for all $x \in S$. The standard basis of F^n, for example, is an orthonormal set. Let $f : F^n \to F^n$ be a linear mapping, and let $\{v_1, \ldots, v_n\}$ be an orthonormal basis of F^n. If $y \in F^n$, then $g_y : F^n \to F$, $x \mapsto \langle f(x), y \rangle$, is a linear map, and the element $f^*(y) = \sum_1^n v_i g_y(v_i)$ is uniquely determined by the condition $g_y(x) = \langle x, f^*(y) \rangle$ for every $x \in F^n$. Consequently, there is a mapping $f^* : F^n \to F^n$ given by $\langle f(x), y \rangle = \langle x, f^*(y) \rangle$ for all $x, y \in F^n$. It is easily checked that $f^* : F^n \to F^n$ is also linear, called the *adjoint* of f. For each $j = 1, \ldots, n$, there exist scalars $a_{ij} \in F$, $i = 1, \ldots, n$, such that $f(v_j) = \sum_1^n v_i a_{ij}$. Thus, we obtain the matrix $A = (a_{ij})$ (over F) called the matrix of f relative to the ordered basis $\{v_i\}$. One readily verifies that $\overline{A}^t$ is the matrix of f^* relative to the basis $\{v_i\}$.

If f is self-adjoint (i.e., $f^* = f$) and $\langle f(x), x \rangle = 0$ for all $x \in F^n$, then

$$0 = \langle f(x + ya), x + ya \rangle = \langle f(x), ya \rangle + \langle f(ya), x \rangle$$
$$= \langle f(x), y \rangle a + \overline{\langle f(x), y \rangle} a$$

for all $x, y \in F^n$ and $a \in F$. By taking suitable values of a, we find that $\langle f(x), y \rangle = 0$ for all $x, y \in F^n$, and therefore $f = 0$. Hence, we deduce that a linear map $f : F^n \to F^n$ is an isometry if and only if

$$\|f(x)\| = \|x\| \iff \langle f(x), f(y) \rangle = \langle x, y \rangle \iff$$
$$f^* \circ f = 1 \iff \overline{A}^t A = I.$$

It follows that the elements of $O(n)$, $U(n)$, and $Sp(n)$ correspond precisely to the linear isometries of $\mathbb{R}^n$, $\mathbb{C}^n$, and $\mathbb{H}^n$, respectively. Accordingly, $O(n)$ may be considered as the group of the linear isometries of $\mathbb{R}^n$, $U(n)$ as the group of the linear isometries of $\mathbb{C}^n$, and $Sp(n)$ as the group of the linear isometries of $\mathbb{H}^n$. The other way around, the linear isometries of $\mathbb{R}^n$, $\mathbb{C}^n$, and $\mathbb{H}^n$ are referred to as the *orthogonal, unitary, or symplectic transformations*, respectively.

For $F = \mathbb{R}$ and $\mathbb{C}$, there is also a subgroup of $GL(n, F)$ which consists of all matrices with determinant 1; this is called the *special linear group over F* and denoted by $SL(n, F)$. It is closed in $GL(n, F)$, since $1 \in F$ is closed and the determinant function is continuous. Consequently, $SO(n) = O(n) \cap SL(n, \mathbb{R})$, called the *special orthogonal group*, is a closed subgroup of $O(n)$, and $SU(n) = U(n) \cap SL(n, \mathbb{C})$, called the *special unitary group*, is a closed subgroup of $U(n)$. It follows that $SO(n)$ and $SU(n)$ are compact groups for every n. Their connectedness property shall be discussed in the next chapter.

Turning to the general discussion, we prove the following.

Proposition 12.2.1 *Let G be a topological group and $H \subseteq G$ be a subgroup. Then, the closure $\overline{H}$ of H in G is a subgroup of G. Moreover, if H is normal in G, then so is $\overline{H}$.*

Proof Let $x, y \in \overline{H}$ and let U be a nbd of xy. By the continuity of the multiplication function $\mu \colon G \times G \to G$, there exist nbds V of x and W of y such that $VW \subseteq U$. Since $V \cap H \neq \varnothing \neq W \cap H$, U intersects H. We see from this that $xy \in \overline{H}$. Similarly, $x^{-1} \in \overline{H}$, and $\overline{H}$ is a subgroup of G.

To prove the second part, suppose that H is normal in G. Given an element $g \in G$, let $\gamma \colon G \to G$ be the conjugation map determined by g, that is, $\gamma(x) = gxg^{-1}$. By the continuity of γ, we have $g\overline{H}g^{-1} = \gamma(\overline{H}) \subseteq \overline{\gamma(H)} \subseteq \overline{H}$. Therefore, $\overline{H}$ is normal in G.   ◇

Proposition 12.2.2 *Let G be a Hausdorff topological group and H be a subgroup of G. If H is commutative, then so is the subgroup $\overline{H}$.*

Proof Let μ denote the multiplication map in G, and define a function $\mu' \colon \overline{H} \times \overline{H} \to G$ by $\mu'(x, y) = yx = \mu(y, x)$. Then, μ' is continuous and, by our hypothesis, agrees with μ on $H \times H$. Since $\mu|(\overline{H} \times \overline{H})$ is continuous and $H \times H$ is dense in $\overline{H} \times \overline{H}$, the conclusion follows from Corollary 4.4.3.   ◇

Interestingly, all proper closed subgroups of the group $\mathbb{R}$ of real numbers and of the circle group $\mathbb{S}^1$ are cyclic.

Example 12.2.4 A nontrivial proper closed subgroup of $\mathbb{R}$ is infinite cyclic. Let $H \neq \{0\}$ be a proper closed subgroup of $\mathbb{R}$. Put $c = \inf \{h \in H \mid h > 0\}$. Obviously, c is a limit point of H. We claim that $c \neq 0$. If $c = 0$, then every element of H is its limit point, since the translation by an element of H is a homeomorphism of $\mathbb{R}$ onto itself and H is invariant under such mappings. We further observe that every real number is a limit point of H. Suppose that $x > 0$ and $x \notin H$. Then, for each real number $r > 0$, we find an $h \in H$ such that $0 < h < r$. Let $n > 0$ be the least integer such that $x < nh$. Then, $x - r < nh < x + r$, and therefore x is a limit point of H. Since H is invariant under the homeomorphism $x \mapsto -x$ of $\mathbb{R}$ onto itself, we see that every real number $x < 0$ is also a limit point of H. Thus, we have $H = \overline{H} = \mathbb{R}$, contrary to the hypothesis, and hence our claim. Clearly, $nc \in H$ for all $n \in \mathbb{Z}$. If there is an $h \in H$ which is not a multiple of c, then we find an integer n such that $nc < h < (n+1)c$. Since H is a subgroup, we have $h - nc \in H$. This contradicts the definition of c, for $0 < h - nc < c$. Therefore, $H = \{nc \mid n \in \mathbb{Z}\}$, as desired.

Example 12.2.5 A proper closed subgroup of $\mathbb{S}^1$ is finite and cyclic. Let H be a proper closed subgroup of $\mathbb{S}^1$. We first show that H is a discrete subset of $\mathbb{S}^1$. If H is not discrete, then $1 \in \mathbb{S}^1$ would be a limit point of H. Consequently, for each integer $n > 0$, the open set $U_n = \{e^{i\theta} \mid 0 < \theta < 1/n\}$ contains a point $z \in H$. Now, given an open arc C of $\mathbb{S}^1$ of length λ, choose an integer $n > 0$ so that $1/n < \lambda$. Then, find an element $z \in H$ such that $0 < \arg(z) < 1/n$. Let $x \in \mathbb{S}^1$ be the mid-point of C and m be the largest integer such that $\arg(z^m) \leq \arg(x)$. Then, $\arg(z^{m+1}) > \arg(x)$. If neither of z^m and z^{m+1} belongs to C, then we have $\arg(z^{m+1}) - \arg(z^m) \geq \lambda$. This contradicts the choice of z, and it follows that C intersects H. Since the open arcs of $\mathbb{S}^1$ form a base for its topology, we deduce that H is dense in $\mathbb{S}^1$. However, this is not true, since H is a closed and proper subset of $\mathbb{S}^1$. Thus, we see that H is discrete. Then, being a closed discrete subset of a compact space, H must be finite. Furthermore, if $H \neq \{1\}$, then we find an element $z = e^{i\theta} \in H$ such that $0 < \theta \leq \pi$ and $|z - 1|$ is minimum. We assert that $H = \langle z \rangle$. In fact, if there is an element x in H, which is not a power of z, then there exists an integer $n > 0$ such that $|z^{-n}x - 1| < |z - 1|$ and $0 < \arg(z^{-n}x) \leq \pi$. This is contrary to the definition of z, and the assertion holds.

For connectedness property, we see that the component of a topological group G containing a given element g is the translate of the component of the identity element of G by g, since the left translation $x \mapsto gx$, (resp. the right translation $x \mapsto xg$) in G is a homeomorphism of G onto itself. Accordingly, the components of G are just the left (or right) cosets of the component of the identity element in G. A notable fact about this component is described in the following.

Proposition 12.2.3 *The component of the identity element of a topological group is a closed normal subgroup.*

Proof Let G be a topological group with the identity element e, and let G_0 be the component of e in G. Since components are closed, G_0 is closed in G. Clearly, the set G_0^{-1} contains e and is connected, by the continuity of the inversion function. Hence, we have $G_0^{-1} \subseteq G_0$, which implies that $G_0 = \left(G_0^{-1}\right)^{-1} \subseteq G_0^{-1}$. So $G_0^{-1} = G_0$. Next, suppose that $x, y \in G_0$. Then, xG_0 obviously contains both x and xy. Being continuous image of G_0 under the translation map λ_x, xG_0 is connected. Therefore, we have $xG_0 \subseteq G_0$, since G_0 is the component of x in G. Thus, $xy \in G_0$, and we conclude that G_0 is a subgroup of G. To see the normality of G_0, consider an element $x \in G$. Since the conjugation map $y \mapsto x^{-1}yx$ is a homeomorphism of G, $x^{-1}G_0x$ is connected. It is also obvious that $e \in x^{-1}G_0x$. So, by the definition of G_0 again, we have $x^{-1}G_0x \subseteq G_0$, and it follows that G_0 is normal in G. $\diamond$

Example 12.2.6 The identity component of the general linear group $GL(n, \mathbb{R})$ is the open subgroup $GL_+(n, \mathbb{R})$ consisting of all real $n \times n$ matrices with positive determinants and the other component is the set $GL_-(n, \mathbb{R})$ of all real $n \times n$ matrices having negative determinants. As noted in the previous section, $GL(n, \mathbb{R})$ is not connected, and it is obviously the disjoint union of $GL_+(n, \mathbb{R})$ and $GL_-(n, \mathbb{R})$. Since $GL(n, \mathbb{R})$ is locally connected, its components are actually path components. So it suffices to show that both the sets $GL_+(n, \mathbb{R})$ and $GL_-(n, \mathbb{R})$ are path-connected.

To this end, we first show that its subgroup $SL(n, \mathbb{R})$ consisting of all matrices with determinant 1 is path-connected. Let $E_i(\lambda)$ denote the elementary matrix obtained by multiplying the ith row of the identity matrix I by λ ($\neq 0$), and let $E_{ij}(\lambda)$ denote the elementary matrix obtained by adding λ times the jth row of I to the ith row. Every matrix $A \in SL(n, \mathbb{R})$ can be written as a product of elementary matrices $E_{ij}(\lambda)$ or $E_i(\lambda)$, and in such a factorization of A if $E_i(\lambda)$ is one of the factors, then so is $E_j(\lambda^{-1})$ for some j (verify). For each $0 \leq t \leq 1$, put $\phi(t) = (1 - t)\lambda + t$ if $\lambda > 0$, and $\phi(t) = (1 - t)\lambda - t$ if $\lambda < 0$. Then, $t \mapsto E_i(\phi(t))$ is a path joining $E_i(\lambda)$ to $E_i(\pm 1)$, according as $\lambda > 0$ or $\lambda < 0$. Since the matrix multiplication is continuous, $t \mapsto E_i(\phi(t)) E_j(1/\phi(t))$ is a path in $SL(n, \mathbb{R})$ joining $E_i(\lambda) E_j(\lambda^{-1})$ to I. It is also clear that $t \mapsto E_{ij}((1 - t)\lambda), 0 \leq t \leq 1$, is a path in $SL(n, \mathbb{R})$ joining $E_{ij}(\lambda)$ to I. Thus, we see that every matrix $A \in SL(n, \mathbb{R})$ can be joined to I by a path in $SL(n, \mathbb{R})$, and therefore $SL(n, \mathbb{R})$ is path-connected. Now, if $A \in GL_+(n, \mathbb{R})$, then $\delta^{-1}A \in SL(n, \mathbb{R})$, where $\delta = \det(A)$. So we have a path f in $SL(n, \mathbb{R})$ joining $\delta^{-1}A$ and I. Then, $t \mapsto \delta f(t)$ is a path in $GL_+(n, \mathbb{R})$ joining A and δI. Also, $t \mapsto ((1 - t)\delta + t) I$ is a path in $GL_+(n, \mathbb{R})$ joining δI and I. Hence, we see that $GL_+(n, \mathbb{R})$ is path-connected.

Next, we observe that $GL_-(n, \mathbb{R})$ is the image of $GL_+(n, \mathbb{R})$ under the left translation λ_E, where E is the elementary matrix $E_1(-1)$. Therefore, it is also path connected.

As in the case of $GL_+(n, \mathbb{R})$, one proves that the general linear groups $GL(n, \mathbb{C})$ and $GL(n, \mathbb{H})$ are path-connected. The special linear group $SL(n, \mathbb{C})$ is also path-connected.

We close this section with the following result for closed subgroups of a compact group, which will be required later.

Proposition 12.2.4 *Let G be a compact Hausdorff group and $H \subset G$ a closed subgroup. Then, for any element $g \in G$, $gHg^{-1} = H$ if and only if $gHg^{-1} \subseteq H$.*

Proof Suppose that $gHg^{-1} \subseteq H$ for some $g \in G$, and consider the mapping ϕ: $G \times G \to G$ defined by $\phi(x, y) = xyx^{-1}$. Clearly, ϕ is continuous, and if $A = \{g^n \mid n = 0, 1, \ldots\}$, then $\phi(A \times H) \subseteq H$. Therefore, $\phi(\overline{A} \times H) \subseteq H$. If we can show that $g^{-1} \in \overline{A}$, then we have $g^{-1}Hg \subseteq H \Rightarrow H \subseteq gHg^{-1}$, and the proposition follows. To this end, let $B = \{g^n \mid n \in \mathbb{Z}\}$. Obviously, B is a subgroup of G, and hence so is $\overline{B}$. If the identity element e of G is an isolated point of $\overline{B}$, then $\overline{B}$ is discrete. Being compact and discrete, $\overline{B}$ is finite; consequently, we have $g^n = e$ for some integer $n > 0$. Thus, in this case, $g^{-1} \in A$ itself. If e is not an isolated point of $\overline{B}$, then for any symmetric nbd V of e in G, there is an integer $n > 0$ such that $g^n \in V$. Then, $g^{n-1} \in g^{-1}V \cap A$. Since the sets $g^{-1}(V)$ form a nbd basis at g^{-1}, we deduce that $g^{-1} \in \overline{A}$, as desired.   ◇

Actually, the proof of the preceding proposition shows that $\overline{A} = \overline{B}$. Thus, we see that, if g is an element of a compact Hausdorff group G, then the closure of the set $\{g^n \mid n = 0, 1, \ldots\}$ is a subgroup of G. It should be noted that neither this statement nor the preceding proposition is true without the condition of compactness on G.

Exercises

1. Prove: (a) $\mathbb{R}$ has no proper open subgroup.

   (b) A connected subgroup of $\mathbb{R}$ is either $\{0\}$ or $\mathbb{R}$.

   (c) Every nontrivial discrete subgroup of $\mathbb{R}$ is infinite cyclic.

2. Prove that a subgroup of a topological group G is discrete $\Leftrightarrow$ it has an isolated point.

3. Let G be a topological group with the identity element e. Prove the closure of $\{e\}$ in G has the trivial topology.

4. Let G be a finite connected topological group. Show that G has the trivial topology.

5. Let G be a topological group. Prove that a subgroup of G with nonempty interior is open.

6. Prove that a subgroup H of a topological group G is either closed or $\overline{H} - H$ is dense in $\overline{H}$.

7. Let G be a connected topological group with the identity element e and U be an open nbd of e. Show that $G = \bigcup_{n \geq 1} U^n$ (that is, every element of G is a finite product of elements of U, and thus G is generated by U).

8. Let G be a connected topological group with the identity element e. If a subgroup $H \subseteq G$ contains an open nbd of e, then $G = H$.

9. • Let H be a discrete subgroup of a Hausdorff group G. Show that H is closed in G. Deduce that $H \cap K$ is finite for every compact subset $K \subseteq G$.

10. Let G be a topological group and H a subgroup of G. If U is a nbd of e in G such that $H \cap \overline{U}$ is closed in G, show that H is closed in G.

11. Prove that every locally compact subgroup of a Hausdorff topological group is closed.

12. Prove that the intersection of all nontrivial closed (or open) subgroups of a topological group G is a normal subgroup of G.

13. Let H be a compact subgroup of a Hausdorff topological group G with $(G : H)$, the index of H in G, finite. Show that there exists a normal subgroup N of G such that $N \subseteq H$ and $(G : N)$ is finite.

14. Let G be a topological group and $H \subset G$ a dense subgroup. If N is a normal subgroup of H, show that $\overline{N}$ is a normal subgroup of G.

15. If H is closed in the topological group G, prove that its normalizer $N(H) = \{g \in G \mid gHg^{-1} = H\}$ is closed in G.

16. Let G be a Hausdorff topological group with the identity element e. Prove that $Z(G) = \{z \in G \mid zx = xz$ for all $x \in G\}$ is a closed normal subgroup of G. (This is called the *center* of G.)

17. Prove: (a) If a matrix M in $O(2)$ commutes with every matrix in $SO(2)$, then $M \in SO(2)$;

   (b) $Z(SO(2k)) = \{1, -1\}, k > 1$;

   (c) $Z(SO(2k+1)) = \{1\}, k \geq 1$.

18. Prove: (a) $Z(U(n)) \cong \mathbb{S}^1$;

   (b) $Z(SU(n))$ is a cyclic group of order n;

   (c) $Z(Sp(n)) = \{1, -1\}$.

19. Let G be a connected topological group and H be a totally disconnected normal subgroup of G. Show that H is contained in the center of G. (Such a subgroup of G is called *central*.)

20. Let G be a Hausdorff topological group with the identity e. If every nbd U of e contains an open subgroup of G, show that G is totally disconnected.

21. Let G be a topological group, K a compact subset and U an open subset of G such that $K \subseteq U$. Prove that there exists an open nbd W of e such that $KW \subseteq U$.

22. Let G be a Hausdorff topological group and V be a compact and open nbd of e. Prove that V contains a compact-open subgroup H.

23. Let G be a locally compact Hausdorff and totally disconnected group. Prove that the open subgroups of G form a local base at e.

24. Let G be a topological group with the identity element e. Show that the path component of e is a normal subgroup of G, and the cosets of this subgroup are precisely the path components of G.

12.3   Quotient Groups and Isomorphisms

Quotient Groups

Let H be a subgroup of a topological group G. Denote the set of all left cosets xH, $x \in G$, by G/H, and let $\pi: G \to G/H$ be the natural map $x \mapsto xH$. The set G/H with the quotient topology determined by π is called the *left coset* (or *factor*) space of G by H. Similarly, the space of right cosets Hx is defined. In general, the space of left cosets of H in G is different from the space of right cosets of H in G, but the two spaces are homeomorphic (the mapping $xH \mapsto Hx^{-1}$ gives a homeomorphism). For $x \in G$, the translation map $\lambda_x \colon G \to G$ induces a homeomorphism of G/H onto itself; consequently, G/H is a homogeneous space. By definition of the quotient topology, it is obvious that, with this topology for G/H, the map $\pi \colon G \to G/H$ is continuous. It is also open, for if U is an open subset of G, then $\pi^{-1}\,(\pi\,(U)) = UH$ is open in G, and hence $\pi(U)$ is open in G/H. Thus, we have proved the first part of

Proposition 12.3.1 *Let G be a topological group and $H \subset G$ be a subgroup. Then, the natural map $\pi \colon G \to G/H$ is continuous and open in the quotient topology for G/H. If H is compact, then π is also closed.*

Proof For the second part, notice that if H is compact and $F \subseteq G$ is closed, then $\pi^{-1}\,(\pi\,(F)) = FH$ is closed in G, by Proposition 12.1.4. So $\pi\,(F)$ is closed in G/H. ◇

Clearly, if H is an open subgroup of a topological group G, then each coset of H in G is open, and the complement $G - H$ is the union of cosets of H. So H is also closed in G. Conversely, if H is a closed subgroup of G having a finite index, then it is also open. Notice further that the factor space G/H is discrete if H is open in G. We also have the following.

Proposition 12.3.2 *Let G be a topological group and $H \subset G$ be a subgroup. The factor space G/H is Hausdorff if and only if H is closed in G.*

Proof If G/H is Hausdorff, then it is clear that H is closed in G, by the continuity of the map $\pi \colon G \to G/H$. Conversely, suppose that H is closed in G and $\pi(x) \neq \pi(y)$, $x, y \in G$. Then, $e \notin x^{-1}yH$, where e is the identity element of G. Consequently, there is an open nbd U of e in G such that $U \cap x^{-1}yH = \emptyset$. Then, by Proposition 12.1.2, we find open nbds V and W of e in G such that $x^{-1}Vx \subseteq W$ and $W^{-1}W \subseteq U$. Observe that VxH and VyH are disjoint nbds of xH and yH in G. So $\pi(Vx)$ and $\pi(Vy)$ are disjoint nbds of $\pi(x)$ and $\pi(y)$ in G/H. ◇

If N is a normal subgroup of the topological group G, then we know that the set G/N of left (or right) cosets of N in G is endowed with a group structure. The group G/N together with the quotient topology determined by the canonical projection $\pi \colon G \to G/N$ is a topological group. Recall that the multiplication function $\mu' \colon G/N \times G/N \to G/N$ is defined by $\mu'(xN, yN) = \mu(x, y)N$, where $\mu \colon G \times G \to G$ is the multiplication mapping $(x, y) \mapsto xy$. Consequently, the following diagram of continuous functions commutes:

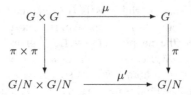

that is, $\mu' \circ (\pi \times \pi) = \mu \circ \pi$. So we have $(\pi \times \pi)^{-1} \mu'^{-1} (U^*) = \mu^{-1} \pi^{-1} (U^*)$ for every $U^* \subseteq G/N$. Notice that $\pi \times \pi$ is an identification, since π is open. As μ and π are continuous, we see that $\mu'^{-1}(U^*)$ is open, whenever U^* is open in G/N. So μ' is continuous.

Since the inversion function $\iota' \colon G/N \to G/N$ satisfies the equation $\iota'(xN) = x^{-1}N = \iota(x)N$, where ι is the inversion mapping $\iota \colon G \to G$, we have the following commutative diagram of continuous functions:

$$
\begin{array}{ccc}
G & \xrightarrow{\ \iota\ } & G \\
\downarrow{\scriptstyle \pi} & & \downarrow{\scriptstyle \pi} \\
G/N & \xrightarrow{\ \iota'\ } & G/N
\end{array}
$$

Hence, as above, we deduce that ι' is continuous.

Therefore, G/N a topological group with the quotient topology. We call it the *quotient* or the *factor group* of G by the normal subgroup N.

By Proposition 12.3.2, the quotient of a topological group G modulo a closed normal subgroup N is a Hausdorff topological group under the induced multiplication and topology. And, by Proposition 12.3.1, we see that most of the topological properties of G passed on to G/N. Note the particular case of G/G_0, where G_0 is the component of the identity element e of G. In this case, we also have the following.

Proposition 12.3.3 *Let G be a topological group and G_0 be the component of the identity element e of G. Then, the quotient group G/G_0 is totally disconnected.*

Proof Let C be a connected subset of G/G_0. We show that C is a singleton set; equivalently, $\pi^{-1}(C)$ is a coset of G_0 in G, where $\pi \colon G \to G/G_0$ is the canonical map. Since the components of G are cosets of G_0 in G, it suffices to prove that $\pi^{-1}(C)$ is connected. Assume on the contrary that $\pi^{-1}(C)$ is disconnected, and let $\pi^{-1}(C) = A \cup B$ be a disconnection. Then, both A and B are nonempty open subsets of $\pi^{-1}(C)$ and $A \cap B = \varnothing$. Accordingly, there exist open sets U and V in G such that $A = U \cap \pi^{-1}(C)$ and $B = V \cap \pi^{-1}(C)$. Since π is open, we see that both $\pi(A) = \pi(U) \cap C$ and $\pi(B) = \pi(V) \cap C$ are nonempty open subsets of C and $C = \pi(A) \cup \pi(B)$. As C is connected, we must have $\pi(A) \cap \pi(B) \neq \varnothing$. So there exist points $a \in A$ and $b \in B$ such that $\pi(a) = \pi(b)$. Then, $E = aG_0 = bG_0 \subseteq \pi^{-1}(C)$ and $E = (aG_0 \cap A) \cup (bG_0 \cap B)$ is a disconnection of E, a contradiction. Therefore, $\pi^{-1}(C)$ is connected, and the proposition follows. $\diamond$

For certain topological properties, we have partial converse.

Proposition 12.3.4 *Let G be a topological group and H be a connected subgroup of G such that the coset space G/H is also connected. Then, G is connected.*

Proof Suppose, on the contrary, that $G = U \cup V$ is a disconnection of G. By Proposition 12.3.1, the quotient map $\pi \colon G \to G/H$, $\pi(g) = gH$, is open. So both $\pi(U)$ and $\pi(V)$ are open in G/H. Since G/H is connected, we find an element $g \in G$ such that $\pi(g) \in \pi(U) \cap \pi(V)$. Then, $U \cap gH \neq \varnothing \neq V \cap gH$. As $gH \subseteq U \cup V$, we have a disconnection of gH. But gH is connected, being a continuous image of the connected set H (by hypothesis), and thus we obtain a contradiction. $\diamondsuit$

Proposition 12.3.5 *Suppose that H is a compact subgroup of a Hausdorff topological group G such that the quotient space G/H is also compact. Then G is compact.*

Proof By Proposition 12.3.1, the quotient map $\pi \colon G \to G/H$ is closed. In fact, π is a proper map, for each coset gH is obviously compact. Now, by "Proposition 6.5.4", we conclude that G is compact. $\diamondsuit$

For analogous result concerning the property of local compactness, refer to Montgomery and Zippin [9].

Isomorphisms

Definition 12.3.6 A *homomorphism* of a topological group G into another topological group G' is a function $f \colon G \to G'$ that is continuous as well as an algebraic homomorphism.

A homomorphism $f \colon G \to G'$ is called

(a) a *monomorphism* if it is injective,
(b) an *epimorphism* if it is surjective, and
(c) an *isomorphism* if it is a homeomorphism.

Thus, an isomorphism between two topological groups G and G' is an algebraic homomorphism $G \to G'$ which is also a homeomorphism. We call G and G' isomorphic (written $G \cong G'$) if there is an isomorphism $G \to G'$.

It is obvious that an inclusion map of a subgroup of the topological group G is a monomorphism, and if N is a normal subgroup of a topological group G, then the natural projection $\pi \colon G \to G/N$ is an epimorphism.

In the previous section, we introduced topological groups $O(n)$, $U(n)$ and $Sp(n)$. These are related by monomorphisms $U(n) \to O(2n)$ and $Sp(n) \to U(2n)$. Since $\mathbb{R}$ is a subfield of $\mathbb{C}$, we can regard $\mathbb{C}^n$ as a linear space over $\mathbb{R}$. Then, the mapping $\psi \colon \mathbb{C}^n \to \mathbb{R}^{2n}$, $(a_1 + \iota b_1, \ldots, a_n + \iota b_n) \mapsto (a_1, b_1, \ldots, a_n, b_n)$ is a linear isomorphism. Consequently, any $\mathbb{C}$-linear map $f \colon \mathbb{C}^n \to \mathbb{C}^n$ defines an $\mathbb{R}$-linear map $\lambda_n(f) \colon \mathbb{R}^{2n} \to \mathbb{R}^{2n}$, in the obvious way. Clearly, $\lambda_n(gf) = \lambda_n(g) \circ \lambda_n(f)$. If the matrix of f in the standard basis of $\mathbb{C}^n$ has z_{rs} as the (rs)th entry, then the matrix of $\lambda_n(f)$ in the standard basis of $\mathbb{R}^n$ has the matrix

$$\begin{bmatrix} a_{rs} & -b_{rs} \\ b_{rs} & a_{rs} \end{bmatrix}$$

at the (rs)-block, where $z_{rs} = a_{rs} + \iota b_{rs}$. So there is a monomorphism $\lambda_n : GL$ $(n, \mathbb{C}) \to GL(2n, \mathbb{R})$. The matrices in the image of λ_n are referred to as the complex-linear real matrices. The isomorphism ψ preserves the norms, and therefore $\lambda_n(f)$ is an isometry whenever f is so. Hence, the restriction of λ_n to $U(n)$ gives a monomorphism $U(n) \to O(2n)$. Since $U(n)$ is compact, we see that it is isomorphic, as a topological group, to the group of complex-linear real orthogonal matrices.

To find a monomorphism $Sp(n) \to U(2n)$, notice that each quaternion can be written as $z + wj$, $z, w \in \mathbb{C}$. The mapping $z + wj \mapsto (z, \overline{w})$ defines a bijection $\psi : \mathbb{H}^n \to \mathbb{C}^{2n}$, which is $\mathbb{C}$-linear when $\mathbb{H}^n$ is considered as a right linear space over $\mathbb{C}$. Accordingly, each linear map $\mathbb{H}^n \to \mathbb{H}^n$ determines a linear map $\mathbb{C}^{2n} \to \mathbb{C}^{2n}$, and by replacing the element $z + wj$ of an $n \times n$ matrix A over $\mathbb{H}$ with the 2×2 matrix

$$\begin{bmatrix} z & -w \\ \overline{w} & \overline{z} \end{bmatrix},$$

we obtain a $2n \times 2n$ matrix over $\mathbb{C}$. Thus, there results in a monomorphism $\lambda_n :$ $GL(n, \mathbb{H}) \to GL(2n, \mathbb{C})$. Observe that $\lambda_n(A) \in U(2n)$ if $A \in Sp(n)$, hence λ_n defines a monomorphism $Sp(n) \to U(2n)$. The matrices in the image of $Sp(n)$ under λ_n are referred to as the quaternionic-linear complex matrices. Since $Sp(n)$ is compact, we see that it is isomorphic, as a topological group, to the subgroup of the quaternionic-linear complex matrices.

The following examples show that two topological groups may have essentially the same algebraic structures but different topological structures, and vice versa.

Example 12.3.1 The group $(\mathbb{R}, +)$ can be given the usual topology as well as the discrete topology to produce two different topological groups.

Example 12.3.2 Let G be the set of all matrices $\begin{bmatrix} a & 0 \\ 0 & b \end{bmatrix}$ and G' be the set of all matrices $\begin{bmatrix} e^a & b \\ 0 & e^{-b} \end{bmatrix}$, where $a, b \in \mathbb{R}$. Then, G is an abelian group under the addition of matrices, and G' is a nonabelian group under the multiplication of matrices. With the usual topologies, the underlying spaces of the topological groups G and G' are homeomorphic to the Euclidean space $\mathbb{R}^2$.

As in ordinary group theory, the *kernel* of f is defined as the inverse image of the identity element of G', and is denoted by $\ker(f)$. The homomorphism f is a monomorphism if $\ker(f) = \{e\}$, where $e \in G$ is the identity element, and an epimorphism if its image $f(G)$ equals G'. It is easily verified that $K = \ker(f)$ is a normal subgroup of G, and the induced homomorphism $\bar{f} \colon G/K \to G'$, $xK \mapsto f(x)$, is a monomorphism of topological groups. The continuity of $\bar{f}$ follows from the factorization $\bar{f}\pi = f$ and the fact that the natural projection $\pi \colon G \to G/K$ is an identification map. The remaining properties of $\bar{f}$ are well known from group theory. Furthermore, if f is an open epimorphism, then $\bar{f}$ is clearly an isomorphism. Conversely, if $\bar{f}$ is an isomorphism (of topological groups), then f must be an open epimorphism, for $\pi \colon G \to G/K$ is open. Thus, we have established the following

Proposition 12.3.7 *Let $f \colon G \to G'$ be a homomorphism between topological groups. Then, the induced homomorphism $\bar{f} \colon G/\ker(f) \to G'$ is an isomorphism if and only if f is an open epimorphism.*

Corollary 12.3.8 *A homomorphism of a compact topological group onto a Hausdorff topological group is open.*

Proof Let G be a compact topological group, G' a Hausdorff topological group and $f \colon G \to G'$ be an epimorphism. Then, $G/\ker(f)$ is compact, and hence the induced map $\bar{f} \colon G/\ker(f) \to G'$ is an isomorphism between topological groups. By the preceding proposition, f is open. ◇

Example 12.3.3 The exponential map $p \colon \mathbb{R}^1 \to \mathbb{S}^1$, $t \mapsto e^{2\pi i t}$, is an open epimorphism of topological groups with the kernel $\mathbb{Z}$, the group of integers. Therefore, the quotient group $\mathbb{R}^1/\mathbb{Z}$ is isomorphic to $\mathbb{S}^1$.

The next proposition often simplifies the problem of checking the continuity or openness of algebraic homomorphism between topological groups.

Proposition 12.3.9 *Let G and G' be topological groups with the identity elements e and e', and let $f \colon G \to G'$ be a homomorphism in the algebraic sense. Then,*

(a) *f is continuous if it is continuous at e and*
(b) *f is open if it carries nbds of e to nbds of e' in G'.*

Proof (a): Let $x \in G$, and let V be an open nbd of $f(x) = x'$ in G'. Then, $\rho_{x'^{-1}}(V)$ is an open nbd of e' in G'. As $f(e) = e'$, we find an open nbd U of e such that $f(U) \subseteq \rho_{x'^{-1}}(V)$. This implies that $f(\rho_x(U)) = \rho_{x'}f(U) \subseteq V$. Since $\rho_x(U)$ is an open nbd of x, f is continuous at x.

(b): Let U be an open subset of G, and $x \in U$. Then, $\rho_{x^{-1}}(U)$ is an open nbd of e. By our hypothesis, $f(\rho_{x^{-1}}(U))$ is a nbd of e'. If $f(x) = x'$, then $\rho_{x'^{-1}}(f(U)) = f(\rho_{x^{-1}}(U))$, and therefore $f(U)$ is a nbd of x'. Since $x \in U$ is arbitrary, $f(U)$ is open. ◇

As an immediate application of the preceding proposition, we see that the function $f \colon \mathbb{C} \to \mathbb{C}$, $z \mapsto z^n$, is an open mapping for every integer $n > 0$, since the image of an open disc $B(0; \delta)$ in $\mathbb{C}$ under f is clearly the open disc $B(0; \delta^n)$. We use this fact to show that the determinant function $\det \colon \mathcal{M}(n, \mathbb{C}) \to \mathbb{C}$ is open in the usual topologies on both spaces. Let U be an open subset of $\mathcal{M}(n, \mathbb{C})$ and $A \in U$. Then, there exists a real $r > 0$ such that $B(A; r) \subseteq U$. If $A = 0$, then the open ball of radius $\epsilon = \left(r/\sqrt{n}\right)^n$ about $0 \in \mathbb{C}$ is contained in $\det(U)$, for the scalar matrix

$$\begin{bmatrix} \sqrt[n]{z} & 0 & \cdots & 0 \\ 0 & \sqrt[n]{z} & \cdots & 0 \\ \vdots & \vdots & \ddots & \vdots \\ 0 & \cdots & \cdots & \sqrt[n]{z} \end{bmatrix}$$

belongs to U when $|z| < \epsilon$. If $A \neq 0$, then we have $zA \in U$ for all $z \in \mathbb{C}$ such that $|z - 1| < r/\|A\|$. Obviously, $\det(zA) = z^n\det(A)$. Since the function $z \mapsto z^n$ is an open mapping of $\mathbb{C}$ into itself, $W = \{z^n : |z - 1| < r/\|A\|\}$ is an open set in $\mathbb{C}$ containing 1. So there is an open ball about 1 (in $\mathbb{C}$) contained in W. It follows that the set of numbers $z^n\det(A)$, as z ranges over the set $\{z : |z - 1| < r/\|A\|\}$, contains an open ball about $\det(A)$. This ball (with center $\det(A)$) is contained in the image of U under the function det, and thus $\det(U)$ is a nbd of $\det(A)$. This shows that $\det(U)$ is open, as desired.

Similarly, the determinant function $\mathcal{M}(n, \mathbb{R}) \to \mathbb{R}$ can be shown to be open.

We know that the restriction of the function $\det : \mathcal{M}(n, F) \to F$, $F = \mathbb{R}$ or $\mathbb{C}$, to $GL(n, F)$ is a homomorphism of this group onto F_0, where F_0 is the multiplicative group of the field F. Since $GL(n, F)$ is an open subset of $\mathcal{M}(n, F)$, we see that $\det : GL(n, F) \to F_0$ is an open epimorphism. The kernel of this homomorhism is obviously the special linear group $SL(n, F) = \{A \in GL(n, F) \mid \det(A) = 1\}$, and so the quotient group $GL(n, F)/SL(n, F)$ is isomorphic to F_0, as a topological group.

For applications of Proposition 12.3.7, it is convenient to know certain conditions under which the homomorphism f is open. In this regard, we have the following.

Theorem 12.3.10 *Let G and G' be locally compact Hausdorff groups, and $f : G \to G'$ be an epimorphism. If G is also Lindelöf, then f is an open map (and hence $G' \cong G/\ker(f)$).*

Proof By Proposition 12.3.9, it is enough to show that $f(N)$ is a nbd of the identity element e' of G' for every nbd N of the identity element e of G. Observe that this can easily be deduced by proving that the interior of $f(U)$ is nonempty for every open nbd U of e in G. For, if N is nbd of e in, then there exists an open nbd V of e such that $VV^{-1} \subseteq N$. Now, if there exists an element y in the interior of $f(V)$ and $y = f(x)$, where $x \in V$, then $f(V)y^{-1}$ is a nbd of e' and $f(V)y^{-1} = f(Vx^{-1}) \subseteq f(N)$. So $f(N)$ is a nbd of e'.

To prove that $f(U)^\circ$ is nonempty for every open set $U \subseteq G$ containing e, we find an open set $V \subseteq G$ such that the closure $\overline{V}$ of V is compact and $e \in V \subseteq \overline{V} \subset U$, for G is locally compact Hausdorff. Then, the family $\{Vx \mid x \in G\}$ is an open covering of G. Since G is Lindelöf, this open covering has a countable subcovering. So we find points $x_1, x_2, \ldots$ in G such that $G = \bigcup_1^\infty V x_n$. Put $K_n = f\left(\overline{V} x_n\right)$ for every n. Then, each K_n is compact, and hence closed in G'. Also, $G' = \bigcup_1^\infty K_n$. We claim that $(K_n)^\circ \neq \varnothing$ for some n. Assume the contrary, and choose a nonempty open set W_0 of G' such that $\overline{W}_0$ is compact. By our assumption, there is a point $y_1 \in W_0 - K_1$. Then, we find an open nbd W_1 of y_1 such that $\overline{W}_1 \subseteq W_0 - K_1$, and $\overline{W}_1$ is compact. Again, by our assumption, there is a point $y_2 \in W_1 - K_2$. Consequently, there is an open nbd W_2 of y_2 such that $\overline{W}_2$ is compact, and $\overline{W}_2 \subseteq W_1 - K_2$. Continuing in this way, we obtain open sets W_n, $n = 1, 2, \ldots$, such that $\overline{W}_n$ is compact, and $\overline{W}_n \cap K_n = \varnothing$ and $\overline{W}_{n+1} \subseteq W_n$. Since the intersection of every finite subfamily of $\overline{W}_n$ is nonempty, we have $\bigcap_1^\infty \overline{W}_n \neq \varnothing$ (by Exercise 5.1.19). But, this is impossible because $G' = \bigcup K_n$ and K_n does not meet $\overline{W}_n$. Hence, our claim. Now, if $(K_n)^\circ \neq \varnothing$, then $f\left(\overline{V}\right) = K_n f\left(x_n^{-1}\right)$ also has a nonempty interior. Since $\overline{V} \subseteq U$, we have $f(U)^\circ \neq \varnothing$. $\diamondsuit$

The condition in the preceding theorem that G be a Lindelöf space cannot be dropped. For, if G is the group of real numbers with the discrete topology and G' is the same group with the usual topology, then the identity map $G \to G'$ is a bijective homomorphism but it is obviously not open. Moreover, the condition of local compactness on G' is also essential, as shown by the following.

Example 12.3.4 Let $\alpha > 0$ be an irrational number, and consider the subgroup $X = \mathbb{Z} + \alpha \mathbb{Z} \subset \mathbb{R}$. Obviously, the restriction of the quotient map $\mathbb{R} \to \mathbb{R}/\mathbb{Z}$ to $\alpha \mathbb{Z}$ is an algebraic isomorphism between $\alpha \mathbb{Z}$ and $X/\mathbb{Z}$. However, these are not homeomorphic, since $\alpha \mathbb{Z}$ is a discrete space while $X/\mathbb{Z}$ is dense in the quotient space $\mathbb{R}/\mathbb{Z} \approx \mathbb{S}^1$. To see the latter statement, observe that if X has a least positive number c, say, then X is generated by c. Hence, $\alpha = mc$ and $1 = nc$ for some $m, n \in \mathbb{Z}$, and we have $\alpha = m/n$, a contradiction. Therefore, there is no least positive number in X; consequently, we can find a strictly decreasing sequence of positive numbers $x_1 > x_2 > \cdots$ in X. Now, given an open interval (a, b), we choose an integer $n > 0$ such that $0 < x_n - x_{n+1} < b - a$. Then, for $u = x_n - x_{n+1}$, there exists an integer $m > 0$ such that $m - 1 < a/u < m$. Clearly, $a < mu < b$. Since X is a subgroup of $\mathbb{R}$, we have $mu \in X$. Thus, we see that X is dense in $\mathbb{R}$, and hence $X/\mathbb{Z}$ is dense in the quotient space $\mathbb{R}/\mathbb{Z}$. Notice that $X/\mathbb{Z}$ is not locally compact, by Theorem 5.4.3.

Recall that the *second isomorphism theorem* of group theory states that if G is a group, and $M, N \subseteq G$ are subgroups and N is normal in G, then the factor group MN/N is isomorphic to $M/(M \cap N)$. The preceding examples also show that the analogue of this theorem for topological groups is not true. However, if G is a second countable, locally compact Hausdorff topological group, and the subgroups M, N and MN are all closed in G, then, by Theorem 12.3.10, $MN/N \cong M/(M \cap N)$ as topological groups.

The analogue of the *third isomorphism theorem* of group theory is valid for topological groups, and we leave the proof to the reader.

We close this section by realizing the geometry of groups $SO(n)$ and $SU(n)$ for small values of n. Note that $SO(1)$ and $SU(1)$ are the trivial groups.

Proposition 12.3.11 $SO(2) \cong \mathbb{S}^1$ *and* $SU(2) \cong \mathbb{S}^3$ *as topological groups.*

Proof Clearly, the multiplication by each $z \in \mathbb{S}^1$ determines a linear isometry μ_z: $\mathbb{C} \to \mathbb{C}$. Also, the canonical vector space isomorphism $\psi: \mathbb{C} \to \mathbb{R}^2$ preserves the norms, and therefore the composition $\psi \circ \mu_z \circ \psi^{-1}$ is a linear isometry $\lambda_z: \mathbb{R}^2 \to \mathbb{R}^2$ for each $z \in \mathbb{S}^1$. Hence, there is a monomorphism $\eta: \mathbb{S}^1 \to O(2)$ which takes $z \in \mathbb{S}^1$ into the matrix of λ_z relative to the standard basis of $\mathbb{R}^2$. If $z = \cos\theta + \imath \sin\theta$, then the matrix of λ_z is

$$\begin{bmatrix} \cos\theta & -\sin\theta \\ \sin\theta & \cos\theta \end{bmatrix}$$

which obviously belongs to $SO(2)$. On the other hand, if $A \in SO(2)$, then it is easily seen that the matrix A is of this form. Consequently, η is the desired isomorphism between $\mathbb{S}^1$ and $SO(2)$.

To see an isomorphism between topological groups $\mathbb{S}^3$ and $SU(2)$, we write $q \in \mathbb{H}$ as $q = z_q + w_q J$, where z_q, w_q are in $\mathbb{C}$. The canonical map $\psi: z_q + w_q J \mapsto \left(z_q, \overline{w_q}\right)$ is an isomorphism of the (right) vector space $\mathbb{H}$ over $\mathbb{C}$ onto $\mathbb{C}^2$. Also, it is norm preserving, and hence an isometry. Observe that the left multiplication $\mu_x: \mathbb{H} \to \mathbb{H}$, $q \mapsto xq$, is a linear isometry for each $x \in \mathbb{S}^3$. Consequently, there is a $\mathbb{C}$-linear isometry $\lambda_x: \mathbb{C}^2 \to \mathbb{C}^2$ given by $\lambda_x = \psi \circ \mu_x \circ \psi^{-1}$. Thus, we have a mapping $\eta: \mathbb{S}^3 \to U(2)$ defined by $\eta(x) = [\lambda_x]$, the matrix of λ_x in the standard basis of $\mathbb{C}^2$. It is easily verified that η is a monomorphism of topological groups. For $x = z_x + w_x J$, the matrix $[\lambda_x]$ is $\begin{bmatrix} z_x & -w_x \\ \overline{w_x} & \overline{z_x} \end{bmatrix}$, which belongs to $SU(2)$ for all $x \in \mathbb{S}^3$. The other way round, if the matrix $A = \begin{bmatrix} a & b \\ c & d \end{bmatrix}$ belongs to $SU(2)$, then $\begin{bmatrix} \overline{a} & \overline{c} \\ \overline{b} & \overline{d} \end{bmatrix} = \begin{bmatrix} d & -b \\ -c & a \end{bmatrix}$. So $d = \overline{a}$, $c = -\overline{b}$ and $1 = \det(A) = |a|^2 + |b|^2$. It follows that $x = a - bJ \in \mathbb{S}^3$, and we have $\eta(x) = A$. Thus, η is a monomorphism of $\mathbb{S}^3$ onto $SU(2)$, and this completes the proof.   ◇

Proposition 12.3.12 *The topological group $SO(3)$ is homeomorphic to the real projective 3-space $\mathbb{R}P^3$.*

Proof Since the space $\mathbb{R}P^3$ is just the quotient group $\mathbb{S}^3/\{-1, 1\}$, we show that $SO(3) \cong \mathbb{S}^3/\{1, -1\}$ as topological groups. Let $\mathbb{S}^3$ be regarded as the group of unit quaternions. Then, we define a continuous homomorphism of $\mathbb{S}^3$ onto $SO(3)$. Given $x \in \mathbb{S}^3$, consider the mapping $f_x: \mathbb{H} \to \mathbb{H}$ defined by $f_x(q) = xq\overline{x}$. This is an $\mathbb{R}$-isomorphism of $\mathbb{H}$ onto itself, for $q \mapsto \overline{x}qx$ is its inverse. Since $\|x\| = 1$, and the norm of the product is the product of the norms, $\|f_x(q)\| = \|q\|$. Hence, the transformation f_x is orthogonal; in fact, it is a special orthogonal transformation of $\mathbb{R}^4 \cong \mathbb{H}$ (verify). We can identify $\mathbb{R}^3$ with the subspace of $\mathbb{H}$ consisting of quaternions q whose real part is zero. Then, for every $q \in \mathbb{R}^3$, we have $2Re(xq\overline{x}) = xq\overline{x} + x\overline{q}\overline{x} = x(q + \overline{q})\overline{x} = 0$. So $xq\overline{x} \in \mathbb{R}^3$, and the restriction of f_x to $\mathbb{R}^3$ defines an isomorphism $g_x: \mathbb{R}^3 \to \mathbb{R}^3$, $q \mapsto xq\overline{x}$. Since $f_x(1) = 1$, and $\mathbb{R}^3$ is orthogonal to the vector 1, it follows that $\det g_x = \det f_x = 1$. Thus, g_x is a special orthogonal transformation of $\mathbb{R}^3$. Let $[g_x]$ denote the matrix of g_x in the basis

$\{\iota, \jmath, k\}$ of $\mathbb{R}^3$. Then, we have a mapping $\phi: \mathbb{S}^3 \to SO\,(3)$ defined by $\phi\,(x) = [g_x]$. If $x, y \in \mathbb{S}^3$, then $g_x\,(q) - g_y\,(q) = (x - y)\,q\bar{x} + yq\,(\bar{x} - \bar{y})$, which implies that $\|g_x\,(q) - g_y\,(q)\| \le \|x - y\|\,\|q\|$ for every $q \in \mathbb{R}^3$. From this, we deduce that $\|[g_x] - [g_y]\| \le 2\sqrt{3}\,\|x - y\|$, and the continuity of ϕ follows. This preserves multiplications, for $g_{xy}\,(q) = xyq\overline{xy} = xyq\bar{y}\bar{x} = \left(g_x \circ g_y\right)(q)$. Thus, ϕ is a (continuous) homomorphism of $\mathbb{S}^3$ into $SO\,(3)$. Since $\mathbb{S}^3$ is compact and $SO\,(3)$ Hausdorff, we have a continuous closed injection $\mathbb{S}^3/\ker(\phi) \to SO(3)$ induced by ϕ.

Next, observe that a quaternion commutes with the pure quaternions ι, $\jmath$ and k if and only if it is real. Accordingly, $g_x = 1 \iff x = \pm 1$, and so $\ker(\phi)$ is $\{-1, 1\}$.

It remains to show that ϕ is surjective. Let H_ι, $H_\jmath$ and H_k be the subgroups of $SO\,(3)$ leaving fixed the quaternions ι, $\jmath$ and k, respectively. We observe that any element $A \in SO\,(3)$ is a product of elements from these subgroups. The vector Ak can be rotated into the (ι, k)-plane by a rotation B in H_k. Then, we can send $(BA)k$ to k by a rotation C in $H_\jmath$. Thus, $(CBA)k = k$ so that $CBA = B'$ belongs to H_k. This implies that $A = B^{-1}C^{-1}B'$. Since ϕ is a homomorphism, it suffices to show that each of these subgroups H_ι, $H_\jmath$ and H_k is contained in the image of ϕ. If A in H_ι, then

$$A = \begin{bmatrix} 1 & 0 & 0 \\ 0 & \cos\theta & -\sin\theta \\ 0 & \sin\theta & \cos\theta \end{bmatrix}$$

for some $0 \le \theta < 2\pi$. Now, for $x = \cos\theta/2 + \iota\sin\theta/2 \in \mathbb{S}^3$, we have $g_x\,(\iota) = \iota$, $g_x\,(\jmath) = \jmath\cos\theta + k\sin\theta$ and $g_x\,(k) = -\jmath\sin\theta + k\cos\theta$ so that $[g_x] = A$. Similarly, it can be shown that every element in $H_\jmath$ and H_k has a preimage in $\mathbb{S}^3$ under ϕ, and we see that ϕ is surjective. $\diamond$

Exercises

1. Let H be a subgroup of a topological group G. Prove that G/H has the indiscrete topology $\Leftrightarrow H$ is dense in G.

2. Let H be a subgroup of a topological group G. Show that the coset space G/H is T_3.

3. Let G be a locally compact topological group. Prove that the identity component of G is the intersection of all open subgroups, and deduce that G is connected $\Leftrightarrow$ it has no proper open subgroups.

4. Prove that $\mathbb{S}^1$ is a closed subgroup of $\mathbb{S}^3$. Also, show that $\mathbb{S}^3/\mathbb{S}^1$ is homeomorphic to $\mathbb{S}^2$. Is $\mathbb{S}^1$ normal in $\mathbb{S}^3$?

5. Show that there is no monomorphism of $\mathbb{S}^1$ into $\mathbb{R}^1$.

6. Prove that the quotient group $\mathbb{C}_0/\mathbb{S}^1$ is isomorphic to the group $\mathbb{R}_+$ of positive reals.

7. Prove that the group of all translations of the topological group $\mathbb{S}^1$ furnished with the compact-open topology is isomorphic to $\mathbb{S}^1$.

8. Show that the group of automorphisms of the circle group $\mathbb{S}^1$ is isomorphic to $\mathbb{Z}_2$.

9. Prove that the determinant function $\det : \mathscr{M}(n, \mathbb{R}) \to \mathbb{R}$ is open.

10. Let $G \subset GL(n, \mathbb{R})$ be a compact group. Prove that every element of G has determinant $+1$ or -1. Is G contained in $O(n)$?

11. Prove every discrete subgroup of $O(2)$ is cyclic or dihedral.

12. Let $C(I)$ be the topological group in Exercise 12.1.2. Prove that the mapping $e_t : C(I) \to \mathbb{R}$ given by $e_t(f) = f(t)$ is a open epimorphism for each $t \in I$, and hence $C(I)/\ker(e_t)$ is isomorphic to $\mathbb{R}$.

13. Let G be a topological group, and M and N be subgroups of G with $M \subset N$. Show that the quotient topology for N/M coincides with the subspace topology induced from G/M.

14. Let G be a topological group. If M and N are normal subgroups of G with $M \subset N$, prove that $(G/M)/(N/M) \cong G/N$ as topological groups.

12.4 Direct Products

Given a family of topological groups G_α, $\alpha \in A$, it is known that their direct product $\prod G_\alpha$ is a group under the coordinatewise group operations: If $x = (x_\alpha)$ and $y = (y_\alpha)$, then $xy = (x_\alpha y_\alpha)$, $x^{-1} = (x_\alpha^{-1})$, and $e = (e_\alpha)$ (where e_α is the identity element of G_α) is the identity element of $\prod G_\alpha$.

We observe that $\prod G_\alpha$ with the product (Tychonoff) topology is a topological group. The multiplication in $\prod G_\alpha$ can be considered as the composition of the function

$$\prod G_\alpha \times \prod G_\alpha \xrightarrow{\eta} \prod (G_\alpha \times G_\alpha) \xrightarrow{\prod \mu_\alpha} \prod G_\alpha,$$

where η takes $((x_\alpha), (y_\alpha))$ into the point whose αth coordinate is (x_α, y_α), and $\mu_\alpha : G_\alpha \times G_\alpha \to G_\alpha$ is the multiplication. The function $\prod \mu_\alpha$ is continuous by Theorem 2.2.10, and the function η is clearly a homeomorphism. So the multiplication in $\prod G_\alpha$ is continuous. The continuity of the inversion function on $\prod G_\alpha$ is also an easy consequence of Theorem 2.2.10. Clearly, each projection $p_\beta : \prod G_\alpha \to G_\beta$ is an open epimorphism. If H is a topological group, then it is easy to see that a function $f : H \to \prod G_\alpha$ is a homomorphism if and only if each composition $p_\alpha \circ f : H \to G_\alpha$ is a homomorphism. In particular, we have monomorphisms $q_\beta : G_\beta \to \prod G_\alpha$ defined by $(q_\beta(x_\beta))_\alpha = e_\alpha$ if $\alpha \neq \beta$, and $(q_\beta(x_\beta))_\beta = x_\beta$.

The direct product of two topological groups G_1 and G_2 is usually denoted by $G_1 \times G_2$. If $p_i: G_1 \times G_2 \to G_i, i = 1, 2$, is the projection and $q_i: G_i \to G_1 \times G_2$ is the canonical injection, then it is obvious that $\ker(p_1) = q_2(G_2)$ and $\ker(p_2) = q_1(G_1)$. By Proposition 12.3.7, there are isomorphisms $(G_1 \times G_2)/q_2(G_2) \cong G_1$ and $(G_1 \times G_2)/q_1(G_1) \cong G_2$ induced by the projections p_1 and p_2, respectively.

There is another fact which is of use at times. Let G be an abelian topological group. Then, $\Delta = \{(x, x) \mid x \in G\}$ is a normal subgroup of the direct product $G \times G$. We observe that the topological group $(G \times G)/\Delta$ is isomorphic to G. Note that the map $\phi: G \times G \to G$, $(x, y) \mapsto xy^{-1}$ has a continuous right inverse $G \to G \times G$, $x \mapsto (x, e)$. Therefore, ϕ is an identification (ref. Exercise 6.2.2). By Theorem 6.2.5, there is a homeomorphism $\psi: (G \times G)/\Delta \to G$ such that $\psi\pi = \phi$, where $\pi: G \times G \to (G \times G)/\Delta$ is the natural projection. It is easily checked that ψ is an algebraic isomorphism.

Recall that in group theory we say that a group G decomposes into the direct product of its subgroups M and N if both M and N are normal in G, $G = MN$ and $M \cap N = \{e\}$. These conditions are equivalent to the requirement that the mapping $M \times N \to G$, $(x, y) \mapsto xy$, is an isomorphism.

For topological groups, we have accordingly the following.

Definition 12.4.1 A topological group G is said to *decompose* into the direct product of its subgroups M and N if the mapping $M \times N \to G$, $(x, y) \mapsto xy$, is an isomorphism of topological groups.

Example 12.4.1 Let $\mathbb{C}_0$ denote the punctured plane $\mathbb{C} - \{0\}$. Then, we have a mapping $\eta: \mathbb{C}_0 \to \mathbb{R}_+ \times \mathbb{S}^1$ defined by $\eta(z) = (|z|, z/|z|)$, where $\mathbb{R}_+$ is the set of positive reals. Note that each element $z \in \mathbb{C}_0$ can be uniquely expressed as $z = |z|(z/|z|)$. Since the absolute value function $z \mapsto |z|$ is an open homomorphism, it follows that η is an isomorphism of topological groups.

Similarly, the punctured quaternionic line $\mathbb{H}_0 = \mathbb{H} - \{0\}$ is isomorphic to $\mathbb{R}_+ \times \mathbb{S}^3$.

Notice that if a topological group G decomposes into the direct product of two subgroups M and N in the algebraic sense (i.e., M and N are normal subgroups of G such that $M \cap N = \{e\}$ and $G = MN$), then the isomorphism $M \times N \to G$, $(x, y) \mapsto xy$, is obviously continuous; however, it need not be a homeomorphism (see Example 12.4.2 below). If G is a compact Hausdorff group and both the subgroups M and N are closed, then this isomorphism is certainly a homeomorphism, and we have a decomposition of the topological group G into the direct product of M and N. By Theorem 12.3.10, it is clear that this proposition is valid for second countable locally compact Hausdorff groups G as well.

Example 12.4.2 Let $\alpha > 0$ be a real number, and consider the subgroups $N = \mathbb{Z} \times \mathbb{Z}$ and $M = \{(x, \alpha x) \mid x \in \mathbb{R}\}$ of $\mathbb{R}^2$. Then, the sum $G = M + N$ is a subgroup of the topological group $\mathbb{R}^2$. If α is an irrational number, then the group G is the direct sum of M and N. But G, as a topological group, fails to decompose into the direct sum of the subgroups M and N (in the sense of Definition 12.4.1). To prove this claim, we first observe that G/M is homeomorphic to $\mathbb{Z} + \alpha \mathbb{Z}$. Clearly, the mapping $f : \mathbb{R}^2 \to \mathbb{R}^1$, $(x, y) \mapsto x - \alpha^{-1} y$ is a continuous open surjection, and hence an identification. It is easily checked that the decomposition space of f is the coset space $\mathbb{R}^2/M$. By Theorem 6.2.5, f induces a homeomorphism $\mathbb{R}^2/M \approx \mathbb{R}^1$, which carries the quotient space G/M onto $\mathbb{Z} + \alpha \mathbb{Z}$. Now, if $G = M \oplus N$ is a decomposition of the topological group G, then the quotient group G/M is isomorphic to the discrete group N. For, the isomorphism $M \times N \cong G$, $(x, y) \mapsto x + y$, carries $M \times \{0\}$ onto M, and so $G/M \cong (M \times N) / (M \times \{0\}) \cong N$, as topological groups. Hence $\mathbb{Z} + \alpha \mathbb{Z}$ is homeomorphic to N, and thus it is a discrete subgroup of $\mathbb{R}$. This implies that $\mathbb{Z} + \alpha \mathbb{Z}$ is closed in $\mathbb{R}$ (by Exercise 12.2.9), contrary to the fact that it is dense (see Example 12.3.4).

Next, we see that a topological group may be homeomorphic to the direct product of its two subgroups, yet fail to decompose into them. As an illustration, consider the group $\mathscr{I}so\,(\mathbb{R}^n)$ of all isometries of the Euclidean space $\mathbb{R}^n$, the composition of mappings being the group multiplication. If $f \in \mathscr{I}so\,(\mathbb{R}^n)$ and $f(0) = b$, then the composition of f and the translation $\rho_{-b} : x \mapsto x - b$ is an isometry of $\mathbb{R}^n$, which fixes the origin 0 of $\mathbb{R}^n$. We show that $g = \rho_{-b} \circ f$ is linear. In fact, we prove

Lemma 12.4.2 *An isometry of $\mathbb{R}^n$, which fixes the origin 0, is a linear map.*

Proof First, observe that if $\langle x, x \rangle = \langle y, y \rangle = \langle x, y \rangle$, then $x = y$; this can be seen by expanding $\langle x - y, x - y \rangle$. Now, if $g : \mathbb{R}^n \to \mathbb{R}^n$ is an isometry, then we have $\|g(x) - g(y)\| = \|x - y\|$ for all $x, y \in \mathbb{R}^n$. If $g(0) = 0$ as well, then obviously $\|g(x)\| = \|x\|$ for every $x \in \mathbb{R}^n$, and hence $\langle g(x), g(y) \rangle = \langle x, y \rangle$ for all $x, y \in \mathbb{R}^n$. Taking $u = x + y$, we obtain the equation $\langle u, u \rangle = \langle u, x \rangle + \langle u, y \rangle = \langle x, x \rangle + 2 \langle x, y \rangle + \langle y, y \rangle$. It follows that

$$\langle g(u), g(u) \rangle = \langle g(u), g(x) + g(y) \rangle = \langle g(x) + g(y), g(x) + g(y) \rangle,$$

and hence $g(x + y) = g(u) = g(x) + g(y)$. Similarly, we find that $g(xa) = g(x)a$, $a \in \mathbb{R}$, and this completes the proof.   ◇

It follows that the mapping $g = \rho_{-b} \circ f$ is an othogonal transformation (refer to Sect. 12.2), and hence f, being the composition of g with the translation $\rho_b : x \mapsto x + b$, is an invertible affine map on $\mathbb{R}^n$. Thus, we see that the group $\mathscr{I}so\,(\mathbb{R}^n)$ is a subgroup of the topological group $Aff_n(\mathbb{R})$, and therefore can be assigned the topology of the product space $GL(n, \mathbb{R}) \times \mathbb{R}^n$ (refer to Exercise 12.1.18). Alternatively, we see that $\mathscr{I}so\,(\mathbb{R}^n)$ can be regarded as a subgroup of $GL\,(n + 1, \mathbb{R})$. For, each isometry f in $\mathscr{I}\,(\mathbb{R}^n)$ determines the matrix

$$M(f) = \begin{bmatrix} A(f) & b \\ 0 & 1 \end{bmatrix},$$

where $b = f(0)$ and $A(f)$ is the matrix of the orthogonal transformation $x \to f(x) - b$. Obviously, $M(f)$ belongs to the group $GL\,(n + 1, \mathbb{R})$. Furthermore, the

Euclidean space $\mathbb{R}^n$ can be identified with the subspace $\mathbb{R}^n \times \{1\} \subset \mathbb{R}^{n+1}$ by the isometry $x \mapsto (x, 1)$. With this identification, we have $M(f)x = f(x)$. The correspondence $f \mapsto M(f)$ is a monomorphism of the group $\mathscr{I}so(\mathbb{R}^n)$ into $GL(n+1, \mathbb{R})$. Therefore, $\mathscr{I}so(\mathbb{R}^n)$ can be identified with a subgroup of $GL(n+1, \mathbb{R})$, and thus it receives the relative topology induced from $GL(n+1, \mathbb{R})$. This topology for $\mathscr{I}so(\mathbb{R}^n)$ is referred to as the *metric topology*. Endowed with this topology, the group $\mathscr{I}so(\mathbb{R}^n)$ is called a *Euclidean group*.

Observe that there is a homeomorphism $\mathscr{I}so(\mathbb{R}^n) \to O(n) \times \mathbb{R}^n$ which maps f into $([g], b)$, where $b = f(0)$ and g is the orthogonal transformation $x \mapsto f(x) - b$. Clearly, the set M of all orthogonal transformations of $\mathbb{R}^n$ forms a subgroup of the euclidean group $\mathscr{I}so(\mathbb{R}^n)$, and the set N of all translations of $\mathbb{R}^n$ is a normal subgroup of $\mathscr{I}so(\mathbb{R}^n)$. It is also obvious that $M \cong O(n)$ and $N \cong \mathbb{R}^n$ under the above homeomorphism. But, the topological group $\mathscr{I}so(\mathbb{R}^n)$ does not decompose into M and N, since the subgroup $M \subset \mathscr{I}so(\mathbb{R}^n)$ is not normal in general. We say that the group $\mathscr{I}so(\mathbb{R}^n)$ is the *semidirect product of $O(n)$ and $\mathbb{R}^n$*.

Exercises

1. Prove: (a) The topological group $\mathbb{R}$ is isomorphic to the topological group $\mathbb{R}_+$ of positive reals.

   (b) The topological group $\mathbb{R}_0 = \mathbb{R} - \{0\}$ is isomorphic to $\mathbb{S}^0 \times \mathbb{R}_+$.

2. Show that the topological group in Exercise 12.1.4 is not isomorphic to the direct product $\prod \mathbb{R}_r$ of copies of $\mathbb{R}$, one copy for each real number r in $\mathbb{R}$, although the underlying groups of both topological groups are the same.

3. Prove that the factor group $\mathbb{R}^n / \mathbb{Z}^n$ is isomorphic to the direct product of n copies of $\mathbb{S}^1$.

4. If g is a homomorphism of $\mathbb{R}$ into itself, prove that $g(x) = xg(1)$ for every $x \in \mathbb{R}$. Deduce that the group of automorphisms of $\mathbb{R}$ is isomorphic to $\mathbb{R} \times \mathbb{Z}_2$.

5. Prove that the orthogonal group $O(n)$ is isomorphic to the group of all isometries of $\mathbb{S}^{n-1}$ onto itself topologized by the metric

$$\rho(f, g) = \sup \{ \| f(x) - g(x) \| : x \in \mathbb{S}^{n-1} \}.$$

6. (a) Prove that $O(n)$ is homeomorphic to $SO(n) \times \mathbb{Z}_2$. Are they isomorphic as topological groups?

   (b) Prove that $U(n)$ is homeomorphic to $SU(n) \times \mathbb{S}^1$. Are they isomorphic as topological groups?

7. Prove that the subgroup of the topological group in Lemma 12.1.5, which consists of all isometries of X onto itself, is compact.

8. Let $F = \mathbb{R}, \mathbb{C}$ or $\mathbb{H}$. Prove that $Aff_n(F)$ is isomorphic to the subgroup

$$\left\{ \begin{bmatrix} A & v \\ 0 & 1 \end{bmatrix} : A \in GL(n, F) \text{ and } v \in F^n \right\}$$

of $GL(n+1, F)$. Deduce that $Aff_n(F)$ considered as a subgroup of $GL(n+1, F)$ is closed. Is it compact?

Chapter 13
Transformation Groups

The study of the symmetries of geometric objects has interested generations of mathematicians for hundreds of years, and groups are intended to analyze symmetries of such objects. In fact, symmetry is merely a self-equivalence of the object, and groups occur concretely in most instances as a family of self-equivalences (automorphisms) of some object such as a topological space, a manifold, a vector space, etc. The elements of such groups are referred to as transformations. We discuss the rudiments of the theory of topological transformation groups in the first section. The second section concerns with geometric motions of the Euclidean spaces. We give here a geometric meaning to the notion of "proper rigid motions" of $\mathbb{R}^n$; in particular, we define rotations of $\mathbb{R}^n$ and show that each element of the group $SO(n)$ determines a rotation (in the new sense).

13.1 Group Actions

Definition 13.1.1 Let G be a topological group and X a space. A *left action of G on X* is a continuous map $\phi: G \times X \to X$ such that

(a) $\phi(g, \phi(h, x)) = \phi(gh, x)$ for all $g, h \in G$ and $x \in X$,
(b) $\phi(e, x) = x$ for all $x \in X$, where $e \in G$ is the identity element.

The space X, together with a left action ϕ of G on X, is called a *left G-space*, and the triple (G, X, ϕ) is called a *topological transformation group*. Often, we denote the topological transformation group (G, X, ϕ) just by the pair (G, X), regarding ϕ as understood. Then it is convenient to denote the image of (g, x) under ϕ by gx. The conditions (a) and (b) are now expressed as $g(hx) = (gh)x$ and $ex = x$.

There is an analogous notion of *right action* of G on X. This is a continuous map $X \times G \to X$, $(x, g) \mapsto xg$, such that $(xg)h = x(gh)$ and $xe = x$. The space X, together with a right action of G on X, is called a *right G-space*. Any right action defines a left action in a canonical way: $gx = xg^{-1}$, and vice versa. Thus, for most purposes, the choice of left or right action is a matter of taste. However, we will see

© Springer Nature Singapore Pte Ltd. 2019
T. B. Singh, *Introduction to Topology*,
https://doi.org/10.1007/978-981-13-6954-4_13

later some situations in which an action appears naturally as a right action. Unless we specify otherwise, (for brevity) by an action of G on X, we shall mean a left action.

Let ϕ be an action of a topological group G on a space X. For each $g \in G$, the map $\phi_g : x \mapsto gx$ is a homeomorphism of X with $\phi_{g^{-1}}$ as its inverse. Therefore the associated function $g \mapsto \phi_g$ defines an algebraic homomorphism G into the group $\mathbf{Homeo}(X)$ of all homeomorphisms of X onto itself (with the composition of mappings as the group operation). This homomorphism is continuous relative to the compact-open topology for $\mathbf{Homeo}\,(X)$ (see Theorem 11.2.13). With this topology, the group $\mathbf{Homeo}\,(X)$ becomes a topological group for many nice spaces X, as we shall see below. Accordingly, an action of the topological group G on such spaces X is essentially a homomorphism of G into the topological group $\mathbf{Homeo}(X)$.

Denote the group $\mathbf{Homeo}\,(X)$ with the compact-open topology by $\mathbf{Homeo}\,(X)_{co}$. If X is locally compact Hausdorff, then the multiplication function

$$\mathbf{Homeo}\,(X)_{co} \times \mathbf{Homeo}\,(X)_{co} \longrightarrow \mathbf{Homeo}\,(X)_{co},$$

$(g, h) \mapsto gh$, is continuous. To see this, let (K, U) be a subbasic open set in $\mathbf{Homeo}\,(X)_{co}$ and suppose that $gh \in (K, U)$. Then $K \subseteq X$ is compact, $U \subseteq X$ is open and $gh\,(K) \subseteq U$. So $h\,(K) \subseteq g^{-1}\,(U)$. Since X is locally compact Hausdorff, for each $x \in K$, there exists an open set V_x in X such that $h\,(x) \in V_x \subseteq \overline{V}_x \subseteq g^{-1}\,(U)$ and $\overline{V}_x$ is compact. By compactness of $h\,(K)$, we find finitely many points $x_1, \ldots, x_n$ in K such that $h\,(K) \subseteq \bigcup_1^n V_{x_i} = V$, say. Then $h \in (K, V)$ and $g \in (\overline{V}, U)$, for $\overline{V}$ is compact and contained in $g^{-1}\,(U)$. Clearly, both (K, V) and $(\overline{V}, U)$ are open in $\mathbf{Homeo}\,(X)_{co}$ and $(\overline{V}, U)\,(K, V) \subseteq (K, U)$. Hence the multiplication in $\mathbf{Homeo}\,(X)_{co}$ is continuous.

Unfortunately, for some locally compact Hausdorff spaces X, the inversion function $\iota : g \mapsto g^{-1}$ on $\mathbf{Homeo}\,(X)_{co}$ is not continuous. However, if X is compact Hausdorff, then it is continuous, since $\iota^{-1}\,(K, U) = (X - U, X - K)$. It follows that $\mathbf{Homeo}\,(X)_{co}$ is a Hausdorff topological group whenever X is a compact Hausdorff space. More generally, we prove

Theorem 13.1.2 *Let X be a locally connected, locally compact Hausdorff space. Then the group $\mathbf{Homeo}\,(X)$ with the compact-open topology is a topological group.*

Proof As observed above, the multiplication operation in $\mathbf{Homeo}\,(X)_{co}$ is continuous for all spaces X. So we only need to establish the continuity of the inversion function $\iota : \mathbf{Homeo}\,(X)_{co} \to \mathbf{Homeo}\,(X)_{co}$, $f \mapsto f^{-1}$. Consider a subbasic open set (K, U) of $\mathbf{Homeo}\,(X)_{co}$ with $f^{-1} \in (K, U)$. Then we have $K \subseteq f(U)$. Since $f(U)$ is open and X is locally connected and locally compact, for each $x \in K$, there exists a connected open set V_x such that $x \in V_x \subseteq \overline{V}_x \subseteq f(U)$ and $\overline{V}_x$ is compact. By the compactness of K, we find finitely many points $x_1, \ldots, x_n$ in K such that $K \subseteq \bigcup_{j=1}^n V_{x_j}$. Clearly, f lies in $\bigcap_{j=1}^n \left(\overline{V}_{x_j}, U\right)^{-1} \subseteq (K, U)^{-1}$. Thus it suffices to show that $\left(\overline{V}, U\right)^{-1}$ is open in the compact-open topology for $\mathbf{Homeo}\,(X)$

for every connected open set V such that $\overline{V}$ is compact. Let $g \in \langle V, U \rangle$ be arbitrary. Since $g\left(\overline{V}\right)$ is compact and $g\left(\overline{V}\right) \subseteq U$, there exists an open set W such that $\overline{W}$ is compact and $g\left(\overline{V}\right) \subseteq W \subseteq \overline{W} \subseteq U$. For the same reason, we find another open set O such that $g\left(\overline{V}\right) \subseteq O \subseteq \overline{O} \subseteq W$. Now, choose a point $x_0 \in g(V)$ and set $N = \langle x_0, V \rangle \cap \langle \overline{W} - O, g^{-1}(U) - \overline{V} \rangle$. Clearly, N is an open nbd of g^{-1}. We observe that $N \subseteq \langle \overline{V}, U \rangle^{-1}$. If $h \in N$, then $h(x_0) \in V$ and $h\left(\overline{W} - O\right) \subseteq g^{-1}(U) - \overline{V}$, which implies that $V \subseteq h(O) \cup h\left(X - \overline{W}\right)$. Since V is connected, we have either $V \cap h(O) = \varnothing$ or $V \subseteq h(O)$. As $x_0 \in g(V) \subseteq O$, we have $h(x_0) \in h(O)$. Thus $V \cap h(O) \neq \varnothing$, and therefore $V \subseteq h(O)$. This forces $\overline{V} \subseteq h\left(\overline{O}\right) \subseteq h(U)$, and so $h \in \left(\overline{V}, U\right)^{-1}$. Since $h \in N$ is arbitrary, we have $N \subseteq \left(\overline{V}, U\right)^{-1}$. Accordingly, $\left(\overline{V}, U\right)^{-1}$ is a nbd of g^{-1}, and hence it is open. It follows that $(K, U)^{-1}$ is open in the space $\mathbf{Homeo}\,(X)_{co}$, and therefore ι is continuous. $\diamond$

As noted earlier, if ϕ is an action of a topological group G on the space X, then the algebraic homomorphism $G \to \mathbf{Homeo}\,(X)_{co}, g \mapsto \phi_g$, is continuous. Conversely, if the topology of $\mathbf{Homeo}\,(X)_{co}$ is admissible, then, by Theorem 11.2.13, a continuous homomorphism $\hat{\phi} \colon G \to \mathbf{Homeo}\,(X)_{co}$ defines an action $\phi \colon G \times X \to X$, given by $\phi(g, x) = \hat{\phi}(g)(x)$. Thus, for locally connected, locally compact, Hausdorff spaces X, an action of a topological group G on X may be defined as a homomorphism of G into the topological group $\mathbf{Homeo}\,(X)_{co}$. We remark that if X is simply a locally compact Hausdorff space, then $\mathbf{Homeo}\,(X)$ becomes a topological group in the topology generated by the subbase, which consists of all sets $\langle K, U \rangle$, and their inverses $(K, U)^{-1}$, where K ranges over all compact subsets of X and U over all its open sets. Thus, the above conclusion remains valid for locally compact Hausdorff spaces, too, regarding $\mathbf{Homeo}\,(X)$ with the new topology as a topological group.

Let (G, X) be a topological transformation group. The action of G on X is called *effective* if the homomorphism $G \to \mathbf{Homeo}\,(X)$ is injective, that is, for each $g \neq e$ in G, there exists a point $x \in X$ such that $gx \neq x$. Clearly, an effective topological transformation group on X is (upto isomorphism) an admissible group of continuous functions of X into itself with composition as the group operation. We say that the action of G on X is *free* if $gx = x$ for any $x \in X$ implies $g = e$, that is, each nontrivial element of G moves every point of X. Notice that a free action is effective.

For each $x \in X$, the set $G_x = \{g \in G \mid gx = x\}$ is clearly a subgroup of G, called the *stabilizer* of x. The subgroup G_x is also referred to as the *isotropy subgroup* (or *stability subgroup*) of G at x. The mapping $G \to X, g \mapsto gx$, is clearly continuous. It is immediate from Corollary 4.4.3 that G_x is closed in G if X is Hausdorff. It is also clear that an action of G on X is free if G_x is trivial for every $x \in X$. If $G_x = G$ for all $x \in X$, then we refer to the action of G on X as *trivial*.

Given a point x in the space X, the set $G(x) = \{gx \mid g \in G\}$ is called the *orbit* of x (under G). Clearly, the orbit of x is the image of the set $G \times \{x\}$ under the given action. The action of G on X is called *transitive* if there is only one orbit, the entire space X itself; equivalently, for every pair of points $x, y \in X$, there is a $g \in G$ such that $gx = y$. Clearly, the isotropy subgroup at gx is gG_xg^{-1}, and if X is transitive, then the stability subgroups G_x, $x \in X$, constitute a complete conjugacy class of subgroups of G.

The following terminology is an important concept in the theory of transformation groups: A point x of a G-space X is called a *fixed point* (or *stationary point*) if $G_x = G$. Notice that a point $x \in X$ is a fixed point of $G \Leftrightarrow G(x) = \{x\}$.

Occasionally, the purely algebraic aspect of a transformation group is needed. Recall that any group can be considered as a topological group with the discrete topology and then it is called a *discrete group*. Accordingly, we can also consider the actions of nontopologized groups on topological spaces. Furthermore, given a group G and a set X, an action of G on X means an action in the sense of Definition 13.1.1, where both G and X are given the discrete topology. In this case, we say that X is a *G-set*. Notice that any G-space can be considered as a G-set by forgetting the topology of G and X. As above, an action of a group G on a set X can be interpreted as a homomorphism G into the permutation group **Symm** (X) of X (referred to as a permutation representation of G).

Example 13.1.1 Any topological group G acts on itself by left translation; more specifically, the mapping $G \times G \to G$, $(g, x) \mapsto gx$, defines an action of G on itself. This is obviously free and transitive. Similarly, the right translations $(x, g) \mapsto xg$ determine a (right) free and transitive action of G on itself.

Another action of G on itself is defined by conjugation: $(g, x) \mapsto gxg^{-1}$. The kernel of this action is clearly the center of G. The mapping $(g, x) \mapsto g^{-1}xg$ of $G \times G$ into G defines a right action.

Example 13.1.2 Let H be a subgroup of a topological group G, and G/H be the space of left cosets xH, $x \in G$. The multiplication mapping μ in G induces a map $\theta: G \times G/H \to G/H$, $(g, xH) \mapsto gxH$, so that the diagram

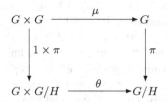

commutes, where $\pi: G \to G/H$ is the natural projection. Hence, we have $\theta^{-1}(O) = (1 \times \pi)(\pi\mu)^{-1}(O)$ for any $O \subseteq G/H$. By Proposition 12.3.1, π is continuous and open in the quotient topology for G/H, and therefore the mapping $1 \times \pi: G \times G \to G \times G/H$ is also continuous and open. It follows that $\theta^{-1}(O)$ is open whenever O is open in G/H. So θ is continuous. It is now easily verified that θ is an action of G on G/H. Clearly, xH is carried to yH by the map $\theta_{yx^{-1}}$, so G is transitive on G/H. Accordingly, the G-space G/H is called a *homogeneous space*. The kernel of this action is $\bigcap_{g \in G} gHg^{-1}$. Notice that Example 13.1.1 is a particular case of the present one.

Example 13.1.3 The natural action of the general linear group $GL\,(n,\mathbb{R})$ on $\mathbb{R}^n$ is defined by $(A, x) \mapsto Ax$, this being multiplication of the $n \times n$ matrix A and the $n \times 1$ matrix $x \in \mathbb{R}^n$ (obtained by writing x as a column matrix). This action is effective, and the origin $0 \in \mathbb{R}^n$ is a fixed point. We observe that there are only two orbits, viz., $\mathbb{R}^n - \{0\}$ and $\{0\}$. If $x \neq 0$ in $\mathbb{R}^n$, then there is a basis $\{b_1, \ldots, b_n\}$ of $\mathbb{R}^n$ with $x = b_1$. Let $\{e_1, \ldots, e_n\}$ be the standard basis of $\mathbb{R}^n$. Then there exist real numbers a_{ij} such that $b_j = \sum_{i=1}^{n} a_{ij} e_i$, $j = 1, \ldots, n$. Clearly, the matrix $A = (a_{ij}) \in GL\,(n,\mathbb{R})$, and $x = Ae_1$. This shows that every $x \neq 0$ is in the orbit of e_1, and thus $\mathbb{R}^n - \{0\}$ is the other orbit.

There are similar actions of $GL\,(n,\mathbb{C})$ on $\mathbb{C}^n$, and $GL\,(n,\mathbb{H})$ on $\mathbb{H}^n$.

Returning to the general case, consider a point x in a G-space X. Then we have the mapping $G \to X, g \mapsto gx$. This is clearly continuous and constant on each coset gG_x, and therefore factors through the coset space G/G_x. Thus there is a continuous mapping $\alpha_x : G/G_x \to X$ given by $\alpha_x\,(gG_x) = gx$. Obviously, α_x is injective and has the image $G\,(x)$. Accordingly, if X is a transitive G-space, then the mapping α_x is a continuous bijection between the coset space G/G_x and X. However, it need not be homeomorphism, as shown by the following examples.

Example 13.1.4 Let $\alpha > 0$ be an irrational number. Then $G = \mathbb{Z}$, the group of integers, acts on the circle $\mathbb{S}^1$ by rotating through $2n\alpha\pi$. By Example 12.3.4, the orbit of $1 \in \mathbb{S}^1$ under G is dense and therefore the orbit of every point $x \in \mathbb{S}^1$ is dense. Consequently, the orbit of no point in $\mathbb{S}^1$ can be a coset space of G, for a discrete subset of $\mathbb{S}^1$ can't be dense in it.

Example 13.1.5 (Irrational Flow) Let $p : \mathbb{R}^1 \to \mathbb{S}^1$ denote the exponential map $r \mapsto e^{2r\pi i}$. Given a fixed number $\alpha \in \mathbb{R}$, we define an action of $G = \mathbb{R}$ on the torus $T = \mathbb{S}^1 \times \mathbb{S}^1$ by

$$(r, (x, y)) \mapsto (p(r)x, p(\alpha r)y).$$

When α is an irrational, the isotropy subgroup at every point of T is the trivial group. For, $(x, y) = (p(r)x, p(\alpha r)y)$ implies that both r and αr are integers. Since α is an irrational number, this is possible only if $r = 0$. Thus the action of G on T in this case is free; it is referred to as an "irrational flow" on the torus. From the multiplication in T, it follows immediately that the orbits through different points of T are translations of the orbit through the point $v = (1, 1)$, and this orbit is the image of the line $L = \{(r, \alpha r) \mid r \in \mathbb{R}\}$ under the canonical map $\pi : \mathbb{R}^2 \to T, (r, s) \mapsto (p\,(r), p\,(s))$. Note that L is a subgroup of $\mathbb{R}^2$. So $\pi\,(L)$ is a subgroup of T. We show that it is dense in T. Observe that $\pi^{-1}\pi\,(L) = L + N = L + H$, where $N = \mathbb{Z} \times \mathbb{Z}$ and $H = \{(0, n - m\alpha) \mid n, m \in \mathbb{Z}\}$. Since $\overline{L + N}$ is a subgroup of $\mathbb{R}^2$ and contains both L and $\overline{H} = \{0\} \times \mathbb{R}$, it contains $L + \overline{H} = \mathbb{R}^2$. It follows that $L + N$ is dense in $\mathbb{R}^2$, and so $\pi\,(L)$ is dense in T. The orbits of this transformation group are referred to as the "flow lines." As the isotropy subgroup of G at the point $v \in T$ is the trivial group, G/G_v is locally compact. But the orbit $G\,(v) = \pi\,(L)$ is not locally compact. For, it is easily checked that $L + N$ does not contain any open ball in $\mathbb{R}^2$; accordingly, every open subset of $\mathbb{R}^2$ contains a limit point of $L + N$, which lies in the complement of $L + N$. It follows that every subset K of $\pi(L)$ with $K^\circ \neq \varnothing$ has limit points in $T - K$. Consequently, the interior of every compact set in $\pi(L)$ is empty, and hence our assertion.

Observe that if α is a rational number, say $\alpha = m/n$, where $(m, n) = 1$, then $\pi(L)$ is closed in T, being the image of I under the continuous map $t \mapsto (p(nt), p(mt))$.

If X is a transitive G-space, then a necessary and sufficient condition for the mapping $\alpha_x : G/G_x \to X$ to be a homeomorphism is that the mapping $G \to X$, $g \mapsto gx$, be open. Another condition for this result is that G be compact and X be Hausdorff. In that case, α_x is closed, and hence a homeomorphism. A transitive G-space X such that α_x is a homeomorphism for every $x \in X$ is called a *topological homogeneous space* (of the topological group G).

We now study two elementary methods of constructing new transformation groups out of old ones. Consider a G-space X and let $H \subset G$ be a subgroup. Then X is also an H-space in an obvious way. In fact, the action of H on X is just the restriction of the action $G \times X \to X$ to $H \times X$. We say that it is obtained by restricting the group G to H. Obviously, the restriction of effective action is effective.

Example 13.1.6 By restricting the natural action of $GL(n, \mathbb{R})$ on $\mathbb{R}^n$ to the orthogonal group $O(n) \subset GL(n, \mathbb{R})$, we obtain an effective action of $O(n)$ on $\mathbb{R}^n$. Unlike Example 13.1.3, there are many orbits in this transformation group. Because orthogonal transformations preserve lengths of vectors, an element of $O(n)$ sends a vector $x \in \mathbb{R}^n$ to a vector of the same length. Conversely, if the nonzero vectors $x, y \in \mathbb{R}^n$ have the same length, then there are orthonormal bases $\{u_i\}$ and $\{v_i\}$ of $\mathbb{R}^n$ with $u_1 = x/\|x\|$ and $v_1 = y/\|y\|$. The linear map $f : u_i \mapsto v_i$ is an orthogonal transformation on $\mathbb{R}^n$. So its matrix A in the standard basis is an element of $O(n)$ and carries x into y, for $\|x\| = \|y\|$. Thus the orbits are concentric spheres (the origin being a sphere of radius zero).

The discussion in the preceding example shows that if $A \in O(n)$ and $x \in \mathbb{S}^{n-1}$, the unit $(n-1)$-sphere, then $Ax \in \mathbb{S}^{n-1}$. Thus there is an action of $O(n)$ on $\mathbb{S}^{n-1}$ obtained by restricting the action of $O(n)$ on $\mathbb{R}^n$ to $\mathbb{S}^{n-1}$. This suggests another method of obtaining transformation groups from a given one.

Definition 13.1.3 Let X be a G-space. A set $Y \subset X$ is called *invariant* under G (or a G-subspace) if $gy \in Y$ for all $g \in G$ and all $y \in Y$.

If X is a G-space and $Y \subseteq X$, we write

$$G(Y) = \{gy \mid g \in G \text{ and } y \in Y\}.$$

With this notation, Y is invariant $\Leftrightarrow G(Y) = Y$. If this is the case, then we have an induced action of G on Y; thus Y becomes a G-space. Note that the orbits in a G-space X are the smallest invariant subsets of X.

Example 13.1.7 The action of the special orthogonal group $SO(n)$ on $\mathbb{S}^{n-1}$ obtained by restricting the standard action of $O(n)$ (by matrix multiplication) is transitive except in the case $n = 1$. For $n = 2$, notice that each element x of $\mathbb{S}^1$ can be written as $x = (\cos\theta, \sin\theta)$ for some θ, and then the matrix $\begin{bmatrix} \cos\theta & -\sin\theta \\ \sin\theta & \cos\theta \end{bmatrix}$ carries the point $1 \in \mathbb{S}^1$ to x. Obviously, all such matrices belong to $SO(2)$, and thus the statement is true in this case. If $n > 2$ and $x \in \mathbb{S}^{n-1}$, then we choose an orthonormal basis $\{f_1 = x, \ldots, f_n\}$ of $\mathbb{R}^n$, and consider the matrix A of this basis relative to the standard basis $\{e_1, \ldots, e_n\}$. It is clear that either A or the matrix B obtained from A by interchanging the last two columns belongs to $SO(n)$, and $x = Ae_1 = Be_1$. This shows that the orbit of e_1 is all of $\mathbb{S}^{n-1}$.

It follows from the preceding example that the orbits of the standard action of $SO(n)$ on $\mathbb{R}^n$, $n > 1$, are concentric spheres.

The above knowledge also helps us to see the connectedness property of $SO(n)$. We show, by induction on n, that these are all connected groups. For $n = 1$, $SO(1) = \{1\}$, which is obviously connected. For $n > 1$, we first establish a relation between $SO(n)$ and $\mathbb{S}^{n-1}$. Since $SO(n)$ acts transitively on $\mathbb{S}^{n-1}$, we have a desired relation once we know the stabilizer of a point in $\mathbb{S}^{n-1}$. Consider the point $e_1 = (1, 0, \ldots, 0)$ of $\mathbb{S}^1$, and suppose that $Ae_1 = e_1$, where $A = (a_{ij}) \in SO(n)$. It is readily checked that $a_{11} = 1$, $a_{i1} = 0 = a_{1i}$ for $2 \le i \le n$. Thus A has the form $A = \begin{bmatrix} 1 & 0 \\ 0 & B \end{bmatrix}$, where $B \in SO(n-1)$. It is also clear that any matrix of this form fixes e_1, and the correspondence $A \leftrightarrow B$ is an isomorphism between the isotropy subgroup of $SO(n)$ at e_1 and $SO(n-1)$. Accordingly, we can identify this subgroup of $SO(n)$ with $SO(n-1)$, and regard $SO(n-1) \subset SO(n)$. Hence the coset space $SO(n)/SO(n-1) \approx \mathbb{S}^{n-1}$. Since the spheres of all positive dimensions are connected and so is $SO(n-1)$ (by induction assumption), Proposition 12.3.4 shows that $SO(n)$ is connected.

In the case of $O(n)$, the coset space $O(n)/O(n-1) \approx \mathbb{S}^{n-1}$, since the action of $O(n)$ on $\mathbb{S}^{n-1}$ is transitive and the isotropy subgroup of $O(n)$ at e_1 is isomorphic to $O(n-1)$. However, the components of $O(n)$ are $SO(n)$ and its complement $\{A \in O(n) \mid \det(A) = -1\}$.

Obviously, $SU(1) = \{1\}$ and $U(1) = \mathbb{S}^1$. As in the case of $SO(n)$, we see that the action of $SU(n)$ $(n > 1)$ on the sphere $\mathbb{S}^{2n-1}$ by matrix multiplication is transitive, and the isotropy subgroup at the point $e_1 = (1, 0, \ldots, 0)$ in $\mathbb{S}^{2n-1}$ is isomorphic to $SU(n-1)$. Consequently, the standard action of $U(n)$ on $\mathbb{S}^{2n-1}$ is also transitive, and the isotropy subgroup of $U(n)$ at e_1 is isomorphic to $U(n-1)$. Therefore $U(n)/U(n-1) \approx \mathbb{S}^{2n-1} \approx SU(n)/SU(n-1)$. Thus the induction applies in both cases, and we conclude by invoking Proposition 12.3.4 that the groups $SU(n)$ and $U(n)$ are connected.

Similarly, we deduce that the symplectic groups $Sp(n)$ are also connected. In fact, $Sp(1) = \mathbb{S}^3$, and it is easily seen that the action of $Sp(n)$ on $\mathbb{S}^{4n-1}$ by matrix multiplication is transitive with the isotropy subgroup at the point $(1, 0, \ldots, 0) \in \mathbb{S}^{4n-1}$

isomorphic to $Sp\,(n-1)$. So $Sp\,(n)\,/Sp\,(n-1)\approx\mathbb{S}^{4n-1}$, and we see by induction that $Sp\,(n)$ is connected for every n.

We now introduce the concept equivalence of G-spaces. Let X and Y be G-spaces with the actions denoted by ϕ and ψ, respectively. A continuous function $f:X\to Y$ is called an *equivariant map* (or a G-*map*) if $f\left(\phi_g\,(x)\right)=\psi_g\,(f\,(x))$ for all $g\in G$ and $x\in X$. If $f:X\to Y$ is an equivarient homeomorphism, then $f^{-1}:Y\to X$ is also equivariant, for $\phi_g\left(f^{-1}\,(f\,(x))\right)=\phi_g\,(x)=f^{-1}\left(\psi_g\,(f\,(x))\right)$. Accordingly, an equivariant map $f:X\to Y$ which is also a homeomorphism is called an *equivalence* of G-spaces, and the actions ϕ and ψ (of G) are said to be equivalent if there exists an equivalence between G-spaces X and Y. Obviously, two actions which are equivalent can not be topologically distinguished from one another, and they are regarded as essentially the same.

If X is a G-space, then the mapping $\alpha_x:G/G_x\to X$, $gG_x\mapsto gx$, is equivariant with respect to the left translation of G on G/G_x for every $x\in X$. Moreover, if X is Hausdorff and a compact group G acts on X transitively, then each α_x is an equivalence of G-spaces. Thus it seems desirable to determine the equivariant maps on the coset spaces of G.

Proposition 13.1.4 *Let G be a topological group and H,K be subgroups of G. Prove that there exists an equivariant map $G/H\to G/K$ if and only if H is conjugate to a subgroup of K.*

Proof Assume that $f:G/H\to G/K$ is a G-map, and $f\,(H)=aK,a\in G$. Then we have $aK=f\,(hH)=hf\,(H)=haK$ for all $h\in H$, by the equivariance of f. So $a^{-1}Ha\subseteq K$. Conversely, suppose that $a^{-1}Ha\subseteq K,a\in G$. Set $f\,(gH)=gaK$ for all $g\in G$. If $g_1H=g_2H$, then $g_1=g_2h$ for some $h\in H$. Also, we have $ha=ak$ for some $k\in K$. Consequently, $g_1aK=g_2haK=g_2akK=g_2aK$, and f is well defined. It is obvious from the definition of f that it is continuous and equivariant.   ◇

Corollary 13.1.5 *If H is a closed subgroup of a compact Hausdorff group G, then each equivariant map $G/H\to G/H$ is a right translation by an element of the normalizer $N\,(H)$ of H in G, and is an equivalence of G-spaces.*

Proof By the proof of the preceding proposition, an equivariant map $f:G/H\to G/H$ is given by $f(gH)=gaH$ for some $a\in G$ satisfying $a^{-1}Ha\subseteq H$. Since G is compact Hausdorff and H is closed in G, $a\in N\,(H)$, by Proposition 12.2.4. So $gaH=gaHa^{-1}a=gHa$, and thus f is the right translation $gH\to gHa$. Clearly, f is an equivalence of the G-space G/H with the inverse $gH\mapsto gHa^{-1}$.   ◇

Orbit Spaces

With each transformation group (G,X), there is associated an equivalence relation. For any two points $x,y\in X$, it is easily checked that the orbits $G\,(x)$ and $G\,(y)$ are either equal or disjoint. Hence there is an equivalence relation $\sim$ on X whose equivalence classes are precisely the orbits of the given action. The relation $\sim$ is given by $x\sim y$ if there exists a $g\in G$ such that $y=gx$.

Definition 13.1.6 The *orbit space* of a G-space X is the quotient space $X/\sim$ whose elements are the orbits of the action, and is denoted by X/G.

Let G be a topological group and $H \subseteq G$ a subgroup. Then the right action $G \times H \to G$, $(g, h) \mapsto gh$, of H on G is free and its orbits are precisely the left cosets of H in G. So the orbit space G/H is the space of left cosets gH. Notice that the notation G/H stands for the same space whether interpreted as an orbit space or as the space of left cosets of H in G. Similarly, the orbit space of the left action $(h, g) \mapsto hg$ of H on G is the space of the right cosets Hg.

It is obvious that if the action is transitive, then the orbit space is a one-point space. The orbit space of each of the transformation groups in Example 13.1.3 is the Sierpinski space. The orbit space of the transformation group in Example 13.1.4 has the trivial topology, and the same is true in the case of Example 13.1.5.

Example 13.1.8 In Example 13.1.6, the mapping $\mathbb{R}^n/O(n) \to [0, \infty)$ which assigns to each sphere its radius is a one-to-one correspondence. This mapping is induced by the continuous open map $\mathbb{R}^n \to [0, \infty)$, $x \mapsto \|x\|$, and hence is a homeomorphism between the orbit space $\mathbb{R}^n/O(n)$ and the ray $[0, \infty)$. We obtain similar results for the actions of $SO(n)$ on $\mathbb{R}^n$ ($n > 1$), $U(n)$ on $\mathbb{C}^n$, and $Sp(n)$ on $\mathbb{H}^n$.

Example 13.1.9 The group $\mathbb{Z}$ of integers acts on $\mathbb{R}^1$ by translation: $(n, x) \mapsto n + x$. The orbit space $\mathbb{R}^1/\mathbb{Z}$ is the circle $\mathbb{S}^1$. The mapping $\mathbb{R}^1 \to \mathbb{S}^1$, $x \mapsto e^{2\pi i x}$, induces a desired homeomorphism.

Example 13.1.10 The group $G = \mathbb{Z} \oplus \mathbb{Z}$ acts on $\mathbb{R}^2$ by $\phi_{(m,n)} : (x, y) \mapsto (m + x, n + y)$. The orbit space $\mathbb{R}^2/G$ is homeomorphic to the torus. The group G may be considered as the group of isometries of $\mathbb{R}^2$ generated by two translations $\lambda_{(1,0)}$ and $\lambda_{(0,1)}$.

Example 13.1.11 The group $G = \mathbb{Z}_2$ acts freely on $\mathbb{S}^n$ by the antipodal map and the orbit space $\mathbb{S}^n/G$ is the real projective space $\mathbb{R}P^n$ (cf. Example 6.1.6).

Example 13.1.12 Let p and q be relatively prime integers, and regard $\mathbb{S}^3$ as the subspace $\{(z_0, z_1) \mid |z_0|^2 + |z_1|^2 = 1\} \subset \mathbb{C}^2$. Then the map $h : \mathbb{S}^3 \to \mathbb{S}^3$ defined by $h(z_0, z_1) = \left(e^{2\pi i/p} z_0, e^{2q\pi i/p} z_1\right)$ is a homeomorphism with $h^p = 1$. Thus h generates a cyclic group of order p and we have a free action of $G = \mathbb{Z}_p$ on $\mathbb{S}^3$ given by $g(z_0, z_1) = h^n(z_0, z_1)$, where g is the residue class of the integer n modulo p. The orbit space of the transformation group $(G, \mathbb{S}^3)$ is called a *lens space* and is denoted by $L(p, q)$.

Returning to the general discussion, consider a G-space X. Then the natural map $\pi : X \to X/G$, $x \mapsto G(x)$, is usually referred to as the *orbit map*. This is continuous, by the definition of the topology for X/G. Moreover, we have

Proposition 13.1.7 *For any G-space X, the orbit map $\pi : X \to X/G$ is open.*

Proof Let $U \subseteq X$ be open. If ϕ denotes the action of G on X, then $gU = \phi_g(U)$ is open in X for every $g \in G$. So $\pi^{-1}\pi(U) = \bigcup_{g \in G} gU$ is open in X. By the definition of the topology on X/G, $\pi(U)$ is open, and π satisfies the desired condition.   ◇

It follows from the preceding proposition that certain topological properties of a G-space X, for example, compactness, connectedness, local compactness, local connectedness, being a first or second countable space, etc. are transmitted to the orbit space X/G. For the actions of compact groups G, some more properties of X such as Hausdorffness, regularity, complete regularity, normality, local compactness, paracompactness, etc. are also transmitted to X/G. This can be seen from

Theorem 13.1.8 *Let G be a compact group, and X be a G-space. Then the orbit map $\pi \colon X \to X/G$ is proper.*

Proof Since G is compact, so is each orbit $G(x)$, $x \in X$. By Theorem 5.5.2, it suffices to prove that π is a closed mapping. For any $A \subseteq X$, we have $\pi^{-1}\pi(A) = G(A)$. So we need to establish that $G(A)$ is closed whenever A is closed in X. Assume that $A \subseteq X$ is closed and let $x \in X - G(A)$ be arbitrary. Then, for each $g \in G$, $g(A) = \{ga \mid a \in A\}$ is closed in X; consequently, $X - g(A)$ is open. By the continuity of the action $G \times X \to X$, we find open nbds N_g of $e \in G$ and U_g of x such that $N_g(U_g) \subseteq X - g(A)$, where $N_g(U_g)$ denotes the image of $N_g \times U_g$ under the action. This implies that U_g does not meet $N_g^{-1} g(A)$. Since G is compact, the open covering $\{N_g^{-1} g \mid g \in G\}$ has finite subcovering. Accordingly, there exist finitely many elements $g_1, \ldots, g_n$ in G such that $G = \bigcup_{i=1}^{n} N_{g_i}^{-1} g_i$. Put $V = \bigcap_{i=1}^{n} U_{g_i}$. Then V is a nbd of x with $V \cap G(A) = \varnothing$. Thus we see that $X - G(A)$ is a nbd of x, and hence $G(A)$ is closed, as desired.   ◇

We end this section by showing that it is sufficient for most purposes to consider only effective actions. If ϕ is an action of G on X and K is the kernel of the associated homomorphism $\hat{\phi} \colon G \to \mathbf{Homeo}(X)$, then G/K acts on X in a canonical way. In fact, we define $\psi \colon G/K \times X \to X$ by setting $\psi(gK, x) = \phi(g, x)$. It is readily verified that ϕ is well defined, and satisfies the conditions (a) and (b) of Definition 13.1.1. To check the continuity of ψ, we note that $\phi = \psi \circ (\pi \times 1)$, where $\pi \colon G \to G/K$ is the natural projection, and 1 is the identity map on X. Consequently, $\psi^{-1}(U) = (\pi \times 1)\phi^{-1}(U)$ for every $U \subseteq X$. Since π is open, it follows that $\psi^{-1}(U)$ is open whenever $U \subseteq X$ is open. So ψ is continuous, and defines an action of G/K on X. This canonical action of G/K on X is clearly effective. The subgroup $K \subseteq G$ is referred to as the "kernel of the action ϕ." We note that if X is Hausdorff, then K is a closed normal subgroup of G. For, if $\langle g_\alpha \rangle$ is a net in K converging to g in G and $x \in X$, then $g_\alpha x \to gx$, by the continuity of the action. So $gx = x$ for all $x \in X$ whence $g \in K$. Also, notice that the kernel K of the action is just $\bigcap_{x \in X} G_x$.

We further show that the orbit space X/G is homeomorphic to the orbit space of the action of G/K on X. To this end, let N be a normal subgroup of G. Then, by restricting the action of G to N, we can consider X as an N-space. Also, G/N

equipped with the quotient structure is a topological group. We observe that there is an induced action of G/N on the orbit space X/N. To see this, we define a function $\theta: G/N \times X/N \to X/N$ by $\theta(gN, N(x)) = N(gx)$. Clearly, θ is single-valued, and we have the commutative diagram

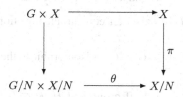

Since the mappings $G \times X \to X$ and $X \to X/N$ are continuous, and the map $G \times X \to G/N \times X/N$ is open and surjective, we see that θ is continuous. It is routine to verify that θ satisfies the conditions of Definition 13.1.1, and thus defines an action. This is referred to as an induced action of G/N on the orbit space X/N. We prove the following

Proposition 13.1.9 *Let X be a G-space and N a normal subgroup of G. Then the orbit space of X/N under the induced action of G/N is homeomorphic to the orbit space X/G.*

Proof Denote the orbit of $N(x)$ under G/N by $[N(x)]$, and define a map $\xi: \frac{X/N}{G/N} \to X/G$ by $\xi([N(x)]) = G(x)$. It makes the diagram

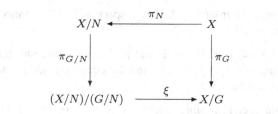

commutative, where π_N, π_G and $\pi_{G/N}$ are all natural projections. Since these maps are continuous, open surjections, it follows that ξ is continuous and open. It is obvious that ξ is a bijection and, therefore, a homeomorphism. $\diamond$

Exercises

1. Let G be a topological group and X be a G-space. For $H \subseteq G$ and $A \subseteq X$, set $H(A) = \{ha \mid h \in H, a \in A\}$. Prove:

   (a) If $H \subseteq G$ and $A \subseteq X$ are open, then $H(A)$ is open in X.

   (b) If $H \subseteq G$ is compact and $A \subseteq X$ is closed, then $H(A)$ is closed in X.

   (c) If $H \subseteq G$ and $A \subseteq X$ are compact, then $H(A)$ is compact.

2. If G is a compact Hausdorff group, prove that every nbd U of e in G contains a nbd V of e which is invariant under conjugation.

3. Let X be a G-space. Prove that $G_{gx} = gG_xg^{-1}$ for every $g \in G$.

4. Let X be a transitive G-set and $x, y \in X$. Prove that

   (a) $G_x = G_y \Leftrightarrow$ there exists an equivariant bijection $f : X \to X$ such that $f(x) = y$, and

   (b) the quotient group $N(G_x)/G_x$ is isomorphic to the group of all equivariant bijections $X \to X$.

5. Let G be a compact Hausdorff group and $H, K \subset G$ be closed subgroups. If there exist equivariant maps $G/H \to G/K$ and $G/K \to G/H$, prove that each is an equivalence and H is conjugate to K.

6. If G is a compact Hausdorff group and $H \subset G$ is a closed subgroup, show that the topological groups **Homeo**$^G (G/H)_{co}$ of all self-equivalences of the G-space G/H and $N(H)/H$ are isomorphic.

7. Prove that the action $(gH, aH) \mapsto gaH$ of $N(H)/H$ on G/H is free.

8. Find actions of $\mathbb{Z}_2$ on the torus such that the orbit space is (a) sphere, (b) cylinder, (c) Klein bottle and (d) torus.

9. Show that $\mathbb{C}P^n$ (resp. $\mathbb{H}P^n$) is the orbit space of a free action of $\mathbb{S}^1$ (resp. $\mathbb{S}^3$) on $\mathbb{S}^{2n+1}$ (resp. $\mathbb{S}^{4n+3}$).

10. Determine the topology of the orbit space of an irrational flow on the torus.

11. Let G be connected group and X be a G-space. If X/G is connected, show that X is also connected.

12. Suppose that X and Y are G-spaces and $f : X \to Y$ is an equivariant map. Show that, for every $x \in X$, $G_x \subseteq G_{f(x)}$ and the equality holds if and only if f is injective on the orbit $G(x)$.

13. Let G be a topological group, and let $f : X \to Y$ be a continuous G-map. Prove:

    (a) f induces a continuous map $\hat{f} : X/G \to Y/G$.

    (b) If f is an equivalence, then so is $\hat{f}$.

14. Let G be a compact group, and X be a locally compact Hausdorff G-space. If $f : X \to X$ is an equivariant map such that $\pi f = \pi$, where $\pi : X \to X/G$ is the orbit map, show that f is an equivalence.

15. Let X be a G-space and A be an invariant subset of X. Prove that the canonical mapping $A/G \to X/G$ is an embedding.

16. Let G be a compact group and X be a G-space. If A is a closed invariant subset of X, prove that each nbd of A contains a G-invariant nbd. (Note the particular case $A = G(x)$.)

17. Let X be a G-space. A *cross section* for the orbit map $\pi: X \to X/G$ is a continuous right inverse of π. A subset $F \subseteq X$ is called a *fundamental domain* of the G-space X if the restriction of π to F is a bijection between F and X/G.

    Let G be a compact group and X be a Hausdorff G-space. Prove that a cross section for the orbit map $\pi: X \to X/G$ exists if and only if there exists a closed fundamental domain of X. Give an example to show that the condition of compactness on G is essential.

18. Let G be a compact group, and X and Y be G-spaces. Suppose that F is a closed fundamental domain of X, and $f': F \to Y$ is a continuous map such that $G_x \subseteq G_{f'(x)}$ for all $x \in F$. Then there is a unique equivariant map $f: X \to Y$ such that $f|F = f'$.

13.2 Geometric Motions

As we know, the position of a moving rigid body is completely determined by describing the location of one of its particles and the orientation of the body. To find the change in the orientation of a moving rigid body, one of the several ways is to study the change in a cartesian set of coordinate axes fixed in the body relative to the coordinate axes parallel to the reference frame fixed in the space and with the same origin as that of the body set of axes. It is obvious that the distance between any pair of particles of the body remains constant throughout the motion of a rigid body (by the definition of a rigid body) and each particle of the body traverses a continuous curve (path). Moreover, any such motion occurs over a period of time. A rigid motion of a Euclidean space (of dimension ≥ 3) is an abstraction of the motion of a rigid body. Accordingly, we introduce the following

Definition 13.2.1 A *proper rigid motion* of the Euclidean space $\mathbb{R}^n$ is a function ϕ from the interval $[0, 1] \subseteq \mathbb{R}$ to the group $\mathscr{I}so(\mathbb{R}^n)$ of isometries of $\mathbb{R}^n$ such that $\phi(0)$ is the translation by $\phi(0)(0)$ and, for each $x \in \mathbb{R}^n$, the mapping $[0, 1] \to \mathbb{R}^n$, $t \mapsto \phi(t)(x)$, is continuous.

Since the elements of $\mathscr{I}so(\mathbb{R}^n)$ are continuous functions $\mathbb{R}^n \to \mathbb{R}^n$, it can also be assigned the relative topology induced from the pointwise topology for $C(\mathbb{R}^n, \mathbb{R}^n)$. It is immediate that ϕ is continuous in this topology for $\mathscr{I}so(\mathbb{R}^n)$. We show that the metric topology of the group $\mathscr{I}so(\mathbb{R}^n)$ (see Sect. 12.4) coincides with the pointwise topology. Recall that, if f is an isometry on $\mathbb{R}^n$, then $f(x) - f(0)$ is an orthogonal linear map. So there exists a $(n \times n)$ matrix $\left(a_{ij}^f\right)$ such that $f(x) - f(0) = \left(a_{ij}^f\right)x$. Writing $f(0) = \left(c_i^f\right)$, etc., the metric on $\mathscr{I}so(\mathbb{R}^n)$ is given by

$$d(g, f) = \sqrt{\sum_{i,j} \left(a_{i,j}^g - a_{i,j}^f\right)^2 + \sum_i \left(c_i^g - c_i^f\right)^2}.$$

Let $e_x : \mathscr{I}so\,(\mathbb{R}^n) \to \mathbb{R}^n$, $f \to f(x)$, be the evaluation map at $x \in \mathbb{R}^n$. Then the sets $e_x^{-1}(U)$, where U ranges over open subsets of $\mathbb{R}^n$ and x varies over points of $\mathbb{R}^n$, form a subbase for the pointwise topology for $\mathscr{I}so\,(\mathbb{R}^n)$. If $f \in e_x^{-1}(U)$, then there is a real $\epsilon > 0$ such that $B(f(x); \epsilon) \subseteq U$. Clearly, $d(g, f) < \epsilon/(1 + \|x\|)$ implies that $\|g(x) - f(x)\| \le \|A(g) - A(f)\| \, \|x\| + \|c^g - c^f\| < \epsilon$. So $g(x) \in U$, and it follows that $e_x^{-1}(U)$ contains an open ball about f. Thus the metric topology for $\mathscr{I}so\,(\mathbb{R}^n)$ is finer than the pointwise topology. To see the converse, consider an open ball $B(f; r)$ about f in $\mathscr{I}so\,(\mathbb{R}^n)$. For each $1 \le j \le n$, denote the unit vectors $(0, \ldots 0, \overset{j}{1}, 0, \ldots, 0)$ in $\mathbb{R}^n$ by u_j and the zero vector by u_0. Then for any real $\epsilon > 0$, the set

$$V = \left\{ g \in \mathscr{I}so\,(\mathbb{R}^n) \mid \|g\,(u_j) - f\,(u_j)\| < \epsilon \text{ for all } 0 \le j \le n \right\}$$

is a basic nbd of f in the pointwise topology for $\mathscr{I}so\,(\mathbb{R}^n)$. Clearly, for a fixed j, we have

$$\sqrt{\sum_{i=1}^{n} \left(a_{ij}^g - a_{ij}^f\right)^2} = \left\| (A(g) - A(f))\, u_j \right\|$$
$$\le \left\| (A(g) - A(f))\, u_j + c^g - c^f \right\| + \left\| c^g - c^f \right\|$$
$$= \left\| g(u_j) - f(u_j) \right\| + \left\| g(0) - f(0) \right\|.$$

Therefore V is contained in $B(f; r)$ when $\epsilon = r/\sqrt{4n+1}$, and we see that the pointwise topology for $\mathscr{I}so\,(\mathbb{R}^n)$ is finer than the metric topology. Thus the two topologies for $\mathscr{I}so\,(\mathbb{R}^n)$ are identical, as desired. Combining this result with the fact that a proper rigid motion is continuous in the pointwise topology for $\mathscr{I}so\,(\mathbb{R}^n)$, we obtain

Proposition 13.2.2 *A proper rigid motion of the Euclidean space $\mathbb{R}^n$ is a continuous function ϕ from the interval $[0, 1]$ to the euclidean group $\mathscr{I}so\,(\mathbb{R}^n)$ such that $\phi(0)$ is the translation by $\phi(0)(0)$.*

As an example, the translation by a vector $a \in \mathbb{R}^n$ can be regarded as the rigid motion of $\mathbb{R}^n$ given by $\phi(t)(x) = x + ta$.

Obviously, the elements of the group $\mathscr{I}so\,(\mathbb{R}^n)$ are homeomorphisms of $\mathbb{R}^n$, so it can also be regarded as a subgroup of the topological group **Homeo**$\,(\mathbb{R}^n)_{co}$. However, we don't obtain a new topological group, for the relative topology for $\mathscr{I}so\,(\mathbb{R}^n)$ induced by the compact-open topology on **Homeo**$\,(\mathbb{R}^n)_{co}$ agrees with the metric topology. Indeed, the compact-open topology is finer than the pointwise topology and, as seen above, the latter coincides with the metric topology for $\mathscr{I}so\,(\mathbb{R}^n)$. On the other hand, the metric topology for $\mathscr{I}so\,(\mathbb{R}^n)$ is finer than the compact-open topology, since the evaluation mapping $e : \mathscr{I}so\,(\mathbb{R}^n) \times \mathbb{R}^n \to \mathbb{R}^n$ is continuous in the metric topology on $\mathscr{I}so\,(\mathbb{R}^n)$, being essentially the matrix multiplication.

Rotations

In the literature, a *rotation* of the euclidean space $\mathbb{R}^n$ is defined as an orthogonal transformation of $\mathbb{R}^n$ with determinant 1 (i.e., an element of the group $SO(n)$), and the group $SO(n)$ is usually referred to as the "rotation group" of $\mathbb{R}^n$. We introduce below a geometric condition to define the term, and provide a rational explanation for the above usage. Recall that rotation literally means a motion of an object in which all its particles follow circles about a common fixed point or a line. Accordingly, we make the following

Definition 13.2.3 A *rotation* of $\mathbb{R}^n$ is a proper rigid motion ϕ such that a point $x_0 \in \mathbb{R}^n$ remains fixed under every transformation $\phi(t)$, $0 \leq t \leq 1$.

If the fixed point x_0 of the motion is chosen as the origin, then each isometry $\phi(t)$ is an orthogonal transformation, by Lemma 12.4.2, and $\phi(0)$ is the identity map. Since the determinant function is continuous and $\det[\phi(0)] = 1$, each $\phi(t)$ is a special orthogonal transformation. Thus the matrix of $\phi(t)$ relative to an orthonormal basis is an element of $SO(n)$. It follows that a *rotation of $\mathbb{R}^n$ about the origin is a continuous map ϕ from the interval $[0, 1]$ to the topological group $SO(n)$, where $SO(n)$ acts on $\mathbb{R}^n$ by the matrix multiplication and $\phi(0)$ is the identity matrix.*

For the plane $\mathbb{R}^2$, the above definition of rotation carries the usual sense of the term. To see this, suppose that $\phi : [0, 1] \rightarrow SO(2)$ is a continuous function such that $\phi(0)$ is the identity matrix. Since each transformation $\phi(t)$ is distance-preserving, the motion of $\mathbb{R}^2$ determined by ϕ can be described by considering the motion of any one of the points of $\mathbb{R}^2$. Let P be an arbitrary point of $\mathbb{R}^2$ other than the origin. If the transformation $\phi(t)$ takes the point P into a point $P(t)$, then $t \mapsto P(t)$ is a continuous function from $[0, 1]$ to $\mathbb{R}^2$ and $\|P(t)\| = \|P\|$. Also, $P(0) = P$, for $\phi(0)$ is the identity matrix. It follows that the point $P(t)$ moves continuously in a circle starting from its initial position. So ϕ determines a rotation of $\mathbb{R}^2$.

On the other hand, a rotation of the plane $\mathbb{R}^2$ about the origin O (in the usual sense) determines a continuous function of the interval $[0, 1]$ into $SO(2)$. To see this, consider such a rotation through an angle $\theta > 0$. Then, for each $0 < \alpha \leq \theta$, there is an isometry f_α which fixes the origin O and sends a point P ($\neq O$) to the point P' such that the measure of the directed angle from the ray through P to the ray through P' is α. If the point P has the coordinates (x, y) with respect to a cartesian coordinate system, then its image P' under f_α has coordinates $(x \cos \alpha - y \sin \alpha, x \sin \alpha + y \cos \alpha)$. It follows that the transformation f_α is just the left multiplication by the matrix

$$\begin{bmatrix} \cos \alpha & -\sin \alpha \\ \sin \alpha & \cos \alpha \end{bmatrix}$$

which has determinant 1. Thus we have the function $t \mapsto [f_{t\theta}]$ from the interval $[0, 1]$ to $SO(2)$, where the transformation f_0 corresponding to the angle zero is the identity mapping of $\mathbb{R}^2$. Clearly, this function is continuous.

The next theorem shows that every element of $SO(n)$ determines a rotation of $\mathbb{R}^n$ in the sense described above.

Theorem 13.2.4 *For every n, the group $SO(n)$ is path-connected.*

Proof The proposition is proved by induction on n. It is known that $SO(1) = \{1\}$ and $SO(2) \cong \mathbb{S}^1$, which are obviously path-connected. Assume that $n > 1$ and $SO(n-1)$ is path-connected. Let $A \in SO(n)$ be arbitrary. It suffices to show that there is a path in $SO(n)$ joining the identity matrix I and A. Let $e_1, \ldots, e_n$ be the standard basis of $\mathbb{R}^n$. If $Ae_1 = e_1$, then $A = \begin{bmatrix} 1 & 0 \\ 0 & A_1 \end{bmatrix}$, where $A_1 \in SO(n-1)$. By our induction assumption, there is a path in $SO(n-1)$ joining I and A_1. The composition of this path with the embedding $SO(n-1) \to SO(n)$, $M \mapsto \begin{bmatrix} 1 & 0 \\ 0 & M \end{bmatrix}$, gives us a desired path in $SO(n)$. So assume that $Ae_1 \neq e_1$. Then the vectors e_1 and Ae_1 determine a plane $\mathscr{P}$. Since $\|Ae_1\| = \|e_1\| = 1$, and $SO(2)$ acts transitively on $\mathbb{S}^1$, there exist matrices $B \in SO(n)$ and $C \in O(n)$ such that $Be_1 = Ae_1$ and

$$
C^{-1}BC = \begin{bmatrix}
\cos\theta & -\sin\theta & 0 & \cdots & 0 \\
\sin\theta & \cos\theta & 0 & \cdots & 0 \\
0 & 0 & 1 & \cdots & 0 \\
\vdots & \vdots & \vdots & \ddots & \vdots \\
0 & 0 & 0 & \cdots & 1
\end{bmatrix}.
$$

Notice that the vectors $Be_2, \ldots, Be_n$ are all orthogonal to $Be_1 = Ae_1$ and $Ae_2, \ldots, Ae_n$ are all orthogonal to Ae_1. Accordingly, they all lie in the $(n-1)$-dimensional euclidean subspace of $\mathbb{R}^n$ orthogonal to Ae_1. The transformation $Be_i \mapsto Ae_i$ is obviously orthogonal and if D denotes the matrix of this transformation relative to the ordered basis $Be_1, \ldots, Be_n$, then D has the form $D = \begin{bmatrix} 1 & 0 \\ 0 & D_1 \end{bmatrix}$ and $BD = A$. So $\det(D) = +1$; consequently, $D_1 \in SO(n-1)$. By our induction assumption, there is a path in $SO(n-1)$ joining I and D_1, and hence a path in $SO(n)$ joining I and D. Since the multiplication in $SO(n)$ is continuous, we have a path in it connecting B to $BD = A$. Moreover, there is a path in $SO(n)$ connecting I to B. To see this, put

$$
B_t = C \begin{bmatrix}
\cos t\theta & -\sin t\theta & 0 & \cdots & 0 \\
\sin t\theta & \cos t\theta & 0 & \cdots & 0 \\
0 & 0 & 1 & \cdots & 0 \\
\vdots & \vdots & \vdots & \ddots & \vdots \\
0 & 0 & 0 & \cdots & 1
\end{bmatrix} C^{-1}
$$

for each $0 \le t \le 1$. Then $t \mapsto B_t$ is a path in $SO(n)$ connecting I to B. The product of this path and the path connecting B to A is a path in $SO(n)$ connecting I to A. $\diamond$

We conclude this subsection with the definition of resultant of two successive rotations of $\mathbb{R}^n$ with the origin as their fixed points: Suppose that $\phi\colon [0, 1] \to SO(n)$ and $\psi\colon [0, 1] \to SO(n)$ are two successive rotations of $\mathbb{R}^n$. Then the resultant of ϕ and ψ is the map $\rho\colon [0, 1] \to SO(n)$ defined by

$$\rho(t) = \begin{cases} \phi(2t) & \text{for } 0 \le t \le 1/2, \text{ and} \\ \psi(2t - 1) \cdot \phi(1) & \text{for } 1/2 \le t \le 1. \end{cases}$$

Reflections

It is noteworthy that rotations and translations in $\mathbb{R}^3$ are physically performable motions. On the other hand, there are rigid motions of $\mathbb{R}^3$ which are not physically performable. To see such an example, consider a hyperplane $H = \{x \in \mathbb{R}^n \mid \langle x, u \rangle = p\}$ (i.e., a translation of an $(n-1)$-dimensional subspace) in $\mathbb{R}^n$, where $\|u\| = 1$. If $x \in \mathbb{R}^n$, then the projection of x into H is a vector v such that $v + pu$ lies in H and $x = v + (p + q)u$ for some $q \in \mathbb{R}$. We define $\phi(x)$ by $\phi(x) = x - 2qu$. Since $\langle v, u \rangle = 0$ and $\langle x, u \rangle = p + q$, we have $\phi(x) = x - 2(\langle x, u \rangle - p)u$. It is easily verified that ϕ is an isometry of $\mathbb{R}^n$ onto itself, called the *reflection* in the plane H. The reflection through the line $\mathbb{R}u$ is given by $\psi(x) = x - 2v = x - 2(x - (p + q)u) = -x + 2(p + q)u = -\phi(x) + 2pu$. See Fig. 13.1.

If the hyperplane H passes through the origin (i.e., $p = 0$), then $\mathbb{R}^n = H \oplus \mathbb{R}u$, where u is the unit vector orthogonal to H. In this case, reflection in the hyperplane H is the transformation $\phi\colon (v, w) \mapsto (v, -w)$. Clearly, ϕ is an orthogonal transformation of $\mathbb{R}^n$. If $\{u_1, u_2, \ldots, u_{n-1}\}$ is an orthonormal basis of H, then the matrix of the reflection ϕ through H in the basis $\{u_1, u_2, \ldots, u_{n-1}, u\}$ is given by

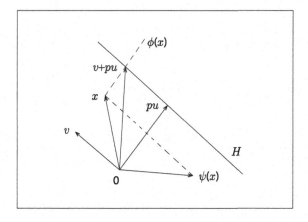

Fig. 13.1 Reflection in a hyperplane

$$A = \begin{bmatrix} 1 & 0 & \cdots & 0 & 0 \\ 0 & 1 & \cdots & 0 & 0 \\ \vdots & \vdots & \ddots & \vdots & \vdots \\ 0 & 0 & \cdots & 1 & 0 \\ 0 & 0 & \cdots & 0 & -1 \end{bmatrix}$$

and the matrix of the reflection ψ through $\mathbb{R}u$ is $-A$. Let $\{e_1, \ldots, e_n\}$ be the standard basis of $\mathbb{R}^n$ and B be the matrix of the basis $\{u_i\}$ with respect to $\{e_i\}$. Then $B \in O(n)$ and the matrix of the reflection ϕ in the basis $\{e_i\}$ is the matrix BAB^{-1}. Conversely, any such matrix fixes the orthogonal complement of the vector Be_n, and can be visualized as a reflection through this hyperplane. It is obvious that a reflection defines a $\mathbb{Z}_2$-action on $\mathbb{R}^n$ with every point of the hyperplane being fixed.

The composition of a reflection through a line and a translation along that line is called a *glide reflection*. Notice that reversing the order of combining gives the same result. We may consider a reflection a special case of glide reflection, where the translation vector is the zero vector. In the euclidean plane R^2, the isometry $(x, y) \mapsto (x + 1, -y)$ consists of the reflection on the x-axis, followed by translation of one unit parallel to it. The isometry group generated by a glide reflection in $\mathbb{R}^2$ is an infinite cyclic group. It fixes a system of parallel lines. Observe that combining two equal glide reflections gives a pure translation with a translation vector that is twice that of the glide reflection, so the even powers of the glide reflection form a translation group.

Chapter 14
The Fundamental Group

A standard problem in topology is the classification of spaces and continuous functions up to homeomorphisms and there are limited tools in general topology to deal with it. For example, one would like to know whether or not euclidean spaces of different dimensions are homeomorphic; alternatively, one may be interested in finding a topological property which can be used to distinguish between a circular disc and an annulus (i.e., a disc with a hole). Notice that topological properties we have studied thus far are not helpful in answering these questions. A solution to such a problem is generally obtained by converting the problem into algebraic questions in such a way that the process always assigns isomorphic groups, rings, etc., to homeomorphic spaces, that is, the associated algebraic structure is a topological invariant. Such an invariant, called the *Fundamental Group* or *Poincaré Group* of the space, was first defined by the great French mathematician Henri Poincaré in 1895. It will be treated in the present chapter.

The construction of the fundamental group of a space is based on the intuitive idea of continuous deformation of a curve. We formalize this concept as "homotopy" in Sect. 14.1 and also study the related notion of deformation of spaces, that is, "homotopy type" of a given space. In Sect. 14.2, we construct the fundamental group and discuss its properties. The third section concerns with computation of the fundamental group of the n-spheres $\mathbb{S}^n$ and some other spaces such as the Möbius band, torus, etc. The knowledge of the fundamental group, particularly of $\mathbb{S}^1$ will be applied to prove a few classical results. In Sect. 14.4, we turn to certain concepts in group theory, which are not generally studied at the undergraduate level. Then, in Sect. 14.5, we return to a powerful theorem about fundamental groups, proved independently by H. Seifert (1931) and E. R. van Kampen (1933). This enables us to compute the fundamental groups of a number of spaces such as the figure 8 space, Klein bottle, etc.

© Springer Nature Singapore Pte Ltd. 2019
T. B. Singh, *Introduction to Topology*,
https://doi.org/10.1007/978-981-13-6954-4_14

14.1   Homotopic Maps

Definition 14.1.1 Let X and Y be topological spaces. Two continuous maps f, g: $X \to Y$ are called homotopic, denoted by $f \simeq g$, if there exists a continuous function

$$H: X \times I \to Y, \quad I = [0, 1],$$

such that $H(x, 0) = f(x)$, $H(x, 1) = g(x)$ for every $x \in X$. The map H is referred to as a *homotopy* connecting f and g, and we often write $H: f \simeq g$.

Given a homotopy $H: X \times I \to Y$, observe that, for each $t \in I$, there is a continuous map $h_t: X \to Y$ defined by $h_t(x) = H(x, t)$ for all $x \in X$. Accordingly, a homotopy connecting f and g gives a one-parameter family of continuous maps $h_t: X \to Y$ indexed by the numbers $t \in I$ with $h_0 = f$ and $h_1 = g$. If we consider the parameter t as representing time, then, as time varies (over I), the point $h_t(x)$, for each $x \in X$, moves continuously from $f(x)$ to $g(x)$. Hence, a homotopy from f to g describes the intuitive idea of continuously deforming f into g, and h_t may be thought of as describing the deformation at time t. If the space $C(X, Y)$ of all continuous maps $X \to Y$ is assigned the compact-open topology, then the function $I \to C(X, Y)$, $t \mapsto h_t$ becomes continuous, and the maps f and g lie in a path component of $C(X, Y)_{co}$. Conversely, by Theorem 11.2.9, each continuous function $h: I \to C(X, Y)_{co}$ defines a homotopy between $h(0)$ and $h(1)$ if X is locally compact Hausdorff. Thus, in this case, two continuous maps $f, g: X \to Y$ are homotopic if and only if they belong to the same path component of the space $C(X, Y)_{co}$. We remark that this is true even for k-spaces X (by Proposition 5.4.13 and Exercise 11.2.24).

Example 14.1.1 Let $Y \subseteq \mathbb{R}^n$ be a convex space and $f, g: X \to Y$ be two continuous maps. Then, $H: X \times I \to Y$, defined by $H(x, t) = (1 - t) f(x) + t g(x)$, is a homotopy connecting f to g. 　　　　　　　　　　　　　　　　　　●

Given two continuous maps $f, g: X \to Y$, it is not always true that f and g are homotopic.

Example 14.1.2 Let $X = Y = \mathbb{S}^n$ (the n-sphere), f be the identity map, and g a constant map. Then, f and g are not homotopic. For $n = 1$, this claim will be proved later but, for $n > 1$, its proof requires more knowledge of algebraic topology than we intend to introduce in this book.

The problem whether or not two given maps $f, g: X \to Y$ are homotopic is actually a special case of the extension problem. For, if we define a map

$$h: Z = (X \times \{0\}) \cup (X \times \{1\}) \to Y$$

by setting $h(x, 0) = f(x)$, $h(x, 1) = g(x)$ for every $x \in X$, then h is continuous, and an extension H of h over $X \times I$ is a homotopy between f and g, and vice versa. So, $f \simeq g$ if and only if h has an extension over $X \times I$.

Proposition 14.1.2 *Let X and Y be spaces, and $C(X, Y)$ the set of all continuous maps $X \to Y$. The homotopy relation $\simeq$ on $C(X, Y)$ is an equivalence relation.*

Proof Reflexivity: Given a continuous map $f: X \to Y$, define $H: X \times I \to Y$ by $H(x, t) = f(x)$ for every $x \in X$ and $t \in I$. It is obvious that $H: f \simeq f$.

Symmetry: Suppose that $F: f \simeq g$, where $f, g: X \to Y$ are continuous. Define $H: X \times I \to Y$ by $H(x, t) = F(x, 1 - t)$ for every $x \in X$ and $t \in I$. Then, $H: g \simeq f$.

Transitivity: Suppose that $f, g, h \in C(X, Y)$ and $F: f \simeq g$ and $G: g \simeq h$. Consider the map $H: X \times I \to Y$ defined by

$$H(x, t) = \begin{cases} F(x, 2t) & \text{for } 0 \le t \le 1/2; \\ G(x, 2t - 1) & \text{for } 1/2 \le t \le 1. \end{cases}$$

By the gluing lemma, we see that H is continuous and it is trivial to check that $H: f \simeq h$.   ◇

Definition 14.1.3 If $f: X \to Y$ is continuous, then its equivalence class $[f] = \{g: X \to Y \mid g$ is continuous and $g \simeq f\}$ is referred to as the *homotopy class* of f.

We will denote by $[X; Y]$ the totality of $[f]$, where $f \in C(X, Y)$. If X is a k-space and $C(X, Y)$ is given the compact-open topology, then $[f]$ is just the path component of f.

Homotopy behaves well with respect to composition of maps in the sense that if $f, g: X \to Y$ are homotopic, then $hf \simeq hg$ for any continuous map $h: Y \to Z$, and $fk \simeq gk$ for any continuous map $k: Z \to X$. In fact, if $H: f \simeq g$ is a homotopy, then $h \circ H: hf \simeq hg$ and $H \circ (k \times 1): fk \simeq gk$, where 1 is the identity map on I. An easy consequence of this is

Proposition 14.1.4 *Composites of homotopic maps are homotopic.*

Proof Suppose that $f_0, f_1: X \to Y$, and $g_0, g_1: Y \to Z$ are homotopic maps. Then, $g_0 f_0 \simeq g_0 f_1$ and $g_0 f_1 \simeq g_1 f_1$, as observed above. By the transitivity of the relation $\simeq$, we have $g_0 f_0 \simeq g_1 f_1$.   ◇

Definition 14.1.5 A continuous map $f: X \to Y$ is called a homotopy equivalence if there is a continuous map $g: Y \to X$ such that $gf \simeq 1_X$ and $fg \simeq 1_Y$. Such a map g, when it exists, is referred to as the homotopy inverse of f and vice versa.

We say that two spaces X and Y have *the same* homotopy type, written $X \simeq Y$, if there exists a homotopy equivalence $X \to Y$. In this case, we also say that X and Y are homotopically equivalent. Obviously, a homeomorphism is a homotopy equivalence, and therefore homeomorphic spaces have the same homotopy type. Of course, the relation of homotopy equivalence is cruder than the topological equivalence, yet it is of central importance in topology because the understanding of homotopy types is a base from which more subtle questions involving topological type can be attacked. Properties which are preserved by homotopy equivalences are called homotopy invariants and much of algebraic topology is concerned with the study of homotopy invariants.

Definition 14.1.6 A space X is called contractible if it is homotopically equivalent to the one-point space.

Proposition 14.1.7 *A space X is contractible if and only if the identity map on X is homotopic to a constant map of X into itself.*

Proof Assume that X is contractible and let $Y = \{y_0\}$ be a one-point space such that $X \simeq Y$. Let $f : X \to Y$ be a homotopy equivalence with homotopy inverse $g : Y \to X$. Then, $1_X \simeq gf$, where gf is a constant map. Conversely, suppose that $1_X \simeq c$, where $c : X \to X$ is a constant map. Let $Y = \{x_0\} = c(X)$. Then the inclusion map $i : Y \hookrightarrow X$ and the map $r : X \to Y$ satisfy $ir = c$ and $ri = 1_Y$. So $X \simeq Y$.   ◇

If X is contractible, then a homotopy connecting the identity map and the constant map of X into itself at x_0 is called a *contraction* of X to x_0.

Example 14.1.3 Any convex subspace of $\mathbb{R}^n$ is contractible, in particular, $\mathbb{R}^n$ itself, the unit cube I^n and the unit disc $\mathbb{D}^n$ are contractible.

Example 14.1.4 The punctured sphere $\mathbb{S}^n - \{pt\}$, being homeomorphic to $\mathbb{R}^n$ for $n > 0$, is contractible. For the same reason, a hemisphere is contractible. These examples show that a contractible space need not be convex.

Example 14.1.5 The "comb space"

$$X = (\{1/n \mid n = 1, 2, \ldots\} \times I) \cup (\{0\} \times I) \cup (I \times \{0\})$$

(refer to Example 3.3.3) is contractible. Let $c : X \to X$ be the constant map at the point $x_0 = (0, 0)$, and consider the map $H : X \times I \to X$ given by

$$H((x, y), t) = \begin{cases} (x, (1 - 2t)\, y) & \text{for } 0 \leq t \leq 1/2, \\ (2(1 - t)\, x, 0) & \text{for } 1/2 \leq t \leq 1. \end{cases}$$

By the gluing lemma, H is continuous and we obviously have $H(x, 0) = x$ and $H(x, 1) = x_0$ for every $x \in X$.

From the above examples, it is immediate that the notion of homotopy equivalence is weaker than that of topological equivalence.

Definition 14.1.8 A continuous map $X \to Y$ is called *null* homotopic or *inessential* if it is homotopic to some constant map $X \to Y$.

With this terminology, a space X is contractible if and only if the identity map on X is null homotopic. Note that a contractible space is necessarily path-connected because if H is a contraction of X to x_0, then, for each $x \in X$, $t \mapsto H(x, t)$ is a path in X joining x to x_0. On the other hand, an annulus (i.e., a region in the plane $\mathbb{R}^2$ bounded by two concentric circles of different radii) and the n-sphere $\mathbb{S}^n$, $n > 0$, are examples of path-connected spaces which are not contractible. We will prove the first assertion later; the proof of the other one, as remarked earlier, needs results from algebraic topology, which are beyond the scope of this book. In this vein, we next prove

Proposition 14.1.9 *Two contractible spaces have the same homotopy type.*

Proof Let X and Y be contractible spaces, and P be a one-point space. Then, $X \simeq P$ and $Y \simeq P$. Suppose that $f: X \to P$ and $g: Y \to P$ are homotopy equivalences with homotopy inverses $f': P \to X$ and $g': P \to Y$. Then, $(f'g)(g'f) = f'(gg')f \simeq f'1_P f = f'f \simeq 1_X$ and, similarly, we also have $(g'f)(f'g) \simeq 1_Y$. Thus, the maps $g'f: X \to Y$ and $f'g: Y \to X$ are homotopy inverse of each other and $X \simeq Y$. $\diamond$

Proposition 14.1.10 *Any two continuous maps of a space X into a contractible space Y are homotopic, and if X is also contractible, then any continuous map $X \to Y$ is a homotopy equivalence.*

Proof By Proposition 14.1.7, there exists a constant map $c: Y \to Y$ such that $1_Y \simeq c$. If $f, g: X \to Y$ are continuous maps, then $f = 1_Y f \simeq cf$, and similarly, $g \simeq cg$. Since $cf = cg$ and $\simeq$ is an equivalence relation, we have $f \simeq g$.

To see the last statement, suppose that X is also contractible and let $\phi: X \to Y$ be a continuous map. Then, by Proposition 14.1.9, there exists a homotopy equivalence $\psi: X \to Y$. As above, we have $\phi \simeq \psi$. Now, if $\psi': Y \to X$ is a homotopy inverse of ψ, then we have $\psi'\phi \simeq \psi'\psi \simeq 1_X$ and $\phi\psi' \simeq \psi\psi' \simeq 1_Y$. Thus, ϕ is a homotopy equivalence, as desired. $\diamond$

It follows that $[X; Y]$ is a singleton set when Y is contractible. Also, the identity map on a contractible space Y is homotopic to any constant map of Y into itself, and any two constant maps of Y into itself are homotopic. However, in general, two null homotopic maps need not be homotopic. For, if Y is a disconnected space, and two points y_0, y_1 belong to distinct path components of Y, then the constant maps $x \mapsto y_0$ and $x \mapsto y_1$ of any space X into Y are obviously not homotopic. On the other hand, if f is a path joining y_0 to y_1, then $H(x, t) = f(t)$ for all $x \in X$ is a homotopy between the constant maps at y_0 and y_1. Thus, it is only when Y is path-connected that all null homotopic maps are homotopic.

A useful property of the null homotopic maps of spheres is described by the following.

Theorem 14.1.11 *A continuous map $f: \mathbb{S}^n \to Y$ is null homotopic if and only if f can be continuously extended to the disc $\mathbb{D}^{n+1}$.*

Proof Suppose that $H: f \simeq c$, where $c: \mathbb{S}^n \to Y$ is the constant map at $y_0 \in Y$. Consider the mapping $\phi: \mathbb{S}^n \times I \to \mathbb{D}^{n+1}$ defined by $\phi(x, t) = (1 - t/2)x$. It is clearly an embedding with the image $C = \mathbb{D}^{n+1} - B$, where $B = B(0; 1/2)$. The inverse $\psi: C \to \mathbb{S}^n \times I$ of ϕ is given by $\psi(z) = (z/\|z\|, 2(1 - \|z\|))$. Notice that ψ maps ∂B onto $\mathbb{S}^n \times 1$, and $\mathbb{S}^n$ onto $\mathbb{S}^n \times 0$; accordingly, the composition $H\psi: C \to Y$ agrees with f on $\mathbb{S}^n$, and is constant on ∂B at y_0. So, we can define a continuous map $F: \mathbb{D}^{n+1} \to Y$ by setting $F(B) = y_0$ and $F|C = H\psi$. Thus, f extends to a continuous map over $\mathbb{D}^{n+1}$.

Conversely, suppose that $g: \mathbb{D}^{n+1} \to Y$ is a continuous extension of f. Then, define $H: \mathbb{S}^n \times I \to Y$ by $H(x, t) = g((1 - t)x + tx_0)$, where $x_0 \in \mathbb{S}^n$ is a fixed point. Clearly, $H(x, 0) = g(x) = f(x)$ and $H(x, 1) = g(x_0)$ for all $x \in \mathbb{S}^n$. If $c:$ $\mathbb{S}^n \to Y$ is a constant map at $g(x_0)$, then $H: f \simeq c$, as desired.   ◇

As an immediate consequence of the preceding theorem, we see that any continuous map of $\mathbb{S}^n$ into a contractible space can be continuously extended over $\mathbb{D}^{n+1}$.

Returning to the general discussion, we see that it is required in many cases to deal with continuous maps having special properties, and homotopies deforming one such map to the other in a certain way. For instance, consider the question of distinguishing between an annulus and a circular disc. An intuitive answer to this question is obtained by considering the possibility, or otherwise, of shrinking to a point a loop drawn on the two spaces. A loop such as f on the annulus (Fig. 14.1a) cannot be shrunk to a point of the annulus without going outside of it, while this is possible for any loop g on the circular disc (Fig. 14.1b).

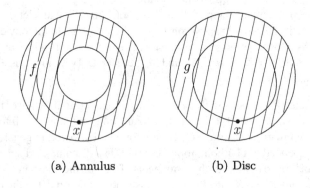

(a) Annulus   (b) Disc

Fig. 14.1 A property of an annulus that distinguishes it from a disc

Technically speaking, by a *loop* in a space X, we mean a path $f: I \to X$ with $f(0) = f(1)$. We also say that f is a *closed path*, and the point $f(0) = f(1)$ is referred to as the *base point of f*. Suppose that the annulus A is bounded by two concentric circles C_1 and C_2 centered at the origin, and having radii 1 and 2, respectively. Consider the loop $f: I \to A$ defined by $f(s) = re^{2\pi is}$, where $1 < r <$ 2. Then, $H: I \times I \to A$ defined by $H(s, t) = re^{2\pi is(1-t)}$ is a homotopy between the loop f and the constant loop based at $(r, 0)$. This example shows that simply asking whether a loop is homotopic to a constant loop is not sufficient to detect holes, and we need a stronger condition than specified in the definition of homotopy. Observe that the terminal points of intermediate paths $s \mapsto H(s, t)$, $t \in I$, in the present case do not stay fixed. Such phenomena are not seen under the following variant of homotopy.

Definition 14.1.12 Let A be a subspace of X, and $f, g: X \to Y$ be continuous maps such that $f|A = g|A$. We say that f is *homotopic to g relative to A* if there exists a continuous map $H: X \times I \to Y$ such that $H(x, 0) = f(x)$, $H(x, 1) = g(x)$ for every $x \in X$, and $H(a, t) = f(a) = g(a)$ for all $a \in A$ and $t \in I$. Such a map H is called *a homotopy from f to g relative to A*, and we write $H: f \simeq g$ rel A to indicate the case.

Example 14.1.6 In Example 14.1.1, if f and g agree on $A \subseteq X$, then the homotopy H is relative to A.

Example 14.1.7 Let $X = Y = \mathbb{D}^2$, $f: X \to Y$ be the identity map and $g: X \to Y$ be the map $z \mapsto -z$. Then, $H: X \times I \to Y$ defined by $H(z, t) = ze^{\pi i t}$ is a homotopy from f to g relative to the origin.

Example 14.1.8 Consider the map $r: X = \mathbb{R}^n - \{0\} \to \mathbb{S}^{n-1}$ defined by $r(x) = x/\|x\|$. If $i: \mathbb{S}^{n-1} \hookrightarrow X$ is the inclusion map, then $1_X \simeq ir$ rel $\mathbb{S}^{n-1}$. Such a homotopy $H: X \times I \to X$ is given by $H(x, t) = (1-t)x + tx/\|x\|$. Also, $X \simeq \mathbb{S}^{n-1}$, since $ri = 1_{\mathbb{S}^{n-1}}$.

Let X and Y be spaces and suppose that $A \subset X$. As in Proposition 14.1.2, we see that the homotopy relative to A is an equivalence relation in the set of all continuous functions $X \to Y$ which agree on A. Moreover, in this terminology, the extended version of Proposition 14.1.4 is

Proposition 14.1.13 *If $f, g: X \to Y$ are homotopic relative to A, $h, k: Y \to Z$ are homotopic relative to B and $f(A) \subseteq B$, then $hf \simeq kg$ rel A.*

The proof is similar to that of Proposition 14.1.4.

The homotopy given in Example 14.1.8 can be seen as a gradual collapsing of the space X onto the subspace $\mathbb{S}^{n-1}$. This sort of homotopy equivalence can thus be visualized geometrically, although homotopy equivalences are hard to visualize, in general. These special maps are introduced in the following.

Definition 14.1.14 Let A be a subspace of X and let $i: A \hookrightarrow X$ be the inclusion map. We say that

(a) A is a *retract* of X if there is a continuous map $r: X \to A$ with $ri = 1_A$. Such a map r is called a *retraction* of X onto A.

(b) A is a *deformation retract* of X if there is a retraction r of X onto A such that $1_X \simeq ir$. A homotopy $H: 1_X \simeq ir$ is called a *deformation retraction* of X onto A.

(c) A is a *strong deformation retract* of X if there is a retraction r of X onto A such that $1_X \simeq ir$ rel A. A homotopy $H: 1_X \simeq ir$ rel A is called a *strong deformation retraction* of X onto A.

The condition of a deformation retraction can be rephrased as follows: A continuous map $H: X \times I \to X$ is a deformation retraction if and only if $H(x, 0) = x$, $H(x, 1) \in A$ for all $x \in X$, and $H(a, 1) = a$ for every $a \in A$. It is a strong deformation retraction if $H(a, t) = a$ for all $t \in I$ and $a \in A$.

Obviously, a one-point subspace of any space is a retract of the space, while a discrete subspace (with more than one point) of a connected space is not. A one-point subspace of a contractible space is a deformation retract of the space. We will see later that $\mathbb{S}^1$ is not contractible, and hence a one-point subspace of $\mathbb{S}^1$ is not its deformation retract. The sphere $\mathbb{S}^{n-1}$, $n \geq 1$, is a strong deformation retract of $\mathbb{R}^n - \{0\}$, as shown by Example 14.1.8. Similarly, it is a strong deformation retraction of the punctured disc $\mathbb{D}^n - \{0\}$. It is clear that if A is a strong deformation retract of X, then A is also a deformation retract of X, and if A is a deformation retract of a space X, then $A \simeq X$. The following examples show that neither of these implications is reversible.

Example 14.1.9 Let X denote the "comb space." We observe that X is not a retract of the unit square I^2; however, these are homotopically equivalent, since both the spaces are contractible. Consider the point $q = (0, 1/2)$. Let $V = X \cap B(q; \epsilon)$, where $\epsilon = 1/4$. If there is a continuous map $r : I^2 \to X$ with $r(q) = q$, then we must have an open ball $B(q; \delta)$ such that $U = I^2 \cap B(q; \delta)$ is mapped into V by r. Since U is connected, $r(U)$ is contained in the set $\{(0, t) \mid 1/4 < t < 3/4\}$, the component of V containing q. However, for n sufficiently large, we have the point $p = (1/n, 1/2)$ in $U \cap X$, which is clearly moved by r. So, r cannot be a retraction of I^2 onto X.

Example 14.1.10 Consider the subspace $Y = \{0\} \times I$ of the "comb space" X. Then, Y is a deformation retract of X, but not a strong deformation retract. The function $H : X \times I \to X$ defined by

$$H((x, y), t) = \begin{cases} (x, (1 - 3t)y) & \text{for } 0 \leq t \leq 1/3, \\ ((2 - 3t)x, 0) & \text{for } 1/3 \leq t \leq 2/3, \\ (0, (3t - 2)y) & \text{for } 2/3 \leq t \leq 1 \end{cases}$$

is a homotopy between 1_X and ir, where $i : Y \hookrightarrow X$ is the inclusion map and $r : (x, y) \mapsto (0, y)$ is the retraction of X onto Y. So, Y is a deformation retract of X. To establish the second claim, assume, on the contrary, that there is a homotopy $F : X \times I \to X$ such that $F(p, 0) = p$, $F(p, 1) \in Y$ for all $p \in X$, and $F(q, t) = q$ for every $q \in Y$ and all $t \in I$. Let q be a point of Y other than the point $(0, 0)$ and let $B(q; \epsilon)$ be an open ball in $\mathbb{R}^2$, which does not meet the set $I \times \{0\} \subset X$. Then, $W = X \cap B(q; \epsilon)$ is a nbd of q in X, and $\{q\} \times I \subset F^{-1}(W)$. By the Tube Lemma 5.1.18, there is an open nbd U of q in X such that $U \times I \subset F^{-1}(W)$. So, for each $p \in U$, $F(\{p\} \times I)$ is contained in the component of W containing p, and this component is the intersection of $B(q; \epsilon)$ with the "tooth" containing p. This contradicts the fact that $F(p, 1) \in Y$ for every $p \in X$, and hence our claim.

The argument given above shows, in particular, that there is no contraction of X to $q = (0, 1)$ leaving the point q fixed, though X is contractible to q.

The simple proof of the following proposition is left to the reader.

Proposition 14.1.15 *A subspace A of a space X is a retract of X if and only if, for any space Y, every continuous map $f : A \to Y$ has a continuous extension over X.*

Notice that A is a retract of X if and only if the identity map on A has an extension over X. The following result shows that any extension question can always be formulated in this particular form.

Proposition 14.1.16 *Let $A \subset X$ be a closed subset and $f : A \to Y$ be a continuous map. Then, f extends over X continuously if and only if Y is a retract of the adjunction space $X \cup_f Y$.*

Proof Let $\pi : X + Y \to X \cup_f Y = Z$ be the quotient map. If $r : Z \to Y$ is a retraction, then the composition $F = r \circ (\pi|X)$ is a continuous extension of f, for $F(x) = r(\pi(x)) = r(f(x)) = f(x)$ when $x \in A$. Conversely, let $F : X \to Y$ be a continuous map such that $F|A = f$. Since Z is the disjoint union of $X - A$ and Y, we can define a function $r : Z \to Y$ by setting

$$r(z) = \begin{cases} F(z) & \text{if } z \in X - A, \text{ and} \\ z & \text{if } z \in Y. \end{cases}$$

Then, the composition $r\pi = g$ is continuous, since $g|X = F$ and $g|Y = 1_Y$. Hence, r is continuous, and it is clear that r is a retraction of Z onto Y. ◇

Exercises

1. Let X be a k-space. Prove that the path components of $C(X, Y)_{co}$ are precisely the homotopy classes.

2. Show that the relation $X \simeq Y$ between topological spaces is an equivalence relation.

3. (a) If $X \simeq Y$ and X is path-connected, show that Y is also path-connected.

   (b) If $X \simeq Y$, show that there is a one-to-one correspondence between their path components.

4. (a) If a continuous map $f : X \to Y$ has a left homotopy inverse g and a right homotopy inverse h, show that f is a homotopy equivalence.

   (b) If one of the continuous maps $f : X \to Y$ and $g : Y \to Z$ and their composition $gf : X \to Z$ are homotopy equivalences, show that the other map is also a homotopy equivalence.

5. Prove: A space X is path-connected $\Leftrightarrow$ every two constant maps of X into itself are homotopic.

6. Let X be a contractible space and Y a path-connected space. Show that any two continuous maps $X \to Y$ are homotopic (and each is null homotopic).

7. (a) If X and Y are contractible, show that $X \times Y$ is also contractible.

   (b) Let X be a contractible space. Show that $X \times Y \simeq Y$ for any space Y.

8. (a) Is the function $H : \mathbb{S}^1 \times I \to \mathbb{S}^1$ defined by $H\left(e^{2\pi \iota s}, t\right) = e^{2\pi \iota s t}$, $(s \neq 1)$, a homotopy between the identity map on $\mathbb{S}^1$ and a constant map?

(b) Suppose that a continuous map $f : \mathbb{S}^1 \to \mathbb{S}^1$ is not homotopic to the identity map. Prove that there is a point $x \in \mathbb{S}^1$ such that $f(x) = -x$.

9. Prove that the identity map on $\mathbb{S}^1$ is homotopic to the antipodal map $x \mapsto -x$. Generalize this result to the sphere $\mathbb{S}^n$, n is odd.

10. If $f, g : X \to \mathbb{S}^n$ are two continuous maps such that $f(x) \neq -g(x)$ for every $x \in X$, show that $f \simeq g$. Deduce that a continuous map $X \to \mathbb{S}^n$ which is not surjective is null homotopic.

11. Let $f : \mathbb{D}^n \to \mathbb{D}^n$ be a homeomorphism such that $f(x) = x$ for every $x \in \mathbb{S}^{n-1}$. Show that there is a homotopy $H : f \simeq 1$ rel $\mathbb{S}^{n-1}$ such that each $h_t : x \mapsto H(x, t)$ is a homeomorphism.

12. If A is a retract of a Hausdorff space X, show that A is closed in X.

13. (a) Show that A is a retract of $X \Leftrightarrow$ there exists a continuous surjection $r : X \to A$ with $r(x) = r(r(x))$ for every $x \in X$.

    (b) Show that a retraction $r : X \to A$ is an identification map.

14. Assume that $X \subset Y \subset Z$. If X is a retract (resp. deformation retract or strong deformation retract) of Y, and Y is a retract (resp. deformation retract or strong deformation retract) of Z, show that X is a retract (resp. deformation retract or strong deformation retract) of Z.

15. If A is a retract (resp. deformation retract or strong deformation retract) of X, and B is a retract (resp. deformation retract or strong deformation retract) of Y, prove that $A \times B$ is a retract (resp. deformation retract or strong deformation retract) of $X \times Y$.

16. Let $A \subset X$ and $f : X \to A$ be a continuous map such that $f|A \simeq 1_A$. If X is contractible, show that A is contractible. Deduce that the retract of a contractible space is contractible.

17. (a) Prove that X is contractible $\Leftrightarrow$ it is a retract of CX, the cone on X.

    (b) Prove that a continuous map $f : X \to Y$ is nullhomotopic $\Leftrightarrow$ f can be continuously extended over CX.

18. (a) Prove that the "center" circle in the Möbius band is its strong deformation retract.

    (b) Show that the circle $\mathbb{S}^1 \times \{0\}$ is a strong deformation retract of the cylinder $\mathbb{S}^1 \times I$.

19. (a) Prove that the disc $\mathbb{D}^n$ is a strong deformation retract of $\mathbb{R}^n$.

    (b) Prove that $(\mathbb{D}^n \times \{0\}) \cup (\mathbb{S}^{n-1} \times I)$ is a strong deformation retract of $\mathbb{D}^n \times I$.

20. Prove that a subspace $A \subset X$ is a deformation retract of X if and only if, for every space Y, each continuous map $f : A \to Y$ can be continuously extended

over X, and any two continuous maps $f, g : X \to Y$ are homotopic whenever $f|A \simeq g|A$.

21. Suppose that A is a strong deformation retract of a compact space X, B is a closed subspace of a Hausdorff space Y, and $f : X \to Y$ is a continuous map such that $f|(X - A)$ is a homeomorphism between $X - A$ and $Y - B$. Show that B is a strong deformation retract of Y.

22. Let $A \subseteq X$ be closed and $f : A \to Y$ be a continuous map. Let T be a locally compact Hausdorff space, and $g : X \times T \to Z$ and $h : Y \times T \to Z$ be continuous maps such that $g(x, t) = h(f(x), t)$ for every $x \in A, t \in T$. Prove that there exists a continuous map $\phi : (X \cup_f Y) \times T \to Z$ such that $(\phi \circ (\pi \times 1))|(X \times T) = g$ and $(\phi \circ (\pi \times 1))|(Y \times T) = h$, where $\pi : X + Y \to X \cup_f Y$ is the quotient map.

23. If $f, g : \mathbb{S}^1 \to X$ are homotopic maps, show that $\mathbb{D}^2 \cup_f X$ and $\mathbb{D}^2 \cup_g X$ have the same homotopy type.

24. • Let X be a triangle (with the interior) having vertices v_0, v_1, v_2. Consider the orientation of the edges in the direction from v_0 to v_1, from v_1 to v_2, and from v_0 to v_2. The quotient space of X obtained by identifying the edges with one another in the given orientation is called the "*dunce cap*" space.

Prove that the "dunce cap" has the homotopy type of a disc (hence, this is a contractible space).

25. (a) Let A be a retract of X and Y a space. Show that, for each continuous map $f : A \to Y$, the subspace Y is a retract of $X \cup_f Y$.

(b) If A is a strong deformation retract of X, show that Y is also a strong deformation retract of $X \cup_f Y$.

26. (a) Let $f : X \to Y$ be a continuous map, and M_f be the mapping cylinder of f. Show that Y, regarded as a subspace of M_f, is a strong deformation retract of M_f.

(b) If f is a homotopy equivalence, show that X is a deformation retract of M_f.

27. Let X and Y be spaces, and λ be the natural injection of Y into $\mathcal{C}(X, Y)_{co}$ (i.e., $\lambda(y) : X \to Y$ is the constant map at y). Show that $\lambda(Y)$ is a retract of $\mathcal{C}(X, Y)_{co}$.

28. Let X be a locally compact Hausdorff contractible space and Y be any space.

(a) Show that $\lambda(Y)$ is a strong deformation retract of $\mathcal{C}(X, Y)_{co}$, where λ is the natural injection of Y into $\mathcal{C}(X, Y)_{co}$.

(b) If $\{x_0\}$ is a strong deformation retract of X, prove that $e_{x_0}^{-1}(y)$ is contractible to $\lambda(y)$ for every $y \in Y$.

14.2 The Fundamental Group

The construction of the fundamental group of a space uses loops in the space. Recall that a loop in a space X is a continuous map $f : I \to X$ with $f(0) = f(1)$. Since I is contractible, any path in X is null homotopic. Therefore, to make better use of paths, we generally consider the homotopies relative to $\{0, 1\}$. A necessary condition for this is that homotopic paths must have the same end points. Intuitively speaking, a homotopy relative to $\{0, 1\}$ between two paths having the same end points deforms one into the other, keeping the end points fixed throughout the deformation (Fig. 14.2).

As remarked in the previous section, "homotopy relative to $\{0, 1\}$" is an equivalence relation on the set of all paths in X having common end points. The resulting equivalence classes are referred to as *homotopy classes or path classes*. If f is a path in X, the homotopy class containing it is denoted by $[f]$. Notice that two paths in the same homotopy class have the same initial point and the same terminal point. So one can define the initial and terminal points of $[f]$ to be those of f.

For $x_0 \in X$, the set of all homotopy classes of loops in X based at x_0 is denoted by $\pi(X, x_0)$. This is the underlying set of the fundamental group of X based at x_0. To turn this set into a group, we must define an operation of "multiplication". This is done via the multiplication of loops, given in

Definition 14.2.1 If f and g are paths in X such that $f(1) = g(0)$, then there is a path $f * g$ in X defined by

$$(f * g)(s) = \begin{cases} f(2s) & \text{for } 0 \leq s \leq 1/2, \\ g(2s - 1) & \text{for } 1/2 \leq s \leq 1. \end{cases}$$

Observe that $f * g$ is just the concatenation of f and g, and it is a path from $f(0)$ to $g(1)$. In particular, if f and g are loops, then $f * g$ is also a loop. So a

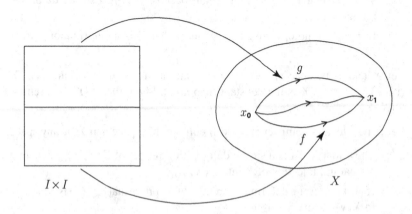

Fig. 14.2 Homotopy relative to $\{0, 1\}$ from a path f to a path g

multiplication can be defined in $\pi(X, x_0)$ by setting $[f] \cdot [g] = [f * g]$. Of course, we must check that this multiplication is well defined. To this end, we prove

Lemma 14.2.2 *Let f, f', g, and g' be paths in X such that $f \simeq f'$ rel $\{0, 1\}$, $g \simeq g'$ rel $\{0, 1\}$ and $f(1) = g(0)$. Then, $f * g \simeq f' * g'$ rel $\{0, 1\}$.*

Proof Since $f'(1) = f(1) = g(0) = g'(0)$, the product $f' * g'$ is defined, and has the same end points as $f * g$. Let $F: f \simeq f'$ and $G: g \simeq g'$ be homotopies relative to $\{0, 1\}$. Then F and G can be glued together to give a homotopy between $f * g$ and $f' * g'$ (see Fig. 14.3). Specifically, we define a mapping $H: I \times I \to X$ by

$$H(s, t) = \begin{cases} F(2s, t) & \text{for } 0 \le s \le 1/2, \\ G(2s - 1, t) & \text{for } 1/2 \le s \le 1. \end{cases}$$

Since $F(1, t) = f(1) = g(0) = G(0, t)$ for all $t \in I$, the gluing lemma applies and we see that H is continuous. We leave it to the reader to verify that $H: f * g \simeq f' * g'$ rel $\{0, 1\}$.   ◇

From the above lemma, it is easily seen that the multiplication in $\pi(X, x_0)$ is well defined. Furthermore, we can prove the following:

Theorem 14.2.3 *The set $\pi(X, x_0)$ forms a group under the multiplication $[f] \cdot [g] = [f * g]$.*

The verification of the group axioms in $\pi(X, x_0)$ is fairly lengthy and it will be carried through via the following three lemmas.

Lemma 14.2.4 *Suppose that f, g, and h are paths in X such that $f(1) = g(0)$ and $g(1) = h(0)$. Then, $(f * g) * h \simeq f * (g * h)$ rel $\{0, 1\}$.*

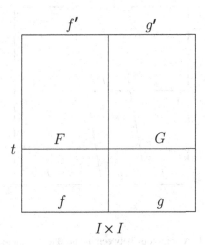

$I \times I$

Fig. 14.3 Product of homotopies F and G

Proof By definition, we have

$$((f * g) * h)(s) = \begin{cases} f(4s) & \text{for } 0 \le s \le 1/4, \\ g(4s - 1) & \text{for } 1/4 \le s \le 1/2, \\ h(2s - 1) & \text{for } 1/2 \le s \le 1; \end{cases}$$

$$(f * (g * h))(s) = \begin{cases} f(2s) & \text{for } 0 \le s \le 1/2, \\ g(4s - 2) & \text{for } 1/2 \le s \le 3/4, \\ h(4s - 3) & \text{for } 3/4 \le s \le 1. \end{cases}$$

These equations provide an intuitive idea for constructing a homotopy connecting $(f * g) * h$ and $f * (g * h)$. We draw the slanted lines in $I \times I$, according to the domains of g in the above formulae, thus dividing $I \times I$ into three closed sets (see Fig. 14.4). Then, we construct suitable continuous functions on each of these pieces and glue them together. In fact, we simply take these functions on horizontal lines AB, BC, and CD at a height t from the base to be reparametrizations of the maps f, g, and h, respectively. Specifically, we define $H : I \times I \to X$ by

$$H(s, t) = \begin{cases} f(4s/(1 + t)) & \text{for } 0 \le s \le (1 + t)/4, \\ g(4s - t - 1) & \text{for } (1 + t)/4 \le s \le (2 + t)/4, \\ h(4s - t - 2)/(2 - t) & \text{for } (2 + t)/4 \le s \le 1. \end{cases}$$

Since f and g agree on the line $t = 4s - 1$, and g and h agree on the line $t = 4s - 2$, the gluing lemma shows that H is continuous. It is now routine to check that $H :$ $(f * g) * h \simeq f * (g * h)$ rel $\{0, 1\}$. $\qquad \diamond$

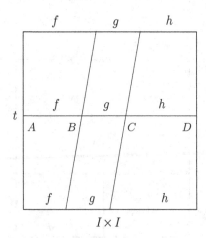

Fig. 14.4   A homotopy between the products of three paths

Lemma 14.2.5 *Let f be a path in X with $f(0) = x_0$, and $f(1) = x_1$. If c_{x_0} and c_{x_1} are the constant paths at x_0 and x_1, respectively, then $c_{x_0} * f \simeq f$ rel $\{0, 1\}$ and $f * c_{x_1} \simeq f$ rel $\{0, 1\}$.*

Proof Considering the definitions of $c_{x_0} * f$ and $f * c_{x_1}$, we divide the square $I \times I$ in each case into a triangle and a trapezium as shown in Fig. 14.5. Then, the desired homotopies are obtained by taking their restrictions on horizontal lines on the squares to be appropriate reparametrizations of f. Explicitly, in the first case, we define $H : I \times I \to X$ by

$$H(s, t) = \begin{cases} x_0 & \text{for } 0 \leq s \leq (1-t)/2, \\ f((2s+t-1)/(1+t)) & \text{for } (1-t)/2 \leq s \leq 1. \end{cases}$$

It is easily checked that $H : c_{x_0} * f \simeq f$ rel $\{0, 1\}$.

In the second case, the required homotopy H is given by

$$H(s, t) = \begin{cases} f(2s/(1+t)) & \text{for } 0 \leq s \leq (1+t)/2, \\ x_1 & \text{for } (1+t)/2 \leq s \leq 1. \end{cases} \qquad \Diamond$$

Clearly, given a path in X, we have another one that traverses it in the opposite direction; specifically, if $f : I \to X$ is a path from x_0 to x_1, then $g : I \to X$ defined by $g(s) = f(1-s)$ is a path from x_1 to x_0. The path g is called the *inverse* (or *reverse*) of f, and denoted by f^{-1}. (Note that f^{-1} is not the inverse of the mapping

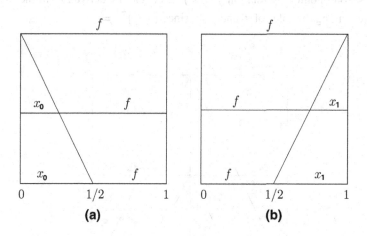

Fig. 14.5 Homotopies: **a** $c_{x_0} * f \simeq f$ and **b** $f * c_{x_1} \simeq f$

f; this just retraces f from its terminal point to its initial point.) With this notation, we have the following.

Lemma 14.2.6 *Let f be a path in X with the initial point x_0 and the terminal point x_1. Then, $f * f^{-1} \simeq c_{x_0}$ rel $\{0, 1\}$ and $f^{-1} * f \simeq c_{x_1}$ rel $\{0, 1\}$, where c_{x_0} and c_{x_1} denote the constant paths at x_0 and x_1, respectively.*

Proof To show the first relation, we define $H : I \times I \to X$ by

$$H(s, t) = \begin{cases} x_0 & \text{for } 0 \le s \le t/2, \\ f(2s - t) & \text{for } t/2 \le s \le 1/2, \\ f(2 - 2s - t) & \text{for } 1/2 \le s \le 1 - t/2, \\ x_0 & \text{for } 1 - t/2 \le s \le 1. \end{cases}$$

Clearly, the gluing lemma applies and H is continuous. It is now trivial to check that $H : f * f^{-1} \simeq c_{x_0}$ rel $\{0, 1\}$.

The intuitive idea behind the construction of the homotopy H is to pull the terminal point of the path f along it to the initial point. Accordingly, on the horizontal line AE at height t above the base in Fig. 14.6, a length $t/2$ at each end should be mapped into x_0 while the segment BC should be mapped into X in the same way as $f * f^{-1}$ maps the interval from 0 to $1/2 - t/2$ on the base, and the segment CD in the same way as $f * f^{-1}$ maps the interval from $1/2 + t/2$ to 1 on the base. Thus, we remain at x_0 during the time AB and DE, and traverse the path f in opposite directions during BD.

The second homotopy relation $f^{-1} * f \simeq c_{x_1}$ can be derived from the first one by interchanging the roles of f and f^{-1}, since $\left(f^{-1}\right)^{-1} = f$.   ◇

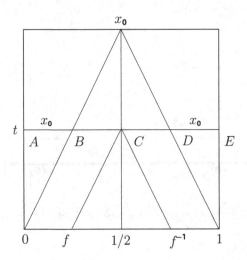

Fig. 14.6 A homotopy connecting $f * f^{-1}$ to the constant path c_{x_0}

Proof of Theorem 14.2.3: If f, g, and h are three loops in X based at x_0, then we have $\big([f] \cdot [g]\big) \cdot [h] = [(f * g) * h]$ and $[f] \cdot \big([g] \cdot [h]\big) = [f * (g * h)]$. Accordingly, the associativity of the multiplication in $\pi(X, x_0)$ follows from the homotopy relation $(f * g) * h \simeq f * (g * h)$ rel $\{0, 1\}$, which is a particular case of Lemma 14.2.4.

If c_{x_0} is the constant loop at x_0, then the homotopy class $\big[c_{x_0}\big]$ acts as the identity element for the multiplication in $\pi(X, x_0)$. For, if f is a loop in X based at x_0, then $f * c_{x_0} \simeq f \simeq c_{x_0} * f$ rel $\{0, 1\}$, by Lemma 14.2.5. So $[f] \cdot \big[c_{x_0}\big] = \big[f * c_{x_0}\big] = [f] = \big[c_{x_0}\big] * [f] = \big[c_{x_0}\big] \cdot [f]$.

For the existence of inverse of an element of $\pi(X, x_0)$, notice that if f is a loop based at x_0, then so is f^{-1}. And, by Lemma 14.2.6, we have $f * f^{-1} \simeq c_{x_0} \simeq f^{-1} * f$ rel $\{0, 1\}$. Hence, $[f] \cdot \big[f^{-1}\big] = \big[c_{x_0}\big] = \big[f^{-1}\big] \cdot [f]$, and $\big[f^{-1}\big]$ is the inverse of $[f]$ in $\pi(X, x_0)$. ◇

The group $\pi(X, x_0)$ is called the *fundamental group* of X at the base point x_0. The class $\big[c_{x_0}\big]$ is the identity element (denoted by 1), and the class $\big[f^{-1}\big]$ is the inverse of $[f]$. In general, the fundamental group of a space X depends on the choice ·of the base point (see Example 2.1.5 below). However, it is independent of the base point when X is path-connected, as we see now.

Theorem 14.2.7 *If X is a path-connected space, then $\pi(X, x_0)$ is isomorphic to $\pi(X, x_1)$ for any two points $x_0, x_1 \in X$.*

Proof Let $h: I \to X$ be a path in X form x_0 to x_1 and k be the path $t \mapsto h(1 - t)$ (i.e., k is the inverse of h). If f is a loop in X based at x_0, then $(k * f) * h$ is a loop based at x_1 (see Fig. 14.7). If f' is another loop in X based at x_0 and $f \simeq f'$ rel $\{0, 1\}$, then $(k * f) * h \simeq (k * f') * h$ rel $\{0, 1\}$, by Lemma 14.2.2. Hence, there is a mapping $\hat{h}: \pi(X, x_0) \to \pi(X, x_1)$ given by $\hat{h}([f]) = [k * f * h]$. By

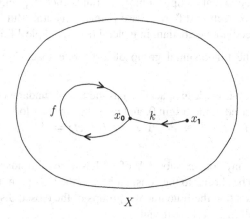

Fig. 14.7 The loop $k * f$ in the proof of Theorem 14.2.7

Lemma 14.2.6, $h * k \simeq c_{x_0}$ rel $\{0, 1\}$, where c_{x_0} is the constant loop at x_0. Therefore, $[h * k]$ is the identity element of $\pi(X, x_0)$ and, for any two elements $[f]$ and $[g]$ in $\pi(X, x_0)$, we have $[f] \cdot [g] = [f] \cdot [h * k] \cdot [g] = [f * h * k * g]$. Hence,

$$\hat{h}([f] \cdot [g]) = [k * f * h * k * g * h] = [k * f * h] \cdot [k * g * h] = \hat{h}([f]) \cdot \hat{h}([g])$$

and it follows that $\hat{h}$ is a homomorphism. Similarly, the path k induces a homomorphism $\hat{k}: \pi(X, x_1) \to \pi(X, x_0)$ given by $\hat{k}([f]) = [h * f * k]$. For any $[f] \in \pi(X, x_0)$, we clearly have

$$\hat{k}\hat{h}([f]) = \hat{k}([k * f * h]) = [h * k * f * h * k] = [h * k] \cdot [f] \cdot [h * k] = [f].$$

So, $\hat{k} \circ \hat{h}$ is the identity map on $\pi(X, x_0)$. By symmetry, $\hat{h} \circ \hat{k}$ is the identity map on $\pi(X, x_1)$, and therefore $\hat{h}$ is an isomorphism with $\hat{k}$ as its inverse.   ◇

By the preceding theorem, the fundamental groups $\pi(X, x_0)$ of a path-connected space X at various base points $x_0 \in X$ may be considered abstractly the same group, and denoted by $\pi(X)$. However, it should be noted that the isomorphism $\hat{h}: \pi(X, x_0) \cong \pi(X, x_1)$ constructed in Theorem 14.2.7 is not canonical, and therefore these groups cannot be identified. To see this, let h_1 and h_2 be two paths in X from x_0 to x_1. Then, there are isomorphisms $\hat{h}_1, \hat{h}_2: \pi(X, x_0) \to \pi(X, x_1)$ induced by h_1 and h_2. The product $h_1 * h_2^{-1}$ is a loop in X based at x_0, and hence determines an element $\alpha = [h_1 * h_2^{-1}]$ of $\pi(X, x_0)$. For any element $[f]$ in $\pi(X, x_0)$, we have

$$\left(\hat{h}_1^{-1} \circ \hat{h}_2\right)([f]) = \left[h_1 * h_2^{-1} * f * h_2 * h_1^{-1}\right] = \alpha[f]\alpha^{-1}.$$

This implies that $\hat{h}_2 = \hat{h}_1 \circ i_\alpha$, where i_α is the inner automorphism of $\pi(X, x_0)$ determined by α. By Lemma 14.2.2, $\hat{h}_1 = \hat{h}_2$ when $h_1 \simeq h_2$ rel $\{0, 1\}$. It follows that the isomorphism induced by a path joining two base points depends on its homotopy class, and if the fundamental group at some point (and hence all points) is abelian, then the isomorphism connecting different base points is natural. But, the fundamental group of a space need not be abelian, in general (see Example 14.5.1).

Example 14.2.1 The fundamental group of a discrete space X is trivial, that is, $\pi(X, x_0) = \{1\}$.

Example 14.2.2 The Euclidean space $\mathbb{R}^n$ has the trivial fundamental group. For, if f is a loop in $\mathbb{R}^n$ based at the origin 0, and c_0 is the constant loop at 0, then $f \simeq c_0$ rel $\{0, 1\}$, in fact, $H: I \times I \to \mathbb{R}^n$ given by $H(s, t) = (1 - t) f(s)$ is a required homotopy.

More generally, any convex subset X of $\mathbb{R}^n$ has a trivial fundamental group, for each loop in X can be shrunk to the constant loop at the base point by a straight-line homotopy. It follows that the fundamental groups of the closed disc $\mathbb{D}^n$, the n-cube I^n, and an open ball in $\mathbb{R}^n$ are all trivial.

The definition of the fundamental group is by no means constructive and the problem of computing group effectively is quite difficult, in general. As we proceed further in the book, we shall see several tools which help in its computation.

We now turn to the invariance properties of the fundamental group. Let $\phi: X \to Y$ be a continuous mapping and $x_0 \in X$. If f is a loop in X based at x_0, then the composite ϕf is a loop in Y based at $y_0 = \phi(x_0)$. Furthermore, if $H: f \simeq f'$ rel $\{0, 1\}$, then $\phi H: \phi f \simeq \phi f'$ rel $\{0, 1\}$. Accordingly, there is a well-defined function $\phi_\#: \pi(X, x_0) \to \pi(Y, y_0)$ given by $\phi_\#[f] = [\phi f]$. By the definition of the product '$*$', we have $\phi \circ (f * g) = (\phi f) * (\phi g)$ for all loops f and g in X based at x_0. So

$$\phi_\# ([f] \cdot [g]) = [\phi(f * g)] = [\phi f] \cdot [\phi g] = \phi_\#[f] \cdot \phi_\#[g].$$

It follows that $\phi_\#$ is a homomorphism, referred to as the *homomorphism induced by* ϕ (on the fundamental groups).

The following "functorial properties" of induced homomorphisms are crucial in applications.

Proposition 14.2.8 (a) *The identity map on X induces the identity automorphism of $\pi(X, x_0)$.*

(b) *If $\phi: X \to Y$ and $\psi: Y \to Z$ are continuous maps, then the composition*
$$\pi(X, x_0) \xrightarrow{\phi_\#} \pi(Y, \phi(x_0)) \xrightarrow{\psi_\#} \pi(Z, \psi\phi(x_0)) \text{ agrees with the homomorphism}$$
$(\psi\phi)_\#$.

The simple proofs are left to the reader.

As an immediate consequence of the preceding proposition, we see that *two homeomorphic spaces have isomorphic fundamental groups.* Thus, the fundamental group is a topological invariant of the space; accordingly, we can use it to distinguish between two spaces by proving that their fundamental groups are not isomorphic. We must add, however, that the fundamental group cannot characterize a topological space. As seen above, two non-homeomorphic spaces (e.g., the Euclidean spaces $\mathbb{R}^1$ and $\mathbb{R}^2$) may well have isomorphic fundamental groups. Actually, the fundamental group is a homotopy invariant of the space, that is, it is preserved by homotopy equivalences. We defer the proof of this assertion for a while, and turn to subspaces of a given space, which have the same fundamental group as the space has.

Proposition 14.2.9 *Let A be a path component of X, and $x_0 \in A$. Then, the homomorphism $i_\#: \pi(A, x_0) \to \pi(X, x_0)$ induced by the inclusion $i: A \hookrightarrow X$ is an isomorphism.*

Proof For a loop g in X based at x_0, we have $g(I) \subseteq A$. So, we can define a loop $f: I \to A$ by setting $f(t) = g(t)$ for all $t \in I$. Obviously, $g = if$ and $i_\#$ is surjective. To see that $i_\#$ is injective, suppose that f_1 and f_2 are two loops in A based at x_0 and $F: if_1 \simeq if_2$ rel $\{0, 1\}$. Since $F(I \times I)$ is path-connected and contains $x_0 = F(0, 0)$, we have $F(I \times I) \subseteq A$. Consequently, F defines a mapping $G: I \times I \to A$. It is obvious that $G: f_1 \simeq f_2$ rel $\{0, 1\}$, and the injectivity of $i_\#$ follows. $\diamond$

From the preceding proposition, it follows that the fundamental group of a space X based at x_0 can give information only about the path component of X containing x_0.

Before coming to next such result, we observe the following property of induced homomorphisms.

Proposition 14.2.10 *Suppose that two continuous maps ϕ and ψ of a space X into a space Y are homotopic relative to a point $x_0 \in X$. Then,*

$$\phi_\# = \psi_\# : \pi(X, x_0) \to \pi(Y, y_0),$$

where $y_0 = \phi(x_0) = \psi(x_0)$.

Proof For every loop f in X based at x_0, we clearly have $\phi \circ f \simeq \psi \circ f$ rel $\{0, 1\}$, and so $\phi_\# = \psi_\#$. $\diamond$

Corollary 14.2.11 *If A is a strong deformation retract of X, then the inclusion $i : A \hookrightarrow X$ induces an isomorphism $\pi(A, x_0) \cong \pi(X, x_0)$, where $x_0 \in A$.*

Proof Let $r : X \to A$ be a retraction such that $1_X \simeq i \circ r$ rel A. Then, by the preceding proposition, $i_\# \circ r_\# = 1_{X\#}$, the identity automorphism of $\pi(X, x_0)$. Moreover, since $r \circ i = 1_A$, we have $r_\# \circ i_\# = 1_{A\#}$, the identity automorphism of $\pi(A, x_0)$. Therefore, $i_\# : \pi(A, x_0) \to \pi(X, x_0)$ is an isomorphism with $r_\#$ as its inverse. $\diamond$

Definition 14.2.12 A path-connected space X is called *simply* connected if $\pi(X) = \{1\}$.

By the preceding corollary, a space containing a point that is its strong deformation retract is simply connected. More generally, we have the following theorem.

Theorem 14.2.13 *A contractible space is simply connected.*

Proof Let X be a contractible space, and let $H : X \times I \to X$ be a contraction of X to x_0. Then, for each $x \in X$, $t \mapsto H(x, t)$ is a path from x to x_0, and therefore X is path-connected. We show that $\pi(X, x_0) = \{1\}$. Recall that the identity element 1 of $\pi(X, x_0)$ is the homotopy class $\left[c_{x_0}\right]$, where c_{x_0} denotes the constant loop at x_0. Also, notice that the function $g : I \to X$, $t \mapsto H(x_0, t)$ is a loop in X based at x_0, so its homotopy class $[g]$ is an element of $\pi(X, x_0)$. We observe that for each loop f in X based at x_0, $f \simeq g * c_{x_0} * g^{-1}$ rel $\{0, 1\}$. A desired homotopy $F : I \times I \to X$ is defined by

$$F(s, t) = \begin{cases} g(4s) & \text{for } 0 \le s \le t/4, \\ H\!\left(f\!\left(\frac{4s-t}{4-3t}\right), t\right) & \text{for } t/4 \le s \le 1 - t/2, \\ g^{-1}(2s - 1) & \text{for } 1 - t/2 \le s \le 1, \end{cases}$$

(see Fig. 14.8). From the above homotopy relation, it is clear that $[f] = [g] \cdot [c_{x_0}] \cdot [g^{-1}] = [g] \cdot [g]^{-1} = 1$, and hence X is simply connected.

By the preceding theorem, the fundamental groups of the "comb space" and the "dunce cap space" (refer to Exercise 14.1.24) are trivial.

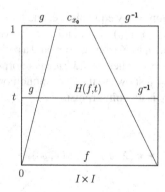

Fig. 14.8 Construction of the homotopy $F: f \simeq (g * c_{x_0}) * g^{-1}$

We further show that two spaces having the same homotopy type have isomorphic fundamental groups, that is, fundamental group is a homotopy invariant. To this end, we observe an important relation between the homomorphisms induced by homotopic maps.

Theorem 14.2.14 *Let $\phi, \psi: X \to Y$ be homotopic maps and suppose that $x_0 \in X$. Then, there is a path g in Y from $\phi(x_0)$ to $\psi(x_0)$ such that the composition*

$$\pi(X, x_0) \xrightarrow{\phi_\#} \pi(Y, \phi(x_0)) \xrightarrow{\hat{g}} \pi(Y, \psi(x_0)),$$

where $\hat{g}$ is the isomorphism $[h] \to [g^{-1} * h * g]$ induced by the path g, agrees with the homomorphism $\psi_\#$.

Proof Let $H: \phi \simeq \psi$ be a homotopy connecting ϕ to ψ and define $g: I \to Y$ by $g(t) = H(x_0, t), t \in I$. Then, g is a path in Y joining $\phi(x_0)$ to $\psi(x_0)$. As seen in the proof of Theorem 14.2.7, g induces the isomorphism $\hat{g}: [h] \to [g^{-1} * h * g]$ from $\pi(Y, \phi(x_0))$ to $\pi(Y, \psi(x_0))$. Thus, it remains to show that $\hat{g}[\phi f] = [\psi f]$ for any loop f in X based at x_0. This clearly holds good if $\phi f \simeq g * (\psi f) * g^{-1}$ rel $\{0, 1\}$. A desired homotopy between the two loops can be constructed as in the previous theorem. Explicitly, we define $F: I \times I \to Y$ by

$$F(s, t) = \begin{cases} g(4s) & \text{for } 0 \leq s \leq t/4, \\ H\left(f\left(\frac{4s-t}{4-3t}\right), t\right) & \text{for } t/4 \leq s \leq 1 - t/2, \\ g^{-1}(2s - 1) & \text{for } 1 - t/2 \leq s \leq 1 \end{cases}$$

The continuity of F is seen by invoking the gluing lemma, and it is easily verified that $F: \phi f \simeq (g * (\psi f)) * g^{-1}$ rel $\{0, 1\}$.   ◇

Corollary 14.2.15 *If $\phi: X \to Y$ is a homotopy equivalence, then for any $x_0 \in X$, $\phi_\#: \pi(X, x_0) \to \pi(Y, \phi(x_0))$ is an isomorphism.*

Proof Suppose that ϕ is a homotopy equivalence, and let $\psi : Y \to X$ be a homotopy inverse of ϕ. Put $y_0 = \phi(x_0)$, $\psi(y_0) = x_1$, and $y_1 = \phi(x_1)$. Since $1_X \simeq \psi \circ \phi$ and $\phi \circ \psi \simeq 1_Y$, there exist a path g in X from x_0 to x_1 and a path h in Y from y_0 to y_1 such that $(\psi\phi)_\# = \hat{g}$ and $\phi_\# \circ \psi_\# = \hat{h}$, where $\hat{g}$ and $\hat{h}$ are isomorphisms (see Fig. 14.9). The equality $\psi_\# \circ \phi_\# = (\psi\phi)_\# = \hat{g}$ shows that $\phi_\#$ is a monomorphism while the equality $\phi_\# \circ \psi_\# = \hat{h}$ implies that $\phi_\#$ is an epimorphism. $\diamond$

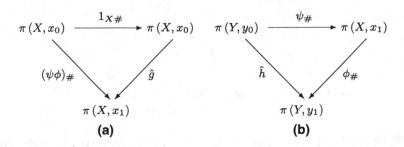

Fig. 14.9   Proof of Corollary 14.2.15

Thus, we have established the homotopy invariance property of the fundamental group: Two path-connected spaces of the same homotopy type have isomorphic fundamental groups. In particular, we see that Corollary 14.2.11 remains valid even if A is just a deformation retract of X. This fact will come in useful in the computation of fundamental groups of some spaces.

Exercises

1. Let $\Omega(X, x_0)$ be the subspace of $\mathcal{C}(I, X)_{co}$ consisting of all loops based at x_0. Prove that $\pi(X, x_0)$ is just the set of all path components in $\Omega(X, x_0)$.

2. Let $f, g : I \to X$ be two paths with initial point x_0 and terminal point x_1. Show:

    (a) $f \simeq g$ rel $\{0, 1\}$ if and only if $f * g^{-1} \simeq c_{x_0}$ rel $\{0, 1\}$.

    (b) If $f \simeq g$ rel $\{0, 1\}$, then $f^{-1} \simeq g^{-1}$ rel $\{0, 1\}$.

3. Find the fundamental group of an indiscrete space.

4. Let x and y be distinct points of a simply connected space X. Prove that there is a unique path class in X with initial point x and terminal point y.

5. Let $r : X \to A$ be a retraction and $i : A \hookrightarrow X$ be the inclusion map. If $i_\# \pi(A)$ is a normal subgroup of $\pi(X)$, show that $\pi(X)$ is the direct product of im $(i_\#)$ and ker $(r_\#)$.

6. If $\phi, \psi : X \to Y$ are homotopic maps and $\phi(x_0) = \psi(x_0)$ for some $x_0 \in X$, prove that the homomorphisms $\phi_\#, \psi_\# : \pi(X, x_0) \to \pi(Y, \phi(x_0))$ differ by an inner automorphism of $\pi(Y, \phi(x_0))$.

7. If $\phi \colon (X, x_0) \to (Y, y_0)$ is (free) null homotopic, show that $\phi_\# \colon \pi(X, x_0) \to \pi(Y, y_0)$ is the trivial homomorphism.

14.3   Fundamental Groups of Spheres

The 0-sphere $\mathbb{S}^0$, being a discrete space, has a trivial fundamental group. Next, we go about determining the fundamental groups $\pi(\mathbb{S}^n)$ for $n \geq 2$, since the computation of $\pi(\mathbb{S}^1)$ is not direct and it involves an approach quite different from the one applicable to other cases. To this end, we prove the following.

Proposition 14.3.1 *Let X be a space and $\{U_\alpha\}$ be an open cover of X such that* (a) *each U_α is simply connected,* (b) *$U_\alpha \cap U_\beta$ is path-connected for any pair of indices α, β, and* (c) *$\bigcap U_\alpha \neq \emptyset$. Then, X is simply connected.*

Proof Choose a point $x_0 \in \bigcap U_\alpha$. Since each point $x \in X$ belongs to some U_α and every U_α is path-connected, there is a path in $U_\alpha \subseteq X$ joining x to x_0. Hence, X is path-connected. We show that $\pi(X, x_0) = \{1\}$. Let f be a loop in X based at x_0. Then, the family $\{f^{-1}(U_\alpha)\}$ is an open cover of the compact metric space I, and therefore has a Lebesgue number δ, say. Consider a subdivision $0 = t_0 < t_1 < \cdots < t_n = 1$ of I so that $t_j - t_{j-1} < \delta$ for every $j = 1, \ldots, n$. Then, f maps the subinterval $[t_{j-1}, t_j]$ into some U_{α_j}, $1 \leq j \leq n$. For notational convenience, we will write U_j for U_{α_j}. Note that the sets U_j and U_k are not necessarily distinct for $j \neq k$. Now, for each j, we define a path $g_j \colon I \to X$ by setting $g_j(s) = f\big((1-s)t_{j-1} + st_j\big)$,

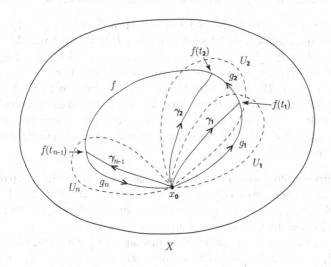

Fig. 14.10 A division of loop f into the paths g_j

$0 \leq s \leq 1$. It is clear that the path g_j lies entirely in U_j. See Fig. 14.10. We observe that $f \simeq (g_1 * \cdots * g_{n-1}) * g_n$ rel $\{0, 1\}$. Divide the interval I into n subintervals by the points $2^{1-n}, 2^{2-n}, \ldots, 2^{-1}$, and consider the map $h : I \to I$ defined by

$$h(s) = \begin{cases} 2^{n-1} s t_1 & \text{for } 0 \leq s \leq 2^{1-n}, \text{ and} \\ (2 - 2^{n-j+1} s) t_{j-1} + (2^{n-j+1} s - 1) t_j & \text{for } 2^{j-n-1} \leq s \leq 2^{j-n}. \end{cases}$$

Then, $fh = (g_1 * \cdots * g_{n-1}) * g_n$, and $H : (s, t) \mapsto (th(s) + (1 - t)s), s, t \in I$, is a homotopy between the identity map on I and h relative to $\{0, 1\}$. So the composition fH is a homotopy between f and fh holding the base point fixed. Notice that, for every $1 \leq j < n$, $f(t_j) \in U_j \cap U_{j+1}$; accordingly, there is a path γ_j from x_0 to $f(t_j) = x_j$ lying entirely in $U_j \cap U_{j+1}$, by our hypothesis. By Lemma 14.2.6, we have $\gamma_j^{-1} * \gamma_j \simeq c_{x_j}$ rel $\{0, 1\}$. So $g_j * g_{j+1} \simeq g_j * \gamma_j^{-1} * \gamma_j * g_{j+1}$ rel $\{0, 1\}$ for every $1 \leq j < n$, and hence

$$f \simeq g_1 * \gamma_1^{-1} * \gamma_1 * g_2 * \cdots * \gamma_{n-1}^{-1} * \gamma_{n-1} * g_n \text{ rel } \{0, 1\}.$$

This implies that

$$[f] = [g_1 * \gamma_1^{-1}] \cdot [\gamma_1 * g_2 * \gamma_2^{-1}] \cdot \cdots \cdot [\gamma_{n-1} * g_n].$$

Notice that each term in the right hand side of this equation is a homotopy class of a loop lying in some U_j, and therefore coincides with the homotopy class $[c_{x_0}]$, for each U_j is simply connected. So, $[f] = [c_{x_0}]$, the identity element of $\pi(X, x_0)$. This proves the proposition. $\diamond$

Corollary 14.3.2 *For $n \geq 2$, the n-sphere $\mathbb{S}^n$ is simply connected.*

Proof Let $p = (0, \ldots, 0, 1) \in \mathbb{S}^n$ and $q = -p$. Then, both $U = \mathbb{S}^n - \{p\}$ and $V = \mathbb{S}^n - \{q\}$ are open subsets of $\mathbb{S}^n$ such that $\mathbb{S}^n = U \cup V$. By Example 2.1.5, the mapping $f : U \to \mathbb{R}^n$ given by $f(x) = \frac{1}{1-x_n}(x_0, \ldots, x_{n-1})$ for $x = (x_0, \ldots, x_n)$ in U is a homeomorphism, and $V \approx U$ under the reflection $(x_0, \ldots, x_{n-1}, x_n) \mapsto (x_0, \ldots, x_{n-1}, -x_n)$. Therefore, both U and V are simply connected. Clearly, the homeomorphism f takes q into the origin 0, and hence it induces a homeomorphism between $U \cap V = \mathbb{S}^n - \{p, q\}$ and $\mathbb{R}^n - \{0\}$. Since the latter space is path-connected for $n > 1$, so is $U \cap V$. Thus, the preceding proposition applies, and the fundamental group of $\mathbb{S}^n$ is trivial. $\diamond$

By the proof of Proposition 14.3.1, we see that the fundamental group of a space X, which has an open covering $\{U_\alpha\}_{\alpha \in A}$ such that $\bigcap U_\alpha \neq \emptyset$ and $U_\alpha \cap U_\beta$ is path-connected for every $\alpha, \beta \in A$, is generated by the images of $\pi(U_\alpha)$ under the homomorphisms induced by the inclusion maps $i_\alpha : U_\alpha \hookrightarrow X$.

We now move on to computation of the fundamental group of the circle $\mathbb{S}^1$. This is intimately related to the study of the exponential map $p : \mathbb{R}^1 \to \mathbb{S}^1, t \mapsto e^{2\pi i t}$. We know that $p(t + t') = p(t) p(t')$ for any $t, t' \in \mathbb{R}^1$ and $p^{-1}(1) = \mathbb{Z}$, the group of integers. If $0 < t < 1$, then $p(t) \neq 1$ and for each $z \in \mathbb{S}^1$, there is unique $t \in [0, 1)$

such that $p(t) = z$. The continuity of trigonometric functions $\sin t$ and $\cos t$ shows that p is continuous. Also, we have

Lemma 14.3.3 *The exponential map $p: \mathbb{R}^1 \to \mathbb{S}^1$ is open.*

Proof Clearly, the open intervals of length less than 1 form a base for $\mathbb{R}^1$, so it suffices to prove that the images of all such open intervals under p are open in $\mathbb{S}^1$. It is obvious that each open interval (a, b) in $\mathbb{R}^1$ with $b - a < 1$ is sent to some open interval contained in $(-1/2, 1/2)$ by the translation map $\tau: t \mapsto t + t_0$, where $t_0 = -(a + 1/2)$. Moreover, for every $a < t < b$, we have $p(t) = p(-t_0)p(\tau(t))$. Since the closed interval $[-1/2, 1/2]$ is compact, the mapping

$$q: [-1/2, 1/2] \to \mathbb{S}^1$$

defined by p is a continuous closed surjection. As $q^{-1}(-1) = \{-1/2, 1/2\}$, it follows that the mapping $t \mapsto p(t)$ is a homeomorphism of the open interval $(-1/2, 1/2)$ onto $U = \mathbb{S}^1 - \{-1\}$. Obviously, U is open in $\mathbb{S}^1$, and therefore $p|(-1/2, 1/2)$ is an open mapping. Since $\tau(a, b)$ is an open subset of $(-1/2, 1/2)$ and the multiplication by $p(-t_0)$ is a homeomorphism of $\mathbb{S}^1$, we conclude that $p(a, b)$ is open in $\mathbb{S}^1$.   ◇

We introduce the following notion to show a crucial property of p: A subset X of $\mathbb{R}^n$ is *starlike with respect to a point $x_0 \in X$* if all the line segments joining x_0 to any other point of X lie entirely in X.

It is obvious that a convex subset of $\mathbb{R}^n$ is starlike with respect to any of its points; however, the converse is not true.

Theorem 14.3.4 *Let X be a compact subspace of a Euclidean space $\mathbb{R}^n$, and suppose that it is starlike with respect to the origin 0. If $f: X \to \mathbb{S}^1$ is a continuous map such that $f(0) = 1$, then, for each integer m, there exists a unique continuous map $\tilde{f}: X \to \mathbb{R}^1$ such that $p\tilde{f} = f$ and $\tilde{f}(0) = m$, where p is the exponential map.*

Proof Since X is compact, f is uniformly continuous. Therefore, there is a real $\delta > 0$ such that $|f(x) - f(x')| < 2$ whenever $\|x - x'\| < \delta$. Since X is bounded, we find a positive integer n such that $\|x\| < n\delta$ for all $x \in X$. Then, for each integer $1 \leq i \leq n$ and every $x \in X$, we have $\|\frac{i}{n}x - \frac{i-1}{n}x\| < \delta$, so $|f(\frac{i}{n}x) - f(\frac{i-1}{n}x)| < 2$. This implies that $f(\frac{i}{n}x)/f(\frac{i-1}{n}x) \neq -1$ for all $x \in X$; consequently, for each i, we have the function

$$g_i: X \to \mathbb{S}^1 - \{-1\}$$

given by $g_i(x) = f(\frac{i}{n}x)/f(\frac{i-1}{n}x)$. It is clear that each g_i is continuous, $g_i(0) = 1$ and $f = g_1 \cdots g_n$. By Lemma 14.3.3, the mapping

$$q: \left(-\tfrac{1}{2}, \tfrac{1}{2}\right) \longrightarrow \mathbb{S}^1 - \{-1\}$$

defined by p is a homeomorphism, and therefore its inverse is a continuous function on $\mathbb{S}^1 - \{-1\}$. Consequently, the mapping $\tilde{f}: X \to \mathbb{R}^1$ given by $\tilde{f}(x) = m + q^{-1}g_1(x) + \cdots + q^{-1}g_n(x)$ is continuous. We also have

$$p\tilde{f}(x) = p(m)p\left(\sum_1^n q^{-1}g_i(x)\right) = \prod_1^n g_i(x) = f(x)$$

for every $x \in X$, and $\tilde{f}(0) = m$, for $q^{-1}(1) = 0$.

To see the uniqueness of $\tilde{f}$, suppose that there is a continuous map $\tilde{g}: X \to \mathbb{R}^1$ satisfying $p\tilde{g} = f$ and $\tilde{g}(0) = m$. Then, the mapping $\tilde{h}: X \to \mathbb{R}^1$ defined by $\tilde{h} = \tilde{g} - \tilde{f}$ is continuous and $p\tilde{h}(x) = 1$ for all $x \in X$. Since $p^{-1}(1) = \mathbb{Z}$, we have $\tilde{h}(x) \in \mathbb{Z}$ for every $x \in X$. Obviously, $\tilde{h}(0) = 0$. Since X is connected and $\tilde{h}$ is continuous, we conclude that $\tilde{h}(x) = 0$ for all $x \in X$. Thus, we have $\tilde{g} = \tilde{f}$, as desired.   ◇

The map $\tilde{f}$ in the above theorem is called a *lifting* of f relative to the exponential map p. The two most important cases of this theorem arise for $X = I$ and $X = I \times I$. It is obvious that the unit interval I is a starlike subset of $\mathbb{R}^1$ with respect to 0, and the unit square $I \times I$ is a starlike subset of $\mathbb{R}^2$ with respect to the point $(0, 0)$. For emphasis, we rephrase the theorem in these two cases:

The Path Lifting Property: If $f: I \to \mathbb{S}^1$ is a path with $f(0) = 1$, then there exists a unique path $\tilde{f}: I \to \mathbb{R}^1$ such that $p\tilde{f} = f$ and $\tilde{f}(0) = 0$.

The Homotopy Lifting Property: If $F: I \times I \to \mathbb{S}^1$ is a homotopy with $F(0, 0) = 1$, then there is a unique homotopy $\tilde{F}: I \times I \to \mathbb{R}^1$ such that $p\tilde{F} = F$ and $\tilde{F}(0, 0) = 0$.

These properties of the exponential map p generalize to a very important class of spaces, namely, the "covering spaces," which will be studied in the next chapter. We shall see there constructive proofs of the above statements.

Definition 14.3.5 Let f be a loop in $\mathbb{S}^1$ based at 1, and $\tilde{f}$ be the lifting of f with the initial point $0 \in \mathbb{R}^1$. *The degree* of f, denoted by $\deg f$, is the number $\tilde{f}(1)$ (the terminal point of $\tilde{f}$).

Let f be a loop in $\mathbb{S}^1$ based at 1. Note that $\deg f$ is always an integer, since $p\tilde{f}(1) = f(1) = 1$. We further observe that $\deg f$ depends only on the homotopy class $[f]$ of f. To see this, suppose that $F: f \simeq g$ rel $\{0, 1\}$. By the Homotopy Lifting Property, there is a homotopy $\tilde{F}: I \times I \to \mathbb{R}^1$ such that $\tilde{F}(0, 0) = 0$ and $p\tilde{F} = F$, where p is the exponential map. We have $p\tilde{F}(0, t) = 1$ for all $t \in I$; accordingly, $\tilde{F}(0, t)$ is an integer for each t. Since $\{0\} \times I \approx I$ is connected, and $\tilde{F}(0, 0) = 0$, we must have $\tilde{F}(0, t) = 0$ for all $t \in I$. Similarly, $\tilde{F}(1, t)$ is also a fixed integer. We define $\tilde{f}(s) = \tilde{F}(s, 0)$ and $\tilde{g}(s) = \tilde{F}(s, 1)$, $s \in I$. Then, $p\tilde{f}(s) = f(s)$, $p\tilde{g}(s) = g(s)$ and $\tilde{f}(0) = 0 = \tilde{g}(0)$. Thus, $\tilde{f}$ and $\tilde{g}$ are the liftings of f and g, respectively, having the same initial point 0. So, $\deg f = \tilde{f}(1) = \tilde{F}(1, 0) = \tilde{F}(1, 1) = \tilde{g}(1) = \deg g$. Hence, we have a function $\deg: \pi(\mathbb{S}^1, 1) \to \mathbb{Z}$ defined by $\deg[f] = \deg f$.

Theorem 14.3.6 *The function* $[f] \mapsto \deg f$ *is an isomorphism of* $\pi(\mathbb{S}^1, 1)$ *onto the group* $\mathbb{Z}$ *of integers.*

Proof To see that the function deg is onto, let n be any integer and consider the path $\tilde{f}_n : I \to \mathbb{R}^1$ defined by $\tilde{f}_n(s) = ns$, $s \in I$. Then, $f_n = p\tilde{f}_n$ is a loop in $\mathbb{S}^1$ based at 1 with deg $f_n = \tilde{f}_n(1) = n$, and so deg is onto.

Next, to prove that deg is one-to-one, suppose that deg $f =$ deg g for two loops f and g in $\mathbb{S}^1$ based at 1. We need to show $f \simeq g$ rel $\{0, 1\}$. Let $\tilde{f}$ and $\tilde{g}$ be the liftings of f and g, respectively, with $\tilde{f}(0) = 0 = \tilde{g}(0)$. Our assumption implies that $\tilde{f}(1) = \tilde{g}(1)$. Consider the mapping $H : I \times I \to \mathbb{R}^1$ defined by $H(s, t) = (1 - t)\tilde{f}(s) + t\tilde{g}(s)$ for all $s, t \in I$. Clearly, $H : \tilde{f} \simeq \tilde{g}$ rel $\{0, 1\}$, and this implies that $pH : f \simeq g$ rel $\{0, 1\}$.

Finally, we prove that deg is a homomorphism. Given loops f and g in $\mathbb{S}^1$ based at 1, let $\tilde{f}$ and $\tilde{g}$, respectively, be their liftings with $\tilde{f}(0) = \tilde{g}(0)$. We show that $\deg (f * g) = \deg f + \deg g = \tilde{f}(1) + \tilde{g}(1)$. To establish this equation, we define a path $\tilde{h} : I \to \mathbb{R}^1$ by

$$\tilde{h}(s) = \begin{cases} \tilde{f}(2s) & \text{for } 0 \leq s \leq 1/2, \\ \tilde{f}(1) + \tilde{g}(2s - 1) & \text{for } 1/2 \leq s \leq 1. \end{cases}$$

It is easily checked that $\tilde{h}$ is a lifting of $f * g$ and $\tilde{h}(0) = 0$. So, $\deg (f * g) = \tilde{h}(1) = \tilde{f}(1) + \tilde{g}(1)$, as desired.   ◇

By the proof of the preceding theorem, it is clear that each loop in $\mathbb{S}^1$ based at 1 is homotopically equivalent to the map $f_n : t \to e^{2\pi int}$ for a unique integer n and, intuitively, f_n wraps the interval I around $\mathbb{S}^1$ $|n|$ times (anticlockwise if $n > 0$ and clockwise if $n < 0$). Thus, the degree of a loop in $\mathbb{S}^1$ based at 1, roughly speaking, tells how many times the path is wrapped around the circle, and is referred to as the "winding number" of the loop.

Example 14.3.1 The fundamental group of the punctured plane $\mathbb{R}^2 - \{pt\}$ is the group of integers $\mathbb{Z}$. For, $\mathbb{S}^1$ is a strong deformation retract of $\mathbb{R}^2 - \{0\}$ so that $\pi\left(\mathbb{R}^2 - \{pt\}\right) \cong \pi\left(\mathbb{S}^1\right) = \mathbb{Z}$, by Corollary 14.2.11.

We can now justify our earlier remark that $\pi(X, x_0)$ generally depends on the choice of the base point x_0.

Example 14.3.2 Consider the subspace X of $\mathbb{R}^2$, which is the union of sets $A = \{(t, \sin 1/ t) \mid 0 < t \leq 1\}$, $B = \{(0, t) \mid -1 \leq t \leq 1\}$, and a circle C lying in the second quadrant and tangent to the y-axis at the point $(0, 1)$ (see Fig. 14.11). It is easily checked that $\pi(X, x_0) = \{1\}$ when $x_0 \in A$, and $\pi(X, x_0) = \mathbb{Z}$ when $x_0 = (0, 1)$. Notice that the space X is connected.

By way of other applications of the knowledge of $\pi\left(\mathbb{S}^1\right)$, we prove the following two classical results.

Theorem 14.3.7 (Brouwer's Fixed-Point Theorem) *A continuous map f of $\mathbb{D}^n$ into itself has at least one fixed point, that is, $f(x) = x$ for some $x \in \mathbb{D}^n$.*

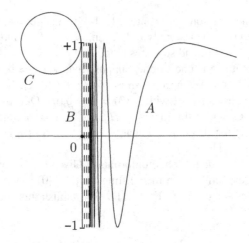

Fig. 14.11 The space X in Example 14.3.2

Proof By our knowledge thus far, we can prove the theorem for $n \leq 2$. Let $f : \mathbb{D}^n \to \mathbb{D}^n$ be a continuous map. Assume that f has no fixed points, that is, $f(x) \neq x$ for each $x \in \mathbb{D}^n$. Let $r(x)$ denote the point of intersection of $\mathbb{S}^{n-1} = \partial \mathbb{D}^n$ with the ray emanating from $f(x)$ and going through the point x. Then, $r(x)$ is uniquely determined, by Lemma 5.3.4. Thus, we have a mapping $r : \mathbb{D}^n \to \mathbb{S}^{n-1}$; explicitly, $r(x) = (1 - \lambda_x) f(x) + \lambda_x x$, where λ_x satisfies the conditions $\lambda_x \geq 1$ and $\|r(x)\| = 1$. One easily sees that

$$\lambda_x = \frac{\|f(x)\|^2 - x \cdot f(x) + \sqrt{\|x - f(x)\|^2 - \|x\|^2 \|f(x)\|^2 + (x \cdot f(x))^2}}{\|x - f(x)\|^2}.$$

Since f is continuous, we see that the mapping $x \mapsto \lambda_x$ is continuous, and therefore so is r. Also, we have $r(x) = x$ for every $x \in \mathbb{S}^{n-1}$, and thus r is a retraction of $\mathbb{D}^n$ onto $\mathbb{S}^{n-1}$. For $n = 1$, this immediately leads to a contradiction, since $\mathbb{D}^1$ is connected and $\mathbb{S}^0$ is not. For $n = 2$, we clearly have $ri = 1$, where $i : \mathbb{S}^1 \hookrightarrow \mathbb{D}^2$ denotes the inclusion map and 1 is the identity map on $\mathbb{S}^1$. Then, for any $x_0 \in \mathbb{S}^1$, the composition

$$\pi\left(\mathbb{S}^1, x_0\right) \xrightarrow{i_\#} \pi\left(\mathbb{D}^2, x_0\right) \xrightarrow{r_\#} \pi\left(\mathbb{S}^1, x_0\right)$$

is the identity automorphism of $\pi\left(\mathbb{S}^1, x_0\right)$. This is impossible, for $\pi\left(\mathbb{D}^2, x_0\right)$ is the trivial group, whereas $\pi\left(\mathbb{S}^1, x_0\right)$ is nontrivial. Therefore, there is a point $x \in \mathbb{D}^2$ such that $f(x) = x$, and the theorem follows.   ◇

In proving the preceding theorem, we have also shown that $\mathbb{S}^{n-1}$ is not a retract of $\mathbb{D}^n$ for $n \leq 2$. In fact, this result holds good for all n and it is known as the Brouwer's

No-retraction Theorem. A proof of this theorem in full generality requires higher dimensional analogues of the fundamental group.

For the other result, we first extend the notion of the degree to any continuous map $g\colon \mathbb{S}^1 \to \mathbb{S}^1$. Note that if g is a continuous mapping of $\mathbb{S}^1$ into itself, then there is a unique continuous mapping g' of $\mathbb{S}^1$ into itself such that $g'(1) = 1$ and $g = g(1)g'$. Obviously, the composition $g'q$ is a loop in $\mathbb{S}^1$ based at 1, where $q\colon I \to \mathbb{S}^1$ is the identification map $t \mapsto e^{2\pi i t}$. This loop in $\mathbb{S}^1$ is uniquely determined by g. Therefore, one can define the *degree* of g to be the integer $\deg g'q$, where $g'(z) = g(z)/g(1)$. In particular, we note that, for each integer n, the mapping $\mathbb{S}^1 \to \mathbb{S}^1$, $z \mapsto z^n$, has degree n.

Proposition 14.3.8 *Two continuous maps of $\mathbb{S}^1$ into itself are homotopic if and only if they have the same degree.*

Proof With the above notation, we first observe that two maps $g, h\colon \mathbb{S}^1 \to \mathbb{S}^1$ are homotopic if and only if $g' \simeq h'$ rel $\{1\}$. If $G\colon g \simeq h$, then the mapping $H\colon \mathbb{S}^1 \times I \to \mathbb{S}^1$, defined by $H(z,t) = G(z,t)/G(1,t)$, is a desired homotopy connecting g' and h'. Conversely, if $g' \simeq h'$, we show that $g \simeq h$. Since the homotopy relation is an equivalence relation, it suffices to prove that $g \simeq g'$. To this end, put $g(1) = e^{i\theta_0}$ and consider the mapping $\mathbb{S}^1 \times I \to \mathbb{S}^1$, $(z,t) \mapsto e^{-it\theta_0}g(z)$. This is clearly a homotopy between g and g'. Next, we see that each continuous map $g\colon \mathbb{S}^1 \to \mathbb{S}^1$ with $g(1) = 1$ determines a loop gq in $\mathbb{S}^1$ based at 1, where $q\colon I \to \mathbb{S}^1$ is the identification map $t \mapsto e^{2\pi i t}$. Furthermore, if two such maps $g, h\colon \mathbb{S}^1 \to \mathbb{S}^1$ are homotopic and $G\colon g \simeq h$ rel $\{1\}$, then $G \circ (q \times 1)\colon I \times I \to \mathbb{S}^1$ is a homotopy relative to $\{0, 1\}$ between the corresponding loops gq and hq. On the other hand, each loop f in $\mathbb{S}^1$ based at 1 induces a continuous map $f'\colon \mathbb{S}^1 \to \mathbb{S}^1$ such that $f'(1) = 1$ and $f = f'q$. Moreover, if f_1 and f_2 are two loops in $\mathbb{S}^1$ based at 1 and $F\colon f_1 \simeq f_2$ rel $\{0, 1\}$, then the function $G\colon \mathbb{S}^1 \times I \to \mathbb{S}^1$, defined by $G(q(s), t) = F(s, t)$, is a homotopy relative to $\{1\}$ between f_1' and f_2', since $q \times 1\colon I \times I \longrightarrow \mathbb{S}^1 \times I$ is an identification. It follows that two continuous maps $g, h\colon \mathbb{S}^1 \to \mathbb{S}^1$ are homotopic if and only if the corresponding loops $g(1)^{-1}gq$ and $h(1)^{-1}hq$ are homotopic relative to $\{0, 1\}$. Finally, by the proof of Theorem 14.3.6, we see that two loops in $\mathbb{S}^1$ based at 1 are homotopic rel $\{0, 1\}$ if and only if they have the same degree, and the proposition follows. ◇

Theorem 14.3.9 (The Fundamental Theorem of Algebra) *A nonconstant polynomial with complex coefficients has at least one complex root.*

Proof Let $f(z) = a_0 + a_1 z + \cdots + a_n z^n$ be a polynomial of degree $n > 0$, where a_i's are complex numbers. We show that $f(z) = 0$ for some $z \in \mathbb{C}$. Obviously, it suffices to consider the case $a_n = 1$. Assume, on the contrary, that $f(z) \neq 0$ for all $z \in \mathbb{C}$. Then, we can consider f as a continuous map of $\mathbb{C}$ into $\mathbb{C}_0 = \mathbb{C} - \{0\}$. For each real $r > 0$, let C_r denote the circle of radius r centered at the origin 0. The mapping $F\colon C_r \times I \to \mathbb{C}_0$ defined by $F(z, t) = f((1-t)z)$, $z \in C_r$, $t \in I$ is a homotopy between $f|C_r$ and the constant map at a_0. Next, we show that, for large r, $f|C_r$ is homotopic to the restriction of the map $g\colon C_0 \to \mathbb{C}_0$, $z \mapsto z^n$, to C_r. Define a mapping $G\colon \mathbb{C} \times I \to \mathbb{C}$ by $G(z, t) = z^n + \sum_0^{n-1}(1-t)a_i z^i$. Clearly, G is continuous. If $G(z, t) = 0$ for a z with $|z| = r$, then we have the following:

$$r^n = \left| -\sum_0^{n-1}(1-t)a_i z^i \right| \leq (1-t)\sum_0^{n-1}|a_i|r^i \leq \sum_0^{n-1}|a_i|r^i.$$

This implies that if $r > \max\{1, \sum_0^{n-1}|a_i|\}$, then $G(z,t) \neq 0$ for every $z \in C_r$ and $t \in I$. Accordingly, for such values of r, G determines a continuous function H: $C_r \to \mathbb{C}_0$. It is easily seen that H is a desired homotopy between $f|C_r$ and $g|C_r$. Since the homotopy relation is an equivalence relation, we deduce that $g|C_r$ is null homotopic. Further, observe that the composition

$$\mathbb{S}^1 \to C_r \xrightarrow{g} \mathbb{C}_0 \to \mathbb{S}^1,$$

where the first mapping is $z \mapsto rz$, and the last mapping is $z \mapsto z/\|z\|$, which coincides with the mapping $h : \mathbb{S}^1 \to \mathbb{S}^1$, $z \mapsto z^n$. Hence, h is freely homotopic to a constant map, and it follows that $n = \deg h = 0$, by Proposition 14.3.8. This contradicts the hypothesis that $\deg f > 0$. $\diamond$

We conclude this section with another tool for computing fundamental groups, which expresses the fundamental group of the product of two spaces as the direct product of the fundamental groups of its factors.

Theorem 14.3.10 *Let X and Y be spaces, and $x_0 \in X$, $y_0 \in Y$. Then,*

$$\pi((X \times Y), (x_0, y_0)) \cong \pi(X, x_0) \times \pi(Y, y_0).$$

Proof Let $p_X : X \times Y \to X$ and $p_Y : X \times Y \to Y$ denote the projection maps. Then, we have a mapping $\phi : \pi((X \times Y), (x_0, y_0)) \longrightarrow \pi(X, x_0) \times \pi(Y, y_0)$ given by $\phi([h]) = ([p_X h], [p_Y h])$ for every loop h in $X \times Y$ based at (x_0, y_0). Obviously, $\phi([h]) = (p_{X\#}([h]), p_{Y\#}([h]))$, where $p_{X\#}$ and $p_{Y\#}$ are the homomorphisms induced by p_X and p_Y, respectively. Hence, ϕ is a homomorphism. We assert that it is an isomorphism. Suppose that f is a loop in X based at x_0, and g is a loop in Y based at y_0. Then, $h : I \to X \times Y$ defined by $h(t) = (f(t), g(t))$ is a loop based at (x_0, y_0), since $p_X h = f$, $p_Y h = g$. Obviously, $\phi([h]) = ([f], [g])$, and thus ϕ is onto. To see that ϕ is one-to-one, suppose that $\phi([h]) = \phi([k])$, h, k are loops in $X \times Y$ based at (x_0, y_0). Then, there exist homotopies $F : p_X h \simeq p_X k$ rel $\{0, 1\}$, and $G : p_Y h \simeq p_Y k$ rel $\{0, 1\}$. Accordingly, $H : I \times I \to X \times Y$, defined by $H(s, t) = (F(s, t), G(s, t))$, is a homotopy between $h = (p_X h, p_Y h)$ and $k = (p_X k, p_Y k)$ rel $\{0, 1\}$. So, $[h] = [k]$ and ϕ is one-to-one. This proves our assertion. $\diamond$

As an immediate consequence of the preceding theorem, we see that the fundamental group of the torus $\mathbb{S}^1 \times \mathbb{S}^1$ is $\mathbb{Z} \oplus \mathbb{Z}$, and that of solid torus $\mathbb{D}^2 \times \mathbb{S}^1$ is $\mathbb{Z}$.

Exercises

1. Determine the fundamental groups of the following spaces:
   (a) Möbius band, (b) cylinder (closed), (c) an annulus, and
   (d) a punctured disc.

2. Let $f: \mathbb{D}^2 \to \mathbb{D}^2$ be a homeomorphism. Show that f maps $\mathbb{S}^1$ onto itself, and $B(0; 1)$ onto itself.

3. If $\phi: \mathbb{S}^1 \to \mathbb{S}^1$ is null homotopic, show that ϕ has a fixed point and $\phi(x) = -x$ for some $x \in \mathbb{S}^1$.

4. Let f be a loop in $\mathbb{S}^1$ based at 1. If f is not surjective, show that $\deg f = 0$. Give an example of a surjective loop f with $\deg f = 0$.

5. Prove that $\mathbb{S}^1$ is not a retract of $\mathbb{D}^2$.

6. Prove that $\mathbb{S}^1 \times \{x\}$ is a retract of the torus $\mathbb{S}^1 \times \mathbb{S}^1$, but it is not a deformation retract for any $x \in \mathbb{S}^1$.

7. Determine the fundamental group of

   (a) a punctured n-space $\mathbb{R}^n - \{pt\}$ $(n > 2)$,

   (b) $\mathbb{S}^m \times \mathbb{S}^n$, $m, n \geq 2$,

   (c) $\mathbb{R}^3$ with a line removed, and

   (d) a punctured n-disc, $n > 2$.

8. Prove that $\mathbb{R}^2$ is not homeomorphic to $\mathbb{R}^n$ for $n \neq 2$.

9. (a) Find a starlike set that is not convex.

   (b) If X is starlike, show that X is simply connected.

10. Let $\phi: (\mathbb{S}^1, 1) \to (X, x_0)$ be a continuous map. Show that ϕ is null homotopic $\Leftrightarrow \phi_\#: \pi(\mathbb{S}^1, 1) \to \pi(X, x_0)$ is the trivial homomorphism.

11. Prove that a continuous map $f: \mathbb{S}^1 \to \mathbb{S}^1$ of degree 1 is homotopic to the identity map.

12. If $f: \mathbb{S}^1 \to \mathbb{S}^1$ is a continuous function such that $f(1) = 1$, show that $f_\#: \pi(\mathbb{S}^1, 1) \to \pi(\mathbb{S}^1, 1)$ is the multiplication by $\deg f$.

14.4 Some Group Theory

In the previous section, we have seen that the fundamental group of the product of two spaces X and Y is isomorphic to the direct product of their fundamental groups. Naturally, one would contemplate computing the fundamental group of the other constructions, such as the union, the wedge, etc., in terms of the fundamental groups of the component spaces. The structure of the fundamental groups of these constructs and the technique involved in their computations entail certain concepts of group theory, which are not generally discussed at the undergraduate level. Therefore, we collect here some necessary background material from group theory to make the discussion of the next section accessible to the reader.

Free Products

Given an indexed family $\{G_\alpha \mid \alpha \in A\}$ of groups, their direct product $\prod G_\alpha$ is a group with the group operation defined componentwise: If $x = (x_\alpha)$ and $y = (y_\alpha)$ are two elements of $\prod G_\alpha$, then $xy = (x_\alpha y_\alpha)$. Obviously, the element $e = (e_\alpha)$ of $\prod G_\alpha$, where e_α is the identity element of G_α, acts as the identity element for this multiplication, and the inverse of x in $\prod G_\alpha$ is the element (x_α^{-1}), where x_α^{-1} is the inverse of x_α in G_α. The group $\prod G_\alpha$ with the projections $p_\beta \colon \prod G_\alpha \to G_\beta$, $\beta \in A$, is characterized by the universal property that for any group H and a family of homomorphisms $f_\alpha \colon H \to G_\alpha$, one for each index α, there exists a unique homomorphism $h \colon H \to \prod G_\alpha$ such that $p_\alpha \circ h = f_\alpha$ for all α.

The direct sum of the G_α, denoted by $\bigoplus G_\alpha$, is the subgroup of $\prod G_\alpha$ consisting of the elements x such that $x_\alpha = e_\alpha$ for all but finitely many indices α. For each index $\beta \in A$, we have a canonical injection $\iota_\beta \colon G_\beta \to \bigoplus G_\alpha$ defined by

$$\left(\iota_\beta(x)\right)_\alpha = \begin{cases} x & \text{for } \alpha = \beta \text{ and} \\ e_\alpha & \text{for } \alpha \neq \beta. \end{cases}$$

If all the groups G_α are abelian, then their direct sum $\bigoplus G_\alpha$ (also the direct product $\prod G_\alpha$) is abelian and, with the family of monomorphisms ι_β, $\beta \in A$, it satisfies the following universal properties: Given an abelian group H and a homomorphism $f_\alpha \colon G_\alpha \to H$ for each $\alpha \in A$, there exists a unique homomorphism $h \colon \bigoplus G_\alpha \to H$ such that $h \circ \iota_\alpha = f_\alpha$. This property, in fact, characterizes the direct sum of the abelian groups.

It should be noted that an analogous theorem for the direct sum of arbitrary (that is, not necessarily abelian) groups does not hold good. This fact, however, leads us to a useful concept, which we study now.

Definition 14.4.1 The *free product* of an indexed family of groups G_α, $\alpha \in A$, is a group F with a family of homomorphisms $\iota_\alpha \colon G_\alpha \to F$, one for each $\alpha \in A$, such that for any group G and any family of homomorphisms $f_\alpha \colon G_\alpha \to G$, $\alpha \in A$, there is a unique homomorphism $h \colon F \to G$ such that $h \circ \iota_\alpha = f_\alpha$, that is, the following triangle commutes for every α.

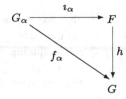

We derive immediately a few consequences of this definition. First, if F is free product of groups G_α, then each homomorphism $\iota_\alpha \colon G_\alpha \to F$ is a monomorphism. For, taking $G = G_\alpha$, $f_\alpha = 1_{G_\alpha}$ and the homomorphism $f_\beta \colon G_\beta \to G$ to be trivial for $\beta \neq \alpha$, we obtain a homomorphism $h \colon F \to G$ such that $h \circ \iota_\alpha =$

1_{G_α} and $h \circ \iota_\beta$ is the trivial homomorphism for every $\beta \neq \alpha$. From the equality $h \circ \iota_\alpha = 1_{G_\alpha}$, it is immediate that ι_α is injective. Second, for $\alpha \neq \beta$, we have $\iota_\alpha(G_\alpha) \cap \iota_\beta(G_\beta) = \{\varepsilon\}$, where ε is the identity element of F. For, if $x_\alpha \in G_\alpha$, $x_\beta \in G_\beta$, and $\iota_\alpha(x_\alpha) = \iota_\beta(x_\beta)$, then, with the above homomorphism h, we have $x_\alpha = (h \circ \iota_\alpha)(x_\alpha) = (h \circ \iota_\beta)(x_\beta) = e_\alpha$, the identity element of G_α. So $\iota_\alpha(x_\alpha) = \varepsilon$, and we conclude that $\iota_\alpha(G_\alpha) \cap \iota_\beta(G_\beta) = \{\varepsilon\}$. Third, F is generated by $\bigcup_\alpha \iota_\alpha(G_\alpha)$. To see this, let H be the subgroup of F generated by $\bigcup_\alpha \iota_\alpha(G_\alpha)$. Then, for each $\alpha \in A$, we have a homomorphism $j_\alpha \colon G_\alpha \to H$ defined by ι_α. By the definition of F, there exists a unique homomorphism $k \colon F \to H$ such that $k \circ \iota_\alpha = j_\alpha$, for every $\alpha \in A$. Obviously, the composition of j_α with the inclusion homomorphism $\ell \colon H \hookrightarrow F$ is ι_α, and thus we have $\ell \circ k \circ \iota_\alpha = \iota_\alpha$ for all α. Since $1_F \circ \iota_\alpha = \iota_\alpha$ is also true for every α, we have $\ell \circ k = 1_F$, by the uniqueness condition in Definition 14.4.1. It follows that ℓ is onto, and hence $H = F$.

The free product of groups G_α, $\alpha \in A$, will be denoted by $\bigstar \{G_\alpha \mid \alpha \in A\}$ or, in the finite case, by $G_1 \star \cdots \star G_n$. We usually identify each group G_β with its image under the monomorphism $\iota_\beta \colon G_\beta \to \bigstar \{G_\alpha \mid \alpha \in A\}$, and regard it as a subgroup of $\bigstar \{G_\alpha \mid \alpha \in A\}$. Then, $\bigstar \{G_\alpha \mid \alpha \in A\}$ is generated by the subgroups G_α. So each element ξ in $\bigstar \{G_\alpha \mid \alpha \in A\}$ can be written as $\xi = x_1 \cdots x_n$, $x_i \in G_{\alpha_i}$. For many purposes, it is important to have a unique factorization of each element ξ of $\bigstar \{G_\alpha \mid \alpha \in A\}$ as a product of elements of the subgroups G_α. Later, we will see the existence of such a factorization for each element other than the identity element of $\bigstar \{G_\alpha \mid \alpha \in A\}$.

As another consequence of the above definition, we see that the free product of the groups G_α, if it exists, is unique up to isomorphism. For, suppose that F together with the homomorphisms $\iota_\alpha \colon G_\alpha \to F$ and F' together with the homomorphisms $\iota'_\alpha \colon G_\alpha \to F'$ are free products of the groups G_α. Then, there are unique homomorphisms $h \colon F \to F'$ and $h' \colon F' \to F$ such that $h \circ \iota_\alpha = \iota'_\alpha$ and $h' \circ \iota'_\alpha = \iota_\alpha$ for all α. Consequently, we have $h' \circ h \circ \iota_\alpha = \iota_\alpha$ and $h \circ h' \circ \iota'_\alpha = \iota'_\alpha$ for every α. Thus, for each α, we have the following commutative diagrams:

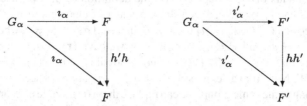

Obviously, these triangles remain commutative when $h'h$ is replaced with the identity isomorphism 1_F on F, and hh' with the identity isomorphism $1_{F'}$ on F'. By the definition of free product, the homomorphisms $F \to F$ and $F' \to F'$ with this property are unique, so we have $h' \circ h = 1_F$ and $h \circ h' = 1_{F'}$. This implies that $h \colon F \to F'$ is an isomorphism with h' as its inverse.

Given a collection $\{G_\alpha \mid \alpha \in A\}$ of groups, a natural question arises about the existence of their free product $\bigstar \{G_\alpha \mid \alpha \in A\}$. The answer to this question is in the affirmative, as shown by the following.

Theorem 14.4.2 *Given a collection $\{G_\alpha \mid \alpha \in A\}$ of groups, their free product exists.*

Proof For each integer $n \geq 0$, we define a *word* of length n in the groups G_α to be a finite sequence (or an ordered n-tuple) $(x_1, \ldots, x_n)$ of elements of the disjoint union $\sum G_\alpha$. If $n = 0$, the word $(x_1, \ldots, x_n)$ is considered to be the empty sequence and denoted by $()$. By a *reduced word* in the G_α, we mean a word such that any two consecutive terms in the word belong to different groups and no term is the identity element of any G_α. The empty word is (vacuously) a reduced word. Let W be the set of all reduced words in the G_α. For each index α and each $x \in G_\alpha$, we define a permutation σ_x of W as follows:

If $x = e_\alpha$, the identity element of G_α, then σ_x is the identity map on W. If $x \neq e_\alpha$ and $w \in W$ is the (nonempty) word $(x_1, \ldots, x_n)$, then we write

$$\sigma_x(w) = \begin{cases} (x, x_1, \ldots, x_n) & \text{if } x_1 \notin G_\alpha, \\ (xx_1, \ldots, x_n) & \text{if } x_1 \in G_\alpha \text{ and } xx_1 \neq e_\alpha, \\ (x_2, \ldots, x_n) & \text{if } x_1 \in G_\alpha \text{ and } xx_1 = e_\alpha. \end{cases}$$

If $w = ()$, then we put $\sigma_x(w) = (x)$.

A simple checking of various cases shows that $\sigma_x \sigma_y = \sigma_{xy}$ for all $x, y \in G_\alpha$. In particular, we have $\sigma_x \sigma_{x^{-1}} = \sigma_{e_\alpha} = \iota$, the identity permutation of W, for every $x \in G_\alpha$. Putting x^{-1} for x in this equation, we get $\sigma_{x^{-1}} \sigma_x = \iota$. So each σ_x is a permutation of W and there is a homomorphism $i_\alpha \colon G_\alpha \to \mathrm{Symm}(W)$ given by $i_\alpha(x) = \sigma_x$. We note that i_α is injective. For, if $x \neq e_\alpha$ in G_α, then $\sigma_x() = (x)$ and $\sigma_{e_\alpha}() = ()$. Denote the image of i_α in $\mathrm{Symm}(W)$ by G'_α and let S be the subgroup of $\mathrm{Symm}(W)$ generated by $\bigcup_\alpha G'_\alpha$. Clearly, each monomorphism i_α may be regarded as a homomorphism of G_α into S. We claim that the group S is the free product of the G_α with respect to the homomorphisms i_α.

To this end, we first find a unique expression for each element of S as a finite product of elements of the groups G'_α. By the definition of S, it is clear that every element $\theta \in S$ can be expressed as a finite product of elements of the groups G'_α. Further, we see that $G'_\alpha \cap G'_\beta = \{\iota\}$ for $\alpha \neq \beta$. In fact, if $\sigma_x = \sigma_y \neq \iota$, where $x \in G_\alpha$, $y \in G_\beta$, then $x \neq e_\alpha$ and $y \neq e_\beta$. Accordingly, we have $(x) = \sigma_x() = \sigma_y() = (y)$, which implies that $x = y \in G_\alpha \cap G_\beta$. This is possible only when $\alpha = \beta$, by the definition of W. Now, if in an expression $\theta = \sigma_{x_1} \cdots \sigma_{x_n}$, two consecutive factors σ_{x_i} and $\sigma_{x_{i+1}}$ belong to the same group, we can replace them by a single element, viz. their product $\sigma_{x_i} \sigma_{x_{i+1}}$, to obtain a similar expression for θ with fewer factors. Moreover, if $\theta \neq \iota$ and a factor in the above expression for θ is ι, then we can delete it from the given expression, obtaining a shorter one. Applying these reduction operations repeatedly, an element $\theta \neq \iota$ can be written so that no two consecutive factors belong to the same group G'_α and no factor is the identity permutation. Such an expression of θ is said to be in the *reduced form*. We show that the expression $\theta = \sigma_{x_1} \cdots \sigma_{x_n}$ in reduced form is unique for any element $\theta \neq \iota$ of S. Notice that, in this case, if $x_j \in G_{\alpha_j}$, then no $x_j = e_{\alpha_j}$ and $\alpha_j \neq \alpha_{j+1}$. So the word $(x_1, \ldots, x_n)$ belongs to W, and we have $\theta() = (x_1, \ldots, x_n)$. Similarly, if $\theta = \sigma_{y_1} \cdots \sigma_{y_m}$ is another expression in reduced form, then $\theta() = (y_1, \ldots, y_m)$. Thus, we have $(x_1, \ldots, x_n) = (y_1, \ldots, y_m)$, which implies that $n = m$ and $x_j = y_j \; \forall \; j$.

Next, to prove our claim, suppose that G is a group and, for each index $\alpha \in A$, let $f_\alpha \colon G_\alpha \to G$ be a homomorphism. We need to construct a unique homomorphism $h \colon S \to G$ such that $h \circ i_\alpha = f_\alpha$ for every α. Notice that if there is a homomorphism $h \colon S \to G$ satisfying $h \circ i_\alpha = f_\alpha$ for all α, and $\theta = \sigma_{x_1} \cdots \sigma_{x_n}$, where $x_j \in G_{\alpha_j}$, is an element of S, then we have

$$h(\theta) = h\left(\sigma_{x_1}\right) \cdots h\left(\sigma_{x_n}\right) = f_{\alpha_1}(x_1) \cdots f_{\alpha_n}(x_n).$$

Accordingly, if $\sigma_{x_1} \cdots \sigma_{x_n}$ is the reduced form of $\theta \neq \iota$, then we define $h(\theta)$ by this equation and put $h(\iota) = e$, the identity element of G. Since the expression $\theta = \sigma_{x_1} \cdots \sigma_{x_n}$ of θ in the reduced form is unique, h is a mapping. Also, the equality $h \circ i_\alpha = f_\alpha$ is obvious for every α. It remains to see that h is a homomorphism. Suppose that $\theta, \zeta \in S$. If θ or ζ is ι, then we obviously have $h(\theta\zeta) = h(\theta)h(\zeta)$. So assume that $\theta \neq \iota \neq \zeta$, and let $\theta = \sigma_{x_1} \cdots \sigma_{x_n}$ and $\zeta = \sigma_{y_1} \cdots \sigma_{y_m}$ be the expressions in reduced form, where $x_k \in G_{\alpha_k}$ and $y_l \in G_{\beta_l}$. Then, the product $\theta\zeta = \sigma_{x_1} \cdots \sigma_{x_n} \sigma_{y_1} \cdots \sigma_{y_m}$ is not in the reduced form in case $\alpha_n = \beta_1$. Therefore, we cannot evaluate h at $\theta\zeta$ directly. To tackle this problem, we consider an auxiliary function ψ from the set $\mathcal{W}$ of all words in the groups G_α to G. Notice that if $w = (z_1, \ldots, z_q)$ is a word in $\mathcal{W}$ with $q > 0$, then there is a unique index $\gamma_j \in A$ such that $z_j \in G_{\gamma_j}$. So w determines a unique element $\vartheta = i_{\gamma_1}(z_1) \cdots i_{\gamma_q}(z_q) = \sigma_{z_1} \cdots \sigma_{z_q}$ of S. Clearly, if $w = (z_1, \ldots, z_q)$ is a reduced word, then the expression $\vartheta = i_{\gamma_1}(z_1) \cdots i_{\gamma_q}(z_q)$ is in the reduced form. We set $\psi(w) = f_{\gamma_1}(z_1) \cdots f_{\gamma_q}(z_q)$ and $\psi() = e$. Then, ψ defines a mapping $\mathcal{W} \to G$ such that $\psi(w) = h(\vartheta)$ if w is a reduced word determining ϑ, and $\psi() = h(\iota)$. Observe that if $w = (z_1, \ldots, z_q) \in \mathcal{W}$, where $z_j \in G_{\gamma_j}$ and $\gamma_j = \gamma_{j+1}$, and the word w' is obtained from w by replacing its consecutive terms z_j and z_{j+1} with their product $z_j z_{j+1}$, then $\psi(w) = \psi(w')$, for $f_{\gamma_j}(z_j) f_{\gamma_{j+1}}(z_{j+1}) = f_{\gamma_j}(z_j z_{j+1})$. And, if a word w' is obtained from w by deleting some $z_j = e_{\gamma_j}$, then also $\psi(w) = \psi(w')$, for $f_{\gamma_j}(z_j) = e$. It follows that if the reduced word w' is obtained from a word $w \in \mathcal{W}$ by applying finitely many times the preceding reduction operations, then we have $\psi(w) = \psi(w')$. Notice further that w and w' determine the same element of S, since each i_γ is a homomorphism. Accordingly, if an element $\vartheta \in S$ is determined by a word w, then $h(\vartheta) = \psi(w)$. Since $\theta\zeta = i_{\alpha_1}(x_1) \cdots i_{\alpha_n}(x_n) i_{\beta_1}(y_1) \cdots i_{\beta_m}(y_m)$, we see that

$$\begin{aligned} h(\theta\zeta) &= \psi(x_1, \ldots, x_n, y_1, \ldots, y_m) \\ &= f_{\alpha_1}(x_1) \cdots f_{\alpha_n}(x_n) f_{\beta_1}(y_1) \cdots f_{\beta_m}(y_m) \\ &= h(\theta)h(\zeta), \end{aligned}$$

and h is a homomorphism. The uniqueness of h is trivial, since S is generated by the images of i_α's. $\diamond$

We now find a unique expression for each element $\xi \neq \varepsilon$ (the identity element) of $\bigstar \{G_\alpha \mid \alpha \in A\}$ as a finite product of elements of the G_α. The notations used here have the meaning assigned to them in the proof of the preceding theorem. Following the procedure adopted for reducing factorizations of elements of the group S, any such expression for ξ can be put in reduced form, wherein no two consecutive factors belong to the same G_α and no factor is the identity element. We observe that this expression of ξ is unique. Assume that $x_1 \cdots x_n = \xi = y_1 \cdots y_m$, where $x_j \in G_{\alpha_j}$ and $y_k \in G_{\beta_k}$ are two factorizations of ξ in reduced form. We need to show that $n = m$ and $x_j = y_j$ for $1 \leq j \leq n$. By the definition of the free product, there exists a unique homomorphism $h' : \bigstar \{G_\alpha \mid \alpha \in A\} \to S$ such that $h'|G_\alpha = i_\alpha$ for every α. So we have $h'(\xi) = i_{\alpha_1}(x_1) \cdots i_{\alpha_n}(x_n) = i_{\beta_1}(y_1) \cdots i_{\beta_m}(y_m)$. Since $i_{\alpha_j}(x_j) = \sigma_{x_j}$ and $i_{\beta_k}(y_k) = \sigma_{y_k}$, it follows that $h'(\xi) = \sigma_{x_1} \cdots \sigma_{x_n} = \sigma_{y_1} \cdots \sigma_{y_m}$. As no $x_j = e_{\alpha_j}$ and $\alpha_j \neq \alpha_{j+1}$, etc., these expressions of $h'(\xi)$ are in reduced form. Therefore, $n = m$ and $x_j = y_j$ for $1 \leq j \leq n$. We note that any expression for the identity element ε of $\bigstar \{G_\alpha \mid \alpha \in A\}$ as a product of elements of the G_α is not in the reduced form.

There is another way to construct the free product of the groups G_α. In fact, one can turn the set W of all reduced words in the G_α itself into a group so that it satisfies the "universal property" stipulated in Definition 14.4.1. The multiplication is defined by concatenation

$$(x_1, \ldots, x_n)(y_1, \ldots, y_n) = (x_1, \ldots, x_n, y_1, \ldots, y_n)$$

and then reduction. The empty word acts as the identity element and the inverse of a word $(x_1, \ldots, x_n)$ in W is the word $(x_n^{-1}, \ldots, x_1^{-1})$. For each group G_α, the required monomorphism $\iota_\alpha : G_\alpha \to W$ is defined by $\iota_\alpha(x) = (x)$. Apparently, this approach is simpler; but, the verification of the axioms for a group in W, particularly that of associativity, is tedious. To avoid this tedium, the former proof for the existence of free products, due to B. L. van der Waerden, has been preferred. Clearly, there is an isomorphism between S and W given by $\theta \mapsto \theta()$ and $\iota \mapsto ()$, the empty word. Under this isomorphism, a word $(x_1, \ldots, x_n)$ in W corresponds to the permutation $\sigma_{x_1} \cdots \sigma_{x_n}$ in S. Observe further that the homomorphism $\bigstar \{G_\alpha \mid \alpha \in A\} \to S$ $x_1 \cdots x_n \mapsto \sigma_{x_1} \cdots \sigma_{x_n}$, is actually an isomorphism. Hence, an element of $\bigstar \{G_\alpha \mid \alpha \in A\}$ having the reduced form $x_1 \cdots x_n$ corresponds to the unique word $(x_1, \ldots, x_n)$ in W. By abuse of the terminology, the product $x_1 \cdots x_n$ will also be called a word; in particular, the empty word is considered the reduced form of ε.

Example 14.4.1 Let $G_1 = \{e_1, g_1\}$ and $G_2 = \{e_2, g_2\}$ be two disjoint copies of the cyclic group $\mathbb{Z}_2$. Then, an element of $\mathbb{Z}_2 \star \mathbb{Z}_2$ other than the identity element has a unique representation as a product of alternating g_1's and g_2's. For example,

$$g_1, g_1 g_2, g_1 g_2 g_1, g_1 g_2 g_1 g_2 \quad \text{and}$$
$$g_2, g_2 g_1, g_2 g_1 g_2, g_2 g_1 g_2 g_1$$

are some of the elements of $\mathbb{Z}_2 \star \mathbb{Z}_2$. Note that $g_1 g_2 \neq g_2 g_1$ and both are of infinite order.

Free Abelian Groups

Recall that a group G is said to be generated by a set $X \subseteq G$ if every element of G can be written as a product of elements of X and their inverses. The empty set generates the trivial group. In case the group G is abelian, the expression of an element $g \in G$ in terms of a generating set $X \neq \emptyset$ simplifies as $g = x_1^{n_1} \cdots x_k^{n_k}$, where $x_i \in X$ and $n_i \in \mathbb{Z}$. In general, one cannot assert that the elements $x_1, \ldots, x_k$ in X and the exponents $n_1, \ldots, n_k$ are uniquely determined by g. An abelian group G is called *free* on a subset $X \subset G$ if every element $g \in G$ different from the identity element e can be uniquely written as $g = x_1^{n_1} \cdots x_k^{n_k}$. If G is free on X, then X generates G and is linearly independent in the sense that a relation $x_1^{n_1} \cdots x_k^{n_k} = e$ with distinct $x_i \in X$ holds if and only if all the exponents n_i are zero. Such a subset of G is called a *basis* for G. If X is a basis for G, then the cyclic subgroup $F_x = \langle x \rangle$ generated by each $x \in X$ is infinite, and G is the direct sum of the subgroups $F_x, x \in X$. It follows that *a free abelian group is the direct sum of isomorphic copies of $\mathbb{Z}$, the group of integers.*

Given a nonempty set X, there exists an abelian group $F(X)$ which is free on X. To construct $F(X)$, we form an infinite cyclic group $\langle x \rangle = \{x^n \mid n \in \mathbb{Z}\}$ for every $x \in X$. The multiplication in $\langle x \rangle$ is defined by $x^n x^m = x^{n+m}$, $1 = x^0$ is taken as the identity element and x^{-n} as the inverse of x^n. It is easily checked that $\langle x \rangle$ is an infinite cyclic group generated by x^1. The element x is identified with x^1, and we say that $\langle x \rangle$ is the infinite cyclic group generated by x. Denote $\langle x \rangle$ by F_x, and let $F(X) = \bigoplus_{x \in X} F_x$ (the external direct sum). Then, $F(X)$ is an abelian group and consists of the functions $f: X \to \bigcup_{x \in X} F_x$ such that $f(x) \in F_x$ and $f(x) = 1$ for all but finitely many x's. For each $x \in X$, consider the function $u_x: X \to \bigcup_{x \in X} F_x$ defined by

$$u_x(y) = \begin{cases} x & \text{if } y = x, \text{ and} \\ 1 \text{ (in } F_y) & \text{for all } y \neq x. \end{cases}$$

Then, $u_x \in F(X)$. We observe that $B = \{u_x \mid x \in X\}$ is a basis of $F(X)$. Let $f \in F(X)$ be arbitrary. Then, $f: X \to \bigcup_{x \in X} F_x$ such that $f(x) \neq 1$ for at most a finite number of x's in X. Suppose that f has the values $x_1^{n_1}, \ldots, x_k^{n_k}$ with nonzero exponents on $x_1, \ldots, x_k$ and is 1 elsewhere. Then, we have $f = u_{x_1}^{n_1} \ldots u_{x_k}^{n_k}$. This shows that $F(X)$ is spanned by B. Next, if $u_{x_1}^{n_1} \cdots u_{x_k}^{n_k} = e$, the identity element of $F(X)$, and $x_1, \ldots, x_k$ are all distinct, then $1 = e(x_i) = \left(u_{x_1}^{n_1} \cdots u_{x_k}^{n_k}\right)(x_i) = x_i^{n_i}$. This implies that $n_i = 0$ for every $i = 1, \ldots, k$ and B is linearly independent. Thus, we see that B is a basis of $F(X)$. Obviously, the canonical mapping $\lambda: X \to F(X)$, $x \mapsto u_x$, is an injection. So we can identify X with $B = \lambda(X)$ and, thus, regard X a subset of $F(X)$. Then, X is a base of $F(X)$.

The free abelian groups are characterized by the following universal mapping property.

Theorem 14.4.3 *If G is a free abelian group with basis X, then any function from X to an abelian group H can be extended to a unique homomorphism of G into H, and conversely.*

Proof Suppose that an abelian group G is free with basis X. Let H be an abelian group and $\phi: X \to H$ be a set function. Then, each g in G distinct from the identity element e can be uniquely written as $g = x_1^{n_1} \cdots x_k^{n_k}$, where the x_i are distinct elements of X and the integers $n_i \neq 0$. We put $\tilde{\phi}(g) = \phi(x_1)^{n_1} \cdots \phi(x_k)^{n_k}$ and $\tilde{\phi}(e) = e'$, the identity element of H. The uniqueness of the expression for g shows that $\tilde{\phi}$ is a well-defined mapping of G into H. It is easily verified that $\tilde{\phi}$ is a homomorphism and $\tilde{\phi}(x) = \phi(x)$ for all $x \in X$. Next, if $\tilde{\psi}: G \to H$ is another homomorphism such that $\tilde{\psi}|X = \phi$, then $\tilde{\psi} = \tilde{\phi}$, since X spans G.

Conversely, suppose that X is a subset of an abelian group G such that, for each abelian group H and each function $\phi: X \to H$, there exists a unique homomorphism $\tilde{\phi}: G \to H$ such that $\tilde{\phi}(x) = \phi(x)$ for all $x \in X$. Let $F(X)$ be the free abelian group generated by X with the canonical injection $\lambda: X \to F(X)$. Taking $H = F(X)$, we get a homomorphism $\tilde{\lambda}: G \to F(X)$ such that $\tilde{\lambda}(x) = \lambda(x)$ for all $x \in X$. If $\mu: X \hookrightarrow G$ is the inclusion map, then there exists a homomorphism $\tilde{\mu}: F(X) \to G$ such that $\tilde{\mu}\lambda(x) = \mu(x)$ for all $x \in X$. So $\tilde{\mu}\tilde{\lambda}(x) = x$ for all $x \in X$. Since the identity automorphism 1_G of G also satisfies these equations, we see that $\tilde{\mu}\tilde{\lambda} = 1_G$. This implies that $\tilde{\lambda}$ is injective. Since $\{\lambda(x) \mid x \in X\}$ generates $F(X)$, $\tilde{\lambda}$ is surjective. Thus, $\tilde{\lambda}$ is an isomorphism with the inverse $\tilde{\mu}$. Since $\{\lambda(x) \mid x \in X\}$ is a basis of $F(X)$, $X = \{\tilde{\mu}\lambda(x) \mid x \in X\}$ is also a basis of G, and thus G is free on X.   ◇

From the preceding theorem, it is clear that every abelian group is a homomorphic image of a free abelian group. For, given an abelian group G, we can always find a set X of generators of G. Then, the homomorphism $F(X) \to G$, which extends the inclusion map $X \hookrightarrow G$, is surjective.

As another consequence of the above theorem, we see that two free abelian groups whose bases have the same cardinality are isomorphic. Let F and F' be two free abelian groups with bases X and X', respectively, and suppose that $|X| = |X'|$. Then, there exists a bijection $\phi: X \to X'$. Clearly, we may regard ϕ as a mapping of X into F'. Then, by the preceding theorem, ϕ extends to a homomorphism $\tilde{\phi}: F \to F'$. Similarly, considering the inverse ψ of ϕ as a mapping of X' into F, we have a homomorphism $\tilde{\psi}: F' \to F$, which is an extension of ψ. Then, we have $(\tilde{\psi} \circ \tilde{\phi})(x) = \psi\phi(x) = x$ for all $x \in X$. Since F is generated by X, it follows that $\tilde{\psi} \circ \tilde{\phi} = 1_F$. Similarly, $\tilde{\phi} \circ \tilde{\psi} = 1_{F'}$, and thus $\tilde{\phi}$ is an isomorphism with $\tilde{\psi}$ as its inverse.

On the other hand, we see that the cardinal number $|X|$ of a basis X of a free abelian group F is uniquely determined. We first observe that if $|X|$ is finite, then every basis of F is finite. For, suppose that Y is another basis of F. Then, each element of X is a product of integral powers of finitely many elements in Y. Accordingly, we find a finite subset $S \subseteq Y$ which also generates F. As Y is a basis of F, we have $Y = S$, and thus it is finite. Now, consider the normal subgroup $N = \{u^2 \mid u \in F\}$ of F. Clearly, the quotient F/N is a vector space over $\mathbb{Z}_2$ with the obvious scalar multiplication, and the set $B = \{xN \mid x \in X\}$ generates F/N over $\mathbb{Z}_2$. If B were linearly dependent over $\mathbb{Z}_2$, then a finite product $x_1 \cdots x_n$ of distinct elements $x_i \in X$ would be square of some element of F. Assume that $x_1 \cdots x_n = y_1^{2k_1} \cdots y_m^{2k_m}$, where $y_j \in X$ and the integers $k_j \neq 0$. This contradicts the linear independence of X over $\mathbb{Z}$. Therefore, B is linearly independent over $\mathbb{Z}_2$, and forms a basis of F/N. Notice that if B is finite, then $|F/N| = 2^{|B|}$ and, if B is infinite, then $|F/N| = |B|$, since the family of all *finite* subsets of B has the same cardinality as B (refer to Theorem A.8.4 in Appendix A). The mapping $x \leftrightarrow xN$ is a bijection between X and B, since X is linear independent over $\mathbb{Z}$. So $|B| = |X|$, and therefore $|X|$ is independent of the choice of the basis X. We call $|X|$ the *rank* of F.

Free Groups

The characterization of free abelian groups described by Theorem 14.4.3 leads us to consider arbitrary groups with the similar property.

Definition 14.4.4 A group F is said to be *free* on a subset $X \subset F$ (or *freely generated by X*) if, for every group G and every function $f : X \to G$, there exists a unique homomorphism $h : F \to G$ such that $h|X = f$. The subset X is referred to as a *basis* for F.

We remark that the requirement of uniqueness on the extensions of maps $X \to G$ is equivalent to condition that X generates F. By the universal mapping property of F (that any function from X to a group G can be extended in just one way to a homomorphism $F \to G$), no relation among the elements of X holds except the trivial relations that are valid for any set of elements in any group. This justifies the use of the adjective "free" to F.

Suppose that F is a free group with basis $X \neq \emptyset$. Then, each mapping f from X to a group G extends to a unique homomorphism $h : F \to G$. Let $x \in X$ be a fixed element. Consider the function $f : X \to \mathbb{Z}$ which sends x to 1 and maps the other elements of X into 0. By our hypothesis, f extends to a homomorphism $\tilde{f} : F \to \mathbb{Z}$. Consequently, the order of x must be infinite; so the cyclic group $\langle x \rangle \subseteq F$ is infinite for every $x \in X$. We observe that F is the free product of the groups $\langle x \rangle$, $x \in X$. Given a family of homomorphisms $h_x : \langle x \rangle \to G$, $x \in X$, we have a function $f : X \to G$ obtained by setting $f(x) = h_x(x)$. Then, there is a unique homomorphism $\phi : F \to G$ such that $\phi(x) = f(x) = h_x(x)$ for every $x \in X$. Since the group $\langle x \rangle$ is generated by x, we have $\phi|\langle x \rangle = h_x$ for every $x \in X$. It follows that F is the free product of infinite cyclic groups $\langle x \rangle$, $x \in X$. In particular, the free group with basis a singleton $\{x\}$ is the infinite cyclic group $\langle x \rangle$; this is the only case where a free group is abelian.

The above property of free groups provides us a clue for the construction of a free group with a given set X as basis. If X is nonempty, then we consider the free product P of the infinite cyclic groups $F_x = \langle x \rangle$, $x \in X$. There is a natural injection $X \to P$ which sends an element $x \in X$ to the word $x \in P$. Thus, X can be regarded as a subset of P, and then P is actually generated by X. We show that the group P is free on X. Let G be a group and $f: X \to G$ a mapping. Then, we have a family of homomorphisms $f_x: F_x \to G, x \in X$, defined by $f_x(x^n) = f(x)^n$. By the universal mapping property for free products, there exists a unique homomorphism $h: P \to G$ which extends each f_x. Then, $f(x) = f_x(x) = h(x)$ for all $x \in X$, and P is free on X. We denote P by F_X. The free group generated by the empty set $\emptyset$ is, by convention, the trivial group $\{1\}$. Thus, we have established the following.

Theorem 14.4.5 *Given a set X, there exists a free group F_X with basis X.*

If X is a basis of a free group F, then each element $w \in F$ can be written as $w = x_1^{n_1} \cdots x_k^{n_k}, x_i \in X$ and $n_i \in \mathbb{Z}$. It can be further expanded as a product of elements of $X \cup X^{-1}$, where $X^{-1} = \{x^{-1} \mid x \in X\} \subset F$. If in this expression of w, certain part is of the form xx^{-1} or $x^{-1}x$, then we can omit it without affecting w. By carrying out all such simplifications for $w \neq 1$ (the identity element of F), we may assume that $w = x_1 \cdots x_m$, where $x_i \in X \cup X^{-1}$ and all $x_i x_{i+1} \neq 1$. Such an expression for w is said to be in reduced form. We shall see in a moment that the expression of each element $w \neq 1$ of F as a product of elements of $X \cup X^{-1}$ in reduced form is unique. We note that X^{-1} is disjoint from X, and $x \leftrightarrow x^{-1}$ is a bijection between X and X^{-1}.

Proposition 14.4.6 *Let G be a group and X be a subset of G such that $X \cap X^{-1} = \emptyset$. Then, the subgroup $\langle X \rangle$ generated by X is free with X as a basis if and only if no product $w = x_1 \cdots x_n$ is the identity element e of G, where $n \geq 1$, $x_i \in X \cup X^{-1}$ and all $x_i x_{i+1} \neq e$.*

Proof Assume first that $\langle X \rangle$ is free with basis X. Then, for each $x \in X$, the cyclic subgroup $\langle x \rangle$ is infinite and $\langle X \rangle$ is the free product of these subgroups. If $w = x_1 \cdots x_n$, where $n \geq 1$, $x_i \in X \cup X^{-1}$ and all $x_i x_{i+1} \neq e$, then w can also be written as $w = x_{j_1}^{k_1} \cdots x_{j_m}^{k_m}$, where $x_{j_i} \in X$, $x_{j_i} \neq x_{j_{i+1}}$ and $k_i \in \mathbb{Z}$ are all nonzero. This expression in $\bigstar \{\langle x \rangle \mid x \in X\}$ is obviously in reduced form, and therefore $w \neq e$.

Conversely, suppose that no such product is e. Consider a set Y that is in one-to-one correspondence with X, and let $f: Y \to X$ be a bijection. Let F_Y be the free group on Y, which exists by Theorem 14.4.5. Then, there exists a homomorphism $h: F_Y \to G$ such that $h|Y = f$. Also, if $u \in F_Y$ is different from the identity element, then we have $u = y_1^{n_1} \cdots y_k^{n_k}$, where $y_i \in Y$, $y_i \neq y_{i+1}$ and $n_i \in \mathbb{Z}$ are all nonzero. So $h(u) = h(y_1)^{n_1} \cdots h(y_k)^{n_k}$ which can be expanded as $x_1 \cdots x_m$, $x_j \in X \cup X^{-1}$. Since $h|Y$ is injective and $y_i \neq y_{i+1}$, $x_j x_{j+1} \neq e$ for all $j = 1, \ldots, m-1$. By our hypothesis, $h(u) \neq e$ and h is a monomorphism. Since $h(Y) = f(Y) = X$, h maps F_Y onto $\langle X \rangle$. Thus, $h: F_Y \to \langle X \rangle$ is an isomorphism carrying Y onto X, and therefore $\langle X \rangle$ is free on X. $\diamond$

As an immediate consequence of the preceding proposition, we see that if X is a basis of a free group F, then the expression of each element $w \neq 1$ of F as a product of elements of $X \cup X^{-1}$ in reduced form is unique. For, if the expressions for two elements v and w of F as products of elements of $X \cup X^{-1}$ in reduced form are distinct, then vw^{-1} has clearly a nonempty expression in reduced form. Hence, $vw^{-1} \neq 1$ which implies that $v \neq w$.

Some authors directly construct the free group F_X on the given set X by taking another set X^{-1} which is disjoint from and equipotent to X. A one-to-one correspondence $X \to X^{-1}$ is indicated by $x \mapsto x^{-1}$. A *word* on $X \cup X^{-1}$ is a formal product $w = y_1 \cdots y_n$, where each $y_i \in X \cup X^{-1}$. An *elementary reduction* of a word w consists of deletion of parts xx^{-1} and $x^{-1}x$, $x \in X$. By applying elementary reductions on w, one can assume that for any $i = 1, \ldots, n-1$, the part $y_i y_{i+1}$ of w is not of the form xx^{-1} or $x^{-1}x$. Then, w is called a *reduced word*. Under the above reduction process, some words, such as xx^{-1}, lead to the *empty word* which has no factors. This is regarded as a reduced word. The set of all reduced words on $X \cup X^{-1}$ is taken as the underlying set for F_X and the binary operation is juxtaposition and then elementary reductions. The empty word acts as the identity element and the inverse of the word $y_1 \cdots y_n$ is defined to be $y_n^{-1} \cdots y_1^{-1}$.

Next, we consider an important relation between the free group and the free abelian group generated by the same set. Recall that if x, y are two elements of a group G, then the element $xyx^{-1}y^{-1}$ in G is denoted by $[x, y]$ and called the *commutator* of x and y. The subgroup of G generated by the set of all commutators in G is called the *commutator subgroup* and denoted by $[G, G] = G'$. It is easily seen that G' is a normal subgroup of G and the factor group G/G' is abelian.

Proposition 14.4.7 *Let F be a free group on a set X. Then, the quotient F/F' is a free abelian group with basis $\{xF' : x \in X\}$.*

Proof Let H be an abelian group and $f : \{xF' : x \in X\} \to H$ be a function. If $\pi : F \to F/F'$ is the natural projection, then the composition $f \circ (\pi|X)$ is a function from X to H. Since F is free on X, there exists a unique homomorphism $\phi : F \to H$ such that $\phi(x) = f\pi(x)$ for every $x \in X$. Clearly, ϕ maps F' into the identity element of H. Hence, there is an induced homomorphism $\bar{\phi} : F/F' \to H$ such that $\bar{\phi} \circ \pi = \phi$. So $\bar{\phi}(xF') = f(xF')$ for every $x \in X$ and, by Theorem 14.4.3, F/F' is a free abelian group with basis $\{xF' : x \in X\}$. $\diamond$

Now, suppose that F_1 and F_2 are free groups with bases X_1 and X_2, respectively, and $F_1 \cong F_2$. Then, the free abelian groups F_1/F_1' and F_2/F_2' are also isomorphic, and therefore have the same rank. By the preceding proposition, the rank of F_i/F_i' is the cardinality of $\{xF_i' : x \in X_i\}$ for $i = 1, 2$. So $\left| \{xF_1' : x \in X_1\} \right| = \left| \{xF_2' : x \in X_2\} \right|$. We observe that $\left| \{xF_i' : x \in X_i\} \right| = |X_i|$. An element of F_i' is a finite product of commutators in F_i, and a commutator in F_i other than the identity element involves at least four elements when expressed as a product in reduced form of elements of $X_i \cup X_i^{-1}$. Consequently, $xy^{-1} \notin F_i'$ for any two distinct elements $x, y \in X_i$. It follows that $x \mapsto xF_i$ is a bijection between X_i and $\{xF_i' : x \in X_i\}$, and thus $|X_1| = |X_2|$. In particular, we see that all bases of a given free group have the same cardinality, called its *rank*.

Interestingly, a free group is determined (upto isomorphism) by its rank. To see this, let F_1 and F_2 be free groups with bases X_1 and X_2, respectively, and suppose that $|X_1| = |X_2|$. Then, there is a bijection $f : X_1 \to X_2$. Let $f_1 : X_1 \to F_2$

and $f_2\colon X_2 \to F_1$ be functions defined by f and f^{-1}, respectively. Then, we have homomorphisms $\phi_1\colon F_1 \to F_2$ and $\phi_2\colon F_2 \to F_1$ such that $\phi_i|X_i = f_i$, $i = 1, 2$. Obviously, the restriction of the composition $\phi_2 \circ \phi_1$ to X_1 is the inclusion map $i_1\colon X_1 \hookrightarrow F_1$. Since the identity map 1_{F_1} also extends i_1, we have $\phi_2 \circ \phi_1 = 1_{F_1}$, by uniqueness of extension. Similarly, $\phi_1 \circ \phi_2 = 1_{F_2}$, and so ϕ_1 is an isomorphism with ϕ_2 as its inverse. Thus, we have proved the following.

Theorem 14.4.8 *Two free groups are isomorphic if and only if they have the same rank.*

As in the case of abelian groups, one can prove the following.

Proposition 14.4.9 *Every group G is the homomorphic image of a free group.*

It is easily seen that the direct sum of two free abelian groups is a free abelian group. Analogously, if two groups G and H are free, then so is their free product $G \star H$. To see this, suppose that G and H are freely generated by the sets $\{x_\alpha \mid \alpha \in A\}$ and $\{y_\beta \mid \beta \in B\}$, respectively. Then, G is the free product of the infinite cyclic groups $G_\alpha = \langle x_\alpha \rangle$, $\alpha \in A$, and H is the free product of the infinite cyclic groups $H_\beta = \langle y_\beta \rangle$, $\beta \in B$. Therefore, it suffices to show that $G \star H$ is the free product of all the G_α and the H_β.

Proposition 14.4.10 *Suppose that G is the free products of the groups G_α, $\alpha \in A$, and H is the free products of the groups H_β, $\beta \in B$. Then, $G \star H$ is the free product of all the G_α and the H_β.*

Proof Let $i_\alpha\colon G_\alpha \to G$ and $i_\beta\colon H_\beta \to H$ be the canonical monomorphisms. Then, given a group K and homomorphisms $f_\alpha\colon G_\alpha \to K$, $\alpha \in A$, and $f_\beta\colon H_\beta \to K$, $\beta \in B$, we have unique homomorphisms $\phi\colon G \to K$ and $\psi\colon H \to K$ such that $\phi \circ i_\alpha = f_\alpha$ for every $\alpha \in A$, and $\psi \circ i_\beta = f_\beta$ for every $\beta \in B$. Let $\nu\colon G \to G \star H$ and $\mu\colon H \to G \star H$ be the canonical monomorphisms. Then, there exists a unique homomorphism $\eta\colon G \star H \to K$ such that $\eta\nu = \phi$ and $\eta\mu = \psi$. So $\eta \circ \nu \circ i_\alpha = \phi \circ i_\alpha = f_\alpha$ for every $\alpha \in A$ and $\eta \circ \mu \circ i_\beta = \psi \circ i_\beta = f_\beta$ for every $\beta \in B$. It follows that $G \star H$ is the free product of the groups $\{G_\alpha \mid \alpha \in A\} + \{H_\beta \mid \beta \in B\}$ relative to the monomorphisms $\nu \circ i_\alpha$, $\mu \circ i_\beta$. $\diamond$

It follows that if G and H are freely generated by the sets $\{x_\alpha \mid \alpha \in A\}$ and $\{y_\beta \mid \beta \in B\}$, respectively, then $G \star H$ is freely generated by the disjoint union of $\{x_\alpha \mid \alpha \in A\}$ and $\{y_\beta \mid \beta \in B\}$.

Generators and Relations

In order to describe a group, we need to list all elements of the group and write down its multiplication table. This description can be shortened by giving a generating set of the group in place of a complete list of its elements, and writing down a set of equations between products of elements of the given generating set, which enables us to construct the multiplication table. For example, consider the group V of plane symmetries of a rectangle (not a square). The nontrivial elements of V are two reflections ρ and σ in the axes of symmetry parallel to the sides and a rotation τ about its center through $180°$. It is called Klein's four group. Clearly, $\tau = \rho \circ \sigma$ so that ρ and σ generate V. Also, $\rho^2 = \sigma^2 = \varepsilon$, the identity element of V, and $\rho\sigma = \sigma\rho$; all other relations in the group, such as $\tau^2 = \varepsilon$, can be derived from these three relations. So V can be described as the group with two generators ρ, σ satisfying the relations $\rho^2 = \sigma^2 = \varepsilon$, $\rho\sigma = \sigma\rho$.

We now discuss an economical way to describe any abstract group. Given a group G, we find a generating set X for G and construct the free group F on X. Let $\phi: F \to G$ be the epimorphism which extends the inclusion map $X \hookrightarrow G$ and N be its kernel. Then $G \cong F/N$. We recall that the elements of F are products in reduced form of elements of X or X^{-1}. Since ϕ is an extension of the inclusion $X \hookrightarrow G$, every element of F is sent by ϕ to the corresponding product in G. In particular, every element of N reduces in G to the identity element. The elements of N other than the identity element are called *relators* for G. If two reduced words s and t in F define the same element in G, then the equation $s = t$ is called a *relation* among the generators X. Observe that N is infinite, unless G is free on X. So it is not possible to list all elements of N. However, we can specify N by a subset $R \subset N$ such that N is the smallest normal subgroup of F containing R (i.e., N is generated by the elements of R and their conjugates). We call R a *set of defining relators* or a *complete set of relators* for the group G. Thus, G can be described by specifying a set X of generators and a set R of defining relators. This description of G is called a *presentation* and we write $G = \langle X|R \rangle$. As noted above, each $r \in R$ reduces in G to the identity element e. The equations $r = e, r \in R$, are referred to as the *defining relations* for G.

We say that a presentation $G = \langle X|R \rangle$ is *finitely generated* (resp. *finitely related*) if X (resp. R) is finite. If G has a presentation which is both finitely generated and finitely related, then we say that it is *finitely presented*. In this case, we write $G = \langle x_1, \ldots, x_n \mid r_1, \ldots, r_m \rangle$ or $G = \langle x_1, \ldots, x_n \mid r_1 = \cdots = r_m = e \rangle$. Sometimes, a group is given by using a hybrid of relators and relations. For example, each of the following

$$\langle a, b \mid a^2, b^2, aba^{-1}b^{-1} \rangle,$$
$$\langle a, b \mid a^2, b^2, ab = ba \rangle \text{ and}$$
$$\langle a, b \mid a^2 = b^2 = e, ab = ba \rangle$$

is interpreted as a presentation of the Klein's four group V. The free group of rank two might be described as the group generated by two elements a, b with the empty set of defining relations. Thus, it has the presentation $\langle a, b \rangle$.

In order to find the group G described by a presentation, we need to construct an isomorphism of G into a known group. The following simple result is quite useful for this purpose.

Theorem 14.4.11 (Dyck) *Let G be group with the presentation $G = \langle X|R \rangle$. If H is a group and there is a map $\theta \colon X \to H$ such that the relations $r(\theta(x)) = 1$ (the identity element of H) holds in H for all $r \in R$, then there exists a unique homomorphism $\phi \colon G \to H$ such that $\phi(x) = \theta(x)$ for every $x \in X$.*

Proof Let Y be a set equipotent to X and $\alpha \colon X \to Y$ is a bijection. If F is the free group on Y and N is the normal subgroup of F generated by the words $\{r(\alpha(x)) \mid r \in R\}$, then $F/N \cong G$ under the isomorphism which sends yN to $\alpha^{-1}(y)$. By the universal mapping property of F, there exists a unique homomorphism $\psi \colon F \to H$ given by $\psi(y) = \theta(\alpha^{-1}(y))$. Clearly, $\psi(r(\alpha(x))) = r(\theta(x)) = 1$. So $r(\alpha(x))$ belongs to $\ker(\psi)$ for every $r \in R$. Since N is generated by the words $r(\alpha(x))$, we have $N \subseteq \ker(\psi)$. Hence, there is an induced homomorphism $\bar{\psi} \colon F/N \to H$. The composition of this homomorphism with the isomorphism $G \cong F/N, x \leftrightarrow \alpha(x)N$, is the desired homomorphism.   ◇

From the preceding theorem, it is clear that two groups having the same presentation are isomorphic; but two completely different presentations can describe the same group. For example, both presentations $\langle a, b \mid a^3 = b^2 = e, ab = ba \rangle$ and $\langle a \mid a^6 = e \rangle$ describe the cyclic group of order 6. Furthermore, there is no effective procedure to determine whether or not the groups given by two finite presentations are isomorphic. Nevertheless, a presentation allows efficient computation in many cases and gives a concrete way of describing abstract groups.

As an illustration, we determine the group described by the presentation $G = \langle x, y \mid xy = yx \rangle$. We observe that it is the free abelian group of rank two. By the preceding theorem, there exists a homomorphism $\phi \colon G \to \mathbb{Z} \times \mathbb{Z}$ such that $\phi(x) = (1, 0)$ and $\phi(y) = (0, 1)$. Since G is generated by the elements x, y, every element of G is a product of powers of x and y. Using the relation $xy = yx$, we can express each element of G in the form $x^m y^n$ and see that G is abelian. Clearly, $\phi(x^m y^n) = (m, n)$ which shows that ϕ is bijective. Thus, $G \cong \mathbb{Z} \times \mathbb{Z}$.

Exercises

1. Let G_α, $\alpha \in A$, be a collection of abelian groups, and let F be their free product with respect to monomorphisms $\iota_\alpha \colon G_\alpha \to F$. If F' is the commutator subgroup of F, show that F/F' is the direct sum of the groups G_α.

2. Suppose that a group G is generated by a family of subgroups G_α, $\alpha \in A$. If, for every $\alpha \neq \beta$, $G_\alpha \cap G_\beta = \{e\}$ (the identity element of G) and the empty word is the only reduced form of e as a finite product of elements of the G_α, show that G is the free product of the groups G_α.

3. Prove that the smallest normal subgroup of a group G containing a subset $X \subset G$ is generated as a subgroup by the set $\{gxg^{-1} \mid g \in G, x \in X\}$.

4. Suppose that G_1 and G_2 are groups and N is the smallest normal subgroup of the free product $G_1 \star G_2$ containing G_1. Prove that $(G_1 \star G_2)/N \cong G_2$.

5. Let F be a free group with basis X and ϕ be a homomorphism of a group G onto F. Suppose that there is a subset $S \subset G$ which is mapped bijectively into X by ϕ. Show that the subgroup $\langle S \rangle$ of G is free on S.

6. Let F be the free group on two elements a and b. Prove that (i) the commutator subgroup F' is not finitely generated, and (ii) the normal subgroup generated by the single commutator $aba^{-1}b^{-1}$ in F is F'.

7. Show that the group D_n of symmetries of a regular n-gon has presentation $\langle a, b \mid a^n = b^2 = (ab)^2 = 1 \rangle$.

8. If G and H are finitely generated (resp. presented) groups, show that $G \star H$ is finitely generated (resp. presented).

14.5 The Seifert–van Kampen Theorem

We now come to a powerful theorem for computing the fundamental group of a path-connected space that can be decomposed as the union of two path-connected, open subsets whose intersection is also path-connected. We begin with the concept of "free products with amalgamation", which is helpful in making the statement of the main theorem rather concise. Let G_1, G_2, and A be groups, and $\eta_1 \colon A \to G_1$, $\eta_2 \colon A \to G_2$ be homomorphisms. The *free product of G_1 and G_2 amalgamated via η_1 and η_2* is defined as the quotient group $(G_1 \star G_2)/N$, where N is the normal subgroup of $G_1 \star G_2$ generated by the words $\eta_1(a)\eta_2(a)^{-1}$, $a \in A$. This is denoted by $G_1 \star_A G_2$. It is clear that there is a homomorphism $A \to G_1 \star_A G_2$ which sends a into $\eta_1(a)N = \eta_2(a)N$. So $G_1 \star_A G_2$ can be thought of as the group of all words in G_1 and G_2 with the relations $\eta_1(a) = \eta_2(a)$, $a \in A$. For each $j = 1, 2$, denote the composition $G_j \xrightarrow{\iota_j} G_1 \star G_2 \xrightarrow{proj} G_1 \star_A G_2$ by ψ_j, where ι_j is the canonical monomorphism $G_j \xrightarrow{\iota_j} G_1 \star G_2$. Then, the amalgamated free product $G_1 \star_A G_2$ with the homomorphisms ψ_1 and ψ_2 is universal for the homomorphisms from G_1 and G_2 whose compositions with η_1 and η_2 agree. For, given homomorphisms $\phi_j \colon G_j \to H$, $j = 1, 2$, such that $\phi_1\eta_1 = \phi_2\eta_2$, there exists a unique homomorphism $\Phi \colon G_1 \star G_2 \to H$ such that $\Phi|G_j = \phi_j$, by the universal property of free products. Clearly, Φ maps each word $\eta_1(a)\eta_2(a)^{-1}$, $a \in A$, to the identity element of H. Therefore, $N \subseteq \ker(\Phi)$, and there is an induced homomorphism $\bar{\Phi} \colon G_1 \star_A G_2 \to H$ such that $\bar{\Phi} \circ \psi_j = \phi_j$ for $j = 1, 2$:

Theorem 14.5.1 (Seifert–van Kampen Theorem) *Suppose that a topological space X is the union of two path-connected open subsets U and V such that their intersection is a nonempty path-connected set. Choose a point $x_0 \in U \cap V$ to be the base point for all fundamental groups under consideration. Then, there is an isomorphism* $\Phi \colon \pi(U) \star_{\pi(U \cap V)} \pi(V) \cong \pi(X)$.

Proof For $S \subseteq X$, denote by η_S the homomorphism $\pi(U \cap V) \to \pi(S)$ induced by the inclusion map $U \cap V \hookrightarrow S$, and let $G = \pi(U) \star_{\pi(U \cap V)} \pi(V)$ be the free product

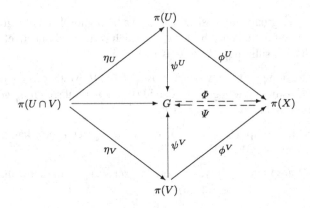

Fig. 14.12 Used in the proof of Theorem 14.5.1

of $\pi(U)$ and $\pi(V)$ amalgamated via homomorphisms η_U and η_V. Let ψ^U be the composition of the monomorphism $\pi(U) \to \pi(U) \star \pi(V)$ and the canonical projection $\pi(U) \star \pi(V) \to G$. Similarly, define $\psi^V : \pi(V) \to G$. Further, let ϕ^S denote the homomorphism $\pi(S) \to \pi(X)$ induced by the inclusion map $S \hookrightarrow X$. Then, it is clear that $\phi^U \circ \eta_U = \phi^{U \cap V} = \phi^V \circ \eta_V$. Therefore, by the universal mapping property of free products with amalgamation, there exists a homomorphism $\Phi \colon G \to \pi(X)$ such that $\Phi \circ \psi^U = \phi^U$ and $\Phi \circ \psi^V = \phi^V$ (see Fig. 14.12). We show that Φ is an isomorphism.

For surjectivity, it clearly suffices to show that $\pi(X)$ is generated by the images of ϕ^U and ϕ^V. To this end, consider a loop f in X based at x_0. By the Lebesgue covering lemma, there exists a subdivision

$$0 = t_0 < t_1 < \cdots < t_n = 1$$

of I such that f maps each subinterval $[t_{i-1}, t_i]$ into either U or V. If two consecutive subintervals $[t_{i-1}, t_i]$ and $[t_i, t_{i+1}]$, say, are mapped by f into the same set U or V, then we omit the common end point t_i and amalgamate these two subintervals to have the subdivision

$$0 = t_0 < t_1 < \cdots < t_{i-1} < t_{i+1} < \cdots < t_n = 1$$

of I. By repeating this process finitely many times, we obtain a subdivision

$$0 = s_0 < s_1 < \cdots < s_m = 1$$

of I such that f takes each subinterval $[s_{j-1}, s_j]$ into either U or V and $f(s_j) \in U \cap V$ for every $j = 1, \ldots, m - 1$. Denote the composite of the linear homeomorphism $I \to [s_{j-1}, s_j]$, $u \mapsto (1 - u)s_{j-1} + us_j$, and the restriction of f to $[s_{j-1}, s_j]$

by f_j for every $j = 1, \ldots, m$. Then, $f \simeq f_1 * \cdots * f_m$ rel$\{0, 1\}$, by the proof of Proposition 14.3.1. Since $U \cap V$ is path-connected, we find a path γ_j from x_0 to $f(s_j)$ in $U \cap V$. Clearly, we have

$$f \simeq f_1 * \left(\gamma_1^{-1} * \gamma_1\right) * f_2 * \cdots * \left(\gamma_{m-1}^{-1} * \gamma_{m-1}\right) * f_m$$
$$\simeq \left(f_1 * \gamma_1^{-1}\right) * \left(\gamma_1 * f_2 * \gamma_2^{-1}\right) * \cdots * \left(\gamma_{m-1} * f_m\right)$$

holding the base point fixed. Thus,

$$[f] = \left[f_1 * \gamma_1^{-1}\right] \cdot \left[\gamma_1 * f_2 * \gamma_2^{-1}\right] \cdots \cdots \left[\gamma_{m-1} * f_m\right].$$

Notice that each of $f_1 * \gamma_1^{-1}$, $\gamma_{m-1} * f_m$, and $\gamma_i * f_{i+1} * \gamma_{i+1}^{-1}$, $1 \leq i \leq i - 1$, is a loop based at x_0 lying completely in U or V. Thus, the homotopy class $[f]$ belongs to the subgroup generated by the images of ϕ^U and ϕ^V, as desired.

For the injectivity of Φ, notice that $\psi^U \circ \eta_U = \psi^V \circ \eta_V$. Under this condition, we prove below (Theorem 14.5.2) that there exists a homomorphism $\Psi : \pi(X) \to G$ such that $\Psi \circ \phi^U = \psi^U$ and $\Psi \circ \phi^V = \psi^V$. This implies that the composition $\Psi \circ \Phi$ agrees with the identity map on the images of $\pi(U)$ and $\pi(V)$ under the homomorphisms ψ^U and ψ^V, respectively. Since G is generated by the union of im $\left(\psi^U\right)$ and im $\left(\psi^V\right)$, we conclude that $\Psi\Phi$ is the identity map on G, and so Φ is injective. $\Diamond$

Theorem 14.5.2 *Assume the hypotheses of Theorem 14.5.1. Then, given two homomorphisms $\psi^U : \pi(U) \to G$ and $\psi^V : \pi(V) \to G$ such that $\psi^U \circ \eta_U = \psi^V \circ \eta_V$, there exists a unique homomorphism $\Psi : \pi(X) \to G$ such that $\Psi \circ \phi^U = \psi^U$ and $\Psi \circ \phi^V = \psi^V$, where the notations η_U, η_V, ϕ^U, and ϕ^V have the above meanings.*

Proof The uniqueness of Ψ is immediate from the fact that $\pi(X)$ is generated by the images of ϕ^U and ϕ^V, and any two homomorphisms $\pi(X) \to G$ satisfying the desired condition agree on the two subgroups.

To prove the existence of Ψ, we first define a mapping Ψ' on the family $\mathcal{L}$ of all loops in X based at x_0 as follows:

For $S = U, V$ and $U \cap V$, if a loop $f \in \mathcal{L}$ lies completely in S, then we denote the homotopy class of f in S by $[f]_S$ and put $\Psi'(f) = \psi^U ([f]_U)$ or $\Psi'(f) = \psi^V ([f]_V)$, according as f lies completely in U or in V. If f lies in $U \cap V$, then we have

$$\psi^U ([f]_U) = \psi^U \eta_U ([f]_{U \cap V}) = \psi^V \eta_V ([f]_{U \cap V}) = \psi^V ([f]_V).$$

Thus, for a loop f in U or in V, $\Psi'(f)$ is a uniquely determined element of G. For any two such loops f and g, it is clear that

(a) $\Psi'(f * g) = \Psi'(f) \cdot \Psi'(g)$ and
(b) if $f \simeq g$ rel $\{0, 1\}$, then $\Psi'(f) = \Psi'(g)$.

Now, let $f \in \mathcal{L}$ be arbitrary. For each $x \in X$, choose a path γ_x from x_0 to x in U or V, preferably in $U \cap V$ when $x \in U \cap V$; in particular, γ_{x_0} will be the constant path c_{x_0} at x_0. Then, choose a subdivision

$$0 = s_0 < s_1 < \cdots < s_m = 1 \tag{14.1}$$

of I such that f maps each subinterval $[s_{i-1}, s_i]$ into U or V. For each $i = 1, \ldots, m$, let f_i be the composite of the linear homeomorphism $I \to [s_{i-1}, s_i]$, $u \mapsto (1 - u)s_{i-1} + us_i$, and the restriction of f to $[s_{i-1}, s_i]$. Then, each f_i is a path lying in U or V and we have $f \simeq f_1 * \cdots * f_m$ rel $\{0, 1\}$. Put $x_i = f(s_i)$ for each $i = 1, \ldots, m$. Obviously, $x_m = x_0$ and $\gamma_{x_i}^{-1} * \gamma_{x_i} \simeq c_{x_i}$ rel $\{0, 1\}$. So

$$\begin{aligned} f &\simeq f_1 * \left(\gamma_{x_1}^{-1} * \gamma_{x_1} \right) * f_2 * \cdots * \left(\gamma_{x_{m-1}}^{-1} * \gamma_{x_{m-1}} \right) * f_m \\ &\simeq \left(\gamma_{x_0} * f_1 * \gamma_{x_1}^{-1} \right) * \cdots * \left(\gamma_{x_{m-1}} * f_m * \gamma_{x_m}^{-1} \right). \end{aligned} \tag{14.2}$$

Observe that each bracketed term in the r.h.s. of (2) is a loop in U or V based at x_0. Thus, by (a),

$$\Psi'(f) = \Psi' \left(\gamma_{x_0} * f_1 * \gamma_{x_1} \right) \cdots \cdot \Psi' \left(\gamma_{x_{m-1}} * f_m * \gamma_{x_m}^{-1} \right)$$

is an element of G. We assert that this element is independent of the choice of a particular subdivision of I. For, if $0 = t_0 < t_1 < \cdots < t_n = 1$ is another subdivision of I such that f maps each subintervals $[t_{j-1}, t_j]$ into U or V, then there is a subdivision of I which has all the points s_i and t_j. So it suffices to show that $\Psi'(f)$ remains unchanged if the subdivision in (1) is refined by dividing a subinterval $[s_{i-1}, s_i]$ into two subintervals $[s_{i-1}, t]$ and $[t, s_i]$. In this case, if g and h are, respectively, the composites of the linear homeomorphisms $I \approx [s_{i-1}, t]$ and $I \approx [t, s_i]$ with the restrictions of f to $[s_{i-1}, t]$ and $[t, s_i]$, then $f_i \simeq g * h$ rel $\{0, 1\}$ and the term $\gamma_{x_{i-1}} * f_i * \gamma_{x_i}^{-1}$ in (2) would be replaced by the product of two loops $\gamma_{x_{i-1}} * g * \gamma^{-1}{}_x$ and $\gamma_x * h * \gamma_{x_i}^{-1}$, where $x = f(t)$. Notice that γ_x lies in U (or V), according as f_i lies in U (or V). So, we have

$$\begin{aligned} \Psi' \left(\gamma_{x_{i-1}} * f_i * \gamma_{x_i}^{-1} \right) &= \Psi' \left(\gamma_{x_{i-1}} * g * h * \gamma_{x_i}^{-1} \right) \\ &= \Psi' \left(\gamma_{x_{i-1}} * g * \gamma_x^{-1} * \gamma_x * h * \gamma_{x_i}^{-1} \right) \\ &= \Psi' \left(\gamma_{x_{i-1}} * g * \gamma_{x^{-1}} \right) \cdot \Psi' \left(\gamma_x * h * \gamma_{x_i}^{-1} \right), \end{aligned}$$

as desired.

Next, we show that Ψ' satisfies the condition (b) for any two homotopic loops f and g in $\mathscr{L}$; this is the trickiest part of the proof. Let F be a homotopy between f and g relative to $\{0, 1\}$. By the Lebesgue covering lemma, we find two subdivisions $0 = s_0 < s_1 < \ldots < s_m = 1$ and $0 = t_0 < t_1 < \ldots < t_n = 1$ of I such that F maps each of the rectangles $[s_{i-1}, s_i] \times [t_{j-1}, t_j]$ into U or V. Then, for each $j = 0, 1, \ldots, n$, we have loops h_j (all based at x_0) defined by $h_j(s) = F(s, t_j)$, $s \in I$. Clearly, the homeomorphism $I \times I \approx I \times [t_{j-1}, t_j]$ followed by the restriction of F to the rectangle $I \times [t_{j-1}, t_j]$ is a homotopy between h_{j-1} and h_j. So, $f = h_0 \simeq h_1 \simeq \cdots \simeq h_n = g$ (rel $\{0, 1\}$), and thus it suffices to prove that $\Psi'(h_{j-1}) = \Psi'(h_j)$. We may, by induction, assume that $n = 1$. Then, for each $i = 1, \ldots, m$, F maps the rectangle $[s_{i-1}, s_i] \times I$ into U or V (see Fig. 14.13).

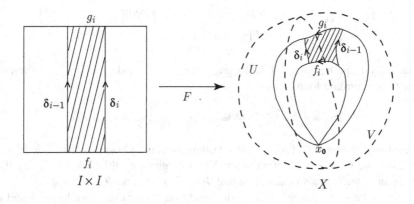

Fig. 14.13 Proof of Theorem 14.5.2

Let f_i and g_i be the paths defined by the restriction of F to $[s_{i-1}, s_i] \times \{0\}$ and $[s_{i-1}, s_i] \times \{1\}$, respectively. Also, put $x_i = F(s_i, 0)$ and $y_i = F(s_i, 1)$. Then, f_i is a path in X from x_{i-1} to x_i, and g_i is a path from y_{i-1} to y_i. As $F(s, 0) = f(s)$ and $F(s, 1) = g(s)$, we see that $f \simeq f_1 * \cdots * f_m$ and $g \simeq g_1 * \cdots * g_m$ rel $\{0, 1\}$. Consequently, we have

$$\left. \begin{aligned} \Psi'(f) &= \Psi'\left(\gamma_{x_0} * f_1 * \gamma_{x_1}^{-1}\right) \cdots \cdots \Psi'\left(\gamma_{x_{m-1}} * f_m * \gamma_{x_m}^{-1}\right) \\ \Psi'(g) &= \Psi'\left(\gamma_{y_0} * g_1 * \gamma_{y_1}^{-1}\right) \cdots \cdots \Psi'\left(\gamma_{y_{m-1}} * g_m * \gamma_{y_m}^{-1}\right) \end{aligned} \right\} \tag{14.3}$$

Moreover, by restricting F to the rectangle $[s_{i-1}, s_i] \times I$, we obtain a homotopy between f_i and g_i; specifically, the mapping $D_i : I \times I \to X$ defined by $D_i(u, t) = F\left((1 - u)s_{i-1} + us_i, t\right)$ is a desired homotopy. Consequently, if δ_i is the path (from x_i to y_i) defined by $\delta_i(t) = F(s_i, t)$, then $f_i \simeq \delta_{i-1} * g_i * \delta_i^{-1}$ rel $\{0, 1\}$. A desired homotopy H_i connecting f_i to $\delta_{i-1} * g_i * \delta_i^{-1}$ is given by

$$H_i(r, t) = \begin{cases} \delta_{i-1}(4r) & \text{for } 0 \leq r \leq t/4, \\ D_i\left(\frac{4r-t}{4-3t}, t\right) & \text{for } t/4 \leq r \leq 1 - t/2, \\ \delta_i^{-1}(2r - 1) & \text{for } 1 - t/2 \leq r \leq 1. \end{cases}$$

Hence, we see that

$$\begin{aligned} f_i &\simeq \delta_{i-1} * \left(\gamma_{y_{i-1}}^{-1} * \gamma_{y_{i-1}}\right) * g_i * \left(\gamma_{y_i}^{-1} * \gamma_{y_i}\right) * \delta_i^{-1} \\ &\simeq \left(\delta_{i-1} * \gamma_{y_{i-1}}^{-1}\right) * \left(\gamma_{y_{i-1}} * g_i * \gamma_{y_i}^{-1}\right) * \left(\gamma_{y_i} * \delta_i^{-1}\right) \quad \text{rel } \{0, 1\}. \end{aligned}$$

Now, by the properties of Ψ' for the loops lying completely in U or V, we find that

$$\Psi'\left(\gamma_{x_{i-1}} * f_i * \gamma_{x_i}^{-1}\right) = \Psi'\left(\gamma_{x_{i-1}} * \delta_{i-1} * \gamma_{y_{i-1}}^{-1}\right) \cdot \Psi'\left(\gamma_{y_{i-1}} * g_i * \gamma_{y_i}^{-1}\right) \cdot$$
$$\Psi'\left(\gamma_{y_i} * \delta_i^{-1} * \gamma_{x_i}^{-1}\right).$$

Notice that the path product of the loops $\gamma_{y_i} * \delta_i^{-1} * \gamma_{x_i}^{-1}$ and $\gamma_{x_i} * \delta_i * \gamma_{y_i}^{-1}$ is homotopic to the constant loop c_{x_0}, so

$$\Psi'\left(\gamma_{y_i} * \delta_i^{-1} * \gamma_{x_i}^{-1}\right) \cdot \Psi'\left(\gamma_{x_i} * \delta_i * \gamma_{y_i}^{-1}\right) = \Psi'(c_{x_0}) = e,$$

the identity element of G. From Eq. (14.3), it now easily follows that $\Psi'(f) = \Psi'(g)$, and therefore there is a function $\Psi : \pi(X) \to G$ given by $\Psi[f] = \Psi'(f)$. By the definition of Ψ', it is also immediate that $\Psi \circ \phi^U = \psi^U$ and $\Psi \circ \phi^V = \psi^V$.

Finally, we show that Ψ' satisfies the condition (a) for every two loops f and g in $\mathscr{L}$, and hence deduce that Ψ is a homomorphism. By the continuity of functions $f(2s)$ on $[0, 1/2]$ and $g(2s - 1)$ on $1/2, 1$, we find a subdivision

$$0 = s_0 < s_1 < \cdots < s_m = 1/2 < s_{m+1} < \cdots < s_n = 1$$

of I such that f maps each subinterval $[2s_{i-1}, 2s_i]$ into U or V for $1 \le i \le m$, and g maps each subinterval $[2s_{i-1} - 1, 2s_i - 1]$ into U or V for $m < i \le n$. Then, $0 = s_0 < s_1 < \cdots < s_n = 1$ is a subdivision of I such that each subinterval $[s_{i-1}, s_i]$ is mapped by $f * g$ into U or V. For $1 \le i \le m$, let f_i be the composition $I \approx [2s_{i-1}, 2s_i] \xrightarrow{f} X$ and, for $m < i \le n$, denote the composition $I \approx [2s_{i-1} - 1, 2s_i - 1] \xrightarrow{g} X$ by g_{i-m}. Then, $f \simeq f_1 * \cdots * f_m$ and $g \simeq g_1 * \cdots * g_{n-m}$ so that

$$f * g \simeq f_1 * \cdots * f_m * g_1 * \cdots * g_{n-m} \quad \text{rel} \{0, 1\}.$$

Put $x_i = f(2s_i), 0 < i \le m$ and $x_i = g(2s_i - 1)$ for $m < i \le n$. Then,

$$\Psi'(f * g) = \Psi'\left(\gamma_{x_0} * f_1 * \gamma_{x_1}^{-1}\right) \cdot \cdots \cdot \Psi'\left(\gamma_{x_{m-1}} * f_m * \gamma_{x_m}^{-1}\right) \cdot$$
$$\Psi'\left(\gamma_{x_m} * g_1 * \gamma_{x_{m+1}}^{-1}\right) \cdot \cdots \cdot \Psi'\left(\gamma_{x_{n-1}} * g_{n-m} * \gamma_{x_n}^{-1}\right)$$
$$= \Psi'(f) \cdot \Psi'(g). \qquad \diamond$$

In most applications of the Seifert–van Kampen theorem, we take one of the sets U, V, or $U \cap V$ to be simply connected. In such cases, we have the following consequences of the theorem.

Corollary 14.5.3 *Let U and V be open path-connected subsets of a topological space X such that $X = U \cup V$. If $U \cap V$ is simply connected, then $\pi(X) \cong \pi(U) \star \pi(V)$.*

Proposition 14.5.4 *Suppose that a topological space $X = U \cup V$ with path-connected open subsets U and V such that $U \cap V \neq \emptyset$ path-connected. If V is simply connected, then $\pi(X) \cong \pi(U)/N$, where N is the normal subgroup of $\pi(U)$ generated by the image of $\pi(U \cap V)$.*

Proof The proof of Theorem 14.5.1 shows that $\pi(X)$ is generated by the image of the homomorphism $\phi^U : \pi(U) \to \pi(X)$, since $\pi(V) = 1$. So ϕ^U is surjective. For the same reason, the composition $\phi^U \circ \eta_U = \phi^V \circ \eta_V$ is the trivial homomorphism, so $\mathrm{im}(\eta_U) \subseteq \ker(\phi^U)$. Consequently, if N is the normal subgroup generated by $\mathrm{im}(\eta_U)$, then $N \subseteq \ker(\phi^U)$. To see the reverse inclusion, take $H = \pi(U)/N$, ψ^U to be the canonical projection $\pi(U) \to H$ and $\psi^V : \pi(V) \to H$ to be the obvious map. Then, by Theorem 14.5.2, there exists a homomorphism $\Psi : \pi(X) \to H$ such that $\Psi \circ \phi^U = \psi^U$. This immediately implies that $\ker(\phi^U) \subseteq N$, and hence the equality holds. $\diamond$

Example 14.5.1 The fundamental group of $\mathbb{S}^1 \vee \cdots \vee \mathbb{S}^1$ (n copies) is the free group on n generators. Suppose that $X = X_1 \vee \cdots \vee X_n$, where each $X_j \approx \mathbb{S}^1$. Let x_j be the base point of X_j for every j and denote the base point $[x_j]$ of X by x_0. First, consider the case $n = 2$. Clearly, the set $W_j = X_j - \{-x_j\}$ is open in X_j. So $U = p\,(X_1 + W_2)$ and $V = p\,(W_1 + X_2)$ are open in X, where $p \colon X_1 + X_2 \to X$ is the quotient map. Since W_j is homeomorphic to the real line $\mathbb{R}$ and its each point is a strong deformation retract, x_j is a strong deformation retract of W_j. Let $F_j \colon W_j \times I \to W_j$ be a strong deformation retraction of W_j onto x_j. Then, the mapping $G_1 \colon (X_1 + W_2) \times I \to X_1 + W_2$ defined by

$$G_1(x, t) = \begin{cases} x & \text{for } x \in X_1, t \in I, \text{ and} \\ F_2(x, t) & \text{for } x \in W_2, t \in I \end{cases}$$

induces a strong deformation retraction of U onto X_1. Similarly, X_2 is a strong deformation retract of V. Also, $H \colon (W_1 + W_2) \times I \to W_1 + W_2$ defined by $H(x, t) = F_j(x, t)$ for $x \in W_j$, $t \in I$ induces a strong deformation retraction of $U \cap V$ onto $x_0 = [x_j] \in X$. Since $U \cap V$ is contractible, Corollary 14.5.3 implies that $\pi(X) \cong \pi(U) \star \pi(V)$. The injections $X_1 \to U$ and $X_2 \to V$, being homotopy equivalences, induce isomorphisms $\pi(X_1, x_1) \cong \pi(U, x_0)$ and $\pi(X_2, x_2) \cong \pi(V, x_0)$. Therefore, $\pi(X) \cong \pi(X_1) \star \pi(X_2)$ which is a free group on two generators.

Now, let $n > 2$ and suppose, by induction, that $X_1 \vee \cdots \vee X_{n-1}$ is a free group on $n - 1$ generators. Let $p \colon \sum X_j \to X$ be the quotient map and, for each $j = 1, \ldots, n$, put $W_j = X_j - \{-x_j\}$. Then, $U = p\,(X_1 + \cdots + X_{n-1} + W_n)$ and $V = p\,(W_1 + \cdots + W_{n-1} + X_n)$ are open path-connected subsets of X such that $X = U \cup V$ and $U \cap V = p\,(W_1 + \cdots + W_n) = \bigvee W_j$. See Fig. 14.14.

We observe that x_0 is a strong deformation retract of $U \cap V$. Suppose that $F_j \colon W_j \times I \to W_j$ is a strong deformation retraction of W_j onto x_j. Then, $H \colon \sum W_j \times I \to \sum W_j$, defined by $H|\,(W_j \times I) = F_j$, is clearly a strong deformation retraction of $\sum W_j$ onto $\{x_1, \ldots, x_n\}$. It induces a strong deformation retraction of $U \cap V$ onto $\{x_0\}$. Thus, $U \cap V$ is simply connected, and hence $\pi(X)$ is the free product of $\pi(U)$ and $\pi(V)$. As above, one sees that $\bigvee_1^{n-1} X_j$ is a strong deformation retract of U and X_n is a strong deformation retract of V. Consequently, $\pi(U)$ is a free group on $n - 1$ generators and $\pi(V)$ is an infinite cyclic group. It follows that $\pi(X)$ is a free group on n generators. Since the isomorphism $\pi\,(\bigvee X_j) \cong \mathbb{Z} \star \cdots \star \mathbb{Z}$ is induced by inclusion of each X_j into $\bigvee X_j$, we can write explicit generators of this free group.

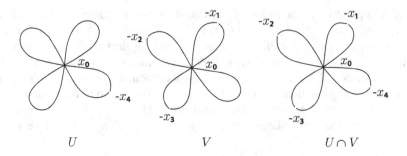

Fig. 14.14 Sets U and V in Example 14.5.1 for $n = 4$

In fact, if f_j denote the standard loop in X_j, then $\pi\left(\bigvee X_j\right)$ is just the free group $\langle [f_1], \ldots, [f_n] \rangle$.

Example 14.5.2 The fundamental group of the Klein bottle K (Example 6.1.5) has a presentation $\langle a, b \mid aba^{-1}b = 1 \rangle$. Recall that K is the quotient space $I^2/\sim$, where $(s, 0) \sim (s, 1)$ and $(0, t) \sim (1, 1 - t)$ for all $s, t \in I$. Choose two positive real numbers $q < r < 1/2$ and consider the closed interval $J = \left[\frac{1}{2} - q, \frac{1}{2} + q\right]$ and the open

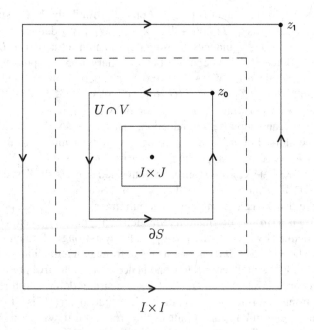

Fig. 14.15 Proof of Example 14.5.2

interval $L = \left(\frac{1}{2} - r, \frac{1}{2} + r\right)$. Put $U = I^2 - J^2$ and $V = L \times L$ (see Fig. 14.15). If $p: I^2 \to K$ is the quotient map, then $p(U)$ and $p(V)$ are path-connected open subsets of K and $K = p(U) \cup p(V)$. Moreover, $p(V) \approx V$ is simply connected and $p(U) \cap p(V) = p\left(V - J^2\right)$ is path-connected. Therefore, by Proposition 14.5.4, $\pi(K, x_0) = \pi(p(U), x_0)/N$, where N is the normal subgroup of $\pi(p(U), x_0)$ generated by the image of $\pi(p(U) \cap p(V), x_0)$ under the homomorphism induced by the inclusion $\overline{\lambda}: p(U) \cap p(V) \hookrightarrow p(U)$.

To determine $\pi(p(U), x_0)$, we first show that $p\left(\partial I^2\right)$ is a strong deformation retract of $p(U)$, where ∂I^2 is the boundary of I^2. Notice that ∂I^2 is a retract of U; a desired retraction is obtained by projecting U onto ∂I^2 from the center of I^2. Specifically, the mapping $\rho: U \to \partial I^2$ given by

$$\rho(s, t) = \begin{cases} \left(\frac{t-s}{2t-1}, 0\right) & \text{for } t \leq s \leq 1 - t \text{ and } t < 1/2 - q, \\ \left(1, \frac{s+t-1}{2s-1}\right) & \text{for } 1 - s \leq t \leq s \text{ and } s > 1/2 + q, \\ \left(\frac{s+t-1}{2t-1}, 1\right) & \text{for } 1 - t \leq s \leq t \text{ and } t > 1/2 + q, \\ \left(0, \frac{s-t}{2s-1}\right) & \text{for } s \leq t \leq 1 - s \text{ and } s < 1/2 - q \end{cases}$$

is a retraction. This induces a continuous map $\bar{\rho}: p(U) \to p\left(\partial I^2\right)$ such that $\bar{\rho}(p|U) = (p|\partial I^2) \circ \rho$ on U. It is readily verified that $\bar{\rho}$ is a retraction of $p(U)$ onto $p\left(\partial I^2\right)$. We observe that $p\left(\partial I^2\right)$ is, in fact, a strong deformation retract of $p(U)$. Notice that the homotopy $F: U \times I \to U$, defined by $F(u, t') = (1 - t')u + t'\rho(u)$, is a strong deformation retraction of U to ∂I^2. Moreover, if $p(u_1) = p(u_2)$ for some $u_1 \neq u_2$ in U, then $u_1, u_2 \in \partial I^2$, by the definition of p. So, for every $t' \in I$, we have $F(u_1, t') = u_1$ and $F(u_2, t') = u_2$, and it follows that the composition pF factors through $p(U) \times I$. Thus, we have a mapping $G: p(U) \times I \to p(U)$ given by $G\left(p(u), t'\right) = pF(u, t')$. Since the mapping $p \times 1: U \times I \to p(U) \times I$ is an identification, G is continuous. It is now easily checked that G is a strong deformation retraction of $p(U)$ to $p\left(\partial I^2\right)$. Hence, $\pi(p(U), x_1) \cong \pi\left(p\left(\partial I^2\right), x_1\right)$. For the latter group, we note that the subspace $p\left(\partial I^2\right) \subset p(U)$ is homeomorphic to the quotient space of ∂I^2 by the equivalence relation obtained by restricting the relation $\sim$ on I^2 to ∂I^2. On the other hand, $\partial I^2/\sim$ is homeomorphic to the "figure 8" space, so we can identify the subspace $p(\partial I^2)$ with the "figure 8" space. Then, by Example 14.5.1, the fundamental group of $\pi\left(p\left(\partial I^2\right), x_1\right)$ is the free group generated by two elements, and hence $\pi(p(U), x_1)$ is also the free group generated by two elements, say, α, β. Of course, we need to change the base point from $x_1 \in p\left(\partial I^2\right)$ to some point $x_0 \in p(U \cap V)$. To this end, let x_1 be the point $p((1, 1))$ and consider the point $x_0 = p(z_0)$, where z_0 is a point in $U \cap V$ on the line segment from the center of $I \times I$ to $z_1 = (1, 1)$. Let γ be the reverse of the path $t' \to G(x_0, t')$ in $p(U)$. Then, $\hat{\gamma}: \pi(p(U), x_1) \to \pi(p(U), x_0), [f] \mapsto [\gamma^{-1} * f * \gamma]$, is an isomorphism. Writing $\hat{\gamma}(\alpha) = a$ and $\hat{\gamma}(\beta) = b$, we see that $\pi(p(U), x_0) = \langle a, b \rangle$.

Finally, we determine the normal subgroup N of $\pi(p(U), x_0)$. Let S denote the square with sides parallel to those of I^2, and whose top right corner is z_0. As above, we see that $\partial S \approx \mathbb{S}^1$ is a strong deformation retract of $U \cap V$. Clearly, $p(U) \cap p(V) \approx U \cap V$, so $\pi(p(U) \cap p(V), x_0)$ is an infinite cyclic group generated by the homotopy class of a loop g which goes once around $p(\partial S)$. Since the retraction ρ maps ∂S homeomorphically onto ∂I^2, we see that $p(\partial S)$ is wrapped twice around $p\left(\partial I^2\right)$ by the retraction $\bar{\rho}$. Thus, the loop $\bar{\rho}\bar{\lambda}g$ goes twice around one of the circles of " figure 8" in the same direction and traverses the other circle in the opposite directions. Accordingly, the homotopy class $\left[\bar{\mu}\bar{\rho}\bar{\lambda}g\right] \in \pi\left(p(U), x_1\right)$ is the word $\alpha\beta\alpha^{-1}\beta$, where $\bar{\mu}\colon p\left(\partial I^2\right) \hookrightarrow p(U)$ is the inclusion map. Since G is a homotopy between $1_{p(U)}$ and $\bar{\mu}\bar{\rho}$, the composition $\hat{\gamma} \circ \bar{\mu}_\# \circ \bar{\rho}_\#$ is the identity isomorphism on $\pi\left(p(U), x_0\right)$ (cf. Theorem 14.2.14). Therefore, $\left[\bar{\lambda}g\right] = \hat{\gamma}\left[\bar{\mu}\bar{\rho}\bar{\lambda}g\right]$ is the word $aba^{-1}b$, and it follows that N is generated by the word $aba^{-1}b$. Thus, $\pi\left(K, x_0\right) = \left\langle a, b \mid aba^{-1}b = 1\right\rangle$.

The technique used in the above computation can be employed to prove a more general result as shown below:

Theorem 14.5.5 *Let X be the space obtained by attaching $\mathbb{D}^2$ to a path-connected Hausdorff space Y by a continuous map $f\colon \mathbb{S}^1 \to Y$. Suppose that $z_0 \in \mathbb{S}^1$ and $y_0 = f(z_0)$. If $j\colon Y \to X$ is the canonical embedding, then $\pi\left(X, j(y_0)\right)$ is isomorphic to the quotient group $\pi\left(Y, y_0\right)/ker(j_\#)$, and $ker(j_\#)$ is the smallest normal subgroup containing the image of the homomorphism $f_\#\colon \pi\left(\mathbb{S}^1, z_0\right) \to \pi\left(Y, y_0\right)$.*

Proof Let $p\colon \left(\mathbb{D}^2 + Y\right) \to X$ be the quotient map, and 0 denote the center of $\mathbb{D}^2$. By Theorem 6.5.4, X is Hausdorff, so $U = X - \{p(0)\}$ is open in X. Also, U is path-connected, since both $\mathbb{D}^2 - \{0\}$ and Y are path-connected and $p(y) \neq p(0)$ for all $y \in Y$. Next, put $V = X - j(Y) = p\left(\text{int}\left(\mathbb{D}^2\right)\right)$. Obviously, the restriction of p to $\text{int}\left(\mathbb{D}^2\right)$ defines a homeomorphism between $\text{int}\left(\mathbb{D}^2\right)$ and V, so V is simply connected. Moreover, it is open in X, since $j(Y)$ is closed. It is also clear that $X = U \cup V$ and $U \cap V$ is path-connected. By the proof of Proposition 14.5.4, the homomorphism $k_{\#1}\colon \pi\left(U, x_1\right) \to \pi\left(X, x_1\right)$ induced by the inclusion map $k\colon U \to X$ is surjective, and $\ker\left(k_{\#1}\right)$ is the smallest normal subgroup of $\pi\left(U, x_1\right)$ containing the image of $\pi\left(U \cap V, x_1\right)$ under the homomorphism induced by the inclusion $\ell\colon U \cap V \hookrightarrow U$. We determine $\pi(U)$ by showing that $i(Y)$, where $i\colon Y \to U$ is the map defined by j, is a strong deformation retract of U. Consider a strong deformation retraction F of $\mathbb{D}^2 - \{0\}$ to $\mathbb{S}^1$, and define the function $H\colon U \times I \to U$ by

$$H(x, t) = \begin{cases} pF(z, t) & \text{if } x = p(z) \text{ for some } z \in \mathbb{D}^2 \text{ and} \\ x & \text{if } x = p(y) \text{ for some } y \in Y. \end{cases}$$

Since $p(Y)$ and $p\left(\mathbb{D}^2 - \{0\}\right)$ are closed in U and $pF(z, t) = p(z) = p(y)$ for $z \in \mathbb{S}^1$ and $f(z) = y$, the gluing lemma shows that H is continuous. It is now easily verified that H satisfies the other conditions to be a strong deformation retraction of U to $i(Y)$. Hence, we deduce that the homomorphism $i_\#\colon \pi\left(Y, y_0\right) \to \pi\left(U, i(y_0)\right)$ induced by i is an isomorphism. Furthermore, we see that the homomorphism $k_{\#0}\colon \pi\left(U, x_0\right) \to \pi\left(X, x_0\right)$ induced by k, where $x_0 = j(y_0)$, is surjective. For, if ϕ is a path in U from x_1 to x_0 and $\psi = k\phi$, then we have the following commutative diagram:

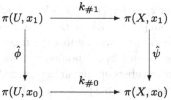

where $\hat{\phi}$ and $\hat{\psi}$ denote the isomorphisms defined by the paths ϕ and ψ, respectively. As the top arrow in the above diagram is surjective, so is the bottom arrow. Accordingly, the homomorphism $j_{\#} : \pi(Y, y_0) \to \pi(X, x_0)$ is surjective, for $j = k \circ i$. It is also clear that the kernel of $j_{\#}$ is the pullback of the kernel of $k_{\#0}$ under the isomorphism $i_{\#}$. Since $\hat{\psi} \circ k_{\#1} = k_{\#0} \circ \hat{\phi}$ and $\hat{\psi}$ is an isomorphism, ker $(k_{\#0})$ is the image of ker $(k_{\#1})$ under the isomorphism $\hat{\phi}$.

As we know, ker $(k_{1\#})$ is the normal subgroup of $\pi(U, x_1)$ generated by the image of $\pi(U \cap V, x_1)$ under the homomorphism $\ell_{\#}$. Further, notice that $\pi(U \cap V, x_1)$ is an infinite cyclic group, since $U \cap V \approx \text{int}(\mathbb{D}^2 - \{0\}) \simeq \mathbb{S}^1$. To find a generator of this group, consider the loop γ in int $(\mathbb{D}^2 - \{0\})$ based at a point z_1 on the radius through z_0, which goes once around the circle passing through z_1 and centered at 0

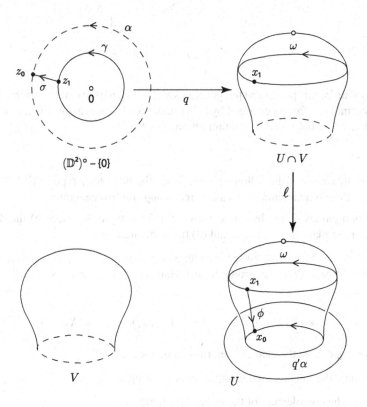

Fig. 14.16 Proof of Theorem 14.5.5

(see Fig. 14.16). Then, with $x_1 = p(z_1)$, $\pi(U \cap V, x_1)$ generated by the homotopy class of $\omega = q\gamma$, where $q : \text{int}(\mathbb{D}^2 - \{0\}) \to U \cap V$ is the homeomorphism defined by p. Regarding ω as a loop in U, we see that the image of $\pi(U \cap V, x_1)$ under the homomorphism $\ell_\#$ is generated by the homotopy class of ω in U. Accordingly, $\ker(k_{\#1})$ is the smallest normal subgroup containing $[\omega]$, and thus we see that $\ker(k_{\#0})$ is the normal subgroup generated by the $[\phi^{-1} * \omega * \phi]$.

It remains to see the pullback of $\ker(k_{\#0})$ under the homomorphism $i_\#$. To this end, consider the straight-line path σ in $\mathbb{D}^2$ from z_1 to z_0 (Fig. 14.16). Then, $\phi = q'\sigma$, where q' is the mapping of $\mathbb{D}^2 - \{0\}$ into U defined by p, is a path in U from x_1 to x_0 and we have $\phi^{-1} * \omega * \phi = q'\sigma^{-1} * q'\gamma * q'\sigma \simeq q'(\sigma^{-1} * \gamma * \sigma)$ rel$\{0, 1\}$. If $\pi(\mathbb{S}^1, z_0)$ is generated by the class $[\alpha]$, then $\sigma^{-1} * \gamma * \sigma \simeq \alpha^{\pm 1}$ rel $\{0, 1\}$ in $\mathbb{D}^2 - \{0\}$; consequently, $\hat{\phi}[\omega] = [q'\alpha]^{\pm 1}$ in $\pi(U, x_0)$. Notice that $q'|\mathbb{S}^1$ coincides with the composition $\mathbb{S}^1 \xrightarrow{f} Y \xrightarrow{i} U$, so we have the following commutative diagram:

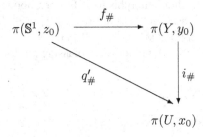

Since $i_\#$ is an isomorphism and $[\alpha]$ generates $\pi(\mathbb{S}^1, z_0)$, we see that the inverse image of the normal subgroup generated by $[q'\alpha]$ under the isomorphism $i_\#$ is the smallest normal subgroup of $\pi(Y, y_0)$ containing $\text{im}(f_\#)$.   ◇

Exercises

1. Show that each of the following spaces has the homotopy type of $\mathbb{S}^1 \vee \mathbb{S}^1$, and hence their fundamental groups are free groups on two generators:

   (a) Two circles joined by a line segment, (b) the figure θ space, (c) the doubly punctured plane $\mathbb{R}^2 - \{p, q\}$, and (d) the punctured torus.

2. Let $X_1, \ldots, X_n$ be pointed spaces with base points $x_1, \ldots, x_n$, respectively. Suppose that each $\{x_i\}$ is a strong deformation retract of an open subset $W_i \subseteq X_i$. Prove that

$$\pi(X_1 \vee \cdots \vee X_n, x_0) \cong \pi(X_1, x_1) \star \cdots \star \pi(X_n, x_n),$$

   where x_0 denote the element determined by the x_i.

3. Compute the fundamental group of the following:

   (a) The complement of the three coordinate axes in $\mathbb{R}^3$.

   (b) The unit 2-sphere $\mathbb{S}^2$ with the unit disc in xy-plane.

(c) The unit 2-sphere $\mathbb{S}^2$ with a diameter.

4. Find a presentation of the fundamental group of the torus.

5. Let $\tau: \mathbb{S}^1 \to \mathbb{S}^1$ be rotation through the angle $2\pi/n, n > 1$, and X be the quotient space of $\mathbb{D}^2$ obtained by identifying the points $x, \tau(x), \ldots, \tau^{n-1}(x)$ for every $x \in \mathbb{S}^1$. Show that $\pi(X)$ is the cyclic group of order n. (Some authors call X the n-fold dunce cap).

6. Determine the fundamental group of the adjunction space of $\mathbb{D}^n$ to a path-connected Hausdorff space Y by a continuous function $f: \mathbb{S}^{n-1} \to Y, n > 2$.

Chapter 15
Covering Spaces

The study of the fundamental group is intimately related to the concept of "covering space". It provides a useful tool for the computation of the fundamental group of various spaces and plays an important role in the study of Riemann surfaces. Moreover, an extremely fruitful notion of "fiber bundle" can be viewed as a generalization of covering spaces. A covering space of a space X is defined by a map over X, called a "covering map". We introduce this concept and discuss its elementary properties in Sect. 15.1. Section 15.2 is devoted to straightforward generalizations of the lifting properties of the exponential map $p: \mathbb{R}^1 \to \mathbb{S}^1$, $p(t) = e^{2\pi i t}$, to covering maps. As a consequence of these generalizations, we obtain a general solution to the lifting problem for these maps in terms of fundamental groups (Theorem 15.2.6). In Sect. 15.3, an action of the fundamental group of the base space of a covering map on the "fiber" over the base point is defined and it is applied to prove the Borsuk–Ulam theorem for $\mathbb{S}^2$. The next three sections mainly concern with the problem of classification of covering spaces of a locally path-connected space. In Sect. 15.4, we define the notion of equivalence for covering spaces of a given space and study the group of covering transformations (self-equivalences) of a covering space. In Sect. 15.5, we shall treat "regular covering maps". It will be seen that such coverings arise as the orbit map of "proper and free" actions of discrete groups. The last section deals with the existence of covering spaces of certain spaces. We first show that every space X satisfying a mild condition has a simply connected covering space, and then the results of the previous two sections are applied to show the existence of a covering space of X with the fundamental group isomorphic to a given subgroup of $\pi(X)$.

© Springer Nature Singapore Pte Ltd. 2019
T. B. Singh, *Introduction to Topology*,
https://doi.org/10.1007/978-981-13-6954-4_15

15.1   Covering Maps

Recall that the key technical tools for computing the fundamental group of the circle were two particular cases of Theorem 14.3.4: The path lifting property and the homotopy lifting property. The essential property of the exponential map used in proving this theorem is captured in the following:

Definition 15.1.1 Let X and $\widetilde{X}$ be topological spaces. A continuous map $p\colon \widetilde{X} \to X$ is called a *covering map* (or *covering projection*) if each point $x \in X$ has an open nbd U such that $p^{-1}(U)$ is a disjoint union of open sets $\widetilde{U}_\alpha$ in $\widetilde{X}$, and $p|\widetilde{U}_\alpha \colon \widetilde{U}_\alpha \to U$ is a homeomorphism for every index α.

If $p\colon \widetilde{X} \to X$ is a covering map, then X is called the *base space* of p, and $\widetilde{X}$ is called a *covering space* of X (relative to p). An open set U satisfying the condition of Definition 15.1.1 is called *admissible* or *evenly covered* and the sets $\widetilde{U}_\alpha$ are referred to as the *sheets over U*. If $x \in X$ and U is an admissible nbd of x, then $p\left(p^{-1}(U)\right) = U$. So $x \in \operatorname{im}(p)$ and p is surjective.

Example 15.1.1 If X and $\widetilde{X}$ are homeomorphic spaces, then any homeomorphism $\widetilde{X} \to X$ is a covering map, so $\widetilde{X}$ is a covering space of X. We use the term "trivial" to refer to such covering maps.

Example 15.1.2 The exponential map $p\colon \mathbb{R}^1 \to \mathbb{S}^1$, $t \mapsto e^{2\pi i t}$, is a covering map. We know that p is continuous and, by Lemma 14.3.3, it is also open. Clearly, $\mathbb{S}^1$ is the union of open sets $U = \mathbb{S}^1 - \{-1\}$ and $V = \mathbb{S}^1 - \{1\}$. Notice that $p^{-1}(U) = \bigcup_{n \in \mathbb{Z}} \widetilde{U}_n$, where $\widetilde{U}_n = \left(n - \frac{1}{2}, n + \frac{1}{2}\right)$. The restriction of p to $\widetilde{U}_n$ is obviously a bijection between $\widetilde{U}_n$ and U. Since p is continuous and open, $p|\widetilde{U}_n$ is a homeomorphism of $\widetilde{U}_n$ onto U. Similarly, each component of $p^{-1}(V) = \bigcup_{n \in \mathbb{Z}} (n, n + 1)$ is

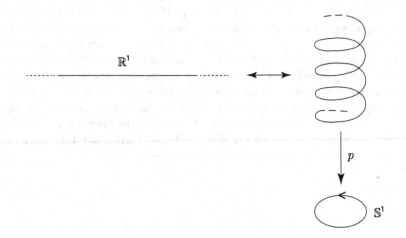

$\mathbb{R}^1$

p

$\mathbb{S}^1$

Fig. 15.1   An infinite spiral as a covering space of $\mathbb{S}^1$

mapped homeomorphically onto V by p. The covering space $\mathbb{R}^1$ may be looked at as an infinite spiral over $\mathbb{S}^1$ (Fig. 15.1).

Example 15.1.3 Let G be a Hausdorff topological group, and H be a discrete subgroup of G. Let G/H denote the space of left cosets (or right cosets) with the quotient topology determined by the natural map $\pi: G \to G/H$. Then π is a covering map. To see this, we first observe that π is an open mapping. If $O \subseteq G$ is open, then Oh is open in G for every $h \in H$, since the right translation $x \mapsto xh$ is a homeomorphism of G onto itself. Thus, $\pi^{-1}\pi(O) = \bigcup_{x \in O} xH = \bigcup_{h \in H} Oh$ is open in G. Since π is an identification map, $\pi(O)$ is open in G/H.

Next, we note that it is enough to prove the existence of an admissible nbd of $\bar{e} \in G/H$, where e is the identity element of G. For, given $g \in G$, the left translation $\lambda_g: G \to G, x \mapsto gx$, is a homeomorphism. This clearly induces a homeomorphism $\theta_g: G/H \to G/H$ given by $\theta_g(\bar{x}) = \overline{gx}$, $\bar{x} \in G/H$. Therefore, it is immediate $\theta_g(U)$ is an admissible nbd of $\bar{g}$ if U is that of $\bar{e}$.

Finally, we show that there is an admissible nbd U of $\bar{e}$ in G/H. Since H is discrete, there exists an open nbd N of e in G such that $N \cap H = \{e\}$. By the continuity of the map $(x, y) \mapsto x^{-1}y$ on $G \times G$, we find an open nbd V of e such that $V^{-1}V \subseteq N$. Set $U = \pi(V)$. Then U is obviously an open nbd of $\bar{e}$ in G/H. We assert that U is evenly covered by π. It is clear that $\pi^{-1}(U) = \bigcup_{h \in H} Vh$. As above, each Vh is open in G. To see that these sets are mutually disjoint, suppose that $Vh \cap Vh' \neq \varnothing$, where $h, h' \in H$. Then there exists $v, v' \in V$ such that $vh = v'h' \Rightarrow v'^{-1}v = h'h^{-1} \in V^{-1}V \cap H \subseteq N \cap H = \{e\}$. So $h = h'$, and the sets Vh are pairwise disjoint. It remains to show that π maps Vh homeomorphically onto U. Being the restriction of a continuous open map to an open set, $\pi|Vh: Vh \to U$ is also continuous and open. If $v, v' \in V$ and $\pi(vh) = \pi(v'h)$, then $vh = v'hh'$ for some $h' \in H$. So $v'^{-1}v = hh'h^{-1} \in V^{-1}V \cap H = \{e\}$, and we have $v = v'$. Thus we see that $\pi|Vh$ is injective. Also, we have $\pi(Vh) = U$ for every $h \in H$, since $xhH = xH$ for all $x \in G$.

Example 15.1.4 Consider the topological group $\mathbb{S}^1$, the unit circle. If n is a positive integer, then the map $p_n: \mathbb{S}^1 \to \mathbb{S}^1$ which maps $z \in \mathbb{S}^1$ into z^n is a covering map and $\mathbb{S}^1$ itself is a covering space $\mathbb{S}^1$. To see this, given $z_0 \in \mathbb{S}^1$, let $U = \mathbb{S}^1 - \{-z_0\}$ and $F = \{z \in \mathbb{S}^1 \mid z^n = -z_0\}$. Then U is a nbd of z_0 and $p_n^{-1}(U) = \mathbb{S}^1 - F$. If $z_0 = e^{\iota\theta_0}$, $0 \leq \theta_0 < 2\pi$, then $-z_0 = e^{\iota(\theta_0 - \pi)}$ and $F = \{e^{\iota(\theta_0 + (2j-1)\pi)/n} \mid j = 1, \ldots, n\}$. For each j, put

$$V_j = \{e^{\iota\theta} \mid (2j - 1)\pi < n\theta - \theta_0 < (2j + 1)\pi\}.$$

Then each V_j is obviously an open arc on $\mathbb{S}^1$ (see Fig. 15.2); in fact, it it can be seen as the intersection of $\mathbb{S}^1$ with a suitable open disc in $\mathbb{R}^2$. Moreover, $p_n^{-1}(U) = \bigcup_1^n V_j$. It is also clear that each of the mappings

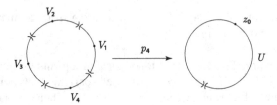

Fig. 15.2 Sheets over an admissible nbd in Example 15.1.4 ($n = 4$)

$$V_j \to (\theta_0 - \pi, \theta_0 + \pi), \quad e^{i\theta} \mapsto n\theta - 2j\pi \quad \text{and}$$
$$(\theta_0 - \pi, \theta_0 + \pi) \longrightarrow \mathbb{S}^1 - \{-z_0\}, \quad \theta \mapsto e^{i\theta}$$

is a homeomorphism, and their composition is the map $z \mapsto z^n$. Thus $p_n | V_j$ is a homeomorphism of V_j onto U.

Example 15.1.5 The plane $\mathbb{R}^2$ is a covering space of the torus $\mathbb{S}^1 \times \mathbb{S}^1$ with the covering projection $(s, t) \mapsto \left(e^{2\pi i s}, e^{2\pi i t}\right)$.

This is a particular case of the following general result.

Proposition 15.1.2 *If* $p \colon \widetilde{X} \to X$ *and* $q \colon \widetilde{Y} \to Y$ *are covering maps, then so is* $p \times q \colon \widetilde{X} \times \widetilde{Y} \to X \times Y, (\tilde{x}, \tilde{y}) \mapsto (p(\tilde{x}), q(\tilde{y}))$.

Proof Given a point (x, y) in $X \times Y$, let U and V be admissible nbds of x and y, respectively. Suppose that $\widetilde{U}_\alpha, \alpha \in A$, are the sheets over U, and $\widetilde{V}_\beta, \beta \in B$, are those over V. Then we clearly have

$$(p \times q)^{-1}(U \times V) = p^{-1}(U) \times q^{-1}(V) = \bigcup\nolimits_{\alpha,\beta}\left(\widetilde{U}_\alpha \times \widetilde{V}_\beta\right).$$

Moreover, $p \times q$ maps each product $\widetilde{U}_\alpha \times \widetilde{V}_\beta$ homeomorphically onto $U \times V$, and thus we see that $U \times V$ is an admissible nbd of (x, y). $\diamond$

The preceding proposition cannot be generalized to arbitrary products, as shown by the following example.

Example 15.1.6 For each integer $n \geq 1$, let $\widetilde{X}_n = \mathbb{R}^1$, $X_n = \mathbb{S}^1$ and $p_n \colon \widetilde{X}_n \to X_n$ be the exponential map $t \mapsto e^{2\pi i t}$. We have already seen that p_n is a covering map. But $\prod p_n \colon \prod \widetilde{X}_n \to \prod X_n$ is not a covering map, for its restriction to any nonempty open subset of $\prod \widetilde{X}_n$ is not injective.

Let $p \colon \widetilde{X} \to X$ be a covering map. The inverse image of a point $x \in X$ under p is called the *fiber over* x. Clearly, the fiber $p^{-1}(x)$ over any point x is a discrete subset of $\widetilde{X}$. The number of points in a fiber is *locally constant* in the sense that each point $x \in X$ has an open nbd U such that the cardinality of $p^{-1}(u)$ is the same for every $u \in U$. For, if U is an admissible open set in X, and $\tilde{U}_\alpha$ is a sheet over U, then $\tilde{U}_\alpha$ contains exactly one point of each fiber $p^{-1}(x)$, $x \in U$, and therefore there is a one-to-one correspondence between $p^{-1}(x)$ and the family of the sheets $\tilde{U}_\alpha$ over U. Since a locally constant function on a connected space is constant, we see that the cardinality of $p^{-1}(x)$ is the same for all $x \in X$ if X is *connected*. In this case, the *number of sheets* (or the *multiplicity*) of p is defined to be the cardinal number of a fiber $p^{-1}(x)$. Covering maps with n (finite) sheets are often called "n-fold" coverings; in particular, a twofold covering is termed a "double covering".

Example 15.1.7 The quotient map $q : \mathbb{S}^2 \to \mathbb{RP}^2$, which identifies each pair of antipodal points is a double covering. We first observe that q is an open map. Let O be any open set in $\mathbb{S}^2$. Since the antipodal map $f : \mathbb{S}^2 \to \mathbb{S}^2$, $f(x) = -x$, is a homeomorphism, $f(O)$ is open in $\mathbb{S}^2$. Obviously, $q^{-1}(q(O)) = O \cup f(O)$ is open, and therefore $q(O)$ is open in $\mathbb{RP}^2$. To show the existence of admissible nbds, let $\xi \in \mathbb{RP}^2$. Suppose that $\xi = q(x) = q(-x)$, and consider an open set O in $\mathbb{S}^2$ containing x which does not contain any pair of antipodal points (O may be taken to be a small disc centered at x). Then $U = q(O)$ is an open set containing ξ, and $q|O : O \to U$ is a bijection. Being the restriction of a continuous open map to an open set, $q|O$ is also continuous and open. Thus $q|O$ is a homeomorphism between O and U. Similarly, q maps $f(O)$ homeomorphically onto U. It is also clear that $q^{-1}(U)$ is the disjoint union of the sets O and $f(O)$. Therefore, U is an admissible nbd of ξ, and thus $\mathbb{S}^2$ is a covering space of $\mathbb{RP}^2$ with q as the covering map.

Similarly, the quotient map $\mathbb{S}^n \to \mathbb{RP}^n$ which identifies the antipodal points is a double covering.

Definition 15.1.3 A continuous map $f : Y \to X$ is called a *local homeomorphism* if each point $y \in Y$ has an open nbd V such that $f(V)$ is open in X and $f|V$ is a homeomorphism of V onto $f(V)$.

With this terminology, a covering map $p : \tilde{X} \to X$ is a local homeomorphism. Consequently, any local property of X, such as local compactness, local path-connectedness, etc., is inherited by $\tilde{X}$. It follows from the following proposition that p is an identification. Thus, the base space of a covering map is a quotient space of its covering space.

Proposition 15.1.4 *A local homeomorphism is an open mapping.*

Proof Let $p : Y \to X$ be a local homeomorphism, and V be an open subset of Y. If $x \in p(V)$, then we find a $y \in V$ such that $p(y) = x$. By our hypothesis, there exists an open neighborhood U of x in X, and an open nbd W of y in Y such that p maps W homeomorphically onto U. Since $V \cap W$ is open in W, and U is open in X, $p(V \cap W)$ is open in X. Obviously, $x \in p(V \cap W) \subseteq p(V)$. Thus, $p(V)$ is a nbd of x, and the proposition follows. $\diamond$

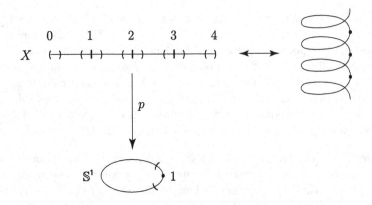

Fig. 15.3 The mapping p in Example 15.1.8 for $n = 4$

However, a local homeomorphism need not be a covering map, as shown by the following.

Example 15.1.8 Let $n > 1$ be an integer and X be the open interval $(0, n)$ with the usual topology. Consider the map $p: X \to \mathbb{S}^1$ given by $p(t) = e^{2\pi i t}$. The space X may be regarded as an open finite spiral over $\mathbb{S}^1$. Being the restriction of a local homeomorphism to an open subset, p is a local homeomorphism. Clearly, p is a surjection; but it is not a covering map because $1 \in \mathbb{S}^1$ has no admissible nbd (Fig. 15.3).

The next result reduces the study of covering maps with locally path-connected base spaces to the case, where they can also assumed connected (equivalently, path-connected).

Proposition 15.1.5 *Let X be a locally path-connected space. Then a continuous map $p: \widetilde{X} \to X$ is a covering map if and only if, for each path component C of X, $p|p^{-1}(C): p^{-1}(C) \to C$ is a covering map.*

Proof Suppose that $p: \widetilde{X} \to X$ is a covering map and $x \in C$. Let U be an admissible open nbd of $x \in X$, and V be the path component of U containing x. Then $V \subseteq C$, for C is a path component of X. Since X is locally path-connected, V is open in X, and therefore open in C. Clearly, V is evenly covered by $q = p|p^{-1}(C)$, and q is a covering map.

Conversely, assume that the map $q: p^{-1}(C) \to C, x \mapsto p(x)$, is a covering map for each path component C of X. Given $x \in X$, let C be the path component of X which contains x. By our hypothesis, there is an open nbd U of x in C, which is evenly covered by q. Since X is locally path-connected, the path component C is open in X, and hence U is open in X. Also, every open subset of $p^{-1}(C)$ is open in $\widetilde{X}$. It is now clear that U is p-admissible, and p is a covering map. $\diamond$

Moreover, in the case of connected and locally path-connected spaces, it suffices to study only their connected covering spaces. This is immediate from the next result.

Proposition 15.1.6 Let $p: \widetilde{X} \to X$ be a covering map. If X is locally path-connected and $\widetilde{C}$ a path component of $\widetilde{X}$, then $p(\widetilde{C})$ is a path component of X and $p|\widetilde{C}: \widetilde{C} \to p(\widetilde{C})$ is a covering map.

Proof Suppose that X is locally path-connected, and let $\widetilde{C}$ be a path component of $\widetilde{X}$. To prove that $p(\widetilde{C})$ is a path component of X, it suffices to establish that it is a component, since the components and path components of a locally path-connected space are identical. It is obvious that $p(\widetilde{C})$ is connected. We observe that $p(\widetilde{C})$ is clopen in X. Let $x \in \mathrm{cl}(p(\widetilde{C}))$. Since X is locally path-connected, there is an admissible path-connected open nbd U of x (ref. Exercise 9). Accordingly, each sheet $\widetilde{U}$ over U is path-connected. We have $\widetilde{C} \cap p^{-1}(U) \neq \varnothing$, for $U \cap p(\widetilde{C}) \neq \varnothing$. So $\widetilde{U} \cap \widetilde{C} \neq \varnothing$ for some sheet $\widetilde{U}$ over U. As $\widetilde{C}$ is a path component of $\widetilde{X}$, we have $\widetilde{U} \subseteq \widetilde{C}$. So $U = p(\widetilde{U}) \subseteq p(\widetilde{C})$ and $x \in \mathrm{int}(p(\widetilde{C}))$. Accordingly, $\mathrm{cl}(p(\widetilde{C})) \subseteq \mathrm{int}(p(\widetilde{C}))$, and it follows that $p(\widetilde{C})$ is both closed and open. Therefore $p(\widetilde{C})$ is a component of X, as desired.

Now, we show that $q = p|\widetilde{C}: \widetilde{C} \to p(\widetilde{C})$ is a covering map. Let $x \in p(\widetilde{C})$ and U be an admissible path-connected open nbd of x in X. Then $U \subseteq p(\widetilde{C})$. If $\widetilde{U}$ is a sheet over U and $\widetilde{U} \cap \widetilde{C} \neq \varnothing$, then $\widetilde{U} \subseteq \widetilde{C}$. It follows that $q^{-1}(U)$ is the disjoint union of those sheets $\widetilde{U}$ over U which intersect $\widetilde{C}$. Thus U is evenly covered by q, and q is a covering map. $\diamond$

We remark that, unlike Proposition 15.1.5, the converse of the preceding proposition is false.

Example 15.1.9 Let X be a countable product of unit circles $\mathbb{S}^1$ and, for each integer $n \geq 1$, $\widetilde{X}_n = \mathbb{R}^n \times X$. Then the mapping $p_n: \widetilde{X}_n \to X$ defined by

$$p_n(t_1, \ldots, t_n, z_1, z_2, \ldots) = \left(e^{2\pi \imath t_1}, \ldots, e^{2\pi \imath t_n}, z_1, z_2, \ldots\right)$$

is clearly a covering map. Let $p: \sum \widetilde{X}_n \to X$ be defined by $p|\widetilde{X}_n = p_n$. Note that the path components of $\sum \widetilde{X}_n$ are the sets $\widetilde{X}_n$. If an open set $O \subseteq X$ is evenly covered by p, then there is a basic connected open set $G \subseteq O$ which is also evenly covered by p, since X is locally connected. Suppose that $G = \bigcap_1^k \pi_i^{-1}(U_i)$, where $\pi_i: X \to \mathbb{S}^1$ is the projection onto the ith factor of X and the $U_i \subseteq \mathbb{S}^1$ are open sets. Since G is connected, so are the U_i. Let $\widetilde{\pi}_i$ be the projection onto the ith factor of $\widetilde{X}_n$ and $ex: \mathbb{R}^1 \to \mathbb{S}^1$ denote the map $t \to e^{2\pi \imath t}$. Then, for $n > k$, $p_n^{-1}(G) = \bigcap_1^k (ex \circ \widetilde{\pi}_i)^{-1}(U_i)$. In this case, the $(k+1)$st factor of each component of $p_n^{-1}(G)$ is $\mathbb{R}^1$, and hence p_n is not injective on any component of $p_n^{-1}(G)$. Accordingly, G is not evenly covered by p_n. On the other hand, G must be evenly covered by each p_n, since $G \subseteq O$. This contradiction shows that no open subset of X is admissible under p, and p is not a covering map.

Exercises

1. Let $\widetilde{X}$ be the product of a space X with a discrete space. Show that the projection $\widetilde{X} \to X$ is a covering map.

2. Let X be the wedge of two circles.

    (a) Prove that the subspace $\widetilde{X} = \{(x, y) \in \mathbb{R}^2 : x$ or y is an integer$\}$ is a covering space of the subspace $X \subset \mathbb{R}^2$.

    (b) Is the quotient map $\pi: \mathbb{S}^1 \to X$ which identifies the points 1 and -1 a covering map?

3. • Let A, B, and C denote the circles of radius 1 centered at 0, 2, and 4, respectively, in the complex plane $\mathbb{C}$. Let $X = A \cup B$ and $\widetilde{X} = A \cup B \cup C$ be subspaces of $\mathbb{C}$. Define a mapping $p: \widetilde{X} \to X$ by setting $p(z) = z$ for $z \in A$, $p(z) = 2 - (z-2)^2$ for $z \in B$ and $p(z) = 4 - z$ for $z \in C$. Show that p is a covering map.

4. Prove that the mapping $z \mapsto z^n$ of $\mathbb{C} - \{0\}$ onto itself, where $n > 0$, is a covering map.

5. Consider the exponential function $\exp: \mathbb{C} \to \mathbb{C}$ given by $\exp(z) = \sum_0^\infty z^n/n$. Show that $\mathbb{C}$ is a covering space of $\mathbb{C} - \{0\}$ relative to exp.

6. Show that the infinite cylinder $\mathbb{R}^1 \times \mathbb{S}^1$ is a covering space of the torus $\mathbb{S}^1 \times \mathbb{S}^1$.

7. Show that there is a two-sheeted covering of the Klein bottle by the torus.

8. Let X be a connected space. Show that a locally constant function on X is constant.

9. • Let $p: \widetilde{X} \to X$ be a covering map. If X is locally path-connected, show that each point of X has a path-connected open nbd V such that each path component of $p^{-1}(V)$ is mapped homeomorphically onto V by p.

10. Let $p_\nu: \widetilde{X}_\nu \to X_\nu$, $\nu \in N$, be a collection of covering maps. Show that the mapping $p: \sum_\nu \widetilde{X}_\nu \longrightarrow \sum_\nu X_\nu$, defined by $p|\widetilde{X}_\nu = p_\nu$, is a covering map.

11. Show, by an example, that the composite of two covering maps need not be a covering map.

12. Suppose that $p: Y \to X$ and $q: Z \to Y$ are covering maps such that $p^{-1}(x)$ is finite for every $x \in X$. Show that the composite pq is a covering map.

13. If N is finite and, for each $\nu \in N$, $p_\nu: \widetilde{X}_\nu \to X$ is a covering map, prove that the function $q: \sum_\nu \widetilde{X}_\nu \to X$, defined by $q|\widetilde{X}_\nu = p_\nu$, is a covering map.

14. Let $p: \widetilde{X} \to X$ be a covering map. If X is Hausdorff, show that $\widetilde{X}$ is also Hausdorff.

15. Let $p: \widetilde{X} \to X$ be a covering map, and suppose that X is connected. Prove that $\widetilde{X}$ is compact if and only if X is compact and the multiplicity of p is finite.

16. Give examples of two local homeomorphisms $X \to Y$ in one of which X is Hausdorff but not Y and in the other Y is Hausdorff but not X.

17. Let $f: X \to Y$ be a local homeomorphism, and suppose that X is compact. Prove:

   (a) $f^{-1}(y)$ is finite for every $y \in Y$.

   (b) If Y is connected and both X and Y are Hausdorff, show that f is a covering map.

15.2 The Lifting Problem

Given continuous maps $p: \widetilde{X} \to X$ and $f: Y \to X$, an important problem in topology is to determine whether or not there exists a continuous map $\tilde{f}: Y \to \widetilde{X}$ such that $p \circ \tilde{f} = f$, that is, $\tilde{f}$ makes the following triangle commutative:

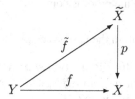

This is referred to as the *lifting problem* for the given map f. If there is such a map $\tilde{f}$, then we say that f can be lifted to $\tilde{f}$ in $\widetilde{X}$, and call $\tilde{f}$ a *lifting* (or *lift*) of f relative to p.

If $p: \widetilde{X} \to X$ is a covering map, then we have a general solution to the lifting problem. We begin by solving this problem for continuous maps $I \to X$, and use the answer to show that every homotopy of a continuous map from a locally connected space Y to X, which has a lifting $Y \to \widetilde{X}$, can also be lifted. These are straightforward generalizations of the path lifting property and the homotopy lifting property of the exponential map $\mathbb{R}^1 \to \mathbb{S}^1$ for covering maps. Finally, we will establish a simple criterion for the existence of liftings of continuous maps from a connected, locally path-connected space into the base space of p purely in terms of the fundamental groups of the spaces involved.

We first prove the following

Theorem 15.2.1 (The Unique Lifting Property) *Let* $p: \widetilde{X} \to X$ *be a covering map, and let* $f: Y \to X$ *be a continuous mapping of a connected space* Y. *Given* $\tilde{x}_0 \in \widetilde{X}$ *and* $y_0 \in Y$ *such that* $f(y_0) = p(\tilde{x}_0)$, *there is at most one continuous map* $\tilde{f}: Y \to \widetilde{X}$ *with* $p\tilde{f} = f$ *and* $\tilde{f}(y_0) = \tilde{x}_0$.

Proof Assume that there are two continuous maps $\tilde{f}_1, \tilde{f}_2 : Y \to \tilde{X}$ such that $p\tilde{f}_1 = f = p\tilde{f}_2$ and $\tilde{f}_1(y_0) = \tilde{x}_0 = \tilde{f}_2(y_0)$. Let

$$A = \left\{ y \in Y \mid \tilde{f}_1(y) = \tilde{f}_2(y) \right\} \text{ and}$$
$$B = \left\{ y \in Y \mid \tilde{f}_1(y) \neq \tilde{f}_2(y) \right\}.$$

It is obvious that $Y = A \cup B$, $A \cap B = \varnothing$. We show that A and B, both, are open. This forces that $B = \varnothing$, since Y is connected and $A \neq \varnothing$ ($y_0 \in A$). So $A = Y$ and $\tilde{f}_1 = \tilde{f}_2$.

Given $a \in A$, let U be an admissible open nbd of $f(a)$ and $\tilde{U}$ be the sheet over U containing $\tilde{f}_1(a) = \tilde{f}_2(a)$. Then $O = \tilde{f}_1^{-1}(\tilde{U}) \cap \tilde{f}_2^{-1}(\tilde{U})$ is an open nbd of a in Y. If $y \in O$, then $\tilde{f}_1(y)$ and $\tilde{f}_2(y)$ belong to $\tilde{U}$. So $p\tilde{f}_1(y) = f(y) = p\tilde{f}_2(y)$, which implies that $\tilde{f}_1(y) = \tilde{f}_2(y)$, since $p|\tilde{U}$ is a homeomorphism. It follows that $y \in A$, and $O \subseteq A$. Thus A is a nbd of each of its points, and therefore open. To show that B is open, let $b \in B$ be arbitrary. Consider an admissible open nbd U of $f(b)$. Since $\tilde{f}_1(b) \neq \tilde{f}_2(b)$, there are distinct sheets $\tilde{U}_1$ and $\tilde{U}_2$ over U containing $\tilde{f}_1(b)$ and $\tilde{f}_2(b)$, respectively. Obviously, $N = \tilde{f}_1^{-1}(\tilde{U}_1) \cap \tilde{f}_2^{-1}(\tilde{U}_2)$ is an open nbd of b. Since $\tilde{U}_1$ and $\tilde{U}_2$ are disjoint, $\tilde{f}_1(y) \neq \tilde{f}_2(y)$ for every $y \in N$, so we have $N \subseteq B$. Therefore B is also open, and the proof is complete. $\diamond$

Next, we show that any path in the base space of a covering map can be lifted.

Theorem 15.2.2 (The Path Lifting Property) *Let $p : \tilde{X} \to X$ be a covering map and let $f : I \to X$ be a path with the origin x_0. If $\tilde{x}_0 \in p^{-1}(x_0)$, then there exists a unique path $\tilde{f} : I \to \tilde{X}$ with $\tilde{f}(0) = \tilde{x}_0$ and $p\tilde{f} = f$.*

Proof Suppose that $[a, b] \subseteq I$ and U is an admissible nbd of $x = f(a)$ such that $f([a, b]) \subseteq U$. If $\tilde{x} \in p^{-1}(x)$, then $\tilde{x}$ lies in a unique sheet, say $\tilde{U}$. It is easy to see that $\tilde{g} : [a, b] \to \tilde{X}$ defined by $\tilde{g} = \left(p|\tilde{U}\right)^{-1} \circ (f|[a, b])$ is continuous and satisfies $p\tilde{g} = f|[a, b]$, and $\tilde{g}(a) = \tilde{x}$.

Now, for each $t \in I$, let U_t be an admissible nbd of $f(t)$. Then $\left\{ f^{-1}(U_t) : t \in I \right\}$ is an open covering of I. Since I is a compact metric space, this open cover of I has a Lebesgue number δ, say. Consider a subdivision of I with points

$$0 = t_0 < t_1 < \cdots < t_n = 1$$

such that $t_i - t_{i-1} < \delta$ for $1 \leq i \leq n$. Then each subinterval $[t_{i-1}, t_i]$ is mapped by f into some U_t. By our initial remarks, there is a continuous map $\tilde{g}_1 : [0, t_1] \to \tilde{X}$ with $p\tilde{g}_1 = f|[0, t_1]$ and $\tilde{g}_1(0) = \tilde{x}_0$. Similarly, there is a continuous map $\tilde{g}_2 : [t_1, t_2] \to \tilde{X}$ with $p\tilde{g}_2 = f|[t_1, t_2]$ and $\tilde{g}_2(t_1) = \tilde{g}_1(t_1)$; indeed for each $1 \leq i \leq n$, there is a continuous map $\tilde{g}_i : [t_{i-1}, t_i] \to \tilde{X}$ with $p\tilde{g}_i = f|[t_{i-1}, t_i]$ and $\tilde{g}_i(t_{i-1}) = \tilde{g}_{i-1}(t_{i-1})$. By the gluing lemma, we may combine the functions $\tilde{g}_i$ into a continuous map $\tilde{f} : I \to \tilde{X}$, where $\tilde{f}(t) = \tilde{g}_i(t)$ for $t_{i-1} \leq t \leq t_i$. The path $\tilde{f}$ clearly lifts f and $\tilde{f}(0) = \tilde{x}_0$. Its uniqueness follows from the unique lifting property of p. $\diamond$

We now show that homotopies connecting maps of a locally connected space into the base space of a covering map can be lifted. It should, however, be noted that this result holds good for homotopies between maps from any space (see Spanier [13] Theorem 3 on p. 67).

Theorem 15.2.3 (The Homotopy Lifting Property) *Let $p: \widetilde{X} \to X$ be a covering map, and let $\tilde{f}: Y \to \widetilde{X}$ and $F: Y \times I \to X$ be continuous maps such that $F(y, 0) = p\tilde{f}(y)$ for every $y \in Y$. If Y is locally connected, then there exists a unique continuous map $\widetilde{F}: Y \times I \to \widetilde{X}$ such that $\widetilde{F}(y, 0) = \tilde{f}(y)$ for every $y \in Y$ and $p\widetilde{F} = F$, that is, $\widetilde{F}$ makes the following diagram*

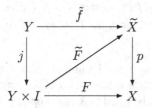

commutative, where j sends y to $(y, 0)$. Moreover, if F is a homotopy rel A for some $A \subseteq Y$, then so is $\widetilde{F}$.

Proof If a homotopy $\widetilde{F}$ having the desired properties exists, then, for each $y \in Y$, $\tilde{f}_y: t \mapsto \widetilde{F}(y, t)$ is a lifting of the path $t \mapsto F(y, t)$ starting at $\tilde{f}(y)$ and, by Theorem 15.2.2, such a lift exists and is unique. So we define $\widetilde{F}(y, t) = \tilde{f}_y(t)$. Then we have $\widetilde{F}(y, 0) = \tilde{f}(y)$ and $p\widetilde{F}(y, t) = p\tilde{f}_y(t) = F(y, t)$ for every $y \in Y$ and $t \in I$. To see the continuity of $\widetilde{F}$, let $\mathcal{U}$ be an open covering of X by admissible open sets. Then $\left\{ F^{-1}(U) \mid U \in \mathcal{U} \right\}$ is an open covering of $Y \times I$. Choose a covering $\mathcal{V}$ of $Y \times I$ consisting of basic open sets each of which is contained in some $F^{-1}(U)$, $U \in \mathcal{U}$. For each $y \in Y$, $\{y\} \times I$ is compact, and therefore the open covering $\{V \cap (\{y\} \times I) \mid V \in \mathcal{V}\}$ of $\{y\} \times I$ has a Lebesgue number ϵ, say. We choose a subdivision

$$0 = t_0 < t_1 < \cdots < t_n = 1$$

of I such that $t_i - t_{i-1} < \epsilon$ for every $i = 1, \ldots, n$. Then there is a member V_i of $\mathcal{V}$ such that $\{y\} \times [t_{i-1}, t_i] \subseteq V_i$. Let N_y be the intersection of the first factors of the V_i. Then $N_y \times [t_{i-1}, t_i]$ is clearly contained in V_i for every i. So F maps $N_y \times [t_{i-1}, t_i]$ into some member of $\mathcal{U}$. Since Y is locally connected, we may assume that N_y is connected. We show, by induction on i, that $\widetilde{F}$ is continuous on each $N_y \times [t_{i-1}, t_i]$. Assume that $\widetilde{F}|\left(N_y \times \{t_{i-1}\}\right)$ is continuous. Then $\widetilde{F}\left(N_y \times \{t_{i-1}\}\right)$ is connected, for N_y is so. As observed above, there is a $U \in \mathcal{U}$ such that $F\left(N_y \times [t_{i-1}, t_i]\right) \subseteq U$. Being a connected set, $\widetilde{F}\left(N_y \times \{t_{i-1}\}\right)$ is contained in a sheet $\widetilde{U}$ (say) over U. Now, if $z \in N_y$ and $t_{i-1} \leq t \leq t_i$, then we have $p(\tilde{f}_z(t)) = F(z, t) \in U$. Since the subinterval $[t_{i-1}, t_i]$ is connected and $\tilde{f}_z(t_{i-1}) = \widetilde{F}(z, t_{i-1}) \in \widetilde{U}$, we have $\tilde{f}_z([t_{i-1}, t_i]) \subseteq \widetilde{U}$, and therefore $\widetilde{F}\left(N_y \times [t_{i-1}, t_i]\right) \subseteq \widetilde{U}$. Since $p|\widetilde{U}$ is a homeomorphism of $\widetilde{U}$ onto U and $p\widetilde{F}(z, t) = F(z, t)$, we see that $\widetilde{F}(z, t) = \left(p|\widetilde{U}\right)^{-1}(F(z, t))$ for every $z \in N_y$

and $t_{i-1} \leq t \leq t_i$. This implies that $\widetilde{F}$ is continuous on $N_y \times [t_{i-1}, t_i]$. For $i = 1$, we note that $\widetilde{F}$ agrees on $N_y \times \{t_0\}$ with the continuous map $\tilde{f}$, so the above argument also applies in this case. Hence, we see that $\widetilde{F}$ is continuous on $N_y \times [t_0, t_1]$ and this completes the induction argument. Since $N_y \times I$ is a finite union of closed sets $N_y \times [t_{i-1}, t_i]$, $1 \leq i \leq n$, the Gluing lemma shows that $\widetilde{F}$ is continuous on $N_y \times I$. Obviously, the family $\{N_y \times I : y \in Y\}$ is an open covering of $Y \times I$, and hence $\widetilde{F}$ is continuous (refer to Exercise 2.1.8). The uniqueness of $\widetilde{F}$ follows from our initial observation that any homotopy of $\tilde{f}$ lifting F provides unique liftings of the paths $t \mapsto F(y, t)$, $y \in Y$.

To see the last statement, suppose that $a \in A$ and $F(a, t) = p\tilde{f}(a)$ for every $t \in I$. Then $\widetilde{F}(a, t) \in p^{-1}(p\tilde{f}(a))$ for every $t \in I$. Since $\widetilde{F}(\{a\} \times I)$ is connected, and $p^{-1}(p\tilde{f}(a))$ is discrete, we have $\widetilde{F}(a, t) = \tilde{f}(a)$ for every $t \in I$ and each $a \in A$, as desired. $\diamond$

Observe that if $f, g : Y \to X$ are homotopic and f has a lifting relative to the covering map $p : \widetilde{X} \to X$, then g also has a lifting in $\widetilde{X}$. Thus, whether or not a continuous map $Y \to X$ can be lifted to $\widetilde{X}$ is a property of the homotopy class of the map. Also, notice that the "Path Lifting Property" of a covering map can be derived from the "Homotopy Lifting Property" by taking Y to be the one-point space.

Theorem 15.2.4 (The Monodromy Theorem) *Let $p : \widetilde{X} \to X$ be a covering map. Suppose that f and g are paths in X and $f \simeq g$ rel $\{0, 1\}$. If $\tilde{f}$ and $\tilde{g}$ are the liftings of f and g, respectively, with $\tilde{f}(0) = \tilde{g}(0)$, then $\tilde{f}(1) = \tilde{g}(1)$ and $\tilde{f} \simeq \tilde{g}$ rel $\{0, 1\}$.*

Proof Let $F : f \simeq g$ rel $\{0, 1\}$. By Theorem 15.2.3, there exists a unique continuous map $\widetilde{F} : I \times I \to \widetilde{X}$ with $p\widetilde{F} = F$ and $\widetilde{F}(s, 0) = \tilde{f}(s)$, $\widetilde{F}(0, t) = \tilde{f}(0)$ and $\widetilde{F}(1, t) = \tilde{f}(1)$ for every $s, t \in I$. The function $\tilde{g}_1 : I \to \widetilde{X}$ defined by $\tilde{g}_1(s) = \widetilde{F}(s, 1)$, $s \in I$, is obviously a continuous lifting of g with the origin $\tilde{g}_1(0) = \tilde{g}(0)$. By Theorem 15.2.1, $\tilde{g}_1 = \tilde{g}$. So $\tilde{g}(1) = \widetilde{F}(1, 1) = \tilde{f}(1)$ and $\widetilde{F} : \tilde{f} \simeq \tilde{g}$ rel $\{0, 1\}$. $\diamond$

Corollary 15.2.5 *Let $p : \widetilde{X} \to X$ be a covering map and $\tilde{x}$ belong to the fiber over $x_0 \in X$.*

(a) *Then the induced homomorphism $p_\# : \pi(\widetilde{X}, \tilde{x}) \to \pi(X, x_0)$ is a monomorphism.*

(b) *If f is a loop in X based at x_0, then the path class $[f]$ belongs to $\mathrm{im}(p_\#)$ if and only if f lifts to a loop in $\widetilde{X}$ based at $\tilde{x}$.*

Proof (a): Let $\tilde{f} : I \to \widetilde{X}$ be a loop in $\widetilde{X}$ based at $\tilde{x}$, and suppose that $p_\#([\tilde{f}]) = [p\tilde{f}]$ is the identity element of $\pi(X, x_0)$. Then $p\tilde{f} \simeq c_{x_0}$ rel $\{0, 1\}$, where c_{x_0} is the constant path at x_0. Since the constant path $c_{\tilde{x}}$ at $\tilde{x}$ is a lifting of c_{x_0} and $\tilde{f}(0) = \tilde{x}$, the Monodromy Theorem implies that $\tilde{f} \simeq c_{\tilde{x}}$ rel $\{0, 1\}$. Hence $[\tilde{f}]$ is the identity element of $\pi(\widetilde{X}, \tilde{x})$, and $p_\#$ is a monomorphism.

(b): If $\tilde{f}$ is a loop in $\tilde{X}$ based at $\tilde{x}$ and $p\tilde{f} = f$, then it is obvious that $[f] \in \text{im}\,(p_\#)$. Conversely, suppose that $[f] \in \text{im}\,(p_\#)$. Then there exists a loop $\tilde{g}$ in $\tilde{X}$ based at $\tilde{x}$ such that $[f] = [p\tilde{g}]$. Therefore there is a homotopy $F: p\tilde{g} \simeq f$ rel $\{0, 1\}$. Let $\tilde{F}:$ $I \times I \to \tilde{X}$ be a homotopy relative to $\{0, 1\}$ such that $p\tilde{F} = F$ and $\tilde{F}(s, 0) = \tilde{g}(s)$, $s \in I$. Then the path $\tilde{f}: I \to \tilde{X}$, defined by $\tilde{f}(s) = \tilde{F}(s, 1)$, is a lift of f and has the same end points as $\tilde{g}$. Thus $\tilde{f}$ is a loop in $\tilde{X}$ based at $\tilde{x}$, and there is no other lifting of f starting at $\tilde{x}$, by Theorem 15.2.1. $\diamond$

The next result provides a general solution to the lifting problem for covering maps in terms of fundamental groups. This is a beautiful instance of the general strategy of algebraic topology: It reduces a topological problem to a purely algebraic question.

Theorem 15.2.6 (The Lifting Theorem) *Let $p: \tilde{X} \to X$ be a covering map, and let Y be a connected and locally path-connected space. If $f: (Y, y_0) \to (X, x_0)$ is continuous, and $\tilde{x}_0 \in p^{-1}(x_0)$, then there exists a lifting $\tilde{f}: (Y, y_0) \to (\tilde{X}, \tilde{x}_0)$ of f if and only if $f_\#\pi(Y, y_0) \subseteq p_\#\pi(\tilde{X}, \tilde{x}_0)$. Moreover, $\tilde{f}$ is unique.*

Proof If there is a lifting $\tilde{f}$ of f with $\tilde{f}(y_0) = \tilde{x}_0$, then $p_\# \circ \tilde{f}_\# = f_\#$, for $p\tilde{f} = f$. So $f_\#\pi(Y, y_0) = (p_\# \circ \tilde{f}_\#)\pi(Y, y_0) \subseteq p_\#\pi(\tilde{X}, \tilde{x}_0)$, and the necessity of the condition follows.

Conversely, suppose that $f_\#\pi(Y, y_0) \subseteq p_\#\pi(\tilde{X}, \tilde{x}_0)$. We construct a fn $\tilde{f}: Y \to \tilde{X}$ as follows. Given $y \in Y$, there is a path h in Y with the origin y_0 and the end y, since Y is path-connected. Accordingly, fh is a path in X with the origin $f(y_0) = x_0$ and the end $f(y)$. By Theorem 15.2.2, there exists a path $\tilde{\phi}$ in $\tilde{X}$ which lifts fh and has the origin $\tilde{x}_0$ (see Fig. 15.4). We observe that the point $\tilde{\phi}(1)$ of $\tilde{X}$ is independent of the choice of path h. Choose another path k in Y from y_0 to y, and let $\tilde{\psi}$ be the path in $\tilde{X}$ that lifts fk and starts at $\tilde{x}_0$. Then $h * k^{-1}$ is a loop in Y based at y_0; consequently, $f \circ (h * k^{-1}) = (fh) * (fk^{-1})$ is a loop in X at x_0. By our hypothesis, $[(fh) * (fk^{-1})] = f_\#([h * k^{-1}])$ belongs to $p_\#\pi(\tilde{X}, \tilde{x}_0)$. So, by Corollary 15.2.5, there exists a loop $\tilde{g}$ in $\tilde{X}$ based at $\tilde{x}_0$ such that $(fh) * (fk^{-1}) = p\tilde{g}$. This implies that $p\tilde{\phi}(2t) = p\tilde{g}(t)$ for $0 \leq t \leq 1/2$, and $p\tilde{\psi}(2 - 2t) = p\tilde{g}(t)$ for $1/2 \leq t \leq 1$. Then, by Theorem 15.2.1, we find that $\tilde{\phi}(t) = \tilde{g}(t/2)$, and $\tilde{\psi}(t) = \tilde{g}(1 - t/2)$ for all $0 \leq t \leq 1$. So $\tilde{\phi}(1) = \tilde{g}(1/2) = \tilde{\psi}(1)$, and we may define a mapping $\tilde{f}: Y \to \tilde{X}$ by setting $\tilde{f}(y) = \tilde{\phi}(1)$. By the definition of the function $\tilde{f}$, we have $p\tilde{f}(y) = p\tilde{\phi}(1) = fh(1) = f(y)$ for every $y \in Y$, and $\tilde{f}(y_0) = \tilde{x}_0$.

To see the continuity of $\tilde{f}$, consider a point $y \in Y$, and let $\tilde{W}$ be an open nbd of $\tilde{f}(y) = \tilde{x}$. Consider an admissible open nbd U of $f(y) = x \in X$. As $\tilde{x} \in p^{-1}(U)$, we find a sheet $\tilde{U}$ over U containing $\tilde{x}$. Replacing $\tilde{W}$ by $\tilde{W} \cap \tilde{U}$, if necessary, we may assume that $\tilde{W} \subseteq \tilde{U}$. Since p is an open map, $W = p(\tilde{W})$ is an open nbd of x with $W \subseteq U$. By the continuity of f, $f^{-1}(W)$ is an open nbd of y in Y. Since Y is locally path-connected, there is a path-connected open nbd V of y with $V \subseteq f^{-1}(W)$. We claim that $\tilde{f}(V) \subseteq \tilde{W}$. Let $h: I \to Y$ be a path in Y from y_0 to y, and let $\tilde{\phi}$ be the lifting of fh with the origin $\tilde{x}_0$. Then $\tilde{\phi}(1) = \tilde{x}$, by the definition of $\tilde{f}$. If $z \in V$, then there is a path $g: I \to V$ from y to z. Clearly, $(fg)(I) \subseteq W \subseteq U$ so

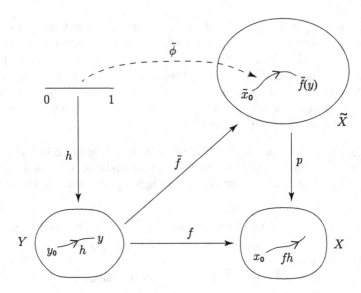

Fig. 15.4 Proof of Theorem 15.2.6

that the composition of fg with the inverse of the homeomorphism $p|\tilde{U} : \tilde{U} \to U$ is
defined. Write $\tilde{\psi} = \left(p|\tilde{U}\right)^{-1} \circ (fg)$. Then $\tilde{\psi}$ is the lifting of fg with $\tilde{\psi}(0) = \tilde{x}$ and
$\tilde{\psi}(1) \in \tilde{W}$. Since $\tilde{\phi}(1) = \tilde{x}$, $\tilde{\phi} * \tilde{\psi}$ is defined. Obviously, $h * g$ is a path in Y from y_0
to z, and $p \circ (\tilde{\phi} * \tilde{\psi}) = (p\tilde{\phi}) * (p\tilde{\psi}) = (fh) * (fg) = f \circ (h * g)$. As $\tilde{\phi} * \tilde{\psi}$ starts
at $\tilde{x}_0$, we have $\tilde{f}(z) = (\tilde{\phi} * \tilde{\psi})(1) = \tilde{\psi}(1) \in \tilde{W}$, and hence our claim. It follows that
$\tilde{f}$ is a lift of f with $\tilde{f}(y_0) = \tilde{x}_0$.

Finally, $\tilde{f}$ is unique, by Theorem 15.2.1. $\diamond$

It should be noted that the condition of local path-connectedness on Y in the
preceding theorem is essential, as shown by the following.

Example 15.2.1 Consider the exponential map $p : \mathbb{R}^1 \to \mathbb{S}^1, t \mapsto e^{2\pi i t}$. Let $\overline{S}$ be the
closed topologist's sine curve (ref. Example 3.1.7) and C be a path in $\mathbb{R}^2$ with the
end points $(0, 1)$ and $(1, \sin 1)$ that intersects $\overline{S}$ only at those two points (Fig. 15.5).
Clearly, $Y = \overline{S} \cup C$ is simply connected. It is also easily seen that the quotient space
$Y/\overline{S}$ is homeomorphic to $\mathbb{S}^1$, and we may assume that $\left[\overline{S}\right] = 1 \in \mathbb{S}^1$. Take the points
$0 \in \mathbb{R}^1, 1 \in \mathbb{S}^1$, and $(0, -1) \in Y$ as base points. Let f be the quotient map $Y \to \mathbb{S}^1$.
If $\tilde{f}$ is constructed as in the proof of the preceding theorem, then the points $(0, y)$,
$-1 \le y \le 1$, are mapped to 0 under $\tilde{f}$ and, by the continuity of path liftings, every
point of the wiggly part S maps to -1. But then $\tilde{f}$ is discontinuous, for the point
$(0, -1)$ is a limit point of S.

As an immediate consequence of the Lifting Theorem, we see that if Y is a locally
path-connected simply connected space, then each continuous map $f : Y \to X$ has

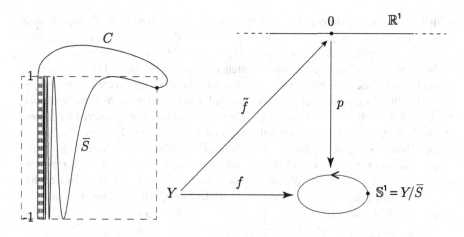

Fig. 15.5 Lifting of a mapping from a space that is not locally path-connected

a unique lifting $\tilde{f}\colon (Y, y_0) \to \left(\widetilde{X}, \tilde{x}_0\right)$ relative to a covering map $p\colon \widetilde{X} \to X$, where $f(y_0) = p(\tilde{x}_0)$.

As another application of the above theorem, we obtain a simple criterion for the existence of fiber preserving continuous maps between connected covering spaces of a locally path-connected space in terms of the fundamental groups of the spaces involved.

Theorem 15.2.7 (The Covering Theorem) *For $i = 1, 2$, let $p_i\colon \widetilde{X}_i \to X$ be covering maps, and suppose that X, $\widetilde{X}_1$ and $\widetilde{X}_2$ are all connected and X is also locally path-connected. If $p_{1\#}\pi\left(\widetilde{X}_1, \tilde{x}_1\right) \subseteq p_{2\#}\pi\left(\widetilde{X}_2, \tilde{x}_2\right)$, where $p_1(\tilde{x}_1) = p_2(\tilde{x}_2) = x_0$, then there exists a unique continuous map $q\colon \widetilde{X}_1 \to \widetilde{X}_2$ such that $p_2 \circ q = p_1$ and $q(\tilde{x}_1) = \tilde{x}_2$, that is, the following diagram of continuous maps commutes:*

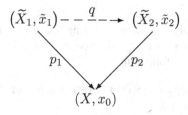

Moreover, q is a covering map.

Proof Since $p_{1\#}\pi\left(\widetilde{X}_1, \tilde{x}_1\right) \subseteq p_{2\#}\pi\left(\widetilde{X}_2, \tilde{x}_2\right)$, the Lifting Theorem shows that there is a unique lifting $q\colon \left(\widetilde{X}_1, \tilde{x}_1\right) \to \left(\widetilde{X}_2, \tilde{x}_2\right)$ of p_1 relative to p_2. Thus, it is the last statement which needs to be established. We first observe that q is surjective. Let $\tilde{y}_2 \in \widetilde{X}_2$ be arbitrary and f be a path in $\widetilde{X}_2$ from $\tilde{x}_2$ to $\tilde{y}_2$. Then $p_2 \circ f$ is a path in X

with the origin x_0. By the path lifting property of p_1, there is a path g in $\widetilde{X}_1$ beginning at $\tilde{x}_1$ such that $p_2 \circ f = p_1 \circ g = p_2 \circ q \circ g$. Since $f(0) = q(\tilde{x}_1) = qg(0)$, we have $f = qg$, by Theorem 15.2.1. So $\tilde{y}_2 = f(1) = q(g(1))$, and q is surjective.

Now, given $\tilde{y}_2 \in \widetilde{X}_2$, we show the existence of an admissible nbd of $\tilde{y}_2$ relative to q. Let U_1 and U_2 be open nbds of $y = p_2(\tilde{y}_2)$, which are evenly covered by p_1 and p_2, respectively. Since X is locally path-connected, there exists a path-connected open set $U \subseteq X$ such that $y \in U \subseteq U_1 \cap U_2$. Clearly, U is also evenly covered by p_1 and p_2, both. Suppose that $p_2^{-1}(U)$ is a disjoint union of open sets $\widetilde{V}_\lambda$, each of which is mapped homeomorphically onto U by p_2. Since U is path-connected, each $\widetilde{V}_\lambda$ is path-connected and, therefore, a path component of $p_2^{-1}(U)$. Let $\widetilde{V}_{\lambda_0}$ be the component of $p_2^{-1}(U)$ that contains $\tilde{y}_2$. We assert that $\widetilde{V}_{\lambda_0}$ is evenly covered by q. Notice that $q^{-1}(\widetilde{V}_{\lambda_0}) \subseteq p_1^{-1}(U)$ and each sheet over U relative to p_1 is path-connected, as in the case of p_2. Write $p_1^{-1}(U) = \bigcup_{\mu \in M} \widetilde{W}_\mu$, where each $\widetilde{W}_\mu$ is a sheet over U relative to p_1. Then, for each $\mu \in M$, $q(\widetilde{W}_\mu) \subseteq p_2^{-1}(U)$, for $p_1 = p_2 q$. Since $q(\widetilde{W}_\mu)$ is path-connected and $\widetilde{V}_{\lambda_0}$ is a path component of $p_2^{-1}(U)$, we have $q(\widetilde{W}_\mu) \subseteq \widetilde{V}_{\lambda_0}$ whenever $q(\widetilde{W}_\mu) \cap \widetilde{V}_{\lambda_0} \neq \varnothing$. Let N be the set of all those indices $\mu \in M$ for which $q(\widetilde{W}_\mu)$ intersects $\widetilde{V}_{\lambda_0}$. Then we have $q^{-1}(\widetilde{V}_{\lambda_0}) = \bigcup \{\widetilde{W}_\mu \mid \mu \in N\}$. Moreover, $q|\widetilde{W}_\mu$ is a homeomorphism of $\widetilde{W}_\mu$ onto $\widetilde{V}_{\lambda_0}$. For, $p_1|\widetilde{W}_\mu = (p_2|\widetilde{V}_{\lambda_0}) \circ (q|\widetilde{W}_\mu)$ and $p_1|\widetilde{W}_\mu$ and $p_2|\widetilde{V}_{\lambda_0}$ are homeomorphisms. This proves our assertion. $\diamond$

Corollary 15.2.8 *If $p: \widetilde{X} \to X$ is a covering map, where X is locally path-connected and $\widetilde{X}$ simply connected, then for any covering map $r: \widetilde{Y} \to X$, there exists a covering map $q: \widetilde{X} \to \widetilde{Y}$ such that $p = r \circ q$, that is, the following diagram of continuous maps is commutative:*

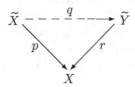

It follows that a simply connected covering space of a locally path-connected space X is a covering space of any other covering space of X. This justifies the adjective "universal" in the following.

Definition 15.2.9 A simply connected covering space of the space X is called a *universal covering space* of X.

For example, the real line $\mathbb{R}^1$ and the plane $\mathbb{R}^2$ are universal covering spaces of the circle $\mathbb{S}^1$ and the torus $\mathbb{S}^1 \times \mathbb{S}^1$, respectively. The n-sphere $\mathbb{S}^n$, $n \geq 2$, is a universal covering space of the real projective space $\mathbb{RP}^n$.

We conclude this section with the remark that a connected and locally path-connected space X does not necessarily have a universal covering space (refer to Example 15.6.1).

Exercises

1. Let $p \colon \widetilde{X} \to X$ be a covering map and $f \colon I \to X$ be a loop based at x_0. Prove:

   (a) If $f \simeq c_{x_0}$ rel $\{0, 1\}$, then any lift of f to a path in $\widetilde{X}$ is a loop homotopic to a constant loop rel $\{0, 1\}$.

   (b) If f lifts to a loop in $\widetilde{X}$ at $\tilde{x}_0$, then any loop homotopic to f rel $\{0, 1\}$ also lifts to a loop in $\widetilde{X}$ at $\tilde{x}_0$.

2. If a space X has a nontrivial path-connected covering space, show that $\pi(X)$ is nontrivial.

3. Given a connected and locally path-connected space X, prove that a continuous map $f \colon X \to \mathbb{S}^1$ can be lifted to a continuous map $\tilde{f} \colon X \to \mathbb{R}^1$ relative to the exponential map $t \mapsto e^{2\pi i t} \Leftrightarrow f$ is null homotopic.

4. Prove that any two continuous maps from a simply connected and locally path-connected space to $\mathbb{S}^1$ are homotopic.

5. Let $p \colon Y \to X$ and $q \colon Z \to Y$ be continuous maps, and assume that the spaces X, Y and Z are all connected and locally path-connected. If the composition $pq \colon Z \to X$ is a covering map, prove that q is a covering map $\Leftrightarrow p$ is so.

15.3   Action of $\pi(X, x_0)$ on the Fiber $p^{-1}(x_0)$

In this section, we adapt the technique used in the computation of $\pi\left(\mathbb{S}^1\right)$ to define an action of the fundamental group $\pi(X, x_0)$ of the base space X of a covering map on the fiber over the base point x_0. This will help presently in computing the fundamental groups of some spaces and proving the famous Borsuk–Ulam theorem for $\mathbb{S}^2$. Its further importance will be seen in the later sections.

Let $\alpha = [f] \in \pi(X, x_0)$, and $\tilde{x} \in F$. Then there is a unique lift $\tilde{f}$ of f with $\tilde{f}(0) = \tilde{x}$. Observe that $\tilde{f}(1) \in F$, for $p\tilde{f}(1) = f(1) = x_0$. So we can define $\tilde{x} \cdot \alpha = \tilde{f}(1)$. By the Monodromy theorem, this definition does not depend on the choice of representative of α. Suppose now that $\beta = [g]$ is another element of $\pi(X, x_0)$, and let $\tilde{g}$ be the lifting of g with $\tilde{g}(0) = \tilde{f}(1)$. Clearly, $\tilde{f} * \tilde{g}$ is the lifting of $f * g$ which starts at $\tilde{x}$ and ends at $\tilde{g}(1)$. Since $\alpha\beta$ is the path class of $f * g$, we have $\tilde{x} \cdot (\alpha\beta) = (\tilde{f} * \tilde{g})(1) = \tilde{g}(1) = (\tilde{x} \cdot \alpha) \cdot \beta$. If $\varepsilon \in \pi(X, x_0)$ is the identity element, then $\tilde{x} \cdot \varepsilon = \tilde{x}$, for the lift of the constant loop at x_0 is the constant loop at $\tilde{x}$. Therefore $\pi(X, x_0)$ acts on the set F by the right (as a group of permutations of F). This action is called the "monodromy" action.

If $\widetilde{X}$ is path-connected, then the above action of $\pi(X, x_0)$ on F is transitive. For, given any two points $\tilde{x}$ and $\tilde{y}$ of F, there is a path $\tilde{f}$ in $\widetilde{X}$ from $\tilde{x}$ to $\tilde{y}$. Now the composition $p\tilde{f}$ is a loop f in X based at x_0 so that $\alpha = [f] \in \pi(X, x_0)$. Obviously, $\tilde{f}$ is the lift of f with the initial point $\tilde{x}$, so $\tilde{x} \cdot \alpha = \tilde{f}(1) = \tilde{y}$.

Proposition 15.3.1 *Let* $p: \tilde{X} \to X$ *be a covering map, where* $\tilde{X}$ *is path-connected. If* $p(\tilde{x}_0) = x_0$, *then the multiplicity of* p *is the index of* $p_\# \pi(\tilde{X}, \tilde{x}_0)$ *in* $\pi(X, x_0)$.

Proof the group $\Gamma = \pi(X, x_0)$ acts transitively on the fiber $F = p^{-1}(x_0)$. Consequently, if $\Gamma_{\tilde{x}_0}$ is the isotropy subgroup of Γ at $\tilde{x}_0$, then there is a bijection $\Gamma / \Gamma_{\tilde{x}_0} \to F$, which takes the coset $\Gamma_{\tilde{x}_0} \alpha$ into $\tilde{x}_0 \cdot \alpha$. We show that $\Gamma_{\tilde{x}_0} = p_\# \pi(\tilde{X}, \tilde{x}_0)$. Let $\alpha \in \Gamma_{\tilde{x}_0}$ and f be a loop in X representing α. If $\tilde{f}$ is the lifting of f with $\tilde{f}(0) = \tilde{x}_0$, then $\tilde{x}_0 = \tilde{x}_0 \cdot \alpha = \tilde{f}(1)$, so $\tilde{f}$ is a loop in $\tilde{X}$ based at $\tilde{x}_0$. Thus we have $[\tilde{f}] \in \pi(\tilde{X}, \tilde{x}_0)$ whence $\alpha = [f] = [p\tilde{f}] \in p_\# \pi(\tilde{X}, \tilde{x}_0)$, and the inclusion $\Gamma_{\tilde{x}_0} \subseteq p_\# \pi(\tilde{X}, \tilde{x}_0)$ follows. To see the reverse inclusion, assume that $\alpha = [p\tilde{g}]$ for some $[\tilde{g}] \in \pi(\tilde{X}, \tilde{x}_0)$. Since $\tilde{g}$ is the lift of $p\tilde{g}$ starting at $\tilde{x}_0$, we have $\tilde{x}_0 \cdot \alpha = \tilde{g}(1) = \tilde{x}_0$. This implies that $\alpha \in \Gamma_{\tilde{x}_0}$, and the equality $\Gamma_{\tilde{x}_0} = \mathrm{im}(p_\#)$ holds.   ◇

It is immediate from the preceding proposition that the multiplicity of a covering map $p: \tilde{X} \to X$ with $\tilde{X}$ simply connected is the order of the group $\pi(X, x_0)$.

Example 15.3.1 For $n \geq 2$, $\pi(\mathbb{R}P^n) \cong \mathbb{Z}/2\mathbb{Z}$. As seen in Example 15.1.7, the quotient map $\mathbb{S}^n \to \mathbb{R}P^n$, which identifies every pair of antipodal points in $\mathbb{S}^n$, is a covering map. The multiplicity of this covering map is obviously 2. Since $\mathbb{S}^n$ is simply connected for $n \geq 2$, the above proposition shows that $|\pi(\mathbb{R}P^n)| = 2$, and therefore $\pi(\mathbb{R}P^n) \cong \mathbb{Z}/2\mathbb{Z}$.

Now, we come to a theorem par excellence.

Theorem 15.3.2 (Borsuk–Ulam Theorem) *If* $f: \mathbb{S}^n \to \mathbb{R}^n$, $n \geq 1$, *is continuous, then there exists a point* $x \in \mathbb{S}^n$ *such that* $f(x) = f(-x)$.

If $f(x) \neq f(-x)$ for every $x \in \mathbb{S}^n$, then the function $g: \mathbb{S}^n \to \mathbb{S}^{n-1}$ defined by $g(x) = [f(x) - f(-x)]/\|f(x) - f(-x)\|$ is continuous and satisfies $g(-x) = -g(x)$. Therefore, it suffices to establish the following.

Theorem 15.3.3 *There is no continuous map* $\mathbb{S}^n \to \mathbb{S}^{n-1}$, $n \geq 1$, *which sends antipodal points to antipodal points.*

With the limited knowledge acquired in this text, we can not prove the theorem in full generality. In fact, we can prove it only for $n = 1, 2$; interestingly, the technique used in proving the case $n = 2$ also works in all other cases.

Proof The case $n = 1$ is obvious, for $\mathbb{S}^1$ is connected but $\mathbb{S}^0$ is not. Now, consider the case $n = 2$ and assume that there is a continuous map $f: \mathbb{S}^2 \to \mathbb{S}^1$ such that $f(-x) = -f(x)$ for all $x \in \mathbb{S}^2$. Let $q_n: \mathbb{S}^n \to \mathbb{R}P^n$, $x \mapsto [x]$, be the quotient map, which identifies the antipodes. Clearly, the map f is relation preserving, so it induces a continuous map $\bar{f}: \mathbb{R}P^2 \to \mathbb{R}P^1$, $[x] \mapsto [f(x)]$. Choose a base point $x_0 \in \mathbb{S}^2$ and write $f(x_0) = y_0$. Since $q_1 f = \bar{f} q_2$, we have the following commutative diagram of groups and homomorphisms, that is, $q_{1\#} \circ f_\# = \bar{f}_\# \circ q_{2\#}$.

$$
\begin{array}{ccc}
\pi\left(\mathbb{S}^2, x_0\right) & \xrightarrow{\ f_\#\ } & \pi\left(\mathbb{S}^1, y_0\right) \\
\Big\downarrow{\scriptstyle q_{2\#}} & & \Big\downarrow{\scriptstyle q_{1\#}} \\
\pi\left(\mathbb{RP}^2, [x_0]\right) & \xrightarrow{\ \bar{f}_\#\ } & \pi\left(\mathbb{RP}^1, [y_0]\right)
\end{array}
$$

Since f is antipode preserving, a path g in $\mathbb{S}^2$ from x_0 to $-x_0$ (which does exist), gives the path fg in $\mathbb{S}^1$ from y_0 to $-y_0$. Thus we have closed paths $q_2 g$ and $q_1 f g$ in $\mathbb{RP}^2$ and $\mathbb{RP}^1$ based at $[x_0]$ and $[y_0]$, respectively. Consider the elements $\alpha = [q_2 g]$ of $\pi\left(\mathbb{RP}^2, [x_0]\right)$ and $\beta = [q_1 f g] \in \pi\left(\mathbb{RP}^1, [y_0]\right)$. Then, by the definition of the monodromy action of the fundamental group $\pi\left(\mathbb{RP}^1, [y_0]\right)$ on the fiber over $[y_0]$, we have $y_0 \cdot \beta = -y_0$, and thus $\beta \neq 1$, the identity element of the group $\pi\left(\mathbb{RP}^1, [y_0]\right)$. Moreover, the equality $q_1 f g = \bar{f} q_2 g$ forces that $\bar{f}_\#(\alpha) = \beta$. This implies that $\bar{f}_\#$ is a nontrivial homomorphism. But the homomorphism $\bar{f}_\# : \pi\left(\mathbb{RP}^2, [x_0]\right) = \mathbb{Z}_2 \longrightarrow \pi\left(\mathbb{RP}^1, [y_0]\right) = \mathbb{Z}$ is obviously trivial. This contradiction proves the theorem for $n = 2$. $\diamond$

Since a continuous function into $\mathbb{R}^n$ is determined by n real-valued continuous functions, the Borsuk–Ulam theorem can be rephrased as follows: If $f_i : \mathbb{S}^n \to \mathbb{R}^1$, $1 \leq i \leq n$, are continuous maps, then there exists a point $x \in \mathbb{S}^n$ such that $f_i(-x) = f_i(x)$ for $i = 1, \ldots, n$. Here is a beautiful consequence of this form of the Borsuk–Ulam theorem. Let $f_1(x)$ denote the temperature, and $f_2(x)$ denote the atmospheric pressure at a particular instant at a point x on the earth's surface. Assuming that both the temperature and the pressure are continuous functions of the point x, we find a pair of diametrically opposite points on the earth's surface having temperature and pressure simultaneously the same.

Exercises

1. Let $p : \widetilde{X} \to X$ be a covering map, where $\widetilde{X}$ is path-connected. Prove that p is a homeomorphism $\Leftrightarrow p_\# \pi\left(\widetilde{X}, \tilde{x}_0\right) = \pi(X, p(\tilde{x}_0))$.

2. Let $p : \widetilde{X} \to X$ be an n-sheeted covering map. If $\widetilde{X}$ is simply connected, and n is a prime, prove that $\pi(X, x_0) \cong \mathbb{Z}/n\mathbb{Z}$.

3. Prove that any continuous map of the projective space $\mathbb{RP}^n$, $n > 1$, into the circle $\mathbb{S}^1$ is homotopic to a constant map.

4. Show that any continuous map $\mathbb{RP}^2 \to \mathbb{RP}^2$ which is nontrivial on the fundamental group can be lifted to a continuous map $\tilde{f} : \mathbb{S}^2 \to \mathbb{S}^2$ such that $\tilde{f}(-x) = -\tilde{f}(x)$ for all $x \in \mathbb{S}^2$.

5. Prove that the following statements are equivalent:

   (a) The Borsuk–Ulam Theorem (15.3.2).

   (b) If $f: \mathbb{S}^n \to \mathbb{R}^n$ is a continuous map such that $f(-x) = -f(x)$ for every $x \in \mathbb{S}^n$, then there is a point $x \in \mathbb{S}^n$ such that $f(x) = 0$.

   (c) Theorem 15.3.3.

6. Prove that no subset of $\mathbb{R}^n$ is homeomorphic to $\mathbb{S}^n$.

7. Prove that there does not exist a continuous injection from $\mathbb{R}^{n+1}$ to $\mathbb{R}^n$ for any $n \geq 1$.

15.4 Equivalence of Covering Spaces

The various possible covering spaces of a given space are classified by means of fiber preserving homeomorphisms, and the existence of such a homeomorphism between two connected covering spaces $\widetilde{X}_1$ and $\widetilde{X}_2$ of a connected and locally path-connected space X is guaranteed by a simple relation between the corresponding subgroups of $\pi(X)$. In view of Propositions 15.1.5 and 15.1.6, this indeed classifies all coverings over a locally path-connected space. Therefore, we shall consider here *only those coverings which have path-connected and locally path-connected base space and path-connected covering space.*

Definition 15.4.1 Two covering spaces $\widetilde{X}_1$ and $\widetilde{X}_2$ of a space X relative to projections $p_i: \widetilde{X}_i \to X$, respectively, are said to be *equivalent* or *isomorphic* if there exists a homeomorphism $h: \widetilde{X}_1 \to \widetilde{X}_2$ such that $p_1 = p_2 \circ h$. The homeomorphism h is called an *equivalence* or *isomorphism* of the covering spaces $\widetilde{X}_1$ and $\widetilde{X}_2$ (also, of the covering maps p_1 and p_2).

Proposition 15.4.2 *Let $p_i: \widetilde{X}_i \to X$, $i = 1, 2$, be covering maps, and suppose that $\tilde{x}_1 \in \widetilde{X}_1$ and $\tilde{x}_2 \in \widetilde{X}_2$ satisfy $p_1(\tilde{x}_1) = p_2(\tilde{x}_2)$. Then there exists an equivalence $h: \widetilde{X}_1 \to \widetilde{X}_2$ such that $h(\tilde{x}_1) = \tilde{x}_2$ if and only if $p_{1\#}\pi(\widetilde{X}_1, \tilde{x}_1) = p_{2\#}\pi(\widetilde{X}_2, \tilde{x}_2)$.*

Proof If $h: \widetilde{X}_1 \to \widetilde{X}_2$ is an equivalence such that $h(\tilde{x}_1) = \tilde{x}_2$, then $h_\#: \pi(\widetilde{X}_1, \tilde{x}_1) \to \pi(\widetilde{X}_2, \tilde{x}_2)$ is an isomorphism and $p_{1\#} = p_{2\#} \circ h_\#$. Therefore $p_{1\#}\pi(\widetilde{X}_1, \tilde{x}_1) = (p_{2\#} \circ h_\#)\pi(\widetilde{X}_1, \tilde{x}_1) = p_{2\#}\pi(\widetilde{X}_2, \tilde{x}_2)$.

For sufficiency, suppose that $p_{1\#}\pi(\widetilde{X}_1, \tilde{x}_1) = p_{2\#}\pi(\widetilde{X}_2, \tilde{x}_2)$. Then, the Lifting Theorem implies that there exist unique continuous maps $h: \widetilde{X}_1 \to \widetilde{X}_2$ and $k: \widetilde{X}_2 \to \widetilde{X}_1$ such that $h(\tilde{x}_1) = \tilde{x}_2$, $p_1 = p_2 \circ h$ and $k(\tilde{x}_2) = \tilde{x}_1$, $p_2 = p_1 \circ k$. Obviously, the composite $kh: \widetilde{X}_1 \to \widetilde{X}_1$ satisfies $kh(\tilde{x}_1) = \tilde{x}_1$ and $p_1 = p_1 \circ (kh)$. By the unique lifting property, we have $kh = 1_{\widetilde{X}_1}$. Similarly, $hk = 1_{\widetilde{X}_2}$, and it follows that h is a homeomorphism with k as its inverse. Thus h is a desired equivalence between $\widetilde{X}_1$ and $\widetilde{X}_2$.   ◇

The preceding proposition, however, does not solve the question of equivalence of two coverings in general. Actually, this is a stronger form of equivalence, that one for which one specifies base points, which, of course, must be preserved under the mappings. It may be possible that a space X does have two equivalent covering spaces $\widetilde{X}_1$ and $\widetilde{X}_2$, but there is no equivalence $\widetilde{X}_1 \to \widetilde{X}_2$ which takes the base point $\tilde{x}_1 \in \widetilde{X}_1$ to the base point $\tilde{x}_2 \in \widetilde{X}_2$, even if $p_1(\tilde{x}_1) = p_2(\tilde{x}_2)$. If $h: \widetilde{X}_1 \to \widetilde{X}_2$ is an equivalence, then $h(\tilde{x}_1)$ clearly belongs to the fiber $p_2^{-1}(p_1(\tilde{x}_1))$ and $p_{1\#}\pi(\widetilde{X}_1, \tilde{x}_1) = p_{2\#}\pi(\widetilde{X}_2, h(\tilde{x}_1))$. Therefore, we need to know how the subgroups $p_\#\pi(\widetilde{X}, \tilde{x}) \subseteq \pi(X, x_0)$ induced by a covering $p: \widetilde{X} \to X$ are related for various choices of $\tilde{x} \in p^{-1}(x_0)$. Note that there are no relations among the subgroups $p_\#\pi(\widetilde{X}, \tilde{x})$ in general, unless the points $\tilde{x}$ can be joined by paths in $\widetilde{X}$. This can be seen, for example, by considering the covering map $p: \mathbb{R}^1 + \mathbb{S}^1 \to \mathbb{S}^1$ defined by $p(x) = e^{2\pi i x}$ for $x \in \mathbb{R}^1$, and $p(x) = x$ for $x \in \mathbb{S}^1$. On the other hand, if $\widetilde{X}$ is path-connected, then we know that $\pi(\widetilde{X}, \tilde{x}) \cong \pi(\widetilde{X}, \tilde{y})$ for any two points $\tilde{x}, \tilde{y} \in \widetilde{X}$. Accordingly, their images under the monomorphism $p_\#$ are isomorphic. The next result shows that any two such subgroups of $\pi(X, x_0)$ can differ by a conjugation only.

Lemma 15.4.3 Let $p: \widetilde{X} \to X$ be a covering map. For $x_0 \in X$, the subgroups $p_\#\pi(\widetilde{X}, \tilde{x}) \subseteq \pi(X, x_0)$, $\tilde{x} \in p^{-1}(x_0)$, constitute a conjugacy class in $\pi(X, x_0)$.

Proof Suppose that $\tilde{x}, \tilde{y} \in p^{-1}(x_0)$. Let $\tilde{f}$ be a path in $\widetilde{X}$ from $\tilde{x}$ to $\tilde{y}$. Then $f = p\tilde{f}$ is a loop in X based at x_0 so that $[f]$ is an element of $\pi(X, x_0)$. There are induced isomorphisms $\eta: \pi(\widetilde{X}, \tilde{x}) \longrightarrow \pi(\widetilde{X}, \tilde{y})$, $[\tilde{g}] \mapsto [\tilde{f}^{-1} * \tilde{g} * \tilde{f}]$, and $\theta: \pi(X, x_0) \longrightarrow \pi(X, x_0)$, $[g] \mapsto [f^{-1} * g * f]$. For any $[\tilde{g}] \in \pi(\widetilde{X}, \tilde{x})$, we have

$$(p_\# \circ \eta)[\tilde{g}] = [p \circ (\tilde{f}^{-1} * \tilde{g} * \tilde{f})] = [f^{-1} * p\tilde{g} * f] = (\theta \circ p_\#)[\tilde{g}],$$

that is, the following diagram of group homomorphisms commutes:

$$
\begin{array}{ccc}
\pi(\widetilde{X}, \tilde{x}) & \xrightarrow{\ \ \eta\ \ } & \pi(\widetilde{X}, \tilde{y}) \\
\Big\downarrow{\scriptstyle p_\#} & & \Big\downarrow{\scriptstyle p_\#} \\
\pi(X, x_0) & \xrightarrow{\ \ \theta\ \ } & \pi(X, x_0)
\end{array}
$$

Since η is surjective and θ is an inner automorphism, we see that $p_\#\pi(\widetilde{X}, \tilde{x})$ and $p_\#\pi(\widetilde{X}, \tilde{y})$ are conjugate subgroups of $\pi(X, x_0)$.

Conversely, for each subgroup S of $\pi(X, x_0)$ in the conjugacy class of the subgroup $p_\#\pi(\widetilde{X}, \tilde{x}_0)$, where $p(\tilde{x}_0) = x_0$, we show that there is a point $\tilde{x}_1 \in p^{-1}(x_0)$ such that $S = p_\#\pi(\widetilde{X}, \tilde{x}_1)$. Suppose that $S = \alpha^{-1} p_\#\pi(\widetilde{X}, \tilde{x}_0)\alpha$, where $\alpha = [f] \in \pi(X, x_0)$. Let $\tilde{f}$ be the lift of f starting at $\tilde{x}_0$. If $\tilde{f}(1) = \tilde{x}_1$, then $p(\tilde{x}_1) = f(1) = x_0$, and so

$\tilde{x}_1 \in p^{-1}(x_0)$. If $\eta: \pi\left(\tilde{X}, \tilde{x}_0\right) \to \pi\left(\tilde{X}, \tilde{x}_1\right)$ is the isomorphism $\left[\tilde{g}\right] \mapsto \left[\tilde{f}^{-1} * \tilde{g} * \tilde{f}\right]$, then we have $\alpha^{-1} \cdot p_\#\left(\left[\tilde{g}\right]\right) \cdot \alpha = (p_\# \circ \eta)\left[\tilde{g}\right]$ for every element $\left[\tilde{g}\right]$ in $\pi\left(\tilde{X}, \tilde{x}_0\right)$. This implies that $S = (p_\# \circ \eta)\,\pi\left(\tilde{X}, \tilde{x}_0\right) = p_\#\pi\left(\tilde{X}, \tilde{x}_1\right)$, as desired.   ◇

The following theorem reduces the problem of equivalence of covering maps to the question about the conjugacy of their corresponding subgroups of the fundamental group of the base space.

Theorem 15.4.4 *Two covering spaces $\tilde{X}_1$ and $\tilde{X}_2$ of a space X relative to covering projections $p_i: \tilde{X}_i \to X$, respectively, are equivalent if and only if the subgroups $p_{1\#}\pi\left(\tilde{X}_1, \tilde{x}_1\right)$ and $p_{2\#}\pi\left(\tilde{X}_2, \tilde{x}_2\right)$, where $p_1(\tilde{x}_1) = p_2(\tilde{x}_2) = x_0$, are conjugate in $\pi(X, x_0)$.*

Proof Suppose that the coverings p_1 and p_2 are equivalent, and let $h: \tilde{X}_1 \to \tilde{X}_2$ be a homeomorphism with $p_1 = p_2 \circ h$. Then we have $p_{1\#}\pi\left(\tilde{X}_1, \tilde{x}_1\right) = (p_{2\#} \circ h_\#)\,\pi\left(\tilde{X}_1, \tilde{x}_1\right) = p_{2\#}\pi\left(\tilde{X}_2, h(\tilde{x}_1)\right)$. As $p_2(\tilde{x}_2) = x_0 = p_2(h(\tilde{x}_1))$, the subgroups $p_{2\#}\pi\left(\tilde{X}_2, h(\tilde{x}_1)\right)$ and $p_{2\#}\pi\left(\tilde{X}_2, \tilde{x}_2\right)$ are conjugate in $\pi(X, x_0)$, by Lemma 15.4.3.

Conversely, assume that the subgroup $p_{1\#}\pi\left(\tilde{X}_1, \tilde{x}_1\right)$ is conjugate in $\pi(X, x_0)$ to the subgroup $p_{2\#}\pi\left(\tilde{X}_2, \tilde{x}_2\right)$. Then, by the preceding lemma again, there is a point $\tilde{y}_2 \in p_2^{-1}(x_0)$ such that $p_{1\#}\pi\left(\tilde{X}_1, \tilde{x}_1\right) = p_{2\#}\pi\left(\tilde{X}_2, \tilde{y}_2\right)$. Therefore, by Proposition 15.4.2, there is a homeomorphism $h: \left(\tilde{X}_1, \tilde{x}_1\right) \to \left(\tilde{X}_2, \tilde{y}_2\right)$ such that $p_1 = p_2 \circ h$. It follows that the covering spaces $\tilde{X}_1$ and $\tilde{X}_2$ of X are equivalent.   ◇

As an application of the above theorem, we determine all nonequivalent connected covering spaces of $\mathbb{S}^1$.

Example 15.4.1 The connected covering spaces of $\mathbb{S}^1$. Since $\pi\left(\mathbb{S}^1\right) \cong \mathbb{Z}$, the non-trivial subgroups of $\pi\left(\mathbb{S}^1\right)$ are isomorphic to infinite cyclic subgroups $n\mathbb{Z} \subseteq \mathbb{Z}$, $n = 1, 2, \ldots$. For each integer $n \geq 1$, we have the covering map $p_n: \mathbb{S}^1 \to \mathbb{S}^1$, $z \mapsto z^n$. If g denotes the loop $t \mapsto e^{2\pi i t}$, $t \in I$, then the homotopy class $[g]$ is a generator of $\pi\left(\mathbb{S}^1, 1\right)$. We observe that $p_n \circ g \simeq g * \cdots * g$ (n-terms) rel $\{0, 1\}$. Let $h: I \to I$ be the homeomorphism which maps the subintervals $\left[0, 2^{1-n}\right]$, $\left[2^{1-n}, 2^{2-n}\right], \ldots, \left[2^{-1}, 1\right]$ linearly onto $[0, 1/n]$, $[1/n, 2/n], \ldots, [(n-1)/n, 1]$, respectively. Then it is easily checked that the composite $p_n \circ g \circ h$ is the product $g * \cdots * g$, and $p_n \circ g \simeq p_n \circ g \circ h$ rel $\{0, 1\}$. So $[g]^n = [g * \cdots * g]$ is a generator of the subgroup $p_{n\#}\pi\left(\mathbb{S}^1, 1\right)$, and it follows that the group $p_{n\#}\left(\pi\left(\mathbb{S}^1, 1\right)\right)$ is isomorphic to $n\mathbb{Z}$. Thus, the covering space corresponding to $n\mathbb{Z}$ is $\mathbb{S}^1$ itself relative to the map p_n. Corresponding to the trivial subgroup of $\mathbb{Z}$, the real line $\mathbb{R}^1$ is the covering space $\mathbb{S}^1$ relative to the map $t \mapsto e^{2\pi i t}$. These are essentially all the connected covering spaces of $\mathbb{S}^1$.

Deck Transformations

Definition 15.4.5 Let $p: \tilde{X} \to X$ be a covering map. A homeomorphism $h: \tilde{X} \to \tilde{X}$ with $ph = p$ is called a *deck transformation* of the covering.

Note that a deck transformation h of a covering $p\colon \widetilde{X} \to X$ permutes the "decks" of $\widetilde{X}$; that is, if U is an admissible open set in X, then h permutes the copies of U in $p^{-1}(U)$. If h is a deck transformation, then $ph^{-1} = phh^{-1} = p$, so h^{-1} is also a deck transformation. Moreover, the composition of two deck transformations of p is obviously a deck transformation and, therefore, the set $\Delta(p)$ of all deck transformations of the covering p forms a group under composition of functions. In the literature, deck transformations are also called *automorphisms* or *covering transformations* of the covering; accordingly, the group $\Delta(p)$ is also referred to as the *automorphism group* or the *covering group* of p.

Proposition 15.4.6 *Let h and h' be two deck transformations of a covering map $p\colon \widetilde{X} \to X$. If $h(\tilde{x}) = h'(\tilde{x})$ for some $\tilde{x} \in \widetilde{X}$, then $h = h'$. In particular, if $h \in \Delta(p)$ has a fixed point in $\widetilde{X}$, then $h = 1_{\widetilde{X}}$.*

Proof Since $\widetilde{X}$ is connected, the proposition follows immediately from the unique lifting property of p. $\diamond$

Example 15.4.2 The group of deck transformations of the covering map $p\colon \mathbb{R}^1 \to \mathbb{S}^1$, $x \mapsto e^{2\pi i x}$ is the group $\mathbb{Z}$ of integers. For each $n \in \mathbb{Z}$, it is clear that the translation $\mathbb{R}^1 \to \mathbb{R}^1$, $x \mapsto x + n$, is a deck transformation. On the other hand, if $h \in \Delta(p)$, then $n = h(0) \in \mathbb{Z}$, and h coincides with the translation $x \mapsto x + n$, by Proposition 15.4.6.

Example 15.4.3 For the covering map $p\colon \mathbb{S}^n \to \mathbb{RP}^n$, $n \geq 2$, the group $\Delta(p)$ consists of just two elements, namely, the identity map of $\mathbb{S}^n$ and the antipodal map.

Let $p\colon \widetilde{X} \to X$ be a covering map. It is obvious that the group $\Delta(p)$ acts on $\widetilde{X}$ (by the left). Under this action of $\Delta(p)$, for each $x_0 \in X$, the fiber $p^{-1}(x_0)$ is invariant. Accordingly, there is a left action of $\Delta(p)$ on $p^{-1}(x_0)$. This action of $\Delta(p)$ on $p^{-1}(x_0)$ commutes with the "monodromy action" of the fundamental group $\pi(X, x_0)$ in the following sense.

Proposition 15.4.7 *Let $p\colon \widetilde{X} \to X$ be a covering map, and $x_0 \in X$ be fixed. If $h \in \Delta(p)$, and $\alpha \in \pi(X, x_0)$, then $h(\tilde{x}) \cdot \alpha = h(\tilde{x} \cdot \alpha)$ for all $\tilde{x} \in p^{-1}(x_0)$.*

Proof Let g be a loop in X (based at x_0) representing α. If $\tilde{g}$ is the lift of g with $\tilde{g}(0) = \tilde{x}$, then $\tilde{x} \cdot \alpha = \tilde{g}(1)$, by the definition of the action of $\pi(X, x_0)$ on $p^{-1}(x_0)$. It is also clear that $h\tilde{g}$ is a lift of g with the initial point $h\tilde{x}$, so $h\tilde{x} \cdot \alpha = h\tilde{g}(1) = h(\tilde{x} \cdot \alpha)$. $\diamond$

For every $\alpha, \beta \in \pi(X, x_0)$ and $\tilde{x} \in p^{-1}(x_0)$, we have

$$(\tilde{x} \cdot \alpha) \cdot \beta = \tilde{x} \cdot \alpha \Leftrightarrow ((\tilde{x} \cdot \alpha) \cdot \beta) \cdot \alpha^{-1} = \tilde{x} \Leftrightarrow \tilde{x} \cdot (\alpha \cdot \beta \cdot \alpha^{-1}) = \tilde{x}.$$

Denoting the group $\pi(X, x_0)$ by Γ, we see that $\Gamma_{\tilde{x} \cdot \alpha} = \alpha^{-1} \Gamma_{\tilde{x}} \alpha$, where $\Gamma_{\tilde{x}}$ and $\Gamma_{\tilde{x} \cdot \alpha}$ are the isotropy subgroups of Γ at $\tilde{x}$ and $\tilde{x} \cdot \alpha$, respectively.

The following concept of group theory will be needed for the next result: If H is a subgroup of a group Γ, the normalizer of H in Γ is the set $N(H)$ of all elements $\gamma \in \Gamma$ such that $\gamma^{-1} H \gamma = H$. This is the largest subgroup of Γ in which H is normal.

Proposition 15.4.8 *Let* $p \colon \widetilde{X} \to X$ *be a covering map. Then, for any two points* $\tilde{x}$ *and* $\tilde{y}$ *in the fiber over a point* $x_0 \in X$, *the following statements are equivalent:*

(a) *There exists an* $h \in \Delta(p)$ *such that* $h(\tilde{x}) = \tilde{y}$.
(b) $p_\# \pi(\widetilde{X}, \tilde{x}) = p_\# \pi(\widetilde{X}, \tilde{y})$.
(c) *There exists an* α *in the normalizer of the subgroup* $p_\# \pi(\widetilde{X}, \tilde{x})$ *of* $\pi(X, x_0)$ *such that* $\tilde{x} \cdot \alpha = \tilde{y}$.

Proof (a) $\Longleftrightarrow$ (b): This is clearly a particular case of Proposition 15.4.2.

(b) $\Rightarrow$ (c): Put $\Gamma = \pi(X, x_0)$. Then, as seen in the proof of Proposition 15.3.1, $p_\# \pi(\widetilde{X}, \tilde{x}) = \Gamma_{\tilde{x}}$ and $p_\# \pi(\widetilde{X}, \tilde{y}) = \Gamma_{\tilde{y}}$. Since $\widetilde{X}$ is path-connected, the action of Γ on $p^{-1}(x_0)$ is transitive. So there exists an $\alpha \in \Gamma$ such that $\tilde{x} \cdot \alpha = \tilde{y}$. Moreover, by (b), $\Gamma_{\tilde{x}} = \Gamma_{\tilde{y}} = \Gamma_{\tilde{x} \cdot \alpha} = \alpha^{-1} \Gamma_{\tilde{x}} \alpha$, and thus $\alpha \in N(\Gamma_{\tilde{x}})$.

(c) $\Rightarrow$ (b): This is trivial, since $\alpha^{-1} \Gamma_{\tilde{x}} \alpha = \Gamma_{\tilde{x} \cdot \alpha}$. $\diamond$

We remark that, unlike the action of the fundamental group $\pi(X)$ on fibers of a covering map $p \colon \widetilde{X} \to X$, the action of the covering group $\Delta(p)$ on fibers of p in general is not transitive. The following consequence of the preceding proposition gives a necessary and sufficient condition for the action of $\Delta(p)$ on a fiber $p^{-1}(x)$ to be transitive.

Corollary 15.4.9 *The group* $\Delta(p)$ *acts transitively on the fiber* $p^{-1}(x_0)$ *if and only if the subgroup* $p_\# \pi(\widetilde{X}, \tilde{x})$ *is normal in* $\pi(X, x_0)$, *where* $p(\tilde{x}) = x_0$.

Proof Suppose that $\Delta(p)$ acts transitively on $p^{-1}(x_0)$, and let α be an arbitrary element of $\pi(X, x_0) = \Gamma$. Then, by definition, $\tilde{x} \cdot \alpha \in p^{-1}(x_0)$ for each $\tilde{x} \in p^{-1}(x_0)$. By our hypothesis, there exists an $h \in \Delta(p)$ such that $h(\tilde{x}) = \tilde{x} \cdot \alpha$. Hence the condition (b) of Proposition 15.4.8 holds for $\tilde{x}$ and $\tilde{y} = \tilde{x} \cdot \alpha$. Since $p_\# \pi(\widetilde{X}, \tilde{x}) = \Gamma_{\tilde{x}}$, we have $\Gamma_{\tilde{x}} = \Gamma_{\tilde{x} \cdot \alpha} = \alpha^{-1} \Gamma_{\tilde{x}} \alpha$, and it follows that $p_\# \pi(\widetilde{X}, \tilde{x})$ is normal in $\pi(X, x_0)$.

Conversely, suppose that $\tilde{x} \in p^{-1}(x_0)$ and $p_\# \pi(\widetilde{X}, \tilde{x})$ is normal in $\Gamma = \pi(X, x_0)$. Given any $\tilde{y} \in p^{-1}(x_0)$, there exists an $\alpha \in \Gamma$ such that $\tilde{x} \cdot \alpha = \tilde{y}$, since Γ acts transitively on $p^{-1}(x)$. Therefore the condition (a) of Proposition 15.4.8 applies, and we find an $h \in \Delta(p)$ such that $h(\tilde{x}) = \tilde{y}$. This shows that $\Delta(p)$ acts transitively on $p^{-1}(x_0)$. $\diamond$

We close this section by establishing an important relationship between the fundamental groups of the covering space and the base space of a covering map and its group of deck transformations.

Theorem 15.4.10 *Let* $p\colon \tilde{X} \to X$ *be a covering map. Then the group* $\Delta\,(p)$ *is iso-morphic to the quotient group* $N\big(p_{\#}\pi(\tilde{X}, \tilde{x})\big)/p_{\#}\pi(\tilde{X}, \tilde{x})$, *where* $N\big(p_{\#}\pi(\tilde{X}, \tilde{x})\big)$ *is the normalizer of* $p_{\#}\pi(\tilde{X}, \tilde{x})$ *in* $\pi\,(X, p\,(\tilde{x}))$.

Proof Write $\Gamma = \pi\,(X, p\,(\tilde{x}))$. Then $p_{\#}\pi(\tilde{X}, \tilde{x})$ is the isotropy subgroup $\Gamma_{\tilde{x}}$. Let $\alpha \in N\,(\Gamma_{\tilde{x}})$ be arbitrary and put $\tilde{x} \cdot \alpha = \tilde{y}$. By Proposition 15.4.8, there exists $h_\alpha \in \Delta\,(p)$ such that $h_\alpha\,(\tilde{x}) = \tilde{y}$ and, by Proposition 15.4.6, this is a unique deck transformation of $\tilde{X}$ which maps $\tilde{x}$ into $\tilde{x} \cdot \alpha$. So there is a mapping $\Psi\colon N\,(\Gamma_{\tilde{x}}) \to \Delta\,(p)$ defined by $\Psi\,(\alpha) = h_\alpha$. We first show that Ψ is a homomorphism. Let $\alpha, \beta \in N\,(\Gamma_{\tilde{x}})$. Then $h_{\alpha\beta}\,(\tilde{x}) = \tilde{x} \cdot (\alpha\beta) = (\tilde{x} \cdot \alpha) \cdot \beta = h_\alpha\,(\tilde{x}) \cdot \beta = h_\alpha\,(\tilde{x} \cdot \beta)$ (by Proposition 15.4.7) $= h_\alpha\,(h_\beta\,(\tilde{x})) = (h_\alpha \circ h_\beta)\,(\tilde{x})$. By Proposition 15.4.6, again, we have $h_{\alpha\beta} = h_\alpha \circ h_\beta$. Thus Ψ is a homomorphism. Next, we observe that Ψ is surjective. Let $h \in \Delta\,(p)$ be arbitrary, and suppose that $h\,(\tilde{x}) = \tilde{y}$. By Proposition 15.4.8, again, there exists $\alpha \in N\,(\Gamma_{\tilde{x}})$ such that $\tilde{y} = \tilde{x} \cdot \alpha = \Psi\,(\alpha)\,(\tilde{x})$. Hence $h = \Psi\,(\alpha)$, and Ψ is surjective. Finally, we have $\alpha \in \ker(\Psi) \Leftrightarrow h_\alpha = 1_{\tilde{X}} \Leftrightarrow h_\alpha\,(\tilde{x}) = \tilde{x} \Leftrightarrow \tilde{x} \cdot \alpha = \tilde{x} \Leftrightarrow \alpha \in \Gamma_{\tilde{x}}$. So $\ker(\Psi) = \Gamma_{\tilde{x}}$, and the theorem follows. $\diamond$

In particular, if $\tilde{X}$ is a simply connected covering space of X, then $\pi\,(X)$ is isomorphic to the group of the deck transformations of the covering map $\tilde{X} \to X$, and its order is equal to the multiplicity of the covering, by Proposition 15.3.1. This result gives a description of the fundamental group of X, which requires no choice of base point. For instance, we have already seen that $\pi\,(\mathbb{S}^1, 1) = \mathbb{Z}$ (Theorem 14.3.6). We also know that the mapping $p\colon \mathbb{R}^1 \to \mathbb{S}^1$, $t \mapsto e^{2\pi it}$, is a covering projection. Since $\mathbb{R}^1$ is simply connected, we have $\pi\,(\mathbb{S}^1) \cong \Delta\,(p) \cong \mathbb{Z}$, by Example 15.4.2.

Exercises

1. For each integer $n \geq 1$, let $p_n\colon \mathbb{S}^1 \to \mathbb{S}^1$ denote the covering map $z \mapsto z^n$. Given two positive integers m and n, show that there is a covering map $q\colon \mathbb{S}^1 \to \mathbb{S}^1$ such that $p_n = p_m \circ q$ if and only if $m|n$. When this is the case, prove that $q = p_k$, where $n = mk$.

2. Let X be a connected, locally path-connected space and $x_0 \in X$. Prove that two covering spaces $(\tilde{X}_i, p_i)$ $(i = 1, 2)$ of X are equivalent if and only if the fibers $p_1^{-1}(x_0)$ and $p_2^{-1}(x_0)$ are isomorphic $\pi(X, x_0)$-set.

3. Using Theorem 15.4.10, determine $\pi\,(\mathbb{R}P^n)$ for $n > 1$ and find all the connected covering spaces of $\mathbb{R}P^n$.

4. Let G be a simply connected Hausdorff topological group and $H \subset G$ a discrete subgroup. Prove that $\pi\,(G/H) \cong H$.

5. Let $p\colon \tilde{X} \to X$ be a covering map, where both X and $\tilde{X}$ are connected and locally path-connected. Show that $\Delta(p)$ is isomorphic to the group of all equivariant bijections of the Γ-set $p^{-1}\,(x_0)$, where $\Gamma = \pi(X, x_0)$.

15.5   Regular Coverings and A Discrete Group Action

Given a covering map $p \colon \widetilde{X} \to X$ with path-connected covering space, we have seen in the last section that any two subgroups $p_{\#}\pi\big(\widetilde{X}, \tilde{x}\big) \subseteq \pi(X, x_0)$ for different base points $\tilde{x} \in p^{-1}(x_0)$ can differ from each other by a conjugation only. The coverings for which all of them coincide are called *regular*. Such coverings are closely related to orbit maps of a particular action of discrete groups. We will study here these coverings and see certain conditions on action of a discrete group so that the orbit map becomes a regular covering.

We shall assume in this section too that a covering has connected locally path-connected base space and connected covering space.

Regular Coverings

Definition 15.5.1 A covering map $p \colon \widetilde{X} \to X$ is called *regular* (or *normal*) if $p_{\#}\pi\big(\widetilde{X}, \tilde{x}\big)$ is a normal subgroup of $\pi(X, p(\tilde{x}))$ for every $\tilde{x} \in \widetilde{X}$.

It is obvious that if $p \colon \widetilde{X} \to X$ is a regular covering, then the subgroup $p_{\#}\pi\big(\widetilde{X}, \tilde{x}\big)$ $\subseteq \pi(X, x_0)$ is independent of the choice of base point $\tilde{x}$ in the fiber over x_0, because the only subgroup conjugate to a normal subgroup H is H itself.

As an example, a covering map $p \colon \widetilde{X} \to X$ with $\widetilde{X}$ simply connected or $\pi(X)$ abelian is clearly regular.

The next result reduces the condition of regularity for a covering map $p \colon \widetilde{X} \to X$ to be satisfied for *some* point $\tilde{x}_0 \in \widetilde{X}$.

Proposition 15.5.2 *A covering map $p \colon \widetilde{X} \to X$ is regular if, for some point $\tilde{x}_0 \in \widetilde{X}$, $p_{\#}\pi\big(\widetilde{X}, \tilde{x}_0\big)$ is normal in $\pi(X, p(\tilde{x}_0))$.*

Proof Assume that $p_{\#}\pi\big(\widetilde{X}, \tilde{x}_0\big)$ is a normal subgroup of $\pi(X, x_0)$, where $x_0 = p(\tilde{x}_0)$. Let $\tilde{x}_1 \in \widetilde{X}$ be arbitrary and put $p(\tilde{x}_1) = x_1$. Since $\widetilde{X}$ is path-connected, there is a path $\tilde{f}$ in $\widetilde{X}$ from $\tilde{x}_0$ to $\tilde{x}_1$. Then $f = p\tilde{f}$ is a path in X from x_0 to x_1, and there are isomorphisms

$$\eta \colon \pi\big(\widetilde{X}, \tilde{x}_0\big) \to \pi\big(\widetilde{X}, \tilde{x}_1\big), \quad [\tilde{g}] \mapsto \left[\tilde{f}^{-1} * \tilde{g} * \tilde{f}\right], \quad \text{and}$$
$$\theta \colon \pi(X, x_0) \to \pi(X, x_1), \quad [g] \mapsto \left[f^{-1} * g * f\right].$$

It is easily checked that $p_{\#} \circ \eta = \theta \circ p_{\#}$, and hence we see that $p_{\#}\pi\big(\widetilde{X}, \tilde{x}_1\big)$ is the image of $p_{\#}\pi\big(\widetilde{X}, \tilde{x}_0\big)$ under θ, for η is surjective. Since $p_{\#}\pi\big(\widetilde{X}, \tilde{x}_0\big)$ is normal in $\pi(X, x_0)$ and θ is an isomorphism, we deduce that $p_{\#}\pi\big(\widetilde{X}, \tilde{x}_1\big)$ is also normal in $\pi(X, x_1)$, as desired. $\diamond$

By Corollary 15.4.9, we also see that *a covering map $p \colon \widetilde{X} \to X$ is regular* if and only if *the group $\Delta(p)$ acts transitively on each fiber of p*

If $p \colon \widetilde{X} \to X$ is a regular covering map, then, by definition, $N\big(p_{\#}\pi\big(\widetilde{X}, \tilde{x}\big)\big) = \pi(X, p(\tilde{x}))$, and hence we deduce from Theorem 15.4.10 the following.

Corollary 15.5.3 *If* $p: \widetilde{X} \to X$ *is a regular covering and* $\tilde{x} \in \widetilde{X}$, *then* $\Delta(p) \cong \pi(X, p(\tilde{x})) / p_{\#}\pi(\widetilde{X}, \tilde{x})$.

Notice that there are coverings such that certain loops in their base space lift to loops as well as paths with distinct endpoints. For example, let X be the wedge of two circles A and B, and let $\widetilde{X}$ be the subspace of $\mathbb{R}^2$, as indicated in Fig. 15.6 below. Consider the mapping p which winds the first and fifth circles of $\widetilde{X}$ once around circle A, the third twice around circle A, and the second and fourth twice around circle B. Then p is a covering map. Let g be the loop in X based at x_0 that traverses the circle A once in counterclockwise direction. Clearly, it lifts to a loop based at $\tilde{x}_0$, while its lifting starting at $\tilde{x}_1$ is not a loop.

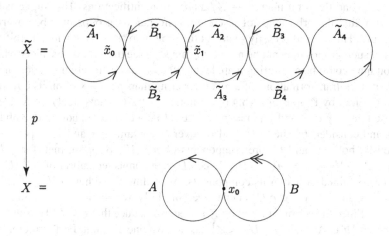

Fig. 15.6 A covering of figure eight space

However, for regular coverings, we have the following

Theorem 15.5.4 *If* $p: \widetilde{X} \to X$ *is a regular covering map, then, for any closed path* g *in* X, *either every lifting of* g *is closed or none is closed. The converse is also true.*

Proof Assume first that p is regular, and let g be a closed path in X based at x having a closed lifting $\tilde{g}$ in $\widetilde{X}$ based at $\tilde{x}$. If $\tilde{h}$ is a lifting of g with $\tilde{h}(0) = \tilde{y}$, then $p(\tilde{x}) = x = p(\tilde{y})$ and the subgroups $p_{\#}\pi(\widetilde{X}, \tilde{x})$ and $p_{\#}\pi(\widetilde{X}, \tilde{y})$ are conjugate in $\pi(X, x)$, by Lemma 15.4.3. Since $p_{\#}\pi(\widetilde{X}, \tilde{x})$ is normal in $\pi(X, x)$, we have $p_{\#}\pi(\widetilde{X}, \tilde{x}) = p_{\#}\pi(\widetilde{X}, \tilde{y})$. Thus $[g] = p_{\#}[\tilde{g}] \in p_{\#}\pi(\widetilde{X}, \tilde{y})$ and, therefore, there exists a loop $\tilde{f}$ in $\widetilde{X}$ based at $\tilde{y}$ such that $p_{\#}[\tilde{f}] = [g]$. It follows that $p\tilde{h} = g \simeq p\tilde{f}$ rel $\{0, 1\}$. By the Monodromy Theorem, we have $\tilde{h}(1) = \tilde{f}(1) = \tilde{y}$, and thus $\tilde{h}$ is a closed path.

Conversely, suppose that the covering map p has the property that either every lifting of a closed path in X is closed or none is closed. To show that p is regular, consider a subgroup $p_{\#}\pi(\widetilde{X}, \tilde{x}) \subseteq \pi(X, x)$. By Lemma 15.4.3 again, a conjugate of $p_{\#}\pi(\widetilde{X}, \tilde{x})$ in $\pi(X, x)$ is the subgroup $p_{\#}\pi(\widetilde{X}, \tilde{y})$ for some $\tilde{y} \in p^{-1}(x)$. So it suffices to prove that $p_\# \pi(\widetilde{X}, \tilde{x}) = p_\# \pi(\widetilde{X}, \tilde{y})$ whenever $p(\tilde{x}) = p(\tilde{y})$. Let $[\tilde{g}] \in \pi(\widetilde{X}, \tilde{x})$. By the Path Lifting Property of p, there is a path $\tilde{h}$ in $\widetilde{X}$ with $p\tilde{h} = p\tilde{g}$ and $\tilde{h}(0) = \tilde{y}$. Since $\tilde{g}$ is a closed lifting of $p\tilde{g}$, $\tilde{h}$ must be closed (by our assumption). Thus $[\tilde{h}] \in \pi(\widetilde{X}, \tilde{y})$ and $p_\#[\tilde{h}] = p_\#[\tilde{g}]$. This implies that $p_\# \pi(\widetilde{X}, \tilde{x}) \subseteq p_\# \pi(\widetilde{X}, \tilde{y})$. Similarly, $p_\# \pi(\widetilde{X}, \tilde{y}) \subseteq p_\# \pi(\widetilde{X}, \tilde{x})$, and the equality holds.   ◇

Proper Actions of Discrete Groups

If $p \colon \widetilde{X} \to X$ is a regular covering map over a locally path-connected space X and $\widetilde{X}$ is connected, then the fibers $p^{-1}(x), x \in X$, are precisely the orbits of $\Delta(p)$, and so the mapping $X \to \widetilde{X}/\Delta(p), x \mapsto p^{-1}(x)$, is a homeomorphism because both the mapping p and the orbit map $\widetilde{X} \to \widetilde{X}/\Delta(p)$ are identifications. This suggests the possibility that the orbit maps of discrete transformation groups may be covering projections. However, it is easily seen that the orbit map associated with an arbitrary action of such groups may not be a covering map. To see the conditions under which an orbit map could be a covering map, let us examine more closely the action of the group of deck transformations of a given covering map $p \colon \widetilde{X} \to X$ on its covering space $\widetilde{X}$. First, by Proposition 15.4.6, we notice that $\Delta(p)$ acts freely on $\widetilde{X}$. Next, suppose that $\tilde{x}, \tilde{y} \in \widetilde{X}$ and $p(\tilde{x}) = p(\tilde{y})$. Let U be a connected admissible nbd of $p(\tilde{x})$, and consider the sheets $U_{\tilde{x}}$ and $U_{\tilde{y}}$ over U containing $\tilde{x}$ and $\tilde{y}$, respectively. Obviously, both $U_{\tilde{x}}$ and $U_{\tilde{y}}$ are components of $p^{-1}(U)$. Suppose that, for $g, h \in \Delta(p)$, $gU_{\tilde{x}} \cap U_{\tilde{y}} \neq \varnothing \neq hU_{\tilde{x}} \cap U_{\tilde{y}}$. Since $gU_{\tilde{x}}$ is a connected subset of $p^{-1}(U)$, it must be contained in one of its components. Accordingly, we have $gU_{\tilde{x}} \subseteq U_{\tilde{y}}$, and it follows that $g\tilde{x} = \tilde{y}$, for $p|U_{\tilde{y}}$ is injective. Similarly, we obtain $h\tilde{x} = \tilde{y}$ and hence $g\tilde{x} = h\tilde{x}$. Since the action of $\Delta(p)$ on $\widetilde{X}$ is free, we deduce that $g = h$. Thus we find that the set $\{h \in \Delta(p) | hU_{\tilde{x}} \cap U_{\tilde{y}} \neq \varnothing\}$ has at most one element. Furthermore, if X is assumed Hausdorff, then we see that this set is empty for $x = p(\tilde{x}) \neq p(\tilde{y}) = y$. Actually, in this case, there are disjoint open nbds V_x and V_y of x and y in X, respectively. Therefore $U_{\tilde{x}} = p^{-1}(V_x)$ and $U_{\tilde{y}} = p^{-1}(V_y)$ are disjoint open nbds of $\tilde{x}$ and $\tilde{y}$ in $\widetilde{X}$, respectively. Clearly, $hU_{\tilde{x}} \subseteq U_{\tilde{x}}$ for every $h \in \Delta(p)$, so we have $hU_{\tilde{x}} \cap U_{\tilde{y}} = \varnothing$. This property of the group of deck transformations of a covering map with the Hausdorff base space is abstracted in the following.

Definition 15.5.5 A discrete group G is said to act *properly* on a space X if for each pair of points $x, y \in X$, there exist open nbds U_x and U_y of x and y in X, respectively, such that $\{g \in G \mid gU_x \cap U_y \neq \varnothing\}$ is finite.

Interestingly, any discrete group G acting freely and properly on a connected Hausdorff space X is the group of deck transformations of a covering projection with the covering space X. To see this, we first note an important property of proper actions of discrete groups.

Lemma 15.5.6 *If a discrete group G acts properly on a Hausdorff space X, then the isotropy group G_x at $x \in X$ is finite and there exists an open nbd U of x such that $gU \cap U = \varnothing$ for $g \notin G_x$.*

Proof Suppose that G is a discrete group and acts properly on a Hausdorff space X. Let $x \in X$ be arbitrary. Then there exists an open nbd U_0 of x such that $H = \{g \in G \mid gU_0 \cap U_0 \neq \varnothing\}$ is finite. Clearly, $G_x \subseteq H$, and therefore it is finite. Now, suppose that $H - G_x = \{h_1, \ldots, h_n\}$ and put $x_i = h_i x$ for every $i = 1, \ldots, n$. Then $x \neq x_i$ for every i. Since X is Hausdorff, we find nbds V_i of x and W_i of x_i in X such that $V_i \cap W_i = \varnothing$. Clearly, each $U_i = V_i \cap h_i^{-1} W_i$ is a nbd of x and satisfies the condition $U_i \cap h_i U_i = \varnothing$. Then $U = \bigcap_{i=0}^{n} U_i$ is an open nbd of x having the desired property. $\diamond$

In particular, if a discrete group G acts freely and properly on a Hausdorff space X, then there exists an open nbd U of x such that $gU \cap U = \varnothing$ for all $g \neq e$, the identity element of G. In the literature, a continuous action of a discrete group having this property is often called *properly discontinuous*.

Theorem 15.5.7 *If a discrete group G acts freely and properly on a connected Hausdorff space X, then the orbit map $q: X \to X/G$ is a regular covering map. Moreover, $G \cong \Delta(q)$, the group of deck transformations of q.*

Proof We have already seen in Sect. 13.1 that q is continuous, open and surjective. To prove that it is a covering projection, we need to find an open covering of X/G by the sets which are evenly covered by q. For each $x \in X$, by the preceding lemma, there exists an open nbd U of x such that $gU \cap U = \varnothing$ for all $g \neq e$, the identity element of G. Set $V = q(U)$. Then V is an open nbd of $q(x)$. We show that it is evenly covered by q. Clearly, $q^{-1}(V)$ is the union of the disjoint open sets gU, $g \in G$. We observe that $q|gU$ is a homeomorphism between gU and V. For every $g \in G$, let θ_g denote the homeomorphism $x \mapsto gx$ of X. Since $(q|gU) \circ (\theta_g|U) = q|U$ and $\theta_g|U$ is a homeomorphism between U and gU for every $g \in G$, it is enough to show that $q|U$ is a homeomorphism between U and V. Obviously, it is continuous, open and maps U onto V. To check the injectivity of q on U, suppose that $q(x) = q(x')$ for $x, x' \in U$. Then $x' = gx$ for some $g \in G$. So $gU \cap U \neq \varnothing$, and hence $g = e$. Consequently, $x = x'$ and $q|U$ is injective. Thus q maps U onto V homeomorphically.

To prove that q is regular, it suffices to show that $\Delta(q)$ is transitive on each fiber of q (by Corollary 15.4.9). By the definition of the orbit space X/G, the group $\Gamma = \{\theta_g \mid g \in G\}$ is transitive on each fiber of q and $\Gamma \subseteq \Delta(q)$, since $q(\theta_g(x)) = q(x)$. In fact, $\Gamma = \Delta(q)$. For, if $h \in \Delta(q)$, then $q(h(x)) = q(x)$ for every $x \in X$. Therefore, given an x, there exists a $g \in G$ such that $h(x) = gx = \theta_g(x)$. Since X is connected, we see that $h = \theta_g$, as desired.

Finally, we have $G \cong \Gamma$, since the action of G on X is effective. $\diamond$

We remark that the proof of Theorem 15.5.7 does not require the condition of local path-connectedness on the space X. Notice also that this result holds good for *properly discontinuous actions* even without the condition of Hausdorffness on X. It will play an important role in the next section in establishing the existence of a covering space of a space X with the given subgroup of $\pi(X)$ as its fundamental group.

Corollary 15.5.8 *If a discrete group G acts freely and properly on a simply connected Hausdorff space X, then $\pi(X/G) \cong G$.*

Proof Let $q: X \to X/G$ denote the orbit map. By the preceding theorem, q is a regular covering projection and G is canonically isomorphic to the group $\Delta(q)$ of deck transformations of q. Accordingly, we can consider the elements of G as the deck transformations of q. Fix a point $x_0 \in X$ and let $\tilde{x}_0 = q(x_0)$. Since X is connected and G is transitive on $q^{-1}(\tilde{x}_0)$, given $\alpha \in \pi(X/G, \tilde{x}_0)$, there exists a unique $g_\alpha \in G$ such that $g_\alpha x_0 = x_0 \cdot \alpha$. So we have a function $\Psi : \pi(X/G, \tilde{x}_0) \to G$ defined by $\Psi(\alpha) = g_\alpha$. By the proof of Theorem 15.4.10, Ψ is a homomorphism with $\ker(\Psi) = q_\# \pi(X, x_0) = \{1\}$. We observe that Ψ is surjective, too. Indeed, given $g \in G$, we find an $\alpha \in \pi(X/G, \tilde{x}_0)$ such that $gx_0 = x_0 \cdot \alpha$, since $\pi(X/G, \tilde{x}_0)$ is transitive on $q^{-1}(\tilde{x}_0)$. By the definition of Ψ, we have $g = \Psi(\alpha)$, and Ψ is surjective. Thus Ψ is an isomorphism. $\diamond$

The above corollary is useful in determining the fundamental groups of many spaces. By way of illustration, we compute the fundamental groups of two spaces, viz., the Klein bottle and the lens space $L(p, q)$.

Example 15.5.1 Consider the following isometries of $\mathbb{R}^2$: The reflection $\rho: (x, y) \mapsto (x, -y)$, the translations $\tau_X : (x, y) \mapsto (x + 1, y)$ and $\tau_Y : (x, y) \mapsto (x, y + 1)$. Then $\sigma = \tau_X \circ \tau_Y \circ \rho$ is a glide reflection of $\mathbb{R}^2$, for $\tau_Y \circ \rho$ is a reflection in the line $y = 1/2$. Let G be the subgroup of the euclidean group $E(2)$ generated by σ and τ_Y. It is easily verified that $\sigma^{-1} \circ \tau_Y \circ \sigma = \tau_Y^{-1}$, so G is a nonabelian group. Obviously, the subgroup generated by $\langle \tau_Y \rangle$ is infinite cyclic and normal in G. We also note that G is spanned by the two parallel glide reflections σ and $\tau_Y\sigma$ which satisfy $\sigma^2 = (\tau_Y\sigma)^2$ (a translation parallel to the x-axis). Since the topology of $E(2)$ coincides with the product topology of $O(2) \times \mathbb{R}^2$, we see that the subgroup of $E(2)$ generated by ρ, τ_X and τ_Y has the discrete topology, and therefore G is a discrete group. Writing $a = \sigma, b = \tau_Y$ and $c = \tau_Y\sigma$, we obtain two different presentations of G: $\langle a, b \mid a^{-1}ba = b^{-1} \rangle$ and $\langle a, c \mid a^2 = c^2 \rangle$. Observe that the first presentation of G is the same as that of $\pi(K)$ derived by using Seifert–van Kampen theorem (see Example 14.5.2).

Next, we show that the action of G on $\mathbb{R}^2$ is properly discontinuous, and hence deduce that the fundamental group of the orbit space $\mathbb{R}^2/G$ is G. Using the relation $\tau_Y \circ \sigma = \sigma \circ \tau_Y^{-1}$, each element g of G can be written as $g = \tau_Y^m \circ \sigma^n$. Consider the open nbd $U = B((x, y); 1/3)$ of (x, y) in $\mathbb{R}^2$. Obviously, we can write $\sigma^n(x, y) = (x + n, z)$. Then $g(x, y) = (x + n, z + m)$, and we see that $gU \cap U = \emptyset$ for every $g \neq e$, as desired.

Finally, we observe that $\mathbb{R}^2/G$ is homeomorphic to the Klein bottle K. It is clear that the image of a unit square S with vertices at the lattice points (the points in the plane $\mathbb{R}^2$ whose coordinates are integers) under each element of G is such a square, and the images of S under all elements of G fill out $\mathbb{R}^2$. Notice that every nontrivial element of G maps a point of the interior of S into the interior of another such square. Also, τ_Y maps the bottom side of $S = I \times I$ homeomorphically onto its top side while σ maps the left side of S homeomorphically onto its right side, as shown in Fig. 15.7. Accordingly, the orbit space $\mathbb{R}^2/G$ is homeomorphic to K.

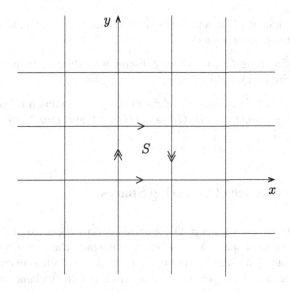

Fig. 15.7 Proof of Example 15.5.1

Example 15.5.2 The fundamental group of the generalized lens space L $(p; q_0, \ldots,$ $q_n)$ is $\mathbb{Z}_p$. Recall that p and the q_i's here are positive integers such that each q_i is relatively prime to p, and L $(p; q_0, \ldots, q_n)$ is defined as the orbit space of a free action of the cyclic group $\mathbb{Z}_p$ on the $(2n + 1)$-sphere $\mathbb{S}^{2n+1}$. In fact, we regard $\mathbb{S}^{2n+1}$ as the subspace $\{(z_0, \ldots, z_n) \mid z_i \in \mathbb{C}^{n+1}$ and $\sum |z_i|^2 = 1\}$ of $\mathbb{C}^{n+1}$, and consider the homeomorphism $h \colon \mathbb{S}^{2n+1} \to \mathbb{S}^{2n+1}$ given by h $(z_0, \ldots, z_n) = (\epsilon^{q_0} z_0, \ldots, \epsilon^{q_n} z_n)$, where $\epsilon = e^{2\pi i/p}$ is a primitive pth root of unity. Clearly, h determines an action of the group $\mathbb{Z}_p$ on $\mathbb{S}^{2n+1}$, which is free and proper. Since $\mathbb{S}^{2n+1}$ is simply connected, Corollary 15.5.8 shows that π $(L$ $(p; q_0, \ldots, q_n))$ is $\mathbb{Z}_p$.

Exercise

1. Let $p \colon \widetilde{X} \to X$ be the covering map of Exercise 15.1.3. Show that p is a regular covering and determine the group Δ (p).

2. If H is a discrete subgroup of a connected Hausdorff group G, prove that the quotient map $G \to G/H$ is a regular covering projection with the covering group H.

3. • Let $p \colon \widetilde{X} \to X$ be a covering projection, where $\widetilde{X}$ is connected and X is locally path-connected. Show that the action of the group $\Delta(p)$ on $\widetilde{X}$ is properly discontinuous.

4. Show that a free action of a finite group on a Hausdorff space is properly discontinuous.

5. Find an example of a free action of an infinite discrete group on some space that is not properly discontinuous.

6. Show that if a group G acts properly discontinuously on a Hausdorff space X, then each orbit $G(x)$ is closed in X.

7. Prove that the set U in Lemma 15.5.6 can be so chosen that it is invariant under G_x and the canonical map $U/G_x \to X/G$ is a homeomorphism onto an open subset of X/G.

15.6   The Existence of Covering Spaces

In Sect. 15.4, we have seen that each connected covering space of a connected and locally path-connected space X picks out a conjugacy class of subgroups of the fundamental group of X. Also, a criterion for equivalence of such covering maps was established there in terms of conjugacy of subgroups of the fundamental group of X induced by the covering maps. In view of these facts, one can expect a classification of the covering spaces of X by means of conjugacy classes of subgroups of $\pi(X)$. Indeed, we only need to show the existence of a covering space of X corresponding to each subgroup of $\pi(X)$.

We first deal with the question of existence of a covering space of X corresponding to the trivial subgroup of $\pi(X)$. The following example shows that not every connected and locally path-connected space has a universal covering space.

Example 15.6.1 Let X be the "Hawaiian earring" (ref. Example 6.4.2). It is clearly connected and locally path-connected. We observe that X has no universal covering space. Assume on the contrary that $\widetilde{X}$ is a universal covering space of X relative to the map p. Let $x_0 \in X$ denote the origin of $\mathbb{R}^2$ and U be a nbd of x_0 that is evenly covered by the covering map p. For $\tilde{x}_0 \in p^{-1}(x_0)$, let $\widetilde{U}$ be the sheet over U containing $\tilde{x}_0$. Since $p|\widetilde{U}$ is a homeomorphism between $\widetilde{U}$ and U, each loop f in U based at x_0 lifts to a loop $\tilde{f}$ in $\widetilde{U}$ based at $\tilde{x}_0$. Since $\widetilde{X}$ is simply connected, there is a base point preserving homotopy between $\tilde{f}$ and the constant path at $\tilde{x}_0$. Hence f is homotopic in X to the constant path at x_0. It follows that the inclusion $i : U \hookrightarrow X$ induces the trivial homomorphism $i_\# : \pi(U, x_0) \to \pi(X, x_0)$. For large n, we have $C_n \subset U$, where C_n is the circle with radius $1/n$ and center $(1/n, 0)$. If $j : C_n \hookrightarrow U$ and $k : C_n \hookrightarrow X$ are the inclusion maps, then we have $k = ij$. It is easily checked that C_n is a retract of X, and therefore the composition

$$\pi(C_n, x_0) \overset{j_\#}{\to} \pi(U, x_0) \overset{i_\#}{\to} \pi(X, x_0)$$

is injective. This implies that $i_\#$ is not trivial, a contradiction. Therefore X has no universal covering space.

By the proof of the above example, we see that if a space X has a universal covering space, then the following condition is necessarily satisfied: Each point $x \in X$ has an open nbd U such that the homomorphism $\pi(U, x) \to \pi(X, x)$ induced by the inclusion map $U \hookrightarrow X$ is trivial (that is, all loops in U are null homotopic in X).

A space X with the above property is called *semilocally simply connected*. With this terminology, we prove the following

Theorem 15.6.1 *A connected, locally path-connected and semilocally simply connected space X has a universal covering space.*

Proof We choose a point $x_0 \in X$ and consider the set $\mathscr{P}$ of all paths in X with origin x_0. For each $f \in \mathscr{P}$, put

$$[f] = \{g \in \mathscr{P} | g(1) = f(1) \text{ and } g \simeq f \text{ rel } \{0, 1\}\}$$

and let $\widetilde{X} = \{[f] \mid f \in \mathscr{P}\}$. We introduce a topology into $\widetilde{X}$ by constructing a basis for it. Since X is locally path-connected, the family $\mathscr{B}$ of all open and path-connected subsets of X form a base. For $U \in \mathscr{B}$ and $f \in \mathscr{P}$ with $f(1) \in U$, let $\langle f, U \rangle$ denote the set of all homotopy classes $[g] \in \widetilde{X}$ such that $g(1) \in U$ and $g \simeq f * \gamma$ rel $\{0, 1\}$ for some path γ in U (see Fig. 15.8).

We show that the family $\widetilde{\mathscr{B}} = \{\langle f, U \rangle \mid U \in \mathscr{B} \text{ and } f \in \mathscr{P}\}$ is a basis for a topology on $\widetilde{X}$. To this end, we first observe the following properties of the sets $\langle f, U \rangle$:

(a) $[f] \in \langle f, U \rangle$.

(b) For U, V in $\mathscr{B}$ and $f \in \mathscr{P}$ with $f(1) \in U \subseteq V$, $\langle f, U \rangle \subseteq \langle f, V \rangle$.

(c) $[g] \in \langle f, U \rangle \Rightarrow \langle g, U \rangle = \langle f, U \rangle$.

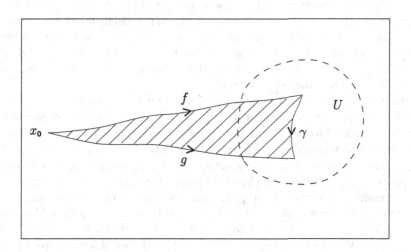

Fig. 15.8 The set $\langle f, U \rangle$ in the proof of Theorem 15.6.1

The proofs of (a) and (b) are obvious. To prove (c), suppose that $[h] \in \langle g, U \rangle$. Then there is a path γ in U such that $g(1) = \gamma(0)$ and $h \simeq g * \gamma$ rel $\{0, 1\}$. By our hypothesis, there is a path δ in U such that $f(1) = \delta(0)$ and $g \simeq f * \delta$ rel $\{0, 1\}$. So $h \simeq (f * \delta) * \gamma \simeq f * (\delta * \gamma)$ rel $\{0, 1\}$. Clearly, the path $\delta * \gamma$ lies in U and we have $[h] \in \langle f, U \rangle$. Thus $\langle g, U \rangle \subseteq \langle f, U \rangle$. For the reverse inclusion, we note that $f \simeq g * \delta^{-1}$ rel $\{0, 1\}$. So the roles of f and g in the above argument can be interchanged, and we obtain $\langle g, U \rangle \supseteq \langle f, U \rangle$. Now, we show that $\mathscr{B}$ satisfies the conditions of Theorem 1.4.5. From (a), it is obvious that the family $\widetilde{\mathscr{B}}$ covers $\widetilde{X}$. Next, suppose that $[h] \in \langle f, U \rangle \cap \langle g, V \rangle$. By (c), we have $\langle f, U \rangle = \langle h, U \rangle$ and $\langle g, V \rangle = \langle h, V \rangle$. Since $\mathscr{B}$ is a basis of X and $h(1) \in U \cap V$, we find a $W \in \mathscr{B}$ such that $h(1) \in W \subseteq U \cap V$. Then, by (b), $[h] \in \langle h, W \rangle \subseteq \langle h, U \rangle \cap \langle h, V \rangle = \langle f, U \rangle \cap \langle g, V \rangle$. Hence $\widetilde{\mathscr{B}}$ is a basis for a topology on $\widetilde{X}$, and we give $\widetilde{X}$ the topology generated by this basis.

Now, we define a mapping $p : \widetilde{X} \to X$ by $p([f]) = f(1)$, and show that p is a covering projection. Since X is path-connected, p is surjective. The continuity of p follows immediately from the fact that if $U \in \mathscr{B}$ and $[f] \in p^{-1}(U)$, then $\langle f, U \rangle$ is a basic nbd of $[f]$ and $p(\langle f, U \rangle) \subseteq U$, by the definitions of p and $\langle f, U \rangle$. Moreover, p is open, for the image of each basic open subset of $\widetilde{X}$ under p is open. In fact, we have $p(\langle f, U \rangle) = U$ for every $f \in \mathscr{P}$ and $U \in \mathscr{B}$. Clearly, we need to show that $U \subseteq p(\langle f, U \rangle)$. Let $x \in U$ be arbitrary. Since U is path-connected, there is a path γ in U with origin $f(1)$ and end x. Then, for $g = f * \gamma$, obviously $[g] \in \langle f, U \rangle$ and $p([g]) = x$. Hence the desired inclusion holds. Next, we show that every point of X has an admissible nbd relative to p. Since X is locally path-connected and semilocally simply connected, for each $x \in X$, there exists a $U \in \mathscr{B}$ such that $x \in U$ and all loops in U are homotopically trivial in X. Clearly, $p^{-1}(U) = \bigcup \{ \langle f, U \rangle \mid f(1) \in U \}$, and any two distinct sets $\langle f, U \rangle$ and $\langle g, U \rangle$ are disjoint, by (c). We further assert that p maps $\langle f, U \rangle$ homeomorphically onto U. Since p is continuous and open, so is its restriction from $\langle f, U \rangle$ to U. Also, it has already observed that $p(\langle f, U \rangle) = U$. To see that p is injective on $\langle f, U \rangle$, suppose that $[g], [h] \in \langle f, U \rangle$ and $g(1) = h(1)$. Then, by (c), $\langle g, U \rangle = \langle f, U \rangle = \langle h, U \rangle$. So $g \in \langle h, U \rangle$, and therefore there exists a path γ in U such that $h(1) = \gamma(0)$ and $g \simeq h * \gamma$ rel $\{0, 1\}$. In particular, we have $\gamma(1) = g(1) = h(1)$, thus γ is a loop in U. By the choice of U, $\gamma \simeq c_{h(1)}$ rel $\{0, 1\}$. Therefore $g \simeq h * \gamma \simeq h * c_{h(1)} \simeq h$ rel $\{0, 1\}$ whence $[g] = [h]$. This proves our assertion, and p is a covering projection.

It remains to show that $\widetilde{X}$ is simply connected. First, we need to prove that $\widetilde{X}$ is path-connected. Consider the point $\tilde{x}_0 = [c_{x_0}]$, where c_{x_0} is the constant path at x_0. We observe that each point of $\widetilde{X}$ can be joined to $\tilde{x}_0$ by a path in $\widetilde{X}$. Let $\tilde{x} = [f] \in \widetilde{X}$ be arbitrary and, for each $t \in I$, define a path f_t in X by $f_t(s) = f(st)$. Then $f_t \in \mathscr{P}$, and we have a function $\tilde{f} : I \to \widetilde{X}$ given by $\tilde{f}(t) = [f_t]$. To see the continuity of $\tilde{f}$, suppose that $t_0 \in I$ and let $\langle g, U \rangle$ be a basic nbd of $\tilde{f}(t_0)$. Then $\langle f_{t_0}, U \rangle = \langle g, U \rangle$. By the continuity of f, there exists an open (or half-open) interval $V \subseteq I$ such that $t_0 \in V$ and $f(V) \subseteq U$. Accordingly, $f((1 - s)t_0 + st) \in U$ for all $s \in I$ and $t \in V$, and we have paths $\gamma_t : I \to U$ defined by $\gamma_t(s) = f((1 - s)t_0 + st)$. Clearly, $f_{t_0}(1) = \gamma_t(0)$ for every $t \in V$. We observe that $f_t \simeq f_{t_0} * \gamma_t$ rel $\{0, 1\}$. Indeed, a homotopy is given by

$$F(s, u) = \begin{cases} f\left(2s((1-u)t + ut_0)/(2-u)\right) & \text{for } 0 \leq s \leq (2-u)/2, \\ f\left((2-2s)t_0 + (2s-1)t\right) & \text{for } (2-u)/2 \leq s \leq 1. \end{cases}$$

Therefore $\tilde{f}(t) = [f_t] \in \langle g, U \rangle$ for all $t \in V$, and it follows that $\tilde{f}$ is continuous at t_0. Obviously, $\tilde{f}(0) = \tilde{x}_0$ and $\tilde{f}(1) = \tilde{x}$, so $\tilde{f}$ is a path in $\tilde{X}$ from $\tilde{x}_0$ to $\tilde{x}$. Now, we show that $\pi(\tilde{X}, \tilde{x}_0) = \{1\}$. Notice that $p(\tilde{x}_0) = x_0$ and if $\Gamma = \pi(X, x_0)$, then $\Gamma_{\tilde{x}_0} = p_\# \pi(\tilde{X}, \tilde{x}_0)$, by the proof of Proposition 15.3.1. Suppose that $\alpha = [g] \in \Gamma$ fixes $\tilde{x}_0$. Then $\tilde{x}_0 = \tilde{x}_0 \cdot \alpha = \tilde{g}(1)$, where $\tilde{g}$ is the lifting of g starting at $\tilde{x}_0$. Clearly, $\tilde{g}$ is given by $\tilde{g}(t) = [g_t]$, where g_t is the path $s \mapsto g(st)$. Thus we have $\tilde{x}_0 = [g] = \alpha$. By definition, $\tilde{x}_0$ is the identity element of $\pi(X, x_0)$, and it follows that $p_\# \pi(\tilde{X}, \tilde{x}_0) = \{1\}$. Since $p_\#$ is a monomorphism, we deduce that $\pi(\tilde{X}, \tilde{x}_0) = \{1\}$. $\diamond$

From Proposition 15.4.2, it is clear that any two universal covering spaces of a connected and locally path-connected space X are equivalent. Accordingly, we often speak of *the* universal covering space of X, when it exists.

Now, suppose that a locally path-connected space X has a universal covering space $\tilde{Y}$. Then each path-connected covering space $\tilde{X}$ of X is the base space of a regular covering map $q: \tilde{Y} \to \tilde{X}$, by Corollary 15.2.8, and it is homeomorphic to the orbit space $\tilde{Y}/\Delta(q)$. By Corollary 15.5.3, $\Delta(q)$ is isomorphic to $\pi(\tilde{X})$ which, as we know, is isomorphic to a subgroup of $\pi(X)$. Accordingly, given a (nontrivial) subgroup H of $\pi(X)$, it seems possible to construct a covering space $\tilde{X}$ of X, via an action of H on the universal covering space of X, such that $\pi(\tilde{X}) = H$. With this end in view, we prove the following.

Lemma 15.6.2 *Let $p: Y \to X$ and $q: Z \to Y$ be continuous maps and suppose that all the three spaces X, Y and Z are connected and locally path-connected. If q and the composition $pq: Z \to X$ are covering maps, then p is also a covering map.*

Proof First, notice that p is a surjection, for pq is so. Next, given a point $x \in X$, find a path-connected open nbd U of x which is evenly covered by pq. We show that U is evenly covered by p. Suppose that V_α, $\alpha \in A$, are the sheets over U relative to pq. Then we have $(pq)^{-1}(U) = \bigcup_\alpha V_\alpha$, so $p^{-1}(U) = \bigcup_\alpha q(V_\alpha)$, for q is surjective. By definition, the V_α,s are open subsets of Z. Since q is a continuous open map and each V_α is homeomorphic to U, we deduce that $q(V_\alpha)$ is a path-connected open subset of Y for every α. We further assert that these sets are closed in $p^{-1}(U)$. To see this, fix an arbitrary index α_0 and consider a point y in the closure of $q(V_{\alpha_0})$ in $p^{-1}(U)$. Since Y is locally path-connected, there exists a path-connected open nbd W of y, which is contained in $p^{-1}(U)$ and is evenly covered by q. Then $q^{-1}(W) \subseteq \bigcup_\alpha V_\alpha$, and so each sheet over W is contained in some V_α, since W is path-connected and the sets V_α are disjoint and open. Moreover, $q^{-1}(W) \cap V_{\alpha_0} \neq \varnothing$, for $y \in W$ is an adherent point of $q(V_{\alpha_0}) \neq \varnothing$. Thus V_{α_0} contains some sheet over W, and hence $y \in W \subseteq q(V_{\alpha_0})$. It follows that $q(V_{\alpha_0})$ is closed in $p^{-1}(U)$, as desired. Therefore each $q(V_\alpha)$ is a path component of $p^{-1}(U)$, being a path-connected clopen subset. Consequently, for every $\alpha, \beta \in A$, we have either $q(V_\alpha) = q(V_\beta)$ or $q(V_\alpha) \cap q(V_\beta) = \varnothing$. Finally, we observe that p maps each $q\,(V_\alpha)$ homeomorphically onto U. Since $(pq)|V_\alpha$ is a homeomorphism between V_α and U, the mapping $q|V_\alpha$ is injective and thus a homeomorphism between V_α and $q\,(V_\alpha)$. From the equality $(pq)|V_\alpha = (p|q(V_\alpha)) \circ (q|V_\alpha)$, we deduce that $p|q\,(V_\alpha)$ is a homeomorphism between $q\,(V_\alpha)$ and U. $\diamond$

Theorem 15.6.3 *Let X be a Hausdorff, connected, locally path-connected and semilocally simply connected space and $x_0 \in X$. Then, for each subgroup $H \subseteq \pi(X, x_0)$, there exists a covering projection $p \colon \widetilde{X} \to X$ such that $p_\# \pi\big(\widetilde{X}, \tilde{x}_0\big) = H$.*

Proof By Theorem 15.6.1, X has a universal covering space $\widetilde{Y}$, which is Hausdorff, since X is so. Let $r \colon \widetilde{Y} \to X$ be the covering projection and $\Delta(r)$ be the group of deck transformations of r. As already seen in the previous section, $\Delta(r)$ acts freely and properly on $\widetilde{Y}$ and $X \approx \widetilde{Y}/\Delta(r)$. Choose a point $\tilde{y}_0 \in r^{-1}(x_0)$. By Theorem 15.4.10, there is an isomorphism $\Psi \colon \pi(X, x_0) \to \Delta(r)$ given by $\Psi(\alpha)\,(\tilde{y}_0) = \tilde{y}_0 \cdot \alpha$. Put $D = \Psi(H)$. Then $D \cong H$ and acts freely and properly on $\widetilde{Y}$, since this is true of $\Delta(r)$. By Theorem 15.5.7, the orbit map $q \colon \widetilde{Y} \to \widetilde{Y}/D$ is a covering projection. We write $\widetilde{X} = \widetilde{Y}/D$ and $\tilde{x}_0 = q\,(\tilde{y}_0)$. Then there is a continuous map $p \colon \widetilde{X} \to \widetilde{Y}/\Delta(r) \approx X$ making the following diagram

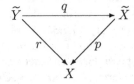

commutative, that is, $r = pq$. Since r and q are covering projections, p is also a covering projection, by the preceding lemma. Obviously, $\tilde{x}_0 \in p^{-1}(x_0)$. We show that $H = p_\# \pi\big(\widetilde{X}, \tilde{x}_0\big)$. Suppose that $\alpha = [f]$, where f is a loop in X based at x_0. Let $\widetilde{F}$ be the lifting of f relative to the covering map r with the origin $\tilde{y}_0$. Then $\widetilde{F}(1) = \tilde{y}_0 \cdot \alpha = \Psi(\alpha)\,(\tilde{y}_0)$. Now, if $\alpha \in H$, then $\Psi(\alpha) \in D$. Consequently, $\widetilde{F}(0)$ and $\widetilde{F}(1)$ are in the same orbit of D, and so $\tilde{f} = q \circ \widetilde{F}$ is a loop in $\widetilde{X}$ based at $\tilde{x}_0$. Then it is obvious that $p\tilde{f} = r\widetilde{F} = f$, and therefore $\alpha \in p_\# \pi\big(\widetilde{X}, \tilde{x}_0\big)$. Conversely, given a loop $\tilde{f}$ in $\widetilde{X}$ based at $\tilde{x}_0$, let $\widetilde{F}$ be the lifting of $\tilde{f}$ relative to the covering map q with the origin $\tilde{y}_0$. Then $p\tilde{f} = f$ (say) is a loop in X based at x_0 and $\widetilde{F}$ is the lifting of f relative to r with the origin $\tilde{y}_0$. Accordingly, for $\alpha = [f] \in \pi(X, x_0)$, we have $\Psi(\alpha)\,(\tilde{y}_0) = \tilde{y}_0 \cdot \alpha = \widetilde{F}(1)$. Since $\tilde{f} = q\widetilde{F}$ is a loop in $\widetilde{X}$, $\tilde{y}_0$ and $\Psi(\alpha)\,(\tilde{y}_0)$ must be in the same orbit of D. So there exists an element $\delta \in D$ such that $\delta\,(\tilde{y}_0) = \Psi(\alpha)\,(\tilde{y}_0)$. By Proposition 15.4.6, $\Psi(\alpha) = \delta \in D$ and we have $\alpha \in H$. Hence $p_\# \pi\big(\widetilde{X}, \tilde{x}_0\big) \subseteq H$, and the equality holds. $\diamond$

We remark that the preceding theorem holds good even without the condition of Hausdorffness on X. In fact, we needed this condition to ensure the Hausdorffness of the universal covering space $\widetilde{Y}$ so that Theorem 15.5.7 applies. However, q would be a regular covering projection even if $\widetilde{Y}$ were not Hausdorff, for the action of $D \subseteq \Delta(r)$ on $\widetilde{Y}$ is properly discontinuous (refer to Exercise 15.5.3). So we conclude that for a connected, locally path-connected and semilocally simply connected space X, there is a one-to-one correspondence between the equivalence classes of covering spaces of X and the conjugacy classes of $\pi(X)$. This correspondence is given by $[\widetilde{X} \xrightarrow{p} X] \longleftrightarrow [p_{\#}\pi(\widetilde{X})]$.

Exercise

1. Give an example of a simply connected space that is not locally path-connected.

2. Prove that an infinite product of circles $\mathbb{S}^1$ has no universal covering space.

3. Let $p: Y \to X$ and $q: Z \to Y$ be continuous maps and assume that the spaces X, Y and Z are all connected and locally path-connected. If p and q are covering maps and X has a universal covering space, show that $pq: Z \to X$ is a covering map.

4. Determine all covering spaces of the lens space $L(p, q)$, up to equivalence.

Appendix A
Set Theory

A.1 Sets

The purpose of this appendix is to introduce some of the basic ideas and terminologies from set theory which are essential to our present work. In this naive treatment, we do not intend to give a complete and precise analysis of set theory, which belongs to the foundations of mathematics and to mathematical logic. Rather, we shall deal with sets on an intuitive basis. We remark that this usage can be formally justified.

Intuitively, a *set* is a collection of objects called *members* of the set. The terms "collection" and "family" are used as synonyms for "set". If an object x is a member of a set X, we write $x \in X$ to express this fact. Implicit in the idea of a set is the notion that a given object either belongs or does not belong to the set. The statement "x is not a member of X" is indicated by $x \notin X$. The objects which make up a set are usually called the *elements* or *points* of the set.

Given two sets A and X, we say that A is a *subset* of X if every element of A is also an element of X. In this case, we write $A \subseteq X$ (read A is *contained* in X) or $X \supseteq A$ (read X *contains* A). The sets A and X are said to be equal, written $A = X$, if $A \subseteq X$ and $X \subseteq A$. We call A a *proper subset* of X (written $A \subset X$) if $A \subseteq X$ but $A \neq X$. The *empty* or *null* set which has no element is denoted by $\varnothing$. We have the inclusion $\varnothing \subseteq X$ for every set X. The set whose only element is x is called a *singleton*, denoted by $\{x\}$. Note that $\varnothing \neq \{\varnothing\}$. If $P(x)$ is a statement about elements in X, that is either true or false for a given element of X, then the subset of all the $x \in X$ for which $P(x)$ is true is denoted by $\{x \in X \mid P(x)\}$ or $\{x \in X : P(x)\}$.

In a particular mathematical discussion, there is usually a set which consists of all primary elements under consideration. This set is referred to as the "universe". To avoid any logical difficulties, all the sets we consider in this section are assumed to be subsets of the universe X.

The *difference* of two sets A and B, denoted by $A - B$, is the set $\{x \in A \mid x \notin B\}$. If $B \subseteq A$, the *complement* of B in A is $A - B$. Notice that the complement operation is defined only when one set is contained in the other, whereas the difference operation does not have such a restriction. The *union* of two sets A and B is the set $A \cup B =$

© Springer Nature Singapore Pte Ltd. 2019
T. B. Singh, *Introduction to Topology*,
https://doi.org/10.1007/978-981-13-6954-4

$\{x \mid x$ belongs to at least one of A, $B\}$. The *intersection* of two sets A and B is the set $A \cap B = \{x \mid x$ belongs to both A and $B\}$. When $A \cap B = \varnothing$, the sets A and B are called *disjoint*; otherwise we say that they *intersect*.

Proposition A.1.1 (a) $X - (X - A) = A$.
  (b) $B \subseteq A \Leftrightarrow X - A \subseteq X - B$.
  (c) $A = B \Leftrightarrow X - A = X - B$.

Proposition A.1.2 $A \cup A = A \cap A$.

Proposition A.1.3 $A \cup B = B \cup A$, $A \cap B = B \cap A$.

Proposition A.1.4 $A \subseteq B \Leftrightarrow A = A \cap B \Leftrightarrow B = A \cup B$.

Proposition A.1.5 (a) $A \cup (B \cup C) = (A \cup B) \cup C$.
  (b) $A \cap (B \cap C) = (A \cap B) \cap C$.

Proposition A.1.6 (a) $A \cap (B \cup C) = (A \cap B) \cup (\cap C)$.
  (b) $A \cup (B \cap C) = (A \cup B) \cap (A \cup C)$.

Proposition A.1.7 (a) $A - B = A \cap (X - B)$.
  (b) $\left. \begin{array}{l} X - (A \cup B) = (X - A) \cap (X - B) \\ X - (A \cap B) = (X - A) \cup (X - B) \end{array} \right\}$ *(De Morgan's laws)*.
  (c) $(A - B) \cup (B - A) = (A \cup B) - (A \cap B)$.
  (d) *If* $X = A \cup B$ *and* $A \cap B = \varnothing$, *then* $B = X - A$.

Let J be a nonempty set, and suppose that a set A_j is given for each $j \in J$. Then the collection of sets $\{A_j \mid j \in J\}$, also written as $\{A_j\}_{j \in J}$, is called an *indexed family* of sets, and J is called an *indexing set* for the family. The *union* of an indexed family $\{A_j \mid j \in J\}$ of subsets of a set X is the set

$$\bigcup_{j \in J} A_j = \{x \in X \mid x \in A_j \text{ for some } j \text{ in } J\},$$

and the intersection is the set

$$\bigcap_{j \in J} A_j = \{x \in X \mid x \in A_j \text{ for every } j \text{ in } J\}.$$

The union of the sets A_j is also denoted by $\bigcup \{A_j \mid j \in J\}$, and their intersection by $\bigcap \{A_j \mid j \in J\}$. If there is no ambiguity about the indexing set, we simply use $\bigcup A_j$ for the union and $\bigcap A_j$ for the intersection. If $J = \{1, \dots, n\}$, $n > 0$ an integer, then we write $\bigcup_{j=1}^{n} A_j$ for the union of $A_1, \dots, A_n$, and $\bigcap_{j=1}^{n} A_j$ for their intersection. Observe that any nonempty collection $\mathscr{C}$ of sets can be considered an indexed family of sets by "self-indexing": The indexing set is $\mathscr{C}$ itself and one assigns to each $S \in \mathscr{C}$ the set S. Accordingly, the foregoing definitions become

$$\bigcup \{S : S \in \mathscr{C}\} = \{x \mid x \in S \text{ for some } S \text{ in } \mathscr{C}\} \quad \text{and}$$

$$\bigcap \{S : S \in \mathscr{C}\} = \{x \mid x \in S \text{ for every } S \text{ in } \mathscr{C}\}.$$

If we allow the collection $\mathscr{C}$ to be the empty set, then, by convention, $\bigcup \{S : S \in \mathscr{C}\} = \varnothing$ and $\bigcap \{S : S \in \mathscr{C}\} = X$, the specified universe of the discourse.

Proposition A.1.8 *Let A_j, $j \in J$, be a family of subsets of a set X. Then we have*

(a) $X - \bigcup_j A_j = \bigcap_j (X - A_j)$, and $X - \bigcap_j A_j = \bigcup_j (X - A_j)$.

(b) *If $K \subset J$, then $\bigcup_{k \in K} A_k \subseteq \bigcup_{j \in J} A_j$ and $\bigcap_{k \in K} A_k \supseteq \bigcap_{j \in J} A_j$.*

A.2 Functions

With each two objects x, y, there corresponds a new object (x, y), called their *ordered pair*. This is another primitive notion that we will use without a formal definition. Ordered pairs are subject to the condition: $(x, y) = (x', y') \Leftrightarrow x = x'$ and $y = y'$. Accordingly, $(x, y) = (y, x) \Leftrightarrow x = y$. The first (resp. second) element of an ordered pair is called the *first* (resp. *second*) *coordinate*. Given two sets X_1 and X_2, their *Cartesian product* $X_1 \times X_2$ is defined to be the set of all ordered pairs (x_1, x_2), where $x_j \in X_j$ for $j = 1, 2$. Thus $X_1 \times X_2 = \{(x_1, x_2) \mid x_j \in X_j \text{ for every } j = 1, 2\}$. Note that $X_1 \times X_2 = \varnothing$ if and only if $X_1 = \varnothing$ or $X_2 = \varnothing$. When both X_1 and X_2 are nonempty, $X_1 \times X_2 = X_2 \times X_1 \Leftrightarrow X_1 = X_2$.

Let X and Y be two sets. A *function* f from the set X to Y (written $f : X \to Y$) is a subset of $X \times Y$ with the following property: for each $x \in X$, there is one and only one $y \in Y$ such that $(x, y) \in f$. A function is also referred to as a *mapping* (or briefly, a map). We write $y = f(x)$ to denote $(x, y) \in f$, and say that y is the image of x under f or the value of f at x. We also say that f maps (or carries) x into y or f sends (or takes) x to y. A function f from X to Y is usually defined by specifying its value at each $x \in X$, and if the value at a typical point $x \in X$ is $f(x)$, we write $x \mapsto f(x)$ to give f. We refer to X as the *domain*, and Y as the *codomain* of f. The set $f(X) = \{f(x) \mid x \in X\}$, also denoted by $\mathrm{im}(f)$, is referred to as the *range* of f.

The *identity* function on X, which sends every element of X to itself, is denoted by 1 or 1_X. A map $c : X \to Y$ which sends every element of X to a single element of Y is called a *constant* function. Notice that the range of a constant function consists of just one element. If $A \subset X$, the function $i : A \hookrightarrow X$, $a \mapsto a$, is called the *inclusion* map of A into X. If $f : X \to Y$ and $A \subset X$, then the *restriction of f to A* is the function $f|A : A \to Y$ defined by $(f|A)(a) = f(a) \; \forall \, a \in A$. The other way around, if $A \subset X$ and $g : A \to Y$ is a function, then an *extension* of g over X is a function $G : X \to Y$ such that $G|A = g$. The inclusion map $i : A \hookrightarrow X$ is the restriction of the identity map 1 on X to A.

Proposition A.2.1 *Let X be a set and $\{A_j \mid j \in J\}$ a family of subsets of X with $X = \bigcup A_j$. If, for each $j \in J$, $f_j : A_j \to Y$ is a function such that $f_j| (A_j \cap A_k) =$*

$f_k|\left(A_j \cap A_k\right)$ for all $j, k \in J$, then there exists a unique function $F: X \to Y$ which extends each f_j.

Proof Given $x \in X$, there exists an index $j \in J$ such that $x \in A_j$. We put $F(x) = f_j(x)$ if $x \in A_j$. If $x \in A_j \cap A_k$, then $f_j(x) = f_k(x)$, by our hypothesis. Thus $F(x)$ is uniquely determined by x, and we have a single-valued function $F: X \to Y$. It is clearly an extension of each f_j. The uniqueness of F follows from the fact that each $x \in X$ belongs to some A_j; consequently, any function $X \to Y$, which agrees with each f_j on A_j, will have to assume the value $f_j(x)$ at x.   ◇

We say that a function $f: X \to Y$ is *surjective* (or a surjection or onto) if $Y = f(X)$. If $f(x) \neq f(x')$ for every $x \neq x'$, then we say that f is *injective* (or an injection or one-to-one). A function is *bijective* (or a *bijection* or a *one-to-one correspondence*) if it is both injective and surjective.

If $f: X \to Y$ and $g: Y \to Z$ are functions, then the function $X \to Z$ which maps x into $g(f(x))$ is called their composition and denoted by $g \circ f$ or simply gf.

Proposition A.2.2 *Suppose that* $f: X \to Y$ *and* $g: Y \to X$ *satisfy* $gf = 1_X$. *Then* f *is injective and* g *is surjective.*

Given a function $f: X \to Y$, a function $g: Y \to X$ such that $gf = 1_X$ and $fg = 1_Y$ is called an *inverse* of f. If such a function g exists, then it is unique and we denote it by f^{-1}. It is clear from the preceding result that f has an inverse if and only if it is a bijection. Also, it is evident that $\left(f^{-1}\right)^{-1} = f$.

Let $f: X \to Y$ be a function. For a set $A \subseteq X$, the subset

$$f(A) = \{f(x) \mid x \in A\}$$

of Y is called the *image* of A under f and, for a set $B \subseteq Y$, the subset

$$f^{-1}(B) = \{x \in X \mid f(x) \in B\}$$

of X is called the *inverse image* of B in X under f.

The *power set* of a set X is the family $\mathscr{P}(X)$ of all subsets of X. A function $f: X \to Y$ induces a function $\mathscr{P}(X) \to \mathscr{P}(Y)$, $A \mapsto f(A)$. It also induces a function $\mathscr{P}(Y) \to \mathscr{P}(X)$, $B \mapsto f^{-1}(B)$. The main properties of these functions are described in the following.

Proposition A.2.3 *Let* $f: X \to Y$ *be a function.*

(a) $A_1 \subseteq A_2 \Rightarrow f(A_1) \subseteq f(A_2)$,
(b) $f(\bigcup_j A_j) = \bigcup_j f(A_j)$,
(c) $f(\bigcap_j A_j) \subseteq \bigcap_j f(A_j)$, *and*
(d) $Y - f(A) \subseteq f(X - A) \Leftrightarrow f$ *is surjective.*

Proposition A.2.4 *Let* $f: X \to Y$ *be a function.*

(a) $B_1 \subseteq B_2 \Rightarrow f^{-1}(B_1) \subseteq f^{-1}(B_2)$,

(b) $f^{-1}(\bigcup_j B_j) = \bigcup_j f^{-1}(B_j)$,

(c) $f^{-1}(\bigcap_j B_j) = \bigcap_j f^{-1}(B_j)$, and

(d) $f^{-1}(Y - B) = X - f^{-1}(B)$.

Proposition A.2.5 *Let $f : X \to Y$ be a function. Then:*

(a) *For each $A \subseteq X$, $f^{-1}(f(A)) \supseteq A$; in particular, $f^{-1}(f(A)) = A$ if f is injective.*

(b) *For each $B \subseteq Y$, $f(f^{-1}(B)) = B \cap f(X)$; in particular, $f(f^{-1}(B)) = B$ if f is surjective.*

A.3 Cartesian Products

In the previous section, we have defined the Cartesian product of two sets. For this operation, the following statements are easily proved.

Proposition A.3.1 (a) $X_1 \times (X_2 \cup X_3) = (X_1 \times X_2) \cup (X_1 \times X_3)$.

(b) $X_1 \times (X_2 \cap X_3) = (X_1 \times X_2) \cap (X_1 \times X_3)$.

(c) $X_1 \times (X_2 - X_3) = (X_1 \times X_2) - (X_1 \times X_3)$.

(d) *For $Y_i \subseteq X_i$, $i = 1, 2$,*

$$(X_1 \times X_2) - (Y_1 \times Y_2) = X_1 \times (X_2 - Y_2) \cup (X_1 - Y_1) \times Y_2 =$$
$$[(X_1 - Y_1) \times (X_2 - Y_2)] \cup [Y_1 \times (X_2 - Y_2) \cup [(X_1 - Y_1) \times Y_2].$$

Proposition A.3.2 *Let X_j, $j \in J$, be a family of subsets of a set X, and let Y_k, $k \in K$, be a family of subsets of a set Y. Then*

(a) $(\bigcup_j X_j) \times (\bigcup_k Y_k) = \bigcup_{j,k} (X_j \times Y_k)$;

(b) $(\bigcap_j X_j) \times (\bigcap_k Y_k) = \bigcap_{j,k} (X_j \times Y_k)$;

With $J = K$,

(c) $(\bigcap_j X_j) \times (\bigcap_j Y_j) = \bigcap_j (X_j \times Y_j)$ *and*

(d) $(\bigcup_j X_j) \times (\bigcup_j Y_j) \supset \bigcup_j (X_j \times Y_j)$.

It is obvious that $(X_1 \times X_2) \times X_3 \neq X_1 \times (X_2 \times X_3)$ when the sets X_1, X_2, and X_3 are all nonempty. However, there is a canonical bijection

$$(X_1 \times X_2) \times X_3 \longleftrightarrow X_1 \times (X_2 \times X_3)$$

given by $(x_1 \times x_2) \times x_3 \leftrightarrow x_1 \times (x_2 \times x_3)$. Therefore, the Cartesian product of n sets $X_1, \ldots, X_n$ $(n > 2)$ may be defined by induction as $X_1 \times \cdots \times X_n = (X_1 \times \ldots \times X_{n-1}) \times X_n$. A typical element x of $X_1 \times \cdots \times X_n$ is written as $x = (x_1, \ldots, x_n)$, where $x_i \in X_i$ for very $i = 1, \ldots, n$ and referred to as the ith coordinate of x. Observe that two elements $x = (x_1, \ldots, x_n)$ and $y = (y_1, \ldots, y_n)$ in $X_1, \ldots, X_n$ are equal if and only if $x_i = y_i$ for very i. Accordingly, an element of $X_1, \ldots, X_n$ is called an ordered n-tuple.

We now extend the definition of the Cartesian product to an arbitrary indexed family of sets $\{X_\alpha \mid \alpha \in A\}$. Of course, the new definition of the product of the sets X_α for $\alpha \in \{1, 2\}$ must reduce to the earlier notion of the Cartesian product of two sets. Observe that each ordered pair (x_1, x_2) in $X_1 \times X_2$ may be considered as defining a function $x \colon \{1, 2\} \to X_1 \cup X_2$ with $x(1) = x_1$ and $x(2) = x_2$. Accordingly, the *Cartesian product* of the X_α is defined to be the set $\prod_\alpha X_\alpha$ (or simply written as $\prod X_\alpha$) of all functions $x \colon A \to \bigcup_\alpha X_\alpha$ such that $x(\alpha) \in X_\alpha$ for each $\alpha \in A$. Occasionally, we denote the product $\prod_\alpha X_\alpha$ by $\prod_{\alpha \in A} X_\alpha$ or $\prod\{X_\alpha \mid \alpha \in A\}$ to avoid any ambiguity about the indexing set. We call X_α the αth *factor* (or *coordinate set*) of $\prod X_\alpha$. If $x \in \prod X_\alpha$, then $x(\alpha)$ is called the αth *coordinate of* x, and is often denoted by x_α.

If the family $\{X_\alpha\}$ has n sets, n a positive integer, then it may be indexed by the set $\{1, \ldots, n\}$. In this case, we have two definitions of its Cartesian product; the one in the present sense will be denoted by $\prod_{i=1}^n X_i$. Notice that the function $\prod_{i=1}^2 X_i \to X_1 \times X_2$, $x \mapsto (x_1, x_2)$, is a bijection. This shows that the notion of the product $\prod_\alpha X_\alpha$ is a reasonable generalization of the notion of the Cartesian product of two sets. Similarly, the mapping $x \mapsto (x_1, \ldots, x_n)$ gives a bijection between $\prod_{i=1}^n X_i \to X_1 \times \cdots \times X_n$, and allows us to identify the two products of $X_1, \ldots, X_n$. We usually call an element $x \in \prod_{i=1}^n X_i$ an ordered n-tuple. In general, an element $x \in \prod_{\alpha \in A} X_\alpha$ is referred to as an A-tuple and written as (x_α). It is obvious that two elements x, y in $\prod_{\alpha \in A} X_\alpha$ are equal if and only if $x_\alpha = y_\alpha$ for all $\alpha \in A$. If one of the sets X_α is empty, then so is $\prod X_\alpha$. On the other hand, if $\{X_\alpha\}$ is a nonempty family of nonempty sets, it is not quite obvious that $\prod_\alpha X_\alpha \neq \varnothing$. In fact, a positive answer to the question of the existence of an element in $\prod_\alpha X_\alpha$ is one of the set-theoretic axioms, known as follows.

Theorem A.3.3 (The Axiom of Choice) *If $\{X_\alpha \mid \alpha \in A\}$ is a nonempty family of nonempty sets, then there exists a function*

$$c \colon A \to \bigcup_\alpha X_\alpha$$

such that $c(\alpha) \in X_\alpha$ for each $\alpha \in A$ (c is called a choice function for the family $\{X_\alpha\}$).

This axiom is logically equivalent to a number of interesting propositions; one such proposition is the following.

Theorem A.3.4 (Zermelo's Postulate) *Let $\{X_\alpha \mid \alpha \in A\}$ be a family of nonempty pairwise disjoint sets. Then there exists a set C consisting of exactly one element from each X_α.*

For, if c is a choice function for the family $\{X_\alpha \mid \alpha \in A\}$ in Theorem A.3.4, then the set $C = c(A)$ is a desired set. Conversely, if $\{X_\alpha \mid \alpha \in A\}$ is a family of nonempty sets, then $Y_\alpha = \{\alpha\} \times X_\alpha$ is nonempty for every α and $Y_\alpha \cap Y_\beta \neq \varnothing$ for all $\alpha \neq \beta$ in A. By the above postulate, there exists a set C which consists of exactly one element from each Y_α. Accordingly, for each $\alpha \in A$, we have a unique $x_\alpha \in X_\alpha$ such that $(\alpha, x_\alpha) \in C$. Obviously, $C \subseteq \bigcup_\alpha Y_\alpha \subseteq A \times (\bigcup_\alpha X_\alpha)$, and we have a function $c \colon A \to \bigcup_\alpha X_\alpha$ defined by $c(\alpha) = x_\alpha$ for all $\alpha \in A$.

Later, we will discuss some more theorems equivalent to the axiom of choice.

It should be noted that the sets X_α in the definition of the Cartesian product $\prod_{\alpha \in A} X_\alpha$ need not to be different from one another; indeed, it may happen that they are all the same set X. In this case, the product $\prod_{\alpha \in A} X_\alpha$ may be called the Cartesian product of A copies of X or the Cartesian Ath power of X, and denoted by X^A. Notice that X^A is just the set of all functions $A \to X$.

Proposition A.3.5 *If $Y_\alpha \subseteq X_\alpha$ for every $\alpha \in A$, then $\prod_\alpha Y_\alpha \subseteq \prod_\alpha X_\alpha$. Conversely, if each $X_\alpha \neq \varnothing$ and $\prod_\alpha Y_\alpha \subseteq \prod_\alpha X_\alpha$, then $Y_\alpha \subseteq X_\alpha$ for every α.*

Proposition A.3.6 *Let $\{X_\alpha \mid \alpha \in A\}$ be a family of nonempty sets, and let $U_\alpha, V_\alpha \subseteq X_\alpha$ for every α. Then we have*

(a) $\prod U_\alpha \cup \prod V_\alpha \subseteq \prod (U_\alpha \cup V_\alpha)$.
(b) $\prod U_\alpha \cap \prod V_\alpha = \prod (U_\alpha \cap V_\alpha)$.

If $\{X_\alpha \mid \alpha \in A\}$ is a family of nonempty sets, then for each $\beta \in A$, we have the mapping $p_\beta \colon \prod_{\alpha \in A} X_\alpha \to X_\beta$ given by $x \mapsto x_\beta$. It is easy to see that each of these maps is surjective. The map p_β is referred to as the *projection* onto the βth factor. For $U_\beta \subseteq X_\beta$, $p_\beta^{-1}(U_\beta)$ is the product $U_\beta \times \prod_{\alpha \neq \beta} X_\alpha$, referred to as a *slab* in $\prod X_\alpha$.

Proposition A.3.7 *If $B \subset A$, then $\bigcap_{\beta \in B} p_\beta^{-1}(U_\beta) = \prod Y_\alpha$, where $Y_\alpha = X_\alpha$ if $\alpha \in A - B$ while $Y_\alpha = U_\alpha$ for $\alpha \in B$.*

A.4 Equivalence Relations

A relation on a set X is a subset $R \subseteq X \times X$. If $(x, y) \in R$, we write $x R y$. The relation $\Delta = \{(x, x) \mid x \in X\}$ is called the identity relation on X. This is also referred to as the diagonal. If R is a relation on X, and $Y \subset X$, then $R \cap (Y \times Y)$ is called the relation induced by R on Y.

Given a set X, a relation R on X is

(a) *reflexive* if $x R x \; \forall x \in X$ (equivalently, $\Delta \subseteq R$),
(b) *symmetric* if $x R y \Rightarrow y R x \; \forall x, y \in X$, and
(c) *transitive* if $x R y$ and $y R z \Rightarrow x R z, \; \forall x, y, z \in X$.

An *equivalence relation* on the set X is a relation which is reflexive, symmetric, and transitive. An equivalence relation is usually denoted by the symbol $\sim$, read "tilde." Suppose that $\sim$ is an equivalence relation on X. Given an element $x \in X$, the set $[x] = \{y \in X \mid y \sim x\}$ is called the *equivalence class* of x. It is clear that

X equals the union of all the equivalence classes, and two equivalence classes are either disjoint or identical. A *partition* of a set X is a collection of nonempty, disjoint subsets of X whose union is X. With this terminology, the family of equivalence classes of X determined by $\sim$ is a partition of X. Conversely, given a partition $\mathscr{E}$ of X, there is an equivalence relation $\sim$ on X such that the equivalence classes of $\sim$ are precisely the sets of $\mathscr{E}$. This relation is obtained by declaring $x \sim y$ if both x and y belong to the same partition set.

Given an equivalence relation $\sim$ on X, the set of all equivalence classes $[x], x \in X$, is called the *quotient set* of X by $\sim$, and is denoted by $X/\sim$. Thus we have another important method of forming new sets from old ones. The map $\pi \colon X \to X/\sim$ defined by $\pi(x) = [x]$ is called the *projection* of X onto $X/\sim$.

If R is a binary relation on a set X, then there is an equivalence relation $\sim$ on X defined by $x \sim y$ if and only if one of the following is true: $x = y, x R y, y R x$, or there exist finitely many points $z_1, \ldots, z_{n+1}$ in X such that $z_1 = x$, $z_{n+1} = y$ and either $z_i R z_{i+1}$ or $z_{i+1} R z_i$ for all $1 \le i \le n$. It is called the *equivalence relation generated by R*.

A.5   Finite and Countable Sets

For counting objects in a set, we use the natural numbers (or positive integers) $1, 2, 3, \ldots$. When it is feasible, the process of counting a set X requires putting it in one-to-one correspondence with a set $\{1, 2, \ldots, n\}$ consisting of the natural numbers from 1 to n. Also, counting of sets in essence serves to determine if one of two given sets has more elements than the other. For this purpose, it may be easier to pair off each member of one set with a member of the other and see if any elements are left over in one of the sets rather than count each. The following definition makes precise the notion of "sets having the same size." We say that two sets X and Y are *equipotent* (or have the same *cardinality*) if there exists a bijection between them. Clearly, the relation of equipotence between sets is an equivalence relation on any given collection of sets.

We shall assume familiarity with the set of natural numbers $\mathbb{N} = \{1, 2, 3, \ldots\}$ (also denoted by $\mathbb{Z}_+$), and the usual arithmetic operations of addition and multiplication in $\mathbb{N}$. We shall also assume the notion of the order relation "less than" $<$ on $\mathbb{N}$ and the "well-ordering property": Every nonempty subset of $\mathbb{N}$ has a smallest element. For any $n \in \mathbb{N}$ the set $\{1, 2, \ldots, n\}$ consisting of the natural numbers from 1 to n will be denoted by $\mathbb{N}_n$. The set $\mathbb{N}_n$ is referred to as an initial segment of $\mathbb{N}$. A set X is *finite* if it is either empty or equipotent to a set $\mathbb{N}_n$ for some $n \in \mathbb{N}$. X is called *infinite* if it is not finite. When X is equipotent to a set $\mathbb{N}_n$ we say that X contains n elements or the *cardinality* of X is n. The cardinality of the empty set is 0. Of course, we must justify that the cardinality of a finite set X is uniquely determined by it. To see this, we first prove the following.

Proposition A.5.1 *For any natural numbers $m < n$, there is no injection from $\mathbb{N}_n$ to $\mathbb{N}_m$.*

Proof We use induction on n to establish the proposition. If $n = 2$, then $m = 1$, and the proposition is obviously true. Now, let $n > 2$ and assume that the proposition is true for $n - 1$. We show that it is true for n. If possible, suppose that there is an injection $f : \mathbb{N}_n \to \mathbb{N}_m$, where $m < n$. If m is not in the image of $\mathbb{N}_{n-1}$, then $f|\mathbb{N}_{n-1}$ is an injection $\mathbb{N}_{n-1} \to \mathbb{N}_{m-1}$, contrary to our inductive assumption. So we may further assume that $f(j) = m$, $j \neq n$. Then $f(n) \neq m$. We define a mapping $g : \mathbb{N}_{n-1} \to \mathbb{N}_{m-1}$ by

$$g(i) = \begin{cases} f(i) & \text{for } i \neq j \\ f(n) & \text{for } i = j. \end{cases}$$

It is clear that g is an injection which, again, contradicts our inductive hypothesis. Therefore, there is no injection $\mathbb{N}_n \to \mathbb{N}_m$.   ◇

The preceding proposition is sometimes called the "Pigeonhole Principle." The contrapositive of the above proposition states that if there is an injection $\mathbb{N}_m \to \mathbb{N}_n$, then $m \leq n$. As an immediate consequence of this, we see that the cardinality of a finite set X is uniquely determined. For, if $f : X \to \mathbb{N}_n$ and $g : X \to \mathbb{N}_m$ are bijections, then the composite $fg^{-1} : \mathbb{N}_m \to \mathbb{N}_n$ is a bijection. Hence, $m \leq n$ and $n \leq m$ which implies that $m = n$. It is now evident that two finite sets X and Y are equipotent if and only if they have the same number of elements. Notice that for infinite sets, the idea of having the same "number of elements" becomes quite vague, whereas the notion of equipotence retains its clarity.

Next, we observe the following useful property of finite sets.

Proposition A.5.2 *An injective mapping from a finite set to itself is also surjective.*

Proof Let X be nonempty finite set, and $f : X \to X$ be injective. Let $x \in X$ be an arbitrary point. Write $x = x_0$, $f(x_0) = x_1$, $f(x_1) = x_2$, and so on. Since X is finite, we must have $x_m = x_n$ for some positive integers $m < n$. Because f is injective, we obtain $x_0 = x_{n-m} \Rightarrow x = f(x_{n-m-1})$. Thus f is surjective.   ◇

Proposition A.5.3 *A proper subset of a finite set X is finite and has cardinality less than that of X.*

Proof If X has n elements, then there is a bijection $X \to \mathbb{N}_n$. Consequently, each subset of X is equipotent to a subset of $\mathbb{N}_n$, and thus it suffices to prove that every proper subset of $\mathbb{N}_n$ is finite and has at most n elements. We use induction on n to show that any proper subset of $\mathbb{N}_n$ is finite and has at most $n - 1$ elements. If $n = 1$, then $\mathbb{N}_1 = \{1\}$ and its only proper subset is the empty set $\varnothing$. Obviously, the proposition is true in this case. Suppose now that every proper subset of $\mathbb{N}_k$ $(k > 1)$ has less than k elements. We show that each proper subset Y of $\mathbb{N}_{k+1}$ has at most k elements. If $k + 1$ does not belong to Y, then either $Y = \mathbb{N}_k$ or it is a proper subset of $\mathbb{N}_k$; in the latter case, Y has at most k elements, by the induction assumption. If $k + 1$ is in Y, then $Y - \{k + 1\} \subset \mathbb{N}_k$. Therefore, either $Y = \{k + 1\}$ or there exists an integer

$m < k$ and a bijection $f : Y - \{k + 1\} \to \mathbb{N}_m$, by the induction assumption again. In the latter case, we define $g : Y \to \mathbb{N}_{m+1}$ by setting $g(y) = f(y)$ for all $y \neq k + 1$, and $g(k + 1) = m + 1$. Clearly g is a bijection. So Y is finite and has $m + 1 \leq k$ elements.   ◇

The contrapositive of this proposition states that if a subset $Y \subseteq X$ is infinite, then so is X.

Theorem A.5.4 *For any nonempty set X, the following statements are equivalent:*

(a) *X is finite.*
(b) *There is a surjection $\mathbb{N}_n \to X$ for some $n \in \mathbb{N}$.*
(c) *There is an injection $X \to \mathbb{N}_n$ for some $n \in \mathbb{N}$.*

Proof (a) $\Rightarrow$ (b): Obvious.

(b) $\Rightarrow$ (c): Let $f : \mathbb{N}_n \to X$ be a surjection. Then $f^{-1}(x) \neq \varnothing$ for every $x \in X$. So, for each $x \in X$, we can choose a number $c(x) \in f^{-1}(x)$. Since $\{f^{-1}(x) \mid x \in X\}$ is a partition of $\mathbb{N}_n$, the function $x \mapsto c(x)$ is an injection from X to $\mathbb{N}_n$.

(c) $\Rightarrow$ (a): If there is an injection $f : X \to \mathbb{N}_n$, then X is equipotent to $f(X) \subseteq \mathbb{N}_n$. By the preceding proposition, there exists a positive integer $m \leq n$ and a bijection $f(X) \to \mathbb{N}_m$. It follows that X is equipotent to $\mathbb{N}_m$ and hence finite.   ◇

Proposition A.5.5 *If X is a finite set, then there is no one-to-one mapping of X onto any of its proper subsets.*

Proof Let X be a finite set and Y a proper subset of X. Suppose that X has n elements. Then there is a bijection $f : X \to \mathbb{N}_n$ for some $n \in \mathbb{N}$. Also, there is a positive integer $m < n$ and a bijection $g : Y \to \mathbb{N}_m$, by Proposition A.5.3. If $h : X \to Y$ is a bijection, then the composite $ghf^{-1} : \mathbb{N}_n \to \mathbb{N}_m$ would be an injection, which contradicts Proposition A.5.1.   ◇

By the preceding proposition, a finite set cannot be equipotent to one of its proper subsets. As $n \mapsto n + 1$ is a bijection between $\mathbb{N}$ and $\mathbb{N} - \{1\}$, the set $\mathbb{N}$ is not finite. In fact, this is the characteristic property of infinite sets. To establish this, we need the following.

Theorem A.5.6 (Principle of Recursive Definition) *Let X be a set and $f : X \to X$ be a function. Given a point $x_0 \in X$, there is a unique function $g : \mathbb{N} \to X$ such that $g(1) = x_0$ and $g(n + 1) = f(g(n))$ for all $n \in \mathbb{N}$.*

This is one of the most useful ways of defining a function on $\mathbb{N}$, which we will take for granted.

Theorem A.5.7 *Let X be a set. The following statements are equivalent:*

(a) *X is infinite.*
(b) *There exists an injection $\mathbb{N} \to X$.*
(c) *X is equipotent to one of its proper subsets.*

Proof (a) $\Rightarrow$ (b): Let $\mathscr{F}$ be the family of all finite subsets of X. By the axiom of choice, there exists a function c from the collection of all nonempty subsets $Y \subseteq X$ to itself such that $c(Y) \in Y$. Since X is not finite, $X - F \neq \varnothing$ for all $F \in \mathscr{F}$. So $c(X - F) \in X - F$. It is obvious that $F \cup \{c(X - F)\}$ is a member of $\mathscr{F}$ for every $F \in \mathscr{F}$. Accordingly, $F \mapsto F \cup \{c(X - F)\}$ is a function of $\mathscr{F}$ into itself. Set $G(1) = \{c(X)\}$ and $G(n + 1) = G(n) \cup \{c(X - G(n))\}$ for every $n \in \mathbb{N}$. By the principle of recursive definition, G is a function $\mathbb{N} \to \mathscr{F}$. We show that the function $\phi \colon \mathbb{N} \to X$ defined by $\phi(n) = c(X - G(n))$ is an injection. It is easily seen, by induction, that $G(m) \subseteq G(n)$ for $m \leq n$. Therefore, for $m < n$, we have $\phi(m) = c(X - G(m)) \in G(m + 1) \subseteq G(n)$, while $\phi(n) \in X - G(n)$.

(b) $\Rightarrow$ (c): Let $\phi \colon \mathbb{N} \to X$ be an injection, and put $Y = \phi(\mathbb{N})$. Consider the mapping $\psi \colon X \to X$ defined by

$$\psi(x) = \begin{cases} \phi\left(1 + \phi^{-1}(x)\right) & \text{if } x \in Y, \text{ and} \\ x & \text{if } x \notin Y. \end{cases}$$

We assert that ψ is a bijection between X and $X - \{\phi(1)\}$. The surjectivity of ψ is clear. To show that it is injective, assume that $\psi(x) = \psi(x')$. Then both x and x' belong to either Y or $X - Y$. If $x, x' \in Y$, then we have $\phi\left(1 + \phi^{-1}(x)\right) = \phi\left(1 + \phi^{-1}(x')\right)$, and it follows from the injectivity of ϕ that $x = x'$. And, if $x, x' \in X - Y$, then $x = x'$, by the definition of ψ.

(c) $\Rightarrow$ (a): This is contrapositive of Proposition A.5.5. $\diamond$

A set X is called *countably infinite* (or *denumerable*, or *enumerable*) if X is equipotent to the set $\mathbb{N}$ of natural numbers. X is called *countable* if it is either finite or denumerable; otherwise, it is called *uncountable*.

For example, the set $\mathbb{Z}$ of all integers is countably infinite, since $f \colon \mathbb{N} \to \mathbb{Z}$, defined by

$$f(n) = \begin{cases} n/2 & \text{if } n \text{ is even, and} \\ -(n - 1)/2 & \text{if } n \text{ is odd,} \end{cases}$$

is a bijection.

We conclude from Theorem A.5.7 that every infinite set contains a countably infinite subset. The following proposition shows that no uncountable set can be a subset of a countable set.

Proposition A.5.8 *Every subset of $\mathbb{N}$ is countable.*

Proof Let $M \subseteq \mathbb{N}$. If M is finite, then it is countable, by definition. If $M = \mathbb{N}$, then M is a countably infinite set, since the identity function from $\mathbb{N}$ to M is a bijection. So assume that M is an infinite proper subset of $\mathbb{N}$. Then we define a function from $\mathbb{N}$ to M by recursion as follows: Let $f(1)$ be the smallest integer in M, which exists by the well-ordering property of $\mathbb{N}$. Suppose that we have chosen integers $f(1), f(2), \ldots, f(k)$. Since M is infinite, $M - f\{1, 2, \ldots, k\}$ is nonempty for every $k \in \mathbb{N}$. We define $f(k + 1)$ to be the smallest integer of this set. By the principle of recursive definition, $f(k)$ is defined for all $k \in \mathbb{N}$. Clearly, $f(1) < f(2) < \cdots$

so that f is injective. Now, we see that f is surjective, too. For each $m \in M$, the set $\{1, 2, \ldots, m\}$ is finite. Since the set $\{f(k) \mid k \in \mathbb{N}\}$ is infinite, there is a $k \in \mathbb{N}$ such that $f(k) > m$. So we can find the smallest integer $l \in \mathbb{N}$ such that $f(l) \geq m$. Then $f(j) < m$ for each $j < l$; accordingly, $m \notin f\{1, \ldots, l-1\}$. By the definition of $f(l)$, we have $f(l) \leq m$, and the equality $m = f(l)$ follows. Thus $f: \mathbb{N} \to M$ is a bijection. $\diamond$

It follows that a subset of $\mathbb{N}$ is either finite or countably infinite. Note that any set which is equipotent to a countable set is countable. Accordingly, any subset of a countable set is countable.

Theorem A.5.9 *For any nonempty set X, the following statements are equivalent:*

(a) *X is countable.*
(b) *There is a surjection $\mathbb{N} \to X$.*
(c) *There is an injection $X \to \mathbb{N}$.*

Proof $(a) \Rightarrow (b)$: If X is countably infinite, then there is a bijection between $\mathbb{N}$ and X, and we are done. If X is finite, then there exists an $n \in \mathbb{N}$ and a bijection $f: \{1, \ldots, n\} \to X$. We define $g: \mathbb{N} \to X$ by setting

$$g(m) = \begin{cases} f(m) & \text{for } 1 \leq m \leq n, \text{ and} \\ f(n) & \text{for } m > n. \end{cases}$$

Clearly, g is a surjection.

$(b) \Rightarrow (c)$: Let $g: \mathbb{N} \to X$ be a surjection. Then $g^{-1}(x)$ is nonempty for every $x \in X$. So it contains a unique smallest integer $h(x)$, say. Then $x \mapsto h(x)$ is a mapping $h: X \to \mathbb{N}$, which is injective, since $g^{-1}(x) \cap g^{-1}(y) \neq \varnothing$ whenever $x \neq y$.

$(c) \Rightarrow (a)$: Suppose that $h: X \to \mathbb{N}$ is an injection. Then $h: X \to h(X)$ is a bijection. By Proposition A.5.8, $h(X)$ is countable, and therefore X is countable. $\diamond$

By definition, a countable set is the range of a (finite or infinite) sequence, and the converse follows from the preceding theorem. Thus the elements of a countable set X can be listed as $x_1, x_2, \ldots$, and such a listing is called an *enumeration* of X. Observe that the range of a function of a countable set is countable, and the domain of an injective function into a countable set is countable.

Lemma A.5.10 *For any finite number factors, $\mathbb{N} \times \cdots \times \mathbb{N}$ is countably infinite.*

Proof Let $2, 3, \ldots, p_k$ be the first k prime numbers, where k is the number of factors in $\mathbb{N} \times \cdots \times \mathbb{N}$. Define a mapping $f: \mathbb{N} \times \cdots \times \mathbb{N} \to \mathbb{N}$ by $f(n_1, n_2, \ldots, n_k) = 2^{n_1} 3^{n_2} \cdots p_k^{n_k}$. By the fundamental theorem of arithmetic, f is injective, and hence $\mathbb{N} \times \cdots \times \mathbb{N}$ is countable. It is obvious that $n \mapsto (n, 1, \ldots, 1)$ is an injection $\mathbb{N} \to \mathbb{N} \times \cdots \times \mathbb{N}$ so that $\mathbb{N} \times \cdots \times \mathbb{N}$ is infinite. $\diamond$

It is immediate from the preceding lemma that a finite product of countable sets X_i is countable, for if $f_i: \mathbb{N} \to X_i$ are surjections for $1 \leq i \leq k$, then so is the mapping

$(n_1, \ldots, n_k) \mapsto (f_1(n_1), \ldots, f_k(n_k))$ of $\mathbb{N} \times \cdots \times \mathbb{N}$ into $X_1 \times \cdots \times X_k$. Notice that if each X_i is nonempty and some X_j is countably infinite, then $\prod_1^k X_i$ is also countably infinite, since $x_j \mapsto \left(x_1^0, \ldots, x_{j-1}^0, x_j, x_{j+1}^0, \ldots, x_k^0\right)$, where $x_i^0 \in X_i$ are fixed elements for $i \neq j$, is an injection $X_j \to \prod_1^k X_i$. We also see that if each X_i is finite, then so is $\prod_1^k X_i$. In fact, if any $X_j = \varnothing$, then $\prod_1^k X_i = \varnothing$, and, if X_i has $m_i > 0$ elements, then $\prod_1^k X_i$ contains $m_1 \cdots m_k$ elements. By induction on k, it suffices to establish this proposition in the case $k = 2$. Let $\mathbb{N}_{m_i} = \{1, 2, \ldots, m_i\}$ and $f_i \colon \mathbb{N}_{m_i} \to X_i$ be bijections, $i = 1, 2$. Then $f_1 \times f_2 \colon \mathbb{N}_{m_1} \times \mathbb{N}_{m_2} \to X_1 \times X_2$, $(a, b) \mapsto (f_1(a), f_2(b))$ is a bijection. Define a mapping $g \colon \mathbb{N}_{m_1} \times \mathbb{N}_{m_2} \to \mathbb{N}_{m_1 m_2}$ by $g(a, b) = (a - 1)m_2 + b$. It is easily verified that g is a bijection, so $X_1 \times X_2$ has $m_1 m_2$ elements.

However, an infinite product of even finite sets is not countable. For example, let $X = \{0, 1\}$, and consider the countable product $X^{\mathbb{N}} = \prod_1^{\infty} X_n$, where each $X_n = X$. Note that an element in $X^{\mathbb{N}}$ is a sequence $s \colon \mathbb{N} \to X$. Let $f \colon \mathbb{N} \to X^{\mathbb{N}}$ be any function. Write $t_n = 1 - f(n)_n$. Then $t = \langle t_n \rangle \in X^{\mathbb{N}}$ and $t \neq f(n)$ for all $n \in \mathbb{N}$. Thus f is not surjective. Since f is an arbitrary function $\mathbb{N} \to X^{\mathbb{N}}$, $X^{\mathbb{N}}$ cannot be countable, by Theorem A.5.9.

This method of proof was first used by G. Cantor, and is known as Cantor's diagonal process. We use this technique to prove the following

Example A.5.1 The set $\mathbb{R}$ of all real numbers is uncountable. Since an uncountable set cannot be a subset of a countable set, it is enough to show that the unit interval $I \subset \mathbb{R}$ is uncountable. By Theorem A.5.9, it suffices to show that there is no surjective mapping $f \colon \mathbb{N} \to I$. Let $f \colon \mathbb{N} \to I$ be any function. We use the decimal representation of real numbers to write $f(n) = 0.a_{n1}a_{n2}\cdots$. Here each a_{ni} is a digit between 0 and 9. This representation of a number is not necessarily unique, but if a number has two different decimal representations, then one of these representations repeats 9s from some place onward and the other repeats 0s from some place onward. We define a new real number r whose decimal representation is $0.b_1 b_2 \cdots$, where $b_n = 3$ if $a_{nn} \neq 3$, and $b_n = 5$ otherwise. It is clear that r has a unique decimal representation and differs from $f(n)$ in the nth decimal place for every $n \in \mathbb{N}$. So $r \neq f(n)$ for all $n \in \mathbb{N}$. Obviously, r belongs to I, and thus f is not surjective.

Proposition A.5.11 *The union of a countable family of countable sets is countable.*

Proof Let A be a countable set, and suppose that for every $\alpha \in A$, E_α is a countable set. Put $X = \bigcup E_\alpha$. We show that X is countable. If $X = \varnothing$, there is nothing to prove. So we assume that $X \neq \varnothing$. Then $A \neq \varnothing$, and we may further assume that $E_\alpha \neq \varnothing$ for every $\alpha \in A$, since the empty set contributes nothing to the union of the E_α. Since A is countable, there is a surjection $\mathbb{N} \to A$, $n \mapsto \alpha_n$. Since E_{α_n} is a nonempty countable set, there is a surjection $f_n \colon \mathbb{N} \to E_{\alpha_n}$. Write $f_n(m) = x_{nm}$ for every $m \in \mathbb{N}$. Then we have a mapping $\phi \colon \mathbb{N} \times \mathbb{N} \to X$ defined by $\phi(n, m) = x_{nm}$. Clearly, ϕ is surjective, for $X = \bigcup_n E_{\alpha_n}$. By Lemma A.5.10, $\mathbb{N} \times \mathbb{N}$ is countable, and hence X is countable. $\Diamond$

If the indexing set A and the sets E_α in the preceding proposition are all finite, then it is easily verified that the union $\bigcup_\alpha E_\alpha$ is finite.

Example A.5.2 The set $\mathbb{Q}$ of all rational numbers is countably infinite. Let $\mathbb{Q}_+$ denote the set of all positive rationals, and $\mathbb{Q}_-$ denote the set of all negative rationals. Then $\mathbb{Q} = \mathbb{Q}_+ \cup \mathbb{Q}_- \cup \{0\}$. Obviously, $\mathbb{Q}_+$ and $\mathbb{Q}_-$ are equipotent, and therefore it suffices to prove that $\mathbb{Q}_+$ is countably infinite. As $\mathbb{N} \subset \mathbb{Q}_+$, $\mathbb{Q}_+$ is infinite. There is a surjective mapping $g : \mathbb{N} \times \mathbb{N} \to \mathbb{Q}_+$ defined by $g(m, n) = m/n$. By Lemma A.5.10, there is a bijection $f : \mathbb{N} \to \mathbb{N} \times \mathbb{N}$, so the composition $gf : \mathbb{N} \to \mathbb{Q}_+$ is surjective. By Theorem A.5.9, $\mathbb{Q}_+$ is countable.

Since $\mathbb{R}$ is uncountable, we see that the set $\mathbb{R} - \mathbb{Q}$ of all irrational numbers is uncountable.

Theorem A.5.12 *The family of all finite subsets of a countable set is countable.*

Proof Let X be a countable set, and $\mathscr{F}(X)$ be the family of all finite subsets of X. If X is a finite set having n elements, then every subset of X is finite, and $\mathscr{F}(X)$ has 2^n members. If X is countably infinite, then there is a bijection between $\mathscr{F}(X)$ and the family $\mathscr{F}(\mathbb{N})$ of all finite subsets of $\mathbb{N}$. So it suffices to prove that $\mathscr{F}(\mathbb{N})$ is countably infinite. It is obvious that $n \mapsto \{n\}$ is an injection $\mathbb{N} \to \mathscr{F}(\mathbb{N})$, so $\mathscr{F}(\mathbb{N})$ is infinite. To see that it is countable, consider the sequence $\langle 2, 3, \ldots, p_k, \ldots \rangle$ of prime numbers. If $F = \{n_1, n_2, \ldots, n_k\} \subset \mathbb{N}$, then the n_i can be indexed so that $n_1 < n_2 < \cdots < n_k$. Put $\nu(F) = 2^{n_1} 3^{n_2} \cdots p_k^{n_k}$. Then $F \mapsto \nu(F)$ defines an injection $\mathscr{F}(\mathbb{N}) - \{\varnothing\} \to \mathbb{N}$, by the fundamental theorem of arithmetic. By Theorem A.5.9, $\mathscr{F}(\mathbb{N}) - \{\varnothing\}$ is countable, and so $\mathscr{F}(\mathbb{N})$ is countable. $\diamond$

It must be noted that the family of all subsets of a countably infinite set is not countable. This follows from the following.

Theorem A.5.13 (Cantor (1883)) *For any set X, there is no surjection $X \to \mathscr{P}(X)$.*

Proof Assume that there is a surjection $f : X \to \mathscr{P}(X)$ and consider the set $S = \{x \in X \mid x \notin f(x)\}$. Then $S \subseteq X$, so there exists an $x \in X$ such that $S = f(x)$. Now, if $x \in S$, then $x \notin f(x)$, and if $x \notin S$, then $x \in f(x)$. Thus, in either case, we obtain a contradiction, and hence the theorem. $\diamond$

We end this section with the following theorem, which will be proved later in Sect. A.8.

Theorem A.5.14 (Bernstein–Schröeder) *Let X and Y be sets. If there exists injections $X \to Y$ and $Y \to X$, then there exists a bijection $X \to Y$.*

A.6 Orderings

Definition A.6.1 Let X be a set. An *order* (or a *simple order*) on X is a binary relation, denoted by $\prec$, such that

(a) if $x, y \in X$, then one and only one of the statements $x = y$, $x \prec y$, $y \prec x$ is true, and

(b) $x \prec y$ and $y \prec z \Rightarrow x \prec z$.

X together with a definite order relation defined in it is called an *ordered set.*

The statement "$x \prec y$" is read as "x precedes y" or "y follows x." We also say that "x is less than y" or "y is greater than x." It is sometimes convenient to write $y \succ x$ to mean $x \prec y$. The notation $x \preceq y$ is used to indicate $x \prec y$ or $x = y$, that is, the negation of $y \prec x$. It should be noticed that an order relation is irreflexive, that is, for no x $x \prec x$. However, if $\prec$ is an order on X, then the relation $\preceq$ satisfies the following conditions:

(a) (Reflexivity) $x \preceq x \ \forall \ x \in X$;

(b) (Antisymmetry) $x \preceq y$ and $y \preceq x \Rightarrow x = y$;

(c) (Transitivity) $x \preceq y$ and $y \preceq z \Rightarrow x \preceq z$; and

(d) (Comparability) $x \preceq y$ or $y \preceq x \ \forall \ x, y \in X$.

A relation having the properties (a)–(c) is called a *partial ordering*. A set together with a definite partial ordering is called a *partially ordered set*. Two elements x, y of a partially ordered set $(X, \preceq)$ are called *comparable* if $x \preceq y$ or $y \preceq x$, and a subset $Y \subseteq X$ in which any two elements are comparable is called a *chain* in X. A partially ordered set that is also a chain is called a *linearly* or *totally ordered set*. Thus we have associated with each simple order a total (or linear) order relation.

Conversely, there is a simple order relation associated with each total ordering. In fact, to any partial order relation $\preceq$ on X, there is associated a unique relation $\prec$ given by

$$x \prec y \Leftrightarrow x \preceq y \text{ and } x \neq y.$$

It is transitive and has the property that (a) for no $x \in X$, the relation $x \prec x$ holds, and (b) if $x \prec y$, then $y \not\prec x$ for every $x, y \in X$. A transitive relation with this property is called a *strict partial ordering*. If $\preceq$ totally orders the set X, then $\prec$ is clearly a simple order (or a strict total ordering) on X. It follows that the notions of a total (resp. partial) order relation and a strict total (resp. strict partial) order relation are interchangeable. If $\preceq$ is a partial order, we use the notation $\prec$ to denote the associated strict partial order, and conversely. The relation $\leq$ is a total ordering on the set $\mathbb{R}$ of real numbers, while the inclusion relation $\subseteq$ on $\mathscr{P}(X)$ is not if X has more than one element. Note that the set inclusion is always a partial ordering on any family of subsets of a set X.

It is obvious that the induced ordering on a subset of a partially ordered set is a partial ordering on that subset. If $(X_1, \preceq_1)$ and $(X_2, \preceq_2)$ are partially (totally) ordered sets, then the *dictionary* (or *lexicographic*) *order* relation on $X_1 \times X_2$ defined by

$$(x_1, x_2) \preceq (y_1, y_2) \text{ if } x_1 \prec_1 y_1 \text{ or if } x_1 = y_1 \text{ and } x_2 \preceq_2 y_2$$

is a partial (total) ordering.

Let $(X, \preceq)$ be a partially ordered set. If $Y \subseteq X$, then an element $b \in X$ is an *upper bound* of Y if for each $y \in Y$, $y \preceq b$. If there exists an upper bound of Y, then we say that Y is bounded above. A *least upper bound* or *supremum* of Y is an element $b_0 \in X$ such that b_0 is an upper bound of Y and if $c \prec b_0$, then c is not an upper bound of Y. Observe that there is at most one such b_0, by antisymmetry, and it may or may not exist. We write $b_0 = \sup Y$ when it exists. Notice that $\sup Y$ may or may not belong to Y. If $\sup Y \in Y$, it is referred to as the *last* (or *largest* or *greatest*) element of Y. Analogously, an element $a \in X$ is a *lower bound* of Y if $a \preceq y$ for all $y \in Y$. If there exists a lower bound of Y, we say that Y is bounded below. A *greatest lower bound* or *infimum* of Y is an element $a_0 \in X$ which is a lower bound of Y and if $a_0 \prec c$, then c is not a lower bound of Y. It is clear that there is at most one such a_0, and it may or may not exist. When it exists, we write $a_0 = \inf Y$. If $\inf Y \in Y$, we call it the *first* (or *smallest* or *least*) element of Y. It is obvious that $\sup Y$ is the first element of the set $\{x \in X \mid y \preceq x \text{ for all } y \in Y\}$, and $\inf Y$ is the last element of the set $\{x \in X \mid x \preceq y \text{ for all } y \in Y\}$.

A partially ordered set X is said to have the least upper bound property if each nonempty subset $Y \subseteq X$ with an upper bound has a least upper bound. Analogously, X is said to have the greatest lower bound property if each nonempty subset $Y \subseteq X$ which is bounded below has a greatest lower bound. If X is a set, then $\mathscr{P}(X)$ ordered by inclusion has the least upper bound property.

Proposition A.6.2 *If an ordered set X has the least upper bound property, then it also has the greatest lower bound property.*

An ordered set X which has the least upper bound (equivalently, the greatest lower bound) property is called *order complete*.

Let $(X, \preceq)$ be a partially ordered set. An element $a \in X$ is called a *minimal element* of X if for every $x \in X$, $x \preceq a \Rightarrow x = a$, that is, no $x \in X$ which is distinct from a precedes a. Similarly, an element $b \in X$ is called a *maximal* element of X if for every $x \in X$, $b \preceq x \Rightarrow x = b$, that is, no $x \in X$ which is distinct from b follows b. It should be noted that if X has a last (first) element, then that element is the unique maximal (minimal) element of X. If $\preceq$ is a total order, a maximal element is the last element, but there are partially ordered sets with unique maximal elements which are not last elements.

If $x, y \in X$ and $x \prec y$, we say that x is a *predecessor* of y (or y is a *successor* of x). If $x \prec y$ and there is no $z \in X$ such that $x \prec z \prec y$, then we say that x is an *immediate predecessor* of y (or y an *immediate successor* of x).

Ordinals

A partially ordered set X is called *well ordered* (or an *ordinal*) if each nonempty subset $A \subseteq X$ has the first element.

The set $\mathbb{N}$ of the positive integers is well ordered by its natural ordering $\leq$. On the other hand, the set $\mathbb{R}$ of real numbers is not well ordered by the usual ordering $\leq$. And, if X has more than one element, then the partial ordering $\subseteq$ in $\mathscr{P}(X)$ is not a well ordering.

Clearly, any subset of a well-ordered set is well ordered in the induced ordering. The empty set $\varnothing$ is considered to be a well-ordered set. If $(X_1, \preceq_1)$ and $(X_2, \preceq_2)$ are well-ordered sets, then the dictionary order relation on $X_1 \times X_2$ is a well ordering.

It was E. Zermelo who first formulated the axiom of choice (though it was being used by mathematicians without explicit formulation) and established that every set can be well ordered (1904). In fact, we have the following.

Theorem A.6.3 *The following statements are equivalent:*

(a) **The Axiom of Choice.**

(b) **Well-Ordering Principle**: *Every set can be well ordered.*

(c) **Zorn's Lemma**: *Let X be a partially ordered set. If each chain in X has an upper bound, then X has a maximal element.*

(d) **Hausdorff Maximal Principle**: *If X is a partially ordered set, then each chain in X is contained in a maximal chain, that is, for each chain C in X, there exists a chain M in X such that $C \subseteq M$ and M is not properly contained in any other chain which contains C.*

We do not wish to prove this theorem and refer the interested reader to the texts by Dugundji [3] and Kelley [6]. We remark that there are several other statements equivalent to the axiom of choice individually.

We note that a well-ordered set $(W, \preceq)$ is totally ordered. For, if $x, y \in W$, then the subset $\{x, y\} \subseteq W$ has a first element; accordingly, either $x \preceq y$ or $y \preceq x$. Moreover, W is order complete. In fact, given a nonempty set $X \subseteq W$ with an upper bound, the first element of the set of all upper bounds of X in W is sup X.

Each element x of W that is not the last element of W has an immediate successor. This is first element of the set $\{y \in W \mid x \prec y\}$, usually denoted by $x + 1$. However, x need not have an immediate predecessor. For example, consider the set $\mathbb{N}$ of positive integers with the usual ordering $\leq$. Choose an element $\infty \notin \mathbb{N}$, and put $W = \mathbb{N} \cup \{\infty\}$. Define a relation $\preceq$ on W by $x \preceq y$ if $x, y \in \mathbb{N}$ and $x \leq y$, $x \preceq \infty$ for all $x \in \mathbb{N}$, and $\infty \preceq \infty$. Then $(W, \preceq)$ is a well-ordered set, and ∞ has no immediate predecessor in W. It is obvious that ∞ is the last element of W. We say that $\mathbb{N} \cup \{\infty\}$ has been obtained from $\mathbb{N}$ by adjoining ∞ as the last element.

For each $x \in W$, the set $S(x) = \{w \in W \mid w \prec x\}$ is called the *section* (or *initial segment*) determined by x. The first element of W is usually denoted by 0; it is obvious that $S(0) = \varnothing$. Each section $S(x)$ in W obviously satisfies the condition: $y \in S(x)$ and $z \preceq y \Rightarrow z \in S(x)$. Conversely, if S is a *proper subset* of W satisfying this condition, then S is a section in W. For, if x is the least element of $W - S$, then $y \prec x \Leftrightarrow y \in S$. So $S = S(x)$. It is clear that a union of sections in W is either a section in W or equals W. Also, it is immediate that an intersection of sections in W is a section in W.

A.6.4 (Principle of Transfinite Induction) Let $(W, \preceq)$ be a well-ordered set. If X is a subset of W such that $S(y) \subseteq X$ implies $y \in X$ for every $y \in W$, then $X = W$.

Proof The first element 0 of W is in X, since $\varnothing = S(0) \subset X$. If $W - X \neq \varnothing$, then it has a first element y_0, say. So $S(y_0) \subseteq X$, which forces $y_0 \in X$, by our hypothesis. This contradiction establishes the theorem. $\diamond$

The preceding theorem is generally used in the following form: For each $x \in W$, let $P(x)$ be a proposition. Suppose that (a) $P(0)$ is true, and (b) for each $x \in W$, $P(x)$ is also true whenever $P(y)$ is true for all $y \in S(x)$. Then $P(x)$ is true for every $x \in W$.

Since each element of $\mathbb{N}$ other than 1 has an immediate predecessor, the induction principle on $\mathbb{N}$ is equivalent to the following statement.

Let $P(n)$ be a proposition defined for each $n \in \mathbb{N}$. If $P(1)$ is true, and for each $n > 1$, the hypothesis "$P(n-1)$ is true" implies that $P(n)$ is true, then $P(n)$ is true for every $n \in \mathbb{N}$.

Let $(W, \preceq)$ and $\left(W', \preceq'\right)$ be well-ordered sets. A mapping $f: W \to W'$ is said to be *order-preserving* if $f(x) \preceq' f(y)$ whenever $x \preceq y$ in W. We call W and W' *isomorphic* (or *of the same order type*) if there is a bijective order-preserving map $f: W \to W'$; such a map f is called an *isomorphism*. An order-preserving injection $f: W \to W'$ is called a *monomorphism*. It should be noted that an *isomorphism* of a partially ordered set X onto a partially ordered set X' is a bijection $f: X \to X'$ such that $x \preceq y \Leftrightarrow f(x) \preceq' f(y)$. However, if X and X' are totally ordered, one needs only the implication $x \preceq y \Rightarrow f(x) \preceq' f(y)$. For, this also implies the reverse implication $f(x) \preceq' f(y) \Rightarrow x \preceq y$: If $x \npreceq y$, then $y \prec x \Rightarrow f(y) \prec' f(x)$, contradicting the antisymmetry of $\preceq'$.

The ordinal $\mathbb{N}$ is isomorphic to the well-ordered set $\omega = \{0, 1, 2, \ldots\}$ of nonnegative integers in its natural order. We note that a well-ordered set may be isomorphic to one of its proper subsets. For example, the well-ordered set $\mathbb{N}$ is isomorphic to the set E of even integers; $n \mapsto 2n$ is a desired isomorphism. However, *no well-ordered set W can be isomorphic to a section in it*; this is immediate from the following fact.

Proposition A.6.5 *If f is a monomorphism of a well-ordered set W into itself, then $f(w) \succeq w$ for every $w \in W$.*

Proof Let $X = \{x \in W \mid f(x) \prec x\}$. If $X \neq \varnothing$, then it has a first element, say x_0. As $x_0 \in X$, we have $f(x_0) \prec x_0$ whence $f(f(x_0)) \prec f(x_0)$. This implies that $f(x_0) \in X$, which contradicts the definition of x_0. Therefore $X = \varnothing$.   ◇

Corollary A.6.6 *Two sections in a well-ordered set W are isomorphic if and only if they are identical.*

Proof Suppose that $S(x)$ and $S(x')$ are isomorphic sections in W and let $f: S(x) \to S(x')$ be an isomorphism. If $S(x) \neq S(x')$, then we have either $x \prec x'$ or $x' \prec x$. By interchanging $S(x)$ and $S(x')$, if necessary, we may assume that $x' \prec x$. Then $x' \in S(x)$ which implies that $f(x') \prec x'$. This contradicts Proposition A.6.5.   ◇

Corollary A.6.7 *The only one-to-one order-preserving mapping of a well-ordered set onto itself is the identity mapping.*

Proof Suppose that W is a well-ordered set and $f: W \to W$ is an isomorphism. If the set $S = \{x \in W \mid f(x) \neq x\}$ is nonempty, then S has a least element a, say. By Proposition A.6.5, we have $a \prec f(a)$. Since f is surjective, $a = f(x)$ for some $x \in W$. So $f(x) \prec f(a) \Rightarrow x \prec a$. Then, by the definition of a, $f(x) = x$ and we obtain $a = x \prec a$, a contradiction. Therefore $S = \varnothing$, and f is the identity mapping.

Theorem A.6.8 *If X and Y are well-ordered sets, then exactly one of the following statements holds.*

(a) *X is isomorphic to Y.*
(b) *X is isomorphic to a section in Y.*
(c) *Y is isomorphic to a section in X.*

Proof We first show that the three possibilities are mutually exclusive. If (a) and (b) occur together, then Y is isomorphic to a section in it, a contradiction. For the same reason, (a) and (c) cannot occur together. If $f : X \to S(y)$ and $g : Y \to S(x)$ are isomorphisms, then $gf : X \to S(x)$ is a monomorphism with $gf(x) \prec x$. This contradicts Proposition A.6.5, so (b) and (c) also cannot occur together.

Now, we prove that one of the above three possibilities does occur. Suppose that neither of (a) and (b) holds, and let Σ be the family of all sections $S(x_\alpha)$ in X such that there exists an isomorphism f_α of $S(x_\alpha)$ onto a section in Y or onto Y itself. For every pair of indices α and β, $S(x_\alpha) \cap S(x_\beta)$ is a section in X, say $S(x_\gamma)$, and $f_\alpha(S(x_\gamma))$ and $f_\beta(S(x_\gamma))$ are clearly isomorphic. If $\alpha \neq \beta$, then one of these is certainly a section in Y, and hence the other is also a section in Y. Therefore, by Corollary A.6.6, we have $f_\alpha(S(x_\gamma)) = f_\beta(S(x_\gamma))$, and then Corollary A.6.7 shows that $f_\alpha(x) = f_\beta(x)$ for all $x \in S(x_\gamma)$. Hence, there is a function

$$f : \bigcup_\alpha S(x_\alpha) \to \bigcup_\alpha f_\alpha(S(x_\alpha))$$

defined by $f(x) = f_\alpha(x)$ if $x \in S(x_\alpha)$. It is easily checked that f is an isomorphism. Moreover, $\bigcup_\alpha S(x_\alpha)$ is either X or a section in it, and $\bigcup_\alpha f_\alpha(S(x_\alpha))$ is either Y or a section in Y. If $\bigcup_\alpha S(x_\alpha) = X$, then we have alternative (a) or (b), contrary to our assumption. So $\bigcup_\alpha S(x_\alpha) = S(x_0)$ for some $x_0 \in X$, and $S(x_0) \in \Sigma$. We assert that $f : S(x_0) \to Y$ is surjective. Assume otherwise. Then we have $f(S(x_0)) = S(y_0)$ for some $y_0 \in Y$. Obviously, we can extend f to an isomorphism $S(x_0) \cup \{x_0\} \to S(y_0) \cup \{y_0\}$ by defining $f(x_0) = y_0$. Note that $S(y_0) \cup \{y_0\}$ is either a section in Y or coincides with Y (if y_0 is the last element of Y). If x_0 were the last element of X, then we would have alternative (a) or (b). Therefore $S(x_0) \cup \{x_0\}$ is a member of the family Σ. Then, from the definition of $S(x_0)$, it follows that $x_0 \in S(x_0)$, a contradiction. Hence, our assertion and the alternative (c) holds.   ◇

Corollary A.6.9 *Any subset of a well-ordered set W is isomorphic to either a section in W or W itself.*

Proof Let X be a subset of W. Then X is a well-ordered set under the induced order. If there is an isomorphism f of W onto a section $S(x_0)$ in X, then there exists a monomorphism $g : X \to X$ defined by f, which satisfies $g(x_0) \prec x_0$. This contradicts Proposition A.6.5. Therefore, one of the two possibilities (a) and (b) in Theorem A.6.8 must hold.   ◇

A.7 Ordinal Numbers

It is evident that the relation of isomorphism between ordinals is an equivalence relation. Ordinal numbers are objects uniquely associated with the isomorphic classes of well-ordered sets. A natural way to define them is to consider these isomorphic classes themselves as the ordinal numbers. The main drawback with this definition is that isomorphic classes of well-ordered sets are unfortunately not *sets*; accordingly, logical contradictions arise when such large classes are collected into sets. To avoid such difficulties, ordinal numbers are defined to be well-ordered sets such that each isomorphic class of well-ordered sets contains exactly one ordinal number.

Definition A.7.1 An *ordinal number* is a well-ordered set $(\mathcal{W}, \preceq)$ such that

(a) if $X \in \mathcal{W}$ and $x \in X$, then $x \in \mathcal{W}$,
(b) for every $X, Y \in \mathcal{W}$, one of the possibilities: $X = Y$, $X \in Y$ or $Y \in X$ holds, and
(c) $X \preceq Y \Leftrightarrow X \in Y$ or $X = Y$.

The fact that ordering $\preceq$ in the above definition is actually a well ordering follows from one of the axioms in set theory:

Every nonempty set X contains an element y such that $x \notin y$ for all $x \in X$. (∗)

As a consequence of this axiom, we see that no nonempty set can be a member of itself, and both the relations $X \in Y$ and $Y \in X$ cannot hold simultaneously for any two sets X, Y. So only one of the three alternatives described in the condition (b) can occur. It is now obvious that $\preceq$ is reflexive and antisymmetric. To see the transitivity of this relation, suppose that $X \preceq Y$ and $Y \preceq Z$ in $\mathcal{W}$. Then we must have one of the relations $X = Z$, $X \in Z$ or $Z \in X$. If $Z \in X$, then we cannot have $X = Y$ or $Y = Z$, for $Y \notin X$ and $Z \notin Y$. So $X \in Y$ and $Y \in Z$. But, this leads to the violation of Axiom (∗) by the set $\{X, Y, Z\}$. Hence $X \preceq Z$. Using the above axiom, we also see that every nonempty subset of $\mathcal{W}$ has a least member.

The empty set $\varnothing$ is obviously an ordinal number. To construct a few more ordinal numbers, observe that if $\mathcal{W}$ is an ordinal number, then so is $\mathcal{W} \cup \{\mathcal{W}\}$. Thus $\{\varnothing\}$, $\{\varnothing, \{\varnothing\}\}$, $\{\varnothing, \{\varnothing\}, \{\varnothing, \{\varnothing\}\}\}$, etc. are first few ordinal numbers. Henceforth, we will generally denote ordinal numbers by small Greek letters.

Proposition A.7.2 *Every element of an ordinal number is an ordinal number.*

Proof Let β be an ordinal number and $\alpha \in \beta$. We show that α is also an ordinal number. Suppose that $x \in y$ and $y \in \alpha$. Then we have either $x = \alpha$ or $x \in \alpha$ or $\alpha \in x$. If $x = \alpha$, then we would have $x \in y$ and $y \in x$, a contradiction. Now, if $\alpha \in x$, then the subset $\{x, y, \alpha\} \subset \beta$ fails to have a least element. So $x \in \alpha$, and the condition Definition A.7.1(a) holds. Next, suppose that $x, y \in \alpha$. Then both x, y are members of β. So one of the possibilities $x = y$, $x \in y$ or $y \in x$ must hold. Thus the condition (b) in Definition A.7.1 is satisfied by α.   ◇

Lemma A.7.3 *If α and β are ordinal numbers, then $\alpha \in \beta$ if and only if $\alpha \subset \beta$.*

Proof If $\alpha \in \beta$, then $x \in \alpha \Rightarrow x \in \beta$, by Definition A.7.1. Moreover, $\alpha = \beta \Rightarrow \alpha \in \alpha$, a contradiction. So we have $\alpha \subset \beta$. Conversely, assume that $\alpha \subset \beta$. Then $\beta - \alpha$ is nonempty and has a least element λ, say. We observe that $\alpha = \lambda$. Suppose that $x \in \lambda$. Then $x \in \beta$, for $\lambda \in \beta$. If $x \notin \alpha$, then we have $\lambda \preceq x$. This implies that either $\lambda = x$ or $\lambda \in x$, and contradicts the fact that λ is an ordinal number, by A.7.2. On the other hand, if $x \in \alpha$, then $x \neq \lambda$ and $\lambda \notin x$, for $\lambda \notin \alpha$. Since both x and λ are members of β, we have $x \in \lambda$. Thus $\alpha = \lambda \in \beta$. $\diamond$

Proposition A.7.4 *If α and β are any two ordinal numbers, then $\alpha \subseteq \beta$ or $\beta \subseteq \alpha$.*

Proof Suppose that α and β are two ordinal numbers. Then $\gamma = \alpha \cap \beta = \{x \mid x \in \alpha$ and $x \in \beta\}$ is an ordinal number. If $\gamma \neq \alpha, \beta$, then $\gamma \subset \alpha$ and $\gamma \subset \beta$. By the preceding lemma, we have $\gamma \in \alpha$ and $\gamma \in \beta$. Thus $\gamma \in \alpha \cap \beta = \gamma$, a contradiction. So $\gamma = \alpha$ or $\gamma = \beta$; accordingly, we have $\alpha \subseteq \beta$ or $\beta \subseteq \alpha$. $\diamond$

Theorem A.7.5 (a) *Every nonempty set of ordinal numbers has a least element with respect to inclusion.*

  (b) *The union of a set of ordinal numbers is an ordinal number.*

Proof (a): Let Σ be a nonempty set of ordinal numbers. Choose $\alpha \in \Sigma$. If $T = \alpha \cap \Sigma$ is empty, then $\sigma \in \Sigma$ implies that $\sigma \not\subset \alpha$, by Lemma A.7.3. Hence $\alpha \subseteq \sigma$, and α is the least element of Σ. If $T \neq \varnothing$, then T contains an element β such that $\beta \in \tau$ or $\beta = \tau$ for every $\tau \in T$, since α is well ordered. Thus $\beta \subseteq \tau$ for every $\tau \in T$. And, if $\sigma \in \Sigma - T$, then we clearly have $\beta \subset \alpha \subseteq \sigma$. Thus, in this case, β is the least element of Σ.

  (b): Let Σ be a set of ordinal numbers. Then $X = \bigcup_{\alpha \in \Sigma} \alpha$ is a set. It is obvious that the condition (a) of Definition A.7.1 holds in X. Furthermore, if $x, y \in X$, then there exist ordinal numbers $\alpha, \beta \in \Sigma$ with $x \in \alpha$ and $y \in \beta$. By Proposition A.7.4, we have $\alpha \subseteq \beta$ or $\beta \subseteq \alpha$. To be specific, suppose that $\alpha \subseteq \beta$. Then $x, y \in \beta$ and, therefore, they satisfy the condition (b) of Definition A.7.1. Lastly, we show that X is well ordered by the relation $x \preceq y \Leftrightarrow x \in y$ or $x = y$. It is obvious that $\preceq$ is a linear ordering on X and, by Proposition A.7.2, every element of X is an ordinal number. Then, for each nonempty subset Y of X, the part (a) of the theorem shows that there is an element $z \in Y$ such that $z \subseteq y$ for all $y \in Y$. By Lemma A.7.3, we deduce that z is the least element of Y. Thus (b) holds. $\diamond$

  Given two ordinal numbers α, β, we write $\alpha \leq \beta$ if and only if $\alpha \subseteq \beta$. Then, by Proposition A.7.4, we have either $\alpha \leq \beta$ or $\beta \leq \alpha$. Moreover, it is immediate from Proposition A.7.5 that any nonempty set of ordinal numbers is well ordered by the relation $\leq$. We also see, by Proposition A.7.2 and Lemma A.7.3, that an ordinal number α consists of all those ordinal numbers which precede it. It follows that *any two distinct ordinal numbers cannot be isomorphic*. For, if $\alpha < \beta$, then α is the section $\{\gamma \in \beta \mid \gamma < \alpha\}$ in β. Therefore, by Proposition A.6.5, α cannot be isomorphic to β.

  Notice that the class $\mathcal{O}$ of all ordinal numbers is not a set. For, if $\mathcal{O}$ were a set, then $X = \bigcup_{\alpha \in \mathcal{O}} \alpha$ would be an ordinal number, by the preceding theorem. Consequently, $T = X \cup \{X\}$ is an ordinal number, and we obtain the contradiction $X < T \leq X$.

Theorem A.7.6 *Every well-ordered set W is isomorphic to a unique ordinal number.*

Proof Clearly, if W is isomorphic to two ordinal numbers α and β, then they themselves are isomorphic, and we must have $\alpha = \beta$.

To find the desired ordinal number, let X be the collection of all elements $x \in W$ for which there are ordinal numbers α_x (depending upon x) such that the section $S(x)$ in W is isomorphic to α_x. Put $\Psi(x) = \alpha_x$. If $x, y \in X$ and $x = y$, then the ordinal numbers α_x and α_y are isomorphic. Hence $\alpha_x = \alpha_y$, and it follows that Ψ is single-valued on X. By an axiom in set theory, we see that X is a set, and so is $\text{im}(\Psi)$. Observe that Ψ is injective, too. For, if $\Psi(x) = \Psi(y)$, then $S(x)$ and $S(y)$ are isomorphic. By Corollary A.6.6, we deduce that $S(x) = S(y)$, which forces $x = y$. Thus $\Psi : X \to \text{im}(\Psi)$ is a bijection. We show that it is, in fact, an isomorphism. Denote the ordering in W by $\preceq$, and suppose that $x, y \in X$ and $y \prec x$. Let $f : S(x) \to \Psi(x)$ be an isomorphism. Then $f(y) < \Psi(x)$, and f induces an isomorphism of $S(y)$ onto $\{\alpha \in \Psi(x) \mid \alpha < f(y)\} = f(y)$. By the definition of Ψ, $\Psi(y) = f(y) < \Psi(x)$. Conversely, suppose that $\beta = \Psi(y) < \Psi(x)$ and let $g : \Psi(x) \to S(x)$ be an isomorphism. Then $\beta \in \Psi(x)$ and $g|\beta$ is an isomorphism of β onto $S(g(\beta))$. So $\Psi(g(\beta)) = \Psi(y)$. Since Ψ is injective, we have $y = g(\beta) \prec x$. This proves our claim.

We next show that $\text{im}(\Psi)$ is an ordinal number. Since the class $\mathcal{O}$ of all ordinal numbers is not a set, we have $\mathcal{O} \neq \text{im}(\Psi)$. So there exists an ordinal number $\nu \notin \text{im}(\Psi)$. If ν contains elements which do not belong to $\text{im}(\Psi)$, then we have such a least element λ, by Theorem A.7.5(a); otherwise, we put $\lambda = \nu$. We observe that $\text{im}(\Psi) = \{\alpha \in \mathcal{O} \mid \alpha < \lambda\} = \lambda$. Clearly, $\alpha < \lambda \Rightarrow \alpha \in \text{im}(\Psi)$. On the other hand, if $\alpha \in \text{im}(\Psi)$, then there exists an $x \in W$ and an isomorphism $g : \Psi(x) = \alpha \to S(x)$. Consequently, for every $\beta < \alpha$, $y = g(\beta) \prec x$ and g induces an isomorphism between β and $S(y)$. So $\beta = \Psi(y) \in \text{im}(\Psi)$, and this implies that $\alpha < \lambda$. Thus have $\text{im}(\Psi) = \lambda$.

To finish the proof, we need to show that $X = W$. If $X \neq W$, then $W - X$ has a least element, say w_0. We observe that $X = S(w_0)$. Suppose $x \in X$, $w \in W$ and $w \prec x$. Then there exists an isomorphism $f : S(x) \to \Psi(x)$, which clearly induces an isomorphism between $S(w)$ and the ordinal number $f(w)$. Therefore, $w \in X$ and we conclude that X is the section $S(w_0)$ in W. Since $\Psi : X \to \lambda$ is an isomorphism, we have $w_0 \in X$, a contradiction. $\Diamond$

By the preceding theorem, it is clear that *each well-ordered set determines a unique ordinal number.*

If an ordinal number α is nonempty, then the least element a of α must be $\varnothing$, for $x \in a$ would imply that $x \in \alpha$ and $x < a$. It follows that $\varnothing$ is the least ordinal number, denoted by 0. We denote the ordinal number $\{\varnothing\}$ by 1, $\{\varnothing, \{\varnothing\}\}$ by 2, $\{\varnothing, \{\varnothing\}, \{\varnothing, \{\varnothing\}\}\}$ by 3, and so on. Notice that the ordinal number denoted by a positive integer n is determined by the well-ordered set $\{0, 1, \ldots, n - 1\}$. Moreover, we see that 1 is the immediate successor of 0, 2 is the immediate successor of 1, etc. In general, for each ordinal number α, $\alpha \cup \{\alpha\}$ is an ordinal number, which is the least ordinal number $> \alpha$. Thus $\alpha \cup \{\alpha\}$ is the immediate successor of α, usually denoted by $\alpha + 1$. However, there are ordinal numbers, such as $\omega = \bigcup_{n=1}^{\infty} n$, which do not have immediate predecessors, and they are called *limit* ordinal numbers. Observe that ω is isomorphic to the well-ordered set of all nonnegative integers with the usual ordering, and each ordinal number $\alpha < \omega$ contains a finite number of elements. We refer to an ordinal number containing a finite number of elements as a *finite* ordinal number. It is obvious that a finite ordinal number coincides with some ordinal number n, and thus ω is the smallest ordinal number greater than every finite ordinal number. Accordingly, ω is called the *first infinite ordinal number*. Using the construction of successors, we obtain the sequence of ordinal numbers $\omega + 1, \omega + 2, \ldots$. By Theorem A.7.5(b), $\bigcup_n (\omega + n)$ is an ordinal number, denoted by $2\omega = \omega + \omega$. Thus, there is a sequence of ordinal numbers $0, 1, \ldots, \omega, \omega + 1, \ldots, 2\omega, 2\omega + 1, \ldots, 2\omega + \omega = 3\omega, \ldots$. The ordinal number $\bigcup_n n\omega$ is denoted by ω^2, $\bigcup_n (\omega^2 + n\omega)$ is denoted by $2\omega^2$, and so on. Notice that these are all countable ordinal numbers.

We conclude this section by showing the existence of an uncountable ordinal number. To this end, it suffices to construct an uncountable well-ordered set, since every well-ordered set determines a unique ordinal number.

Proposition A.7.7 *There is an uncountable ordinal $(W, \preceq)$ with the last element ω_1 such that, for each $w \in W$ other than ω_1, the section $S(w) = \{x \in W \mid x \prec w\}$ is countable.*

Proof Let X be any uncountable set. By the well-ordering principle, there is a well-ordering $\preceq$ for X. If X does not have a last element, we choose an element $\infty \notin X$ and construct a well-ordered set $X \cup \{\infty\}$ by adjoining ∞ to X as the last element. This is done by extending the relation on X to $X \cup \{\infty\}$: Define $x \preceq \infty$ for all $x \in X$. So we can assume that the well-ordered set $(X, \preceq)$ has a last element. Let Y be the set of elements $y \in X$ such that the set $\{x \in X \mid x \preceq y\}$ is uncountable. Then Y is nonempty, and therefore has a least element. If ω_1 denotes the least element of Y, then $W = \{x \in X \mid x \preceq \omega_1\}$ is the desired set. $\diamond$

The ordinal W in the preceding proposition is unique in the sense that, if W' is any ordinal with the same properties, then there is an isomorphism of W onto W'. For, by Theorem A.6.8, we can assume that there is a monomorphism $f : W \to W'$. Then $f(W)$ is clearly uncountable, and $f(\omega_1)$ is its last element. If $f(\omega_1)$ is distinct from the last element ω_1' of W', then $f(W)$ would be countable, being a section $S(w')$ for some $w' < \omega_1'$. This contradiction shows that $f(\omega_1) = \omega_1'$, and so f is an isomorphism of W onto W'.

Also, observe that the section $S(\omega_1)$ in W is uncountable, and any countable ordinal is isomorphic to a section $S(x)$ in W for some $x \prec \omega_1$. Another notable fact about W is that if $A \subset W$ is countable and does not contain ω_1, then $\sup A \prec \omega_1$. To see this, note that the set $X_a = \{x \in W \mid x \preceq a\}$ is countable for each $a \in A$. Since A is countable, so is $Y = \bigcup_a X_a$. Clearly, $W \neq Y$, for W is uncountable. Let w_0 be the least element of $W - Y$. Then we have $y \in Y \Leftrightarrow y \prec w_0$. It follows that w_0 has only a countable number of predecessors, and therefore $w_0 \prec \omega_1$. Clearly, w_0 is an upper bound for A, so sup $A \prec \omega_1$.

The ordinal number determined by the ordinal W in the preceding proposition is called the *first* (or *least*) *uncountable ordinal number*, and is denoted by Ω. The section $\{\alpha \in \Omega \mid \alpha < \Omega\}$, usually denoted by $[0, \Omega)$, is the set of all countable ordinal numbers and has no largest element. For, if $\alpha \in [0, \Omega)$, then its immediate successor $\alpha + 1$ is countable. So $\alpha + 1 < \Omega$, since $[0, \Omega)$ is uncountable. Accordingly, Ω is also a limit ordinal; in fact, there are uncountably many limit ordinals in $[0, \Omega]$.

A.8 Cardinal Numbers

As seen in Sect. A.5, two finite sets X and Y are equipotent if and only if they have the same number of elements. We associate with each set X an object $|X|$ such that two sets X and Y are equipotent (that is, X and Y have the same cardinality) if and only if $|X| = |Y|$.

Isomorphic ordinals are obviously equipotent, and we observe that the converse is also true for *finite* ordinals. We first show that an ordinal X consisting of n elements, n any nonnegative integer, is isomorphic to a section $S(n)$ in ω. For, if $X = \varnothing$, then it is $S(0)$. So assume that $X \neq \varnothing$, and define a mapping $f : X \to \omega$ as follows. Map the first element x_1 of X into the integer 0. If $X - \{x_1\} \neq \varnothing$, denote the first element of $X - \{x_1\}$ by x_2 and put $f(x_2) = 1$. If $X \neq \{x_1, x_2\}$, denote the first element of $X - \{x_1, x_2\}$ by x_3 and put $f(x_3) = 2$. Since X consists of finitely many elements only, this process terminates after finitely many steps. Thus, we obtain a positive integer n such that $X = \{x_1, \ldots, x_n\}$, and there is a mapping $f : X \to \omega$ given by $f(x_i) = i - 1, i = 1, 2, \ldots, n$. Clearly, $x_1 < \cdots < x_n$ so that f is an isomorphism of X onto the section $S(n)$ in ω. It follows that any two finite ordinals both having the same number of elements are isomorphic. However, this may not be true for ordinals having infinitely many elements. For example, if $\omega \cup \{q\}$ is the ordinal obtained from the ordinal ω by adjoining q as the last element, then $\omega \cup \{q\}$ is not isomorphic to ω, by Proposition A.6.5. But $\phi : \omega \cup \{q\} \to \omega$ defined by $\phi(n) = n + 1, \phi(q) = 0$ is a bijection.

By the well-ordering principle, every set X can be well-ordered and, by Theorem A.7.6, it has the cardinality of an ordinal number. We define $|X|$ to be the least ordinal number such that X is equipotent to $|X|$. Clearly, the object $|X|$ is uniquely determined by X, and is called the *cardinal number* of X.

If X is finite and has n elements, then $|X|$ is the ordinal number n. The cardinal number $|\mathbb{N}|$ is denoted by $\aleph_0$, and the cardinal number $|\mathbb{R}|$ is denoted by $\aleph_1$ and also by c, called the *cardinality of the continuum*. The mapping $x \mapsto x/(1 - |x|)$ is a bijection between the open interval $(-1, 1)$ and $\mathbb{R}$, and the mapping $x \mapsto (2x - a - b)/(b - a)$ is a bijection between an open interval (a, b) and $(-1, 1)$. Therefore the cardinality of any open interval (a, b) is c. By Theorem A.5.14, we see that the cardinality of a closed interval or a half open interval is also c.

Proof of Theorem A.5.14: Let $f: X \rightarrow Y$ be an injection. By the definition of cardinal number $|Y|$, there is a bijection $\psi: Y \rightarrow |Y|$. Then the composition ψf is an injection of X into $|Y|$; consequently, there is a well ordering, say, $\preceq$ on X such that ψf is an isomorphism of $(X, \preceq)$ into $|Y|$. If $o(X, \preceq)$ denotes the ordinal number determined by the well-ordered set $(X, \preceq)$, then, by Corollary A.6.9, we conclude that $o(X, \preceq) \le |Y|$. On the other hand, $|X| \le o(X, \preceq)$, by the definition of $|X|$. Since the relation $\le$ between ordinal numbers is transitive, we have $|X| \le |Y|$. Similarly, we also have $|Y| \le |X|$ if there is an injection $Y \rightarrow X$. It follows that $|X| = |Y|$, and hence X and Y are equipotent.

Given two sets X and Y, we have $|X| < |Y|$ if and only if there exists an injection $X \rightarrow Y$ but there is no bijection between X and Y. Accordingly, it is logical to say that X *has fewer elements than* Y when $|X| < |Y|$. As seen in §A.5, there is no surjection from $\mathbb{N}$ to I while there is an injection $\mathbb{N} \rightarrow I, n \rightarrow 1/n$. So $\aleph_0 < c$.

Cardinal Arithmetic

The *sum* of two cardinal numbers α and β, denoted by $\alpha + \beta$, is the cardinality of the disjoint union of two sets A and B, where $|A| = \alpha$ and $|B| = \beta$.

Obviously, $n + \aleph_0 = \aleph_0$, for $\mathbb{N} = \{1, 2, \ldots, n\} \cup \{n + 1, n + 2, \ldots\}$. Since $\mathbb{N}$ is the union of disjoint sets $\{1, 3, 5, \ldots\}$ and $\{2, 4, 6, \ldots\}$, we obtain $\aleph_0 + \aleph_0 = \aleph_0$. Similarly, we derive $\aleph_0 + \cdots + \aleph_0 = \aleph_0$.

Considering the interval $[0, 2)$ as the union of the intervals $[0, 1)$ and $[1, 2)$, we see that $c + c = c$. Therefore, for any integer $n \ge 0$,

$$c \le n + c \le \aleph_0 + c \le c + c = c$$

which implies that $n + c = \aleph_0 + c = c + c = c$.

In fact, for any infinite cardinal number α, we have $\alpha + \aleph_0 = \alpha$. To prove this, we first observe that every infinite set X contains a countably infinite subset. Choose an element $x_1 \in X$. Since $X \ne \{x_1\}$, we can choose an element $x_2 \in X$ such that $x_2 \ne x_1$. We still have $X - \{x_1, x_2\} \ne \varnothing$. Consequently, we can choose an element $x_3 \in X$ such that $x_3 \ne x_1, x_2$. Assume that we have chosen n distinct elements $x_1, x_2, \ldots, x_n$ of X. Since X is infinite, it is possible to choose an $x_{n+1} \in X$ that is distinct from each of the x_i. Repeating this process indefinitely, we obtain a sequence $\langle x_1, x_2, \ldots \rangle$ of distinct points of X. The set $\{x_1, x_2, \ldots\}$ constructed in this way is obviously countably infinite. Now, given an infinite cardinal number α, find a set X with $\alpha = |X|$. Then there exists a countably infinite set $Y \subseteq X$. Writing $\beta = |X - Y|$, we have $\alpha = |Y| + |X - Y| = \aleph_0 + \beta$, so $\alpha + \aleph_0 = \beta + \aleph_0 + \aleph_0 = \beta + \aleph_0 = \alpha$.

Let M be a set and suppose that, for each $m \in M$, α_m is a cardinal number. Then $\sum_{m \in M} \alpha_m$ denotes the cardinal number of the union of pairwise disjoint sets A_m, where $|A_m| = \alpha_m$.

Clearly, the set $\mathbb{N}$ can be written as the disjoint union of countably many countable sets:

$$
\begin{array}{cccc}
1 & 3 & 5 & 7 \cdots \\
2 & 6 & 10 & 14 \cdots \\
4 & 12 & 20 & 28 \cdots \\
8 & 24 & 40 & 56 \cdots \\
\vdots & \vdots & \vdots & \vdots \cdots
\end{array}
$$

So $\aleph_0 + \aleph_0 + \cdots = \aleph_0$.

If $\alpha_1, \alpha_2, \alpha_3, \ldots$, are cardinal numbers such that $1 \leq \alpha_n \leq \aleph_0$ for every $n = 1, 2, 3, \ldots$, then, by Proposition A.5.11, we derive

$$
\alpha_1 + \alpha_2 + \cdots = \aleph_0.
$$

The decomposition of the interval $[0, \infty)$ into the intervals $[n - 1, n)$, where $n = 1, 2, \ldots$, shows that $c + c + \cdots = c$.

The *product* of two cardinal numbers α and β, denoted by $\alpha\beta$, is the cardinality of the Cartesian product of two sets A and B, where $|A| = \alpha$ and $|B| = \beta$. If μ is a cardinal number, then the μth power of α, denoted by α^μ, is the cardinal number of the set A^M, where M is a set with $|M| = \mu$.

Suppose that for every $m \in M$, there is the same cardinal number $\alpha_m = \alpha$. Then $\sum_{m \in M} \alpha_m = \alpha\mu$ and $\prod_{m \in M} \alpha_m = \alpha^\mu$, where $\mu = |M|$.

For the first formula, we note that $A \times M = \bigcup_{m \in M} A_m$, where $A_m = A \times \{m\}$. Obviously, A_m is equivalent to A, and the family $\{A_m \mid m \in M\}$ is pairwise disjoint. So $|A \times M| = \sum_{m \in M} |A_m|$ and we have $\sum_{m \in M} \alpha_m = \alpha\mu$.

With $A_m = A$ for all $m \in M$, we have $\prod_{m \in M} A_m = A^M$, and the second formula follows.

Using the above results, we obtain

$$
\begin{aligned}
n\aleph_0 &= \aleph_0 + \cdots + \aleph_0 = \aleph_0, \\
\aleph_0\aleph_0 &= \aleph_0 + \aleph_0 + \cdots = \aleph_0, \\
nc &= c + \cdots + c = c, \\
\aleph_0 c &= c + c + \cdots = c,
\end{aligned}
$$

where n is any positive integer.

Proposition A.8.1 $c = 2^{\aleph_0}$.

Proof By definition, $2^{\aleph_0}$ is the cardinality of $2^{\mathbb{N}}$, the set of all sequences whose terms are the digits 0 and 1 only (that is, the dyadic sequences). Interestingly, only these two digits are required to write the binary (or dyadic) representation of a real number x in $[0, 1]$; specifically, as the binary expansion, the representation $x = 0.a_1 a_2 \cdots a_n \cdots$ means the number $x = \sum a_n / 2^n$, where each a_n is either 0 or 1. Note that, as in the case of decimal expansion, certain numbers in $[0, 1]$ have two binary expansions (e.g., $1/2 = 0.100 \cdots = 0.011 \cdots$). Hence we deduce that $c \leq |2^{\mathbb{N}}| \leq 2c = c$. ◇

Note that $\aleph_0^2 = \aleph_0\aleph_0 = \aleph_0$ and $\aleph_0^n = \aleph_0 \cdots \aleph_0 = \aleph_0$, by induction on n. Similar results can be proved for any transfinite cardinal number α. With this end in view, we first establish the following.

Proposition A.8.2 *For any infinite cardinal number α, $2\alpha = \alpha$.*

Proof Let X be an infinite set with $|X| = \alpha$. Denote the two-point set $\{0, 1\}$ by 2. Then, for any set A, $2 \times A$ is the union of disjoint sets $\{0\} \times A$ and $\{1\} \times A$ and so we have $2|A| = |A| + |A| = |2 \times A|$. Now, consider the family $\mathscr{F}$ of all pairs (A, f) such that $A \subseteq X$ and $f : A \to 2 \times A$ is a bijection. By Theorem A.5.7, X contains a countably infinite set A and, by Proposition A.5.11, the set $2 \times A$ is also countable. Hence, there is a bijection $f : A \to 2 \times A$, and (A, f) belongs to $\mathscr{F}$. Consider the binary relation $\leq$ on $\mathscr{F}$ defined by $(A, f) \leq (B, g)$ if $A \subseteq B$ and $f = g|B$. It is easily verified that $\leq$ is a partial ordering and every chain in $\mathscr{F}$ has an upper bound in $\mathscr{F}$. So the Zorn's lemma applies, and we get a maximal member (M, h) in $\mathscr{F}$. Then $|M| + |M| = |M|$. We show that $|M| = |X|$. If $X - M$ is infinite, then it contains a countably infinite set B. As above, there exists a bijection g between B and $2 \times B$. Then we have a bijection $k : B \cup M \to 2 \times (B \cup M)$ defined by $k|B = g$ and $k|M = h$. Thus $(B \cup M, k)$ belongs to $\mathscr{F}$, and this contradicts the maximality of M. Therefore, $X - M$ must be finite; accordingly, M is infinite. If $Y \subseteq M$ is a countably infinite set (which does exist), then $|Y \cup (X - M)| = |Y| + |X - M| = |Y|$ and we obtain

$$\begin{aligned} |X| &= \left| Y \cup (X - M) \cup (M - Y) \right| \\ &= \left| Y \cup (X - M) \right| + |M - Y| \\ &= |Y| + |M - Y| = |M|. \end{aligned}$$

$\diamond$

As an immediate consequence of this result, we see that if α is an infinite cardinal number, then, by induction, $n\alpha = \alpha$ for every integer $n > 0$. And, for a cardinal number $\beta \leq \alpha$, $\alpha + \beta = \alpha$, since $\alpha \leq \alpha + \beta \leq 2\alpha = \alpha$.

Proposition A.8.3 *For any infinite cardinal number α, $\alpha^2 = \alpha\alpha = \alpha$.*

Proof Let X be an infinite set with $|X| = \alpha$ and let $\mathscr{F}$ be the family of all pairs (A, f) such that $A \subseteq X$ and $f : A \to A \times A$ is a bijection. Clearly, X contains a countably infinite set A and, by Lemma A.5.10, the set $A \times A$ is also countable. So there is a bijection f between A and $A \times A$, and thus the pair (A, f) belongs to $\mathscr{F}$. We partially order $\mathscr{F}$ by $(A, f) \leq (B, g)$ if $A \subseteq B$ and $f = g|B$. Given a chain $\mathscr{C} = \{A_i, f_i\}$ in $\mathscr{F}$, put $B = \bigcup A_i$ and define $f : B \to B \times B$ by setting $f(x) = f_i(x)$ if $x \in A_i$. Then f is clearly a bijection so that the pair (B, f) belongs to $\mathscr{F}$. It is obvious that (B, f) is an upper bound for the chain $\mathscr{C}$. By the Zorn's lemma, we have a maximal member (M, h) in $\mathscr{F}$. Since $h : M \to M \times M$ is a bijection, $|M| = |M||M|$. We observe that $|X| = |M|$. As $M \subseteq X$, $|M| \leq |X|$. If $|M| < |X|$, then we find that $|M| < |X - M|$. For, $|X - M| \leq |M|$ implies that

$$|X| = |M| + |X - M| \leq |M| + |M| = |M|,$$

by the preceding proposition. This contradicts our assumption that $|M| < |X|$. Let $j : M \to X - M$ be an injection and put $Y = j(M)$. Then $|Y| = |M|$. Since $|Y|$ is infinite, we have $3|Y| = |Y|$, and it follows that there are three disjoint sets A, B, and C contained in Y such that $Y = A \cup B \cup C$ and $|Y| = |A| = |B| = |C|$. So $|A| = |M \times Y|$, $|B| = |Y \times M|$ and $|C| = |Y \times Y|$; accordingly, there is a bijection $k : Y \to (M \times Y) \cup (Y \times M) \cup (Y \times Y)$. Since $M \cap Y = \varnothing$, we have a bijection $M \cup Y \to (M \cup Y) \times (M \cup Y)$ which extends h (and k). This contradicts the maximality of M, and hence $|M| = |X|$. $\diamond$

If α is an infinite cardinal number, then, by induction on n, we see that $\alpha^n = \alpha$ for every integer $n > 0$. Moreover, for any cardinal number $\beta \leq \alpha$, we have $\alpha \leq \alpha\beta \leq \alpha\alpha = \alpha$, and therefore $\alpha\beta = \alpha$. In particular, notice that $\aleph_0\alpha = \alpha$. We use these results to prove a generalization of Theorem A.5.12.

Theorem A.8.4 *Let X be an infinite set and $\mathscr{F}$ be the family of all finite subsets of X. Then $|\mathscr{F}| = |X|$.*

Proof Since the mapping $X \to \mathscr{F}$, $x \mapsto \{x\}$, is an injection, we have $|X| \leq |\mathscr{F}|$. To see the opposite inequality, for each integer $n > 0$, let $\mathscr{F}_n$ denote the family of all those subsets of X, which contain exactly n elements. Now, for each set $F \in \mathscr{F}_n$, choose an element $\phi(F)$ in $X \times \cdots \times X = X^n$ (n copies), which has all its coordinates in F. Of course, there are several choices for $\phi(F)$, we pick any one of these. Then $\phi : \mathscr{F}_n \to X^n$ is an injection, and so $|\mathscr{F}_n| \leq |X^n| = |X|^n = |X|$. Obviously, we have $\mathscr{F} = \bigcup_n \mathscr{F}_n$ and therefore

$$|\mathscr{F}| \leq \sum_n |\mathscr{F}_n| \leq \sum_n |X| = \aleph_0|X| = |X|.$$

$\diamond$

Let X be a set and $A \subseteq X$. The function $f_A : X \to \{0, 1\}$ defined by

$$f_A(x) = \begin{cases} 0 & \text{for } x \notin A, \text{ and} \\ 1 & \text{for } x \in A \end{cases}$$

is called the *characteristic function* of A. The function which is zero everywhere is the characteristic function of the empty set, and the function which is identically 1 on X is the characteristic function of X. The set of all functions $X \to \{0, 1\}$ is denoted by 2^X. Obviously, every element of 2^X is a characteristic function on X.

Proposition A.8.5 *For any set X, there is a bijection between the set $\mathscr{P}(X)$ of all its subsets and 2^X.*

Proof For a set $A \subseteq X$, let f_A denote the characteristic function of A. Then the mapping $\phi : 2^X \to \mathscr{P}(X)$, $f \mapsto f^{-1}(1)$, is clearly surjective. It is also injective. For, if $f \neq g$ are in 2^X, then we have $f(x) \neq g(x)$ for some $x \in X$. It follows that x belongs to one of the sets $f^{-1}(1)$ and $g^{-1}(1)$, but not to the other. So $\phi(f) \neq \phi(g)$, and ϕ is an injection. $\diamond$

From the above proposition, it is clear that $\left| \mathscr{P}(X) \right| = 2^{|X|}$. The function $x \mapsto \{x\}$ is obviously an injection from X into $\mathscr{P}(X)$, but there is no bijection between these sets, by Theorem A.5.13. Therefore $|X| < \left| \mathscr{P}(X) \right|$, and thus we have established the following.

Proposition A.8.6 *For any set X, $|X| < 2^{|X|}$.*

Appendix B
Fields $\mathbb{R}$, $\mathbb{C}$, and $\mathbb{H}$

B.1 The Real Numbers

We shall not concern ourselves here with the construction of the real number system on the basis of a more primitive concept such as the positive integers or the rational numbers. Instead, we assume familiarity with the system $\mathbb{R}$ of real numbers as an ordered field which is complete (that is, it has the least upper bound property). We review and list the essential properties of $\mathbb{R}$, however, in aid of the reader.

With the usual addition and multiplication, the set $\mathbb{R}$ has the following properties.

Theorem B.1.1 (a) $\mathbb{R}$ *is an abelian group under addition, the number* 0 *acts as the neutral element.*

   (b) $\mathbb{R} - \{0\}$ *is an abelian group under multiplication, the number* 1 *acts as the multiplicative identity (unit element).*

   (c) *For all* $a, b, c \in \mathbb{R}$, $a(b+c) = ab + ac$.

A *field* is a set F containing at least two elements $1 \neq 0$ together with two binary operations called *addition* and *multiplication*, denoted by $+$ and $\cdot$ (or juxtaposition), respectively, which satisfy Theorem B.1.1. The element 0 is the identity element for addition, and 1 acts as the multiplicative identity.

With this terminology, the set $\mathbb{R}$ is a field under the usual addition and multiplication. This field has an order relation $<$ which satisfies (a) $a + b < a + c$ if $b < c$, and (b) $ab > 0$ if $a > 0$ and $b > 0$. A field which also has an order relation satisfying these two conditions is called an *ordered field*. The set $\mathbb{Q}$ of all rational numbers is another example of an ordered field.

We call an element a of an ordered field *positive* if $a > 0$, and *negative* if $a < 0$.

Theorem B.1.2 *The following statements are true in every ordered field:*

   (a) $a > 0 \Rightarrow -a < 0$, *and vice versa.*

   (b) $a > 0$ *and* $b < c \Rightarrow ab < ac$,
      $a < 0$ *and* $b < c \Rightarrow ab > ac$.

© Springer Nature Singapore Pte Ltd. 2019
T. B. Singh, *Introduction to Topology*,
https://doi.org/10.1007/978-981-13-6954-4

(c) $a \neq 0 \Rightarrow a^2 > 0$. *In particular* $1 > 0$.
(d) $a > 0 \Rightarrow a^{-1} > 0$,
$\quad a < 0 \Rightarrow a^{-1} < 0$.
(e) $ab > 0 \Rightarrow$ *either both a* > 0 *and b* > 0 *or both a* < 0 *and b* < 0.
(f) $ab < 0 \Rightarrow$ *either both a* < 0 *and b* > 0 *or both a* > 0 *and b* < 0.

The ordered field $\mathbb{R}$ has *the least upper bound property*: If $S \subseteq \mathbb{R}$ is nonempty and bounded above, then sup S exists in $\mathbb{R}$.

This is also called the *completeness property* of $\mathbb{R}$. It follows that $\mathbb{R}$ is a complete ordered field. Moreover, $\mathbb{R}$ has no gaps: If $a < b$, then $c = (a + b)/2$ satisfies $a < c < b$. Thus we see that $\mathbb{R}$ is a linear continuum (refer to Exercise 3.1.11).

Using the completeness property, it can be shown that every nonempty set of real numbers with a lower bound has a infimum. Another important consequence of the completeness property of $\mathbb{R}$ is described by the following.

Theorem B.1.3 (Archimedean Property) *If a real number* $x > 0$, *then given any real number* y, *there exists a positive integer* n *such that* $nx > y$.

Using this property of $\mathbb{R}$, one can prove that, for any two real numbers $x < y$, there is a rational number $r \in \mathbb{Q}$ such that $x < r < y$. This fact is usually stated by saying that $\mathbb{Q}$ is *dense* in $\mathbb{R}$.

The *absolute value* of a real number x, denoted by $|x|$, is defined by $|x| = x$ for $x \geq 0$, and $|x| = -x$ for $x < 0$.

Proposition B.1.4 *For any* $x, y \in \mathbb{R}$, *we have*

(a) $|x| \geq 0$ \qquad\qquad\qquad (b) $|x| = 0 \Leftrightarrow x = 0$.
(c) $|-x| = |x|$. \qquad\qquad\quad (d) $|xy| = |x||y|$.
(e) $|x| \leq y \Leftrightarrow -y \leq x \leq y$. (f) $-|x| \leq x \leq |x|$.
(g) $|x + y| \leq |x| + |y|$. \qquad (h) $||x| - |y|| \leq |x - y|$.
(i) $|x - y| \leq |x| + |y|$.

By (d), $|x| = \sqrt{x^2}$.

B.2 The Complex Numbers

The set $\mathbb{R}^2$ of all ordered pairs (x, y) of real numbers turns into a field under the following addition and multiplication:

$$(x, y) + (x', y') = (x + x', y + y'),$$
$$(x, y)(x', y') = (xx' - yy', xy' + yx').$$

The element $(0, 0)$ acts as the neutral element for addition, and the element $(1, 0)$ plays the role of multiplicative identity. It is routine to check that $\mathbb{R}^2$ is a field under these definitions. It is usually denoted by $\mathbb{C}$, and its elements are referred to as the complex numbers. It is readily verified that $(x, 0) + (y, 0) = (x + y, 0)$, and $(x, 0)(y, 0) = (xy, 0)$. This shows that the complex numbers of the form $(x, 0)$ form a subfield of $\mathbb{C}$, which is isomorphic to $\mathbb{R}$ under the correspondence $x \mapsto (x, 0)$. We can therefore identify this subfield of $\mathbb{C}$ with the real field and regard $\mathbb{R} \subset \mathbb{C}$.

Writing $\iota = (0, 1)$, we have $\iota^2 = -1$ and $(x, y) = (x, 0) + (y, 0)(0, 1) = x + y\iota$, using the identification $x \leftrightarrow (x, 0)$. Thus $\mathbb{C} = \{x + y\iota \mid x, y \in \mathbb{R}\}$, where $\iota^2 = -1$. If $z = x + y\iota$ is a complex number, then we call x the real part of z (denoted by $\mathrm{Re}(z)$), and y the imaginary part of z (denoted by $\mathrm{Im}(z)$).

If $z = x + y\iota \in \mathbb{C}$, its *conjugate* is defined to be the complex number $\bar{z} = x - y\iota$. For any complex numbers z and w, we have

$$\overline{z + w} = \bar{z} + \bar{w}, \quad \overline{zw} = \bar{z}\bar{w}.$$

Observe that $z\bar{z}$ is a positive real number unless $z = 0$.

The *absolute value* $|z|$ of a complex number z is defined to be $\sqrt{z\bar{z}}$ (the nonnegative square root). Clearly, $|z| > 0$ except when $z = 0$, and $|0| = 0$. It is also obvious that $|\mathrm{Re}(z)| \leq |z|$ and $|z| = |\bar{z}|$. If z and w are any two complex numbers, then it is easily checked that $|zw| = |z||w|$ and $|z + w| \leq |z| + |w|$.

B.2.1 (Schwarz Inequality) If $z_1, \ldots, z_n$ and $w_1, \ldots, w_n$ are complex numbers, then we have

$$\left| \sum_1^n z_j \bar{w}_j \right|^2 \leq \left(\sum_1^n |z_j|^2 \right) \left(\sum_1^n |w_j|^2 \right).$$

Proof Put $\alpha = \sum_1^n |z_j|^2$, $\beta = \sum_1^n |w_j|^2$ and $\gamma = \sum_1^n z_j \bar{w}_j$. We need to show that $\alpha\beta - |\gamma|^2 \geq 0$. If $\alpha = 0$ or $\beta = 0$, this is trivial. We therefore assume that $\alpha \neq 0 \neq \beta$. Then $\alpha, \beta > 0$ and we have

$$\sum_1^n |\beta z_j - \gamma w_j|^2 = \sum_1^n (\beta z_j - \gamma w_j)(\beta \bar{z}_j - \overline{\gamma w}_j)$$
$$= \beta^2 \alpha - \beta|\gamma|^2 = \beta(\alpha\beta - |\gamma|^2).$$

The left-hand side of the first equality is obviously nonnegative, and therefore we must have $\alpha\beta - |\gamma|^2 \geq 0$. $\diamond$

B.3 The Quaternions

By using the usual scalar and (nonassociative) vector product in $\mathbb{R}^3$, Hamilton defined (in 1843) a multiplication in $\mathbb{R}^4$, which together with componentwise addition makes it into a skew field. This field has proven to be fundamental in several areas of mathematics and physics. We intend to discuss here some basic facts about it. Recall that a *skew field* (also called a *division ring*) satisfies all field axioms except the commutativity of multiplication.

The mapping $\mathbb{R}^4 \to \mathbb{R} \times \mathbb{R}^3$, $(x_0, x_1, x_2, x_3) \mapsto (x_0, (x_1, x_2, x_3))$, is a bijection. If we define the vector space structure on $\mathbb{R} \times \mathbb{R}^3$ over $\mathbb{R}$ componentwise:

$$(a, x) + (b, y) = (a + b, x + y) \text{ and } c(a, x) = (ca, cx),$$

then this mapping becomes an isomorphism. Consequently, we can identify $\mathbb{R} \times \mathbb{R}^3 = \mathbb{H}$ with $\mathbb{R}^4$, and call its elements *quaternions*. If $q = (a, x)$, we refer to a as the real part of q and x as the vector part of q. There are canonical monomorphisms $\mathbb{R} \to \mathbb{H}, a \mapsto (a, 0)$, and $\mathbb{R}^3 \to \mathbb{H}, x \mapsto (0, x)$, of vector spaces. Hence, we identify the real number a with the quaternion $(a, 0)$, and the vector x with the quaternion $(0, x)$. Then a quaternion (a, x) can be written as $a + x$. Accordingly, for any two quaternions $q = a + x$ and $r = b + y$, we have $q + r = a + b + x + y$, and if c is a real, then $cq = ca + cx$.

If $x, y \in \mathbb{R}^3 \subset \mathbb{H}$, we first define $xy = -x \cdot y + x \times y$, where $\cdot$ is the usual scalar product, and $\times$ is the usual vector product in $\mathbb{R}^3$. Notice that xy is in general an element of $\mathbb{H}$. It is easily checked that this multiplication of vectors in $\mathbb{H}$ is associative. As the multiplication in $\mathbb{H}$ ought to be distributive, we set

$$qr = ab + ay + bx + xy$$

for $q = a + x$ and $r = b + y$. We leave it to the reader to verify the following conditions for all quaternions q, r, s, and real c:

$$q(cr) = c(qr) = (cq)r, \quad q(rs) = (qr)s,$$
$$q(r + s) = qr + qs, \qquad (r + s)q = rq + sq.$$

The quaternion $1 = 1 + 0$ acts as the multiplicative identity: $1q = q = q1$ for every $q \in \mathbb{H}$.

To complete the proof that $\mathbb{H}$ is a skew field with the above addition and multiplication, it remains to verify that every nonzero quaternion q has a multiplicative inverse. With this end in view, we define the *conjugate* of $q = a + x$ as $\bar{q} = a - x$. Observe that $\overline{q + r} = \bar{q} + \bar{r}$, $\overline{cq} = c\bar{q}$, $\overline{qr} = \bar{r}\bar{q}$ for any quaternions q, r, s, and real c. Also, it is straightforward to see that $q\bar{q} = \bar{q}q = a^2 + x \cdot x$ (a real). We define the modulus of q to be $|q| = \sqrt{q\bar{q}}$. Notice that $|q|$ is the Euclidean norm of q when it is considered as an element of $\mathbb{R}^4$. For $q, r \in \mathbb{H}$, we have

$$|qr|^2 = (qr)(\overline{qr}) = qr\bar{r}\bar{q} = q|r|^2\bar{q} = |r|^2q\bar{q} = |r|^2|q|^2,$$

which implies that $|qr| = |q||r|$. Clearly, $|q| = 0 \Leftrightarrow q = 0$ and, if $q \neq 0$, then $q(\bar{q}/|q|^2) = 1 = (\bar{q}/|q|^2)q$. Thus $\bar{q}/|q|^2$ is the inverse of q in $\mathbb{H}$, usually denoted by q^{-1}.

Observe that $x \in \mathbb{R}^3$ is a unit vector $\Leftrightarrow x \cdot x = 1 \Leftrightarrow x^2 = -1$, and two vectors $x, y \in \mathbb{R}^3$ are orthogonal $\Leftrightarrow x \cdot y = 0 \Leftrightarrow xy = -yx$. A *right-handed orthonormal system* in $\mathbb{R}^3$ is an ordered triple $\imath, \jmath, k$ of vectors in $\mathbb{R}^3$ such that $\imath, \jmath, k$ are of unit length, mutually orthogonal and $\imath \times \jmath = k$. So, if $\imath, \jmath, k$ form a right-handed orthonormal system, then $\iota^2 = \jmath^2 = k^2 = -1$ and $\iota \jmath k = -1$. Conversely, these conditions imply that $\iota, \jmath, k$ are of unit length, and $\iota \jmath = k$ whence $\iota \cdot \jmath = 0$ and $\iota \times \jmath = k$. Thus $\iota, \jmath, k$ form a right-handed orthonormal system.

Suppose now that $\iota, \jmath, k$ is a right-handed orthonormal system in $\mathbb{R}^3$. Then any vector $x \in \mathbb{R}^3$ can be written uniquely as $x = x_1\iota + x_2\jmath + x_3k$, $x_i \in \mathbb{R}$; accordingly, any quaternion q can be expressed uniquely as

$$q = q_0 + q_1\iota + q_2\jmath + q_3k, \quad q_i \in \mathbb{R}.$$

Clearly, we have $\iota \jmath = k = -\jmath\iota$, $\jmath k = \iota = -k\jmath$, $k\iota = \jmath = -\iota k$. Using these rules, we obtain the following formula for the product of two elements $q = q_0 + q_1\iota + q_2\jmath + q_3k$, $q' = q_0' + q_1'\iota + q_2'\jmath + q_3'k$ in $\mathbb{H}$:

$$qq' = (q_0q_0' - q_1q_1' - q_2q_2' - q_3q_3') + (q_0q_1' + q_1q_0' + q_2q_3' - q_3q_2')\,\iota + (q_0q_2' + q_2q_0' + q_3q_1' - q_1q_3')\,\jmath + (q_0q_3' + q_3q_0' + q_1q_2' - q_2q_1')\,k.$$

We also note that $\bar{q} = q_0 - q_1\iota - q_2\jmath - q_3k$ and $|q| = \sqrt{(q_0^2 + q_1^2 + q_2^2 + q_3^2)}$.

For any unit vector $x \in \mathbb{R}^3$, the set of quaternions $a + bx, a, b \in \mathbb{R}$, is a subfield of $\mathbb{H}$ isomorphic to $\mathbb{C}$ under the mapping $a + bx \mapsto a + b\iota$. In particular, the subfield of quaternions with no $\jmath$ and k components is identified with $\mathbb{C}$, and we regard $\mathbb{C}$ as a subfield of $\mathbb{H}$. Thus, we have field inclusions $\mathbb{R} \subset \mathbb{C} \subset \mathbb{H}$. It is obvious that any real number commutes with every element of $\mathbb{H}$. Conversely, if a quaternion q commutes with every element of $\mathbb{H}$, then $q \in \mathbb{R}$. We emphasize, however, that the elements of $\mathbb{C}$ do not commute with the elements of $\mathbb{H}$.

References

1. G.E. Bredon, *Topology and Geometry* (Springer, New York, 1993)
2. R. Brown, *Elements of Modern Topology* (McGraw-Hill, London, 1968)
3. J. Dugundji, *Topology* (Allyn and Bacon, Boston, 1965)
4. D.B. Fuks, V.A. Rokhlin, *Beginner's Course in Topology* (Springer, Heidelberg, 1984)
5. I.M. James, *Topological and Uniform Spaces* (Springer, New York, 1987)
6. J.L. Kelley, *General Topology* (van Nostrand, New York, 1955)
7. W.S. Massey, *Algebraic Topology: An Introduction* (Springer, New York, 1967)
8. G. McCarty, *Topology: An Introduction with Application to Topological Groups* (McGraw-Hill, New York, 1967)
9. D. Montgomery, L. Zippin, *Topological Transformation Groups* (Interscience Publishers, New York, 1955)
10. J.R. Munkres, *Topology: A First Course* (Prentice-Hall, NJ, 1974)
11. L. Pontriajagin, *Topological Groups* (Princeton University Press, New York, 1939)
12. I.M. Singer, J.A. Thorpe, *Lecture Notes on Elementary Topology and Geometry* (Scott, Foresman and Company, IL, 1967)
13. E.H. Spanier, *Algebraic Topology* (McGraw-Hill, New York, 1967)
14. L.A. Steen, J.A. Seebach, *Counterexamples in Topology* (Springer, New York, 1978)
15. R.C. Walker, *The Stone-Čech Compactification* (Springer, New York, 1974)
16. S. Willard, *General Topology* (Addison-Wesley, MA, 1970)

© Springer Nature Singapore Pte Ltd. 2019
T. B. Singh, *Introduction to Topology*,
https://doi.org/10.1007/978-981-13-6954-4

Index

© Springer Nature Singapore Pte Ltd. 2019
T. B. Singh, *Introduction to Topology*,
https://doi.org/10.1007/978-981-13-6954-4

Printed in the United States
By Bookmasters